D0936955

ORGANIC AND BIOLOGICAL CHEMISTRY

A Series of Monographs

EDITED BY
LOUIS F. FIESER and MARY FIESER
Harvard University, Cambridge, Mass.

1: MARTIN D. KAMEN: Isotopic Tracers in Biology, 3rd Edition, 1957

2: K. VENKATARAMAN: The Chemistry of Synthetic Dyes, Vol. 1, 1952

3: K. VENKATARAMAN: The Chemistry of Synthetic Dyes, Vol. 2, 1952

4: S. C. NYBURG: X-Ray Analysis of Organic Structures. in preparation

ACADEMIC PRESS INC., PUBLISHERS, NEW YORK

ISOTOPIC TRACERS IN BIOLOGY

An Introduction to Tracer Methodology

by

MARTIN D. KAMEN

Professor of Biochemistry
Graduate Department of Biochemistry
Brandeis University
Waltham, Massachusetts

THIRD EDITION, REVISED, ENLARGED AND RESET

1957
ACADEMIC PRESS INC., PUBLISHERS, NEW YORK

Library of Congress Catalog Card Number: 57-8376

First Printing, 1957
Second Printing, 1959
Third Printing, 1962

PRINTED IN THE UNITED STATES OF AMERICA

PREFACE TO THIRD EDITION

When this book first appeared in 1947*, isotopic tracers were rare—even exotic—in biological laboratories. Hence, the first edition reported and commented on pioneer experiences. Now, the use of tracers is commonplace; so much so that the literature reporting experiments in all branches of biology has proliferated beyond the grasp of any one writer. The text in hand has been revised a number of times to accommodate new and relevant facts, but only one completely new edition has appeared since 1948. A third edition is long overdue.

The objectives of this third edition are the same as those defined in the first edition. The main task in preparing it has been to evaluate new material covering many aspects of a number of fields, some not even mentioned in the first edition. I am tempted to borrow the diplomatic phrase "agonizing re-appraisal" to describe the difficulties inescapable in such a process. Perhaps the most challenging of these was the problem of choosing which among the masses of new work to include. The deciding factor had to be the pedagogic value of the new material, rather than its novelty, in relation to the older material.

Three major changes have been made in this edition. First, the scope of the book has been extended to stable isotopes, as the new title indicates. As a result, the chapters on general methodology have been expanded. New chapters on the elements nitrogen and oxygen have been added. Numerous concrete examples illustrating work with both stable and radioactive tracers have been included; some of these are completely new and others are more detailed than they were in previous editions.

Second, the chapters on nuclear physics and chemistry (I–III) have been rewritten and modernized.

Third, two new chapters have been added. One of these (V) is a short discussion of practical matters which arise when investigators make the transition from book learning to the laboratory. The other chapter (VII) is an extension of an original chapter on biochemical applications. It is included to illustrate one of the most important uses of tracer procedures: their application, with other methods, to solve some old problems in which a kind of impasse had developed.

For the rest, I have adhered to my previous practice of selecting material with which I was personally familiar, whether through actual participation

* M. D. Kamen, "Radioactive Tracers in Biology," 1st ed., 1947; 2nd ed., 1951. Academic Press, New York.

(as in the reminiscences about the discovery of C^{14} given in Chapter X), or through acquaintance with investigators doing the researches described.

As in the second edition, I have included current estimates on tolerance amounts of ingested radioactive isotopes. I wish to emphasize that these estimates are nothing more than educated guesses, the bases for which are indicated either in the text or in the references cited.

I would like to add a note of thanks to my wife, Beka Doherty Kamen, who pulled together and edited the critical first draft. I am also heavily indebted to Mrs. Margot C. Bartsch who shepherded and proofread succeeding drafts, and who typed and edited the final draft. Finally, I must thank the critics of previous editions who made many helpful suggestions in their reviews. I have incorporated a number of these in the present edition.

Many colleagues were most generous in giving permission to use material from their books and articles. Acknowledgments are recorded at appropriate places in the text. I should mention in particular the cooperation I received from Dr. Michel Ter-Pogossian, Associate Professor at the Mallinckrodt Institute of Radiology, who made a number of original drawings and supplied unpublished material which was most helpful in preparation of the section on scintillator detectors in Chapter III.

St. Louis, Missouri
April, 1957

MARTIN D. KAMEN

REMARKS ON SECOND PRINTING

On the occasion of a second printing, opportunity has been afforded to make some corrections and additions following a few suggestions noted by some reviewers who were kind enough to read the text critically. The major addition is Appendix 5 which describes briefly new material on systems and procedures for assay of low-energy β-emitters using liquid scintillation media. The writer is indebted to Dr. Seymour Rothchild for much of the information contained in this Appendix.

Waltham, Massachusetts
December, 1958

MARTIN D. KAMEN

CONTENTS

II. Radiation Characteristics of Tracer Atoms

III. Isotopic Assay

IV. Radiation Hazards

V. Practical Interlude

VI. Survey of Tracer Methodology: Biochemical Aspects, Part I

VII. Survey of Tracer Methods: Biochemical Aspects, Part II

VIII. Survey of Tracer Methodology: Physiological and Medical Aspects

IX. The Isotopes of Hydrogen

X. The Isotopes of Carbon

XI. The Isotopes of Oxygen, Nitrogen, Phosphorus, and Sulfur

XII. Various Radioactive Nuclides of Importance in Biology

Chapter I

ATOMIC NUCLEI, RADIOACTIVITY, AND THE PRODUCTION OF RADIOACTIVE ISOTOPES

1. INTRODUCTORY REMARKS

The application of tracer procedures to biological problems does not depend on detailed knowledge of the physical background of tracer methodology. However, most research workers and students who work with tracers want basic information about the nature of tracers and the fundamentals of atomic and nuclear physics. The introductory chapters of this book are written with this in mind. The following brief discussion of fact and theory about atoms and nuclei can be elaborated by reference to the bibliography included in the text.

2. GENERAL PROPERTIES OF NUCLEI

A. Nuclear Structure

All elements are made up of small ultimate units or *atoms*. Atoms contain a positively charged *nucleus* (radius $<10^{-12}$ cm.) which, although it comprises a very small fraction of the atomic volume (radius $\sim 10^{-8}$ cm.), accounts for practically all the weight of the atom. Most of the atomic volume can be said to be relatively "empty," being occupied by negatively charged *electrons*, which are many thousandfold lighter than the nuclear particles. Nuclear matter, in other words, is enormously more dense than atomic matter; its density approximates 10^8 tons per cubic centimeter. Such fantastic densities imply forces of a type not encountered in common experience. It is necessary to suppose that enormous attractive forces spring into being when matter is compressed to form atomic nuclei.

Present knowledge holds that nuclei consist of particles called *neutrons* and *protons*, for which the collective term is *nucleons*. Nuclei are built up by addition of approximately equal quantities of neutrons and protons, beginning with the lightest nucleus, that of the ordinary hydrogen atom. The nucleus of hydrogen is nothing more than a single proton. Protons and neutrons differ in that protons carry a unit positive electric charge (4.8025×10^{-10} electrostatic unit). Further description of these particles must be deferred for a brief excursion into the theory of nuclear forces.

Experimental evidence available indicates that nuclear forces extend only over very short distances—distances, in fact, much less than those assigned to nuclear dimensions. It is assumed that there is a continuous formation of unstable, short-lived particles smaller than nucleons and that nuclear forces arise in some way from exchange of these particles between the nucleons. It is evident that simple Coulombic (electrostatic) forces cannot be involved because the neutron is not electrically charged. Furthermore, the electrostatic forces in the nucleus would be repulsive rather than attractive, because the protons are all positively charged and would cause the nuclei to fly apart.

From investigations of cosmic-ray particles it is known that nuclear particles of mass intermediate between electrons and protons exist. These are called *mesons*. The first to be discovered was found to have a mass equivalent to 210 electron masses and is now called the μ meson. It was suggested that exchange of these μ mesons between nucleons might account for nuclear forces. This idea was not supported by experiment, however, and was abandoned in favor of the idea that another type of meson, discovered later and called the π meson, is the actual particle involved. Its mass is equivalent to 275 electron masses, and it appears to have many other properties necessary to act as the exchange particle between nucleons. At present, the π meson, which occurs in uncharged, positively charged, and negatively charged forms, is central to most theories of nuclear structure. No adequate theory of nuclear forces has yet been developed.

B. Mass Number and Atomic Number

The number of protons in a nucleus is the *atomic number* of the element and is usually symbolized by Z. Values of Z range from 1 for hydrogen to 101 for the most recently discovered transuranic elements. Z is the integral nuclear positive charge. Its magnitude determines the number of negative electrons required to accord with the observed electrical neutrality of the atom.

The total number of nucleons (neutrons plus protons) in the nucleus is called the *mass number* and is usually symbolized by A. The value of A is always expressed as the whole number nearest the atomic weight of the atom under consideration. (For example, see Section 2-D.) Values for A range from 1 to 255.

The atomic number is usually written as a left subscript and the atomic mass number as a right superscript to the chemical symbol; e.g., ${}_1H^1$, ${}_{11}Na^{23}$, and ${}_{15}P^{32}$ refer to certain atomic species of the elements hydrogen, sodium, and phosphorus, respectively. Sometimes the atomic number is omitted because it can be inferred from the chemical symbol.

C. Isotopes

Because the chemical properties of the atom are determined by the value of the nuclear charge or atomic number, addition of neutrons to any nuclear complex of protons and neutrons changes the mass by an integral amount but does not change the nuclear charge. Since the nuclear charge determines the number of extranuclear electrons, which, in turn, determines the chemistry of the atom, no change occurs in the chemical behavior of the atom when neutrons are added to the atomic nucleus. Consequently there are nuclei, and hence atoms, which vary in nuclear mass but not in chemical nature. These are called *isotopes*. Some elements have only one stable isotope each ($_4Be^9$, $_9F^{19}$, $_{11}Na^{23}$, $_{15}P^{31}$, etc.); others are mixtures of two or more stable isotopes.

Sulfur may be cited as an example. Four isotopes of sulfur with mass numbers 32, 33, 34, and 36 are known. In the nomenclature described above these would be written $_{16}S^{32}$, $_{16}S^{33}$, $_{16}S^{34}$, and $_{16}S^{36}$. Each of these nuclei contains $A = 16$ protons, and $(A - Z) = 16$, 17, 18, and 20 neutrons, respectively. The ratio of the number of neutrons to the number of protons for stable nuclei is very close to unity. It increases with increasing values of Z until at $_{83}Bi^{209}$ there is a ratio of 126/83 or 1.5.

It is also possible for nuclei with the same mass number but different atomic number (*isobars*) to exist. Examples are $_{48}Cd^{113}$ and $_{49}In^{113}$, $_{18}A^{40}$ and $_{20}Ca^{40}$. Finally, it is also possible that nuclei of identical charge and mass number may exist in slightly different configurations or energy states. Such nuclei are called *isomers* (see p. 15).

The terms *isotope, isobar*, and *isomer* refer to particular species of atomic nuclei. The collective term for nuclear species is *nuclide*.[1] Thus H^2, H^1, Li^6, P^{31}, Ca^{40}, and Ca^{42} are all nuclides, but only H^2, H^1 and Ca^{42}, Ca^{40} are isotopes of hydrogen and calcium, respectively.

D. Nuclear Mass

The mass of any nucleus is referred to the mass of the oxygen isotope of weight 16 which is defined as having a mass exactly equal to 16. Thus, the hydrogen nucleus, or proton, has a mass of 1.00758 compared to O^{16}. Its mass number, A, is 1, according to the definition in Section 2-B.

The scale used in describing nuclear mass is called the *physical atomic weight scale*. This is not identical with the *chemical atomic weight scale*, which is used to express atomic weights in chemistry. In the chemical scale, the standard weight is that of the natural form of oxygen, which contains small amounts of the rare isotopes O^{17} and O^{18} and is assigned a mass of exactly 16.0000 . . . , although it actually weighs a bit more than

[1] Kohman, T. P., *Am. J. Phys.* **15**, 356 (1947).

the standard O^{16} atom. The chemical unit is larger than the physical unit by 1.000272 (± 0.000005). The uncertainty in the sixth figure arises from fluctuations in the isotopic content at this degree of precision.

The isotopic masses usually encountered in the literature are based on the physical scale. They are not nuclear but atomic masses and include the masses of the extranuclear electrons in the neutral atoms.

E. Other Fundamental Nuclear Properties

In addition to charge and mass, the nucleus has properties analogous to those associated with electrons in atomic physics: spin, mechanical moment, magnetic moment, and electric moment. All nuclei are also subject to one of two types of statistics, depending on the quantum mechanical description employed. This, in turn, depends on whether a nucleus has an odd or even number of constituent particles, i.e., an odd or even mass number. The existence of these properties, however, is not relevant to tracer methodology and need not be considered further.

The nuclei of major importance for this discussion are the *neutron* (n); the *proton* (p); the *deuteron* (d), which is the heavy hydrogen nucleus ($_1H^2$); and the *alpha* particle (α), which is the helium nucleus ($_2He^4$). The *negative electron* or *negative beta particle* (β^-), and the *positive electron* or *positron* (β^+), although they do not exist in the nucleus, are produced by it in certain nuclear transformations. In addition, the list of nuclear entities include *gamma radiations* (γ), which are high-energy photons (electromagnetic radiation quanta), and *neutrinos*, which are hypothetical particles postulated to occur in those nuclear transformations involving β-ray emissions (see p. 11).

3. SYSTEMATICS OF NUCLEI

A. Introduction

Derived from Einstein's theory of special relativity is the concept of equivalence of mass, M, and energy, E, expressed in the relation $E = Mc^2$, where c is the velocity of light. This relation is fundamental for nuclear physics and has been verified convincingly in nuclear studies. According to this principle, the disappearance of mass is accompanied by the liberation of very large amounts of energy. Before discussing the energy magnitudes involved, it is convenient to introduce at this point the concept of the *electron volt* (ev.) as a unit of energy.

A unit electric charge moving through a potential difference of one international volt acquires a kinetic energy which is spoken of as one "electron volt equivalent." The heat energy to which this corresponds can be calculated in the following manner. Suppose one mole (6.02×10^{23}) of electrons is confined in a space between parallel plates of an electrical condenser charged to 1 volt. The electrons fall into the positively charged plate, their kinetic energy being dissipated as heat. This heat energy in joules is the total charge in coulombs multiplied by the voltage difference across the condenser. One mole of electrons is equivalent to a *faraday*, which is approximately 96,500 coulombs. Hence $96{,}500 \times 1 = 96{,}500$ joules of heat appear. One gram-calorie (g.-cal.) corresponds to 4.18 joules, so that 96,500/4.18 or 23,000 g.-cal./mole of heat represents 1 ev. of kinetic energy per atom. Each electron gains a kinetic energy corresponding to heat motion communicated by $23{,}000/6.02 \times 10^{23}$ g.-cal. Thus, 1 ev. is the equivalent of 1.602×10^{-12} erg. Since heats of chemical reactions usually vary from a few kilocalories to a few hundred kilocalories per mole, it can be seen that the range of chemical energies is included in the range 0 to 10 ev.

The magnitude of the energies involved in nuclear interactions can be shown by application of the mass-energy relation. It is obvious that these energies will be enormous in comparison with ordinary chemical reactions, because, in the latter, no detectable mass loss is observed. In nuclear transmutations, on the other hand, there are very appreciable changes in total mass. It is found that one absolute mass unit (m.u.), $\frac{1}{16}$ of O^{16}, corresponds to 931 millions of electron volts (Mev.).[2] In other words, disappearance of 1 m.u. would liberate approximately 2.1×10^{13} g.-cal. of heat per mole of nuclei involved. The heat liberated in the burning of one mole of sugar to CO_2 and water is only 7.2×10^5 g-cal.

The simplest nuclear reaction is the combination of a neutron and a proton to form a deuteron, i.e.,

$$n + p \rightarrow d + \gamma \tag{1}$$

This reaction is exothermic (releases energy), 2.18 Mev. of energy as electromagnetic (γ) radiation being emitted on fusion of a neutron and the proton. The γ-ray energy representing the difference in mass between the reactants (free neutron and proton) and product nucleus (the deuteron) is called the "binding energy." This quantity is analogous to the heat of chemical reactions. If the deuteron is to be disintegrated into a neutron and proton, energy is required and the reaction is endothermic. The mass of the neutron can be calculated from data of reaction 1 in the following way. The relation between the mass of the neutron, M_n, the mass

[2] The energy equivalent to 1 atomic mass unit (a.m.u.) can be found readily from the Einstein equivalence relation. M, the weight of 1 a.m.u. in grams, is the weight of $\frac{1}{16}$ of the O^{16} atom divided by Avogadro's number, which is the number of atoms in a gram atomic weight. Thus, $M = 1/6.02 \times 10^{23} = 1.661 \times 10^{-24}$ g. Multiplying this by the square of the velocity of light (c^2) gives $1.661 \times 10^{-24} \times (2.998 \times 10^{10})^2$ g. cm.^{2} sec.$^{-2}$, or $E = Mc^2 = 1.493 \times 10^{-3}$ erg. Since 1 ev. $= 1.602 \times 10^{-12}$ erg, 1 a.m.u. $=$ 931 ev.

of the proton, M_p, the mass of the deuteron, M_d, and the binding energy, ΔE, follows immediately from reaction 1 as

$$M_n = M_d - M_p + \Delta E \tag{2}$$

ΔE converted to mass units is 2.18/931, or 0.00234 m.u. When the known masses are substituted for the deuterium atom and the hydrogen atom,

$$M_n = 2.01473 - 1.00813 + 0.00234 = 1.00894 \text{ m.u.} \tag{2a}$$

This type of calculation can be applied to any nucleus, stable or unstable, provided the binding energy for the reaction whereby such a nucleus is formed is known. In most tables, the masses of the neutral atoms, rather than the masses of the nuclei, are given. Thus, the mass of the extranuclear electrons is included. Atomic mass instead of nuclear mass can be used in these calculations, because the electronic masses cancel out whenever stable isotopes are involved. For example, in reaction 1 the one extranuclear electron from deuterium ($_1H^2$) cancels the electron from protium ($_1H^1$).

It also possible to calculate nuclear masses of unstable isotopes from a knowledge of the maximum energy involved in the disintegration. As an example, C^{14} emits a negative electron (β^- particle) with a maximum kinetic energy of 0.15 Mev. This process forms the residual nucleus N^{14}. Thus,

$$C^{14} \rightarrow N^{14} + \beta^- + 0.15 \text{ Mev.} \tag{3}$$

The mass of N^{14} is 14.00754, so the mass of C^{14} is 14.00754 + 0.00016 or 14.00770. Here, again, it should be noted that the atomic mass is used in place of the nuclear mass. This is because the residual nucleus has its positive charge (atomic number) increased by one unit when a negative electron leaves a radioactive nucleus. Thus, another electron is required in the atomic orbit. As far as the over-all mass balance is concerned, all that happens is that an electron leaves the nucleus and joins the product atom. Hence, no change in total number of electrons is involved.

This is not true when a positive β-ray emitter is involved, because one less electron is required for the product atom. One negative electron with a rest mass equivalent to 0.00055 m.u. goes off with the initially emitted positive electron and is lost from the orbital electrons. The masses of the positive and negative electrons are equal; two electron masses should be added to the product nucleus to attain mass balance when atomic masses are used. In the disintegration of $_7N^{13}$ a positron is emitted and $_6C^{13}$ with atomic mass 13.00761 is formed. The maximum energy of the radiations emitted gives the heat of reaction, ΔE, as 1.198 Mev., which is 0.00129 m.u. Hence, the mass of $_7N^{13}$ can be calculated as follows:

$$_7N^{13} \rightarrow {}_6C^{13} + \Delta E + 2\beta \tag{4}$$

$$m_{N^{13}} = 13.00761 + 0.00129 + 0.0011 = 13.0100 \tag{4a}$$

In all these calculations it is assumed that the product nucleus is formed in its lowest (most stable) energy state. If this is not the case, γ radiation corresponding to the transition from the upper energy to the lowest energy state will contribute energy, and hence mass, which must be added to the mass value obtained in the above manner.

B. Isotope Classification and Nuclear Forces

It may be assumed that nonradioactive isotopes represent stable combinations of neutrons and protons. Thus, the nucleus of carbon must contain, in addition to 6 protons, no more than 6 or 7 neutrons. These combinations correspond to the two stable carbon nuclei found in nature, namely, $_6C^{12}$ and $_6C^{13}$. Eight neutrons cause formation of an unstable configuration ($_6C^{14}$) of 14 particles. The stable configuration for 14 particles is one consisting of 7 protons and 7 neutrons ($_7N^{14}$). By changing a neutron into a proton, $_6C^{14}$ is transformed to $_7N^{14}$. This requires emission of a negative β particle, thus:

$$n \rightarrow p + \beta^- \tag{5}$$

(The participation of neutrinos in this process will be neglected in the present discussion; see p. 12.) Suppose 5 neutrons are associated with 6 protons to form $_6C^{11}$. This nucleus represents an unstable configuration of 11 particles, the stable configuration being the naturally occurring nonradioactive isotope of boron, $_5B^{11}$, which consists of 5 protons and 6 neutrons. A proton is transformed into a neutron with consequent positive electron emission to effect the necessary change in composition. Hence $_6C^{11}$ disintegrates by positive electron emission to $_5B^{11}$; thus the process

$$p \rightarrow n + \beta^+ \tag{6}$$

occurs in the $_6C^{11}$ nucleus. Positron emission cannot occur unless the mass difference between parent and daughter nuclei exceeds two electron masses ($\sim$1 Mev.) because both a positive and negative electron are lost in the process. When the relative instability of a configuration with excess protons is not sufficient to supply this energy, an alternative process called "*K* capture" can take place. The reader will note that essentially the same nuclear composition can be obtained in one of two ways—either by adding a negative electron to a nucleus or by removing a positron. A nucleus which should emit a positive electron but is lacking in the necessary energy can reduce its positive charge by capturing an orbital electron from the nearest inner electron shell, the *K* shell. Less frequently, capture of electrons from orbits other than the *K* shell can occur. This process of *K* capture, or orbital electron capture, can take place whether sufficient energy for positron emission is available or not. The prediction of relative probability of K capture or positron emission is one of the interesting problems in nuclear physics.

Thus, for carbon, an increase in the number of neutrons beyond 7 or a decrease below 6 results in unstable nuclei. One of these, C^{14}, transforms to N^{14}; the other, C^{11}, transforms to B^{11}. Likewise, there is a C^{10} which is even more unstable than C^{11} and transforms to B^{10}. It is found in this way that throughout the whole periodic system there is, for any given number of protons, a restricted number of neutrons which will form a stable combination. The binding energy of a single neutron and proton as in ${}_1H^2$ has been found to be approximately 2 Mev., or about 1 Mev. per particle. Throughout most of the periodic table, however, the binding energy per particle, whether neutron or proton, is considerably higher and essentially constant at 7 to 8 Mev. The only forces of this magnitude known from previous physical experience are those operative between the charged components, i.e., the protons. Since these are Coulomb forces they should be entirely repulsive and very large at the small distances of separation between protons in nuclei. Quantitatively, the nuclear radius, R, is given empirically by the relation

$$R = (1.43 \times 10^{-13})A^{1/3} \text{ cm.} \tag{7}$$

where A is the mass number. If it is supposed that Z protons are distributed uniformly throughout a spherical nucleus, the electrostatic energy of repulsion, E_c, is given by

$$E_c = 0.067A^{5/3} \text{ Mev.} \tag{8}$$

Since this force is purely repulsive, it must be supposed that Coulomb's law is not operative at short distances or that a new attractive force becomes effective between nuclear particles at nuclear distances. This attractive force far outweighs the Coulomb repulsion and increases linearly with the number of particles because the binding energy per particle is constant. It is seen from Eq. 8 that the Coulomb repulsion increases as the $\frac{5}{3}$ power. It follows that for large values of A (for heavy elements) the Coulomb energy will become appreciable as compared to the total binding energy. Thus, although the Coulomb energy in ${}_2He^4$ is only $\sim$0.1% of the total nuclear binding energy, it is nearly 25% of the total binding energy in ${}_{82}Pb^{206}$. To help hold the heavier nuclei together when the proton repulsions become large, it is necessary to have more neutrons per proton. In this way more binding energy is obtained without concomitant repulsion energy, because there is no Coulomb repulsion between neutrons. It is plausible to account in this fashion for the deviation in the neutron-proton ratio toward values some 50% higher than unity as the atomic number increases.

As remarked in Section 2-A, nuclear forces extend over small distances (1 to 3 $\times$ 10^{-13} cm.). This is consistent with the observed linear increase in binding energy with mass number which indicates that nuclear particles influence only a near neighbor. Such a force is comparable to the homopolar saturation force in chemical bonds. Because the neutron-proton ratio is nearly unity, it seems that the force between neutrons and protons is the major factor in holding nuclei together. There are also attractive forces between protons (*p-p*) and neutrons (*n-n*), but these are somewhat smaller than the *p-n* force. It is postulated that *n-n* forces exist because there are many elements, particularly the heavier ones, which have neutrons in excess of protons. This *n-n* force is responsible for the extra bonding required to maintain nuclear stability as the Coulomb repulsion due to the protons mounts. Furthermore, stable neighboring isobars exist, an example being $_{50}Sn^{115}$ and $_{49}In^{115}$. The only difference between these nuclei is that a *p-n* pair in tin is replaced by an *n-n* pair in indium.

The existence of a definite *p-p* force practically equal in magnitude to the *n-n* force follows from the case of $_1H^3$ and $_2He^3$, the former being very slightly unstable with respect to the latter. Here an *n-p* pair in $_1H^3$ is replaced by a *p-p* pair to form $_2He^3$. The slight difference in binding energy of these two nuclei is consistent with the notion that the *p-p* attraction must be very similar to the *n-n* attraction.

C. Isotope Ratios

When the isotopic composition of the elements is studied, it is found that the relative abundance of the isotopes of nearly every element varies little, if at all. This is true no matter what sources the elements come from. Samples collected from extraterrestrial sources, such as meteorites, do not vary in isotopic content from those found with terrestrial specimens. The only important variations in isotopic constitution occur in those elements in which radioactive processes are operative. Some evidence exists for slight variations in the isotopic contents of hydrogen, carbon, oxygen, and a number of other elements. These variations are too small to affect all but a very few tracer applications, but they are of great importance in supplying information on geochemical processes and the history of various parts of the earth's surface. A discussion of the isotopy involved in geology and cosmology is beyond the scope of this book. A recent general survey by Kohman and Saito is included in the list of references at the end of this chapter.

For general purposes, however, constant isotopic content is one of the most valid generalizations which can be made with regard to the elements. In fact, it serves as the basis for the tracer method because it affords a means of labeling elements simply by changing the isotopic content. For instance, carbon from any natural source is invariably a mixture of C^{13} and C^{12}. The percentage of the former is always $1.10\% \pm 0.02\%$. Hence, any carbon sample prepared with an isotopic composition in which the percentage of C^{13} is appreciably different can be distinguished from normal carbon and, therefore, constitutes a labeled carbon.

The assay methods for determination of isotope abundance in the case of stable isotopes depend, in general, on the use of the mass spectrograph (see Chapter III). In the special cases of hydrogen and oxygen, density methods based on the use of liquid or gas samples are also employed. It is also possible to vary the isotope content by adding isotopes not previously present, such as radioactive isotopes ($_6C^{11}$ and $_6C^{14}$). Thus, a sample of carbon admixed with C^{11} and C^{14} is radioactive and hence can be distinguished from normal carbon by the methods employed in assay of radioactive bodies (see Chapter III). Since radioactive isotopes of nearly all the elements have been prepared, this method of labeling is quite general.

The most arduous and demanding techniques are required to obtain appreciable separation of isotopes by chemical or physical means. Isotopes therefore may be considered ideal labels or tracers. In general, then, *any element is said to be labeled if its natural isotopic content is altered.* The labeling is accomplished by increasing the relative amount of a rare stable isotope or by adding a radioactive isotope. Either of these two types of isotopes is called a "tracer." Inclusion of tracers in any aggregation of atoms of normal isotopic content produces a labeled sample of the element. Inclusion of labeled atoms in a molecule results in a labeled molecule.

D. Specific Activity and Atomic Per Cent Excess

In all tracer applications, it is necessary to define labeled content. When radioactive tracers are used, the labeled content is referred to as the *specific activity* which is the *radioactivity per unit weight of radioactive material.* One may express the specific activity in any unit of radioactivity—millicuries, counts per minute, etc. The specific labeled content or specific isotopic (labeled) content is related linearly to the specific activity, being the ratio of the number of radioactive atoms to the total number of isotope atoms. For the case of phosphorus cited above, the *specific labeled content* is the number of P^{32} atoms divided by the total number of P atoms (P^{31} and P^{32}), thus:

$$\text{Specific labeled content} = P^{32}/(P^{31} + P^{32})$$

In most tracer samples the amount of P^{32} present is negligible compared to the P^{31}, so that this expression reduces to P^{32}/P^{31}. Sometimes it is possible to prepare almost pure samples of P^{32} as well as other radioactive isotopes. Thus, as will be discussed in Section 6-B, it is possible to pick a transmutation process exploiting a nuclear reaction in which the target nucleus is not identical with the product nucleus. Specifically, for the production of P^{32}, chloride or sulfide targets may be irradiated with neutrons, producing the radioactive phosphorus. The only inactive phosphorus present (P^{31}) occurs as a natural contaminant because the sulfide or chloride is not absolutely chemically pure. This P^{31} contamination can be minimized by careful chemical processing of the target material.

Specific isotopic content when expressed on a percentage basis is known as *isotopic atom per cent*, or, in short, *atom per cent.* This term is used in place of specific activity when referring to stable isotopic tracers such as H^2, N^{15}, and C^{13}. As mentioned earlier in this chapter, C^{13} is present in normal carbon in the amount of 1.00%. Tracer samples contain C^{13} in excess of this amount. Thus, a sample with 2.00 atom per cent C^{13} has 1.00 atom per cent *excess* C^{13}. It is more customary to employ the latter term rather than atom per cent in reporting isotopic content of stable isotopic labeled samples. Thus, *atom per cent excess* is the excess in percentage abundance of isotope in labeled element over that in the normal element.

4. RADIOACTIVITY

A. Types of Radioactive Decay

As stated previously, only certain combinations of neutrons and protons are stable. An excess of one or the other component leads to a redistribution of particles during which the neutron-proton ratio is brought to the proper value for stability. This may be accomplished in several ways, for example: (1) transformation of a neutron into a proton (negative β-ray emission), (2) transformation of a proton into a neutron (positron emission or K capture), (3) emission of an α particle. In all these cases there may also be emission of electromagnetic radiation in the form of γ rays, x-rays, etc. The emission of α particles is confined almost entirely to the heavy elements ($Z > 82$). Radioactivity produced artificially in the light and medium heavy elements ($Z < 82$) is associated almost entirely with the emission of negative or positive electrons. The properties of the various radiations encountered in radioactive decay will be considered in Chapter II.

In Section 4-B the fundamental decay law will be discussed. For the present the reader need bear in mind only that radioactive disintegration follows an exponential law and that every radioactive body has its own characteristic rate of disintegration which remains constant in time.

1. Beta Decay. The emission of negative electrons from atomic nuclei was established early in the history of radioactivity. The emission of positrons was discovered relatively recently. Both kinds of particles are assumed to arise during disintegration and not to be present as such in the nucleus.

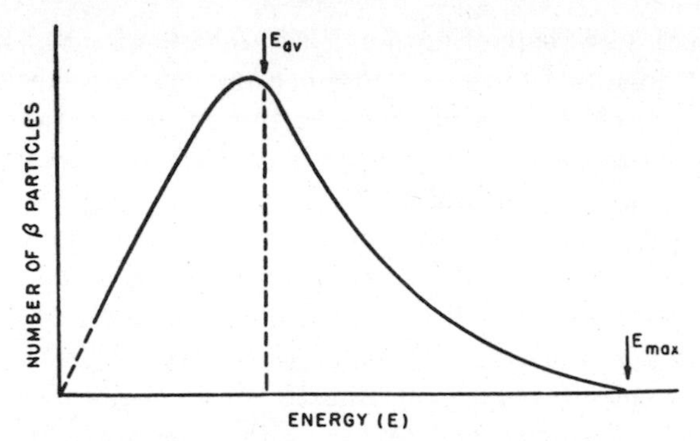

FIG. 1. Typical β-energy spectrum.

Positron emission differs from negative electron emission in one important respect. Positrons have an ephemeral existence only; on coming to rest or being slowed in flight, positrons disappear along with a negative electron. This "annihilation" results in the production of two electromagnetic photons ("annihilation radiation"), each with a kinetic energy equivalent to the mass of the missing electron (0.5 Mev.). Conservation of momentum requires that these two photons depart from the scene of the annihilation in exactly opposite directions. Rarely, one photon carrying nearly all the energy of the annihilation can be emitted. In this case, the momentum and energy of recoil are taken up by the nucleus in the field of which the annihilation takes place. The production of two quanta occurs principally with low-energy positrons.

Conversely, if an electromagnetic photon with an energy equal to or greater than the rest mass of a positron and a negative electron (≥ 1 Mev.) enters the field of a nucleus, a pair of electrons, one positive and one negative, can be produced. Such a process is spoken of as "pair production."

Investigations of the energies of the β particles emitted in nuclear transformations show that there is a continuous distribution in energy from zero energy to the maximum permitted by the nuclear reaction energy. If the number of β particles possessing a given energy ($E \pm dE/2$) is plotted as ordinate against the energy, E, as abscissa, an energy "spectrum" is obtained. A typical β-energy spectrum is shown in Fig. 1. It will be seen that the spectrum shows a maximum value (E_{max}) and an average value (E_{av}) which is the average energy of all the β particles emitted. This average energy is usually about one-third of the maximum energy. It is difficult in practice to determine the exact shape of the β spectrum at either end. At low energies there are low intensities and also complications introduced by interaction of nuclear charge with the emitted electron. At the high-energy end, it is difficult to pick out the exact end point because of straggling and the presence of "bremsstrahlung" (see p. 51). The shape of the curve, i.e., the energy distribution, is not identical for all radioactive isotopes.

The continuous spectrum of β rays has puzzled physicists ever since its discovery in 1914 by Chadwick. It would be expected that β-particle emission—which involves transitions between discrete nuclear energy states—would lead to discrete, not continuous, spectra. Nevertheless, actual calorimetric measurements show that the energy per β particle is the average and not the maximum energy for the β spectrum. It appears that energy is not conserved in β decay. Another difficulty arises when single β-decay events are studied. It is found that the angle between the recoiling nucleus and the emitted β particle is not 180°, as it should be for conservation of momentum in the process, but something less. Thus, momentum also appears not to be conserved.

Because of these and other difficulties, it has been postulated that a third particle, undetected as yet, is emitted during β decay. This particle is called the *neutrino* and has been assigned all the properties needed to avoid the difficulties described in understanding the β-decay process. It is further supposed that a similar particle is emitted in positron decay. By analogy to the β-decay process, this second hypothetical particle is called the *antineutrino*. Further, it is assumed that the process of K capture, which is the alternative to positron emission, probably also involves emission of neutrinos.

Nuclear disintegrations involving β-ray emissions can occur in a variety of ways. The radioactive nucleus may dissipate all its transformation energy in a single transition. In this case a simple β spectrum results. On the other hand, the residual nucleus may possess several energy states to any of which a transition is possible. In this case "complex" β spectra consisting of the several single β-ray spectra may result. Such spectra are accompanied by γ radiation resulting from transitions in the residual nucleus. Gamma radiation of nuclear origin may also be associated with a single β spectrum. Various disintegration schemes representing these possibilities are given diagrammatically in Fig. 2.

It should be noted that no simple relation exists between the energy or range of the β-particles produced during nuclear transformations and the probability of disintegration of β-radiating nuclei. Thus, although Geiger and Nuttall have established that, for α particles, a linear relation exists between the logarithm of the range and the logarithm of the disintegration rate, this is not found to be true in general for β particles. S^{35} and C^{14} emit β rays with the same energy maximum, but they differ by four orders of magnitude in probability of disintegration.

It has been noted (see p. 7) that a mode of decay alternative to positron emission is the K-capture process. This process may be difficult to detect because no nuclear radiation need accompany this kind of transformation. The characteristic radiations are those resulting from the rearrangement of the extranuclear electrons after a nuclear capture of an orbital electron. If

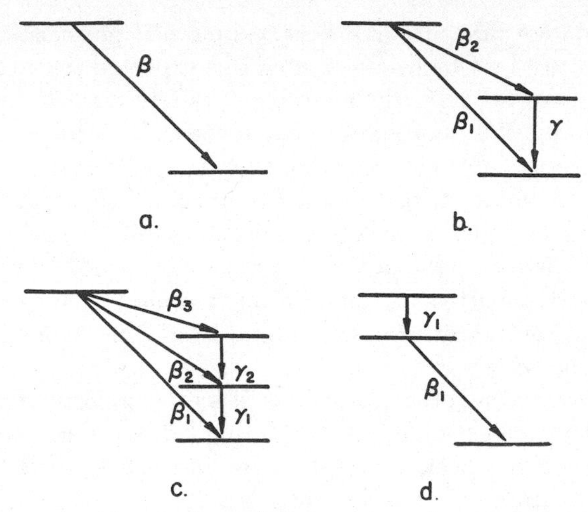

FIG. 2. Some disintegration schemes.

a K electron is captured, there results emission of x-rays after the transition of L, M, N, and other electrons into the vacancy created in the K shell. Thus, it is possible to excite the entire x-ray spectrum of the daughter element which results from K capture. In addition, negative electrons, called "Auger" electrons (named after the French physicist P. Auger) can result from a kind of internal photoelectric effect in which the emission of x-ray photons, say K x-rays, is replaced by emission of L electrons with a kinetic energy equal to the difference in binding energy of the K·and L shells.

2. Gamma Decay. The atomic nucleus can exist in a number of energy states, transitions between which are evidenced by emission of electromagnetic radiation, or photons, in a manner analogous to the production of atomic spectra by transitions of electrons between various atomic energy states. In radioactive decay, the energies observed for γ radiations are much greater than those corresponding to atomic radiation and range up to about 3 Mev. The energy distribution, unlike that in β spectra, corresponds to transitions between discrete energy states, i.e., line spectra rather than continuous spectra.

Just as x-ray emission in K capture may be accompanied by ejection of Auger electrons, so may nuclear γ emission be coincident with or transformed wholly into emission of negative electrons. This process is called "internal conversion." It is not pictured as the emission of a nuclear photon followed by a collision process in which the photon transfers its energy to an extranuclear electron. Rather, it is represented by a direct coupling of the nuclear transition energy with the electron. Whatever the mechanism, there results a negative electron with a discrete kinetic energy given by the initial energy of the nuclear photon from which is subtracted the binding energy of the electron in the atom. Thus, internal conversion electrons are characterized by an energy distribution which is discrete and not continuous, in contrast to nuclear β particles.

It is appropriate to elaborate briefly the phenomenon of nuclear isomerism given passing mention on p. 3. In general, transitions between nuclear energy states occur extremely rapidly, average transition times being of the order 10^{-13} sec. In some cases the excited states are more stable, i.e., metastable, and the nucleus theoretically can exist in such states for much longer times. This possibility leads to the existence of a nuclide in simultaneous isomeric states. Such states with average lives of 10^{-7} sec. to several months are known. An excited isomeric state can decay either by a transition to a lower energy state ("isomeric transition") with γ-ray emission, or to a neighboring isobar by a β decay or K capture. An example is Mn^{52}, which exhibits isomeric transition, K capture, and positron emission.

The phenomenon of nuclear isomerism introduces some limitations in the use of radioactive isotopes as tracers which are discussed in a later section.

3. Alpha Decay. This type of radioactive transformation is of limited importance because no tracer elements of major interest exhibit radioactive isotopes which decay by emission of α particles. It should be remarked only that the α particle is emitted with a discrete energy forming a sharp distribution and that concomitant emission of γ rays with their accompanying secondary radiations is possible when the daughter atom is left in an excited state.

B. Fundamental Decay Law

The rate at which radiation is emitted is a function of nuclear constitution and is not alterable by ordinary chemical or physical means.[3] The process whereby radioactive transformation takes place is governed by chance. Studies of statistical theory and its application to numerous cases of radioactive decay have been made by a number of workers,[4] and it has been shown conclusively that radioactive decay is a statistical process. Hence, it is permissible to assume that the probability of decay at any time is proportional to the number of atoms. Experimentally the rate of decay is seen to follow an exponential course with fluctuations governed by the *Poisson distribution law* (see pp. 96, 97).

[3] Two reports have appeared which indicate that under very special conditions, as in K capture by a light nucleus (Be^7), chemical bonding can affect radioactive decay rate to a small but significant extent. See Leininger, R. F., Segrè, E., and Wiegand, C., *Phys. Rev.* **76,** 897 (1949), and Bouchez, R., Daudel, R., Daudel, P., and Muxart, R., *Compt. rend.* **227,** 525 (1948).

[4] Kohlrausch, F., *Ergeb. exakt. Naturw.* **5,** 197 (1926); Kovarik, A. F., *Phys. Rev.* **13,** 272 (1919); Feather, N., *ibid.* **35,** 705 (1930); Curtis, L. F., *Bur. Standards J. Research* **4,** 595 (1930).

The experimentally observed decay law can be derived statistically[5] if it is assumed that all atoms of a given isotope have the same decay probability and that this probability is not dependent on the age of any particular atom. On this basis, the change, ΔN, in a unit time, Δt, is related to the number of atoms, N, of isotope present by the expression

$$\Delta N/\Delta t = -\lambda N \tag{9}$$

where λ is the *disintegration constant* characteristic of the isotope. For very small intervals of time, dt, the differential expression for Eq. 9 can be written

$$dN/dt = -\lambda N$$

or

$$dN/N = -\lambda\, dt \tag{9a}$$

At time $t = 0$, N_0 atoms are present. At any time t thereafter, the number present, N_t, can be found by integration of Eq. 9a within the limits $t = 0$ to $t = t$ in time, and N_0 to N in the number of atoms. Thus

$$\int_{N_0}^{N} dN/N = -\lambda \int_0^t dt$$

and

$$\ln N/N_0 = -\lambda t \tag{10}$$

or

$$\log N/N_0 = -\lambda t/2.303$$

This can be written

$$N = N_0 e^{-\lambda t} \tag{10a}$$

which is the familiar exponential law for radioactive transformation of a single isotope.

C. Half-Life and Mean Life

The intensity of radioactivity is given by dN/dt. This, in turn, is determined by N, so that, if the intensity drops to one-half its initial value, the number of atoms must also have dropped to half those initially present. The time required for this diminution by half is called the *half-life* and is symbolized by $\tau_{1/2}$. Substitution of $N/N_0 = \frac{1}{2}$ in Eq. 10 gives

$$\ln \tfrac{1}{2} = -\lambda\tau_{1/2} \qquad \text{or} \qquad \ln 2 = \lambda\tau_{1/2} \tag{11}$$

Substitution of 2.303 $\log_{10} 2$ for ln 2 reduces the value for the half-life to

$$\tau_{1/2} = 0.693/\lambda \tag{12}$$

[5] Ruark, A. E., *Phys. Rev.* **44**, 654 (1933).

The actual life of any particular atom can have any value from zero to infinity. However, the average length of time during which any atom exists before disintegration is a definite quantity. If N_0 atoms exist initially, then at time t there will remain $N = N_0e^{-\lambda t}$ atoms. In the next interval, dt, the number of atoms decaying will be $\lambda N dt$ or $\lambda N_0 e^{-\lambda t}\, dt$. All these atoms had life of t. The product of this number of atoms and their life, t, integrated over all time intervals from zero to infinity gives the total time all the atoms lived. This, divided by the number of atoms initially present, N_0, gives the *average* or *mean life*, symbolized by τ. Thus

$$\tau = 1/N_0 \int_0^\infty \lambda N_0 e^{-\lambda t} t \cdot dt = 1/\lambda \tag{13}$$

The average life is seen to be the reciprocal of the disintegration constant and is 1.45 times the half-life. If the logarithm of the radioactivity is plotted against the time (Fig. 3), Eq. 10 shows that a straight line will result, the slope of which is negative and equal to the disintegration constant divided by 2.303.

If two or more isotopes are present in any tracer sample, the resultant decay curve will be a composite curve. Accurate analysis of the components is possible only if the disintegration constants are not too similar. If one of the isotopes is very long-lived, the decay can be followed until all the short-lived component or components have vanished, at which time the curve will become a straight line characteristic of the long-lived component. Extrapolation of this line to zero time will give the initial activity as well as subsequent activity of the long-lived component. By subtracting the ordinates of this line from the corresponding ordinates of the total activity curve, a new curve is obtained representing the activity of the short-lived components. If there is only one of these, i.e., two isotopes to begin with, then the resultant curve will again be a straight line characteristic of

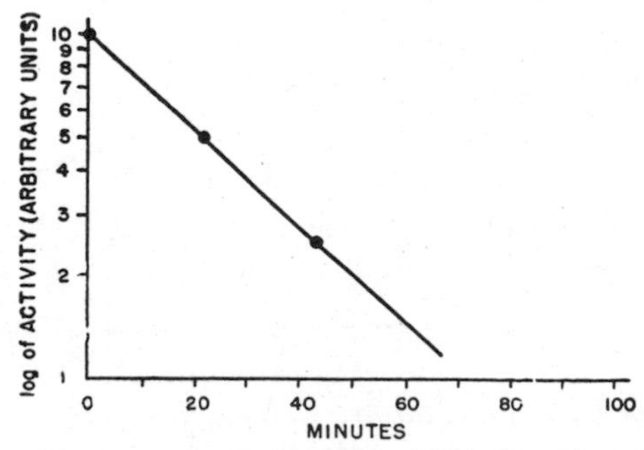

FIG. 3. Decay curve for the isotope C^{11} ($\tau_{1/2} = 21.0$ min.). It will be noted that the intensity of radioactivity drops a factor of 2 every 21 min. Ordinates are plotted on log scale.

the short-lived isotope. If there are many components, a repetition of the subtraction process is necessary, if it is possible to find a straight portion for extrapolation. The procedure is illustrated in Fig. 4 for the two-component case of one isotope with a half-life of 1 day and another with a half-life of 2 hr.

It is also possible to analyze composite decay curves if each isotope emits radiation easily separable from that of the others by differences in absorption properties. Thus, a sample of radiophosphorus ($_{15}P^{32}$, $\tau_{1/2} = 14.3$ days) contaminated with radiosulfur ($_{16}S^{35}$, $\tau_{1/2} = 87.1$ days) will show the proper decay if measured with an instrument which allows radiation to enter through a relatively thick window. Since the sulfur radiation cannot penetrate the window but the phosphorus radiation can, the sulfur radiation will not affect the instrument, and, hence, only the phosphorus decay will be noted.

D. Radioactive Chains

It is of interest to investigate the relations which hold when the parent atom, P, decays to a daughter, D, which in turn is radioactive and decays to a stable atom. It will be convenient to assume that, at time $t = 0$, only the parent, P, is present in initial number P_0. The number of atoms, D, formed in a time dt is given by $\lambda_P P$, where λ_P is the disintegration constant for P. This follows, since $\lambda_P P$ atoms of P have disintegrated in this time

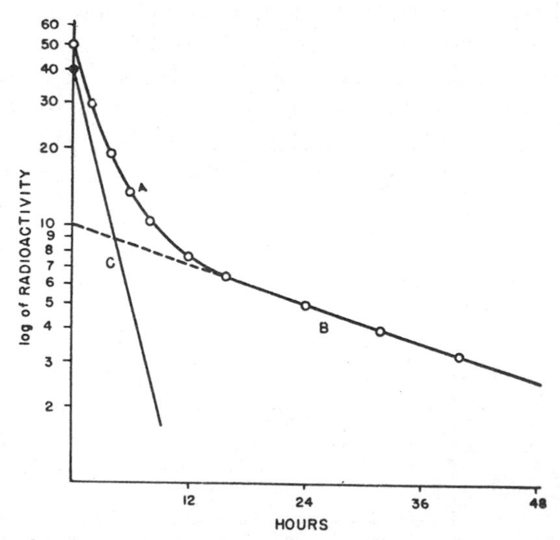

Fig. 4. Composite decay curve for two isotopes, one with $\tau_{1/2} = 1$ day, the other with $\tau_{1/2} = 2$ hr. It is seen that the initial intensity of the short-lived component, C, is four times that of the long-lived component, B. The total decay curve, A, is made up by addition of the two linear relations B and C. C is derived by subtracting the extrapolated portion of B (dotted line) from A. Ordinates are plotted on log scale.

interval to form D. In the same interval of time, the number of daughter atoms lost is given by the number present, D, multiplied by λ_D. Hence the total change in D with time in the interval dt is the difference of the number gained, $\lambda_P P$, and the number lost, $\lambda_D D$, or

$$dD/dt = \lambda_P P - \lambda_D D \tag{14}$$

P is related to the number of parent atoms at zero time, P_0, by the expression $P = P_0 e^{-\lambda_p t}$, so that Eq. 14 becomes

$$dD = (P_0 \lambda_P e^{-\lambda_p t} - D\lambda_D)\, dt \tag{15}$$

The solution of this expression, if it is assumed that $D = 0$ at time $t = 0$, is

$$D = \frac{\lambda_P}{\lambda_D - \lambda_P}(P)(1 - e^{-(\lambda_D - \lambda_P)t}) \tag{16}$$

Equation 16 makes it possible to calculate the number of daughter atoms, D, present at any time t, when there were initially only parent atoms present.

One special case is of interest. Suppose that the daughter atom decays much more rapidly than the parent. This implies that $\lambda_D \gg \lambda_P$. Equation 16 reduces to

$$D = (\lambda_P/\lambda_D)(P)(1 - e^{-\lambda_D t}) \tag{17}$$

If in time t little or no change in the number of parent atoms take place, i.e., the parent atoms are very long-lived and $1/\lambda_D \ll t \ll 1/\lambda_P$, then Eq. 17 reduces to

$$D \rightarrow D_\infty = P_0(\lambda_P/\lambda_D) \tag{18}$$

This expression describes the final number of short-lived daughter atoms in equilibrium with the long-lived parent atoms. In the production of radioactive isotopes the atom-transmuting apparatus (cyclotron, uranium pile, etc.) may be likened to the long-lived parent which is producing the constant number of atoms of daughter isotope, $R = \lambda_P P$. This term, R, depends on the operating characteristics of the apparatus rather than radioactive decay but is in all other respects the equivalent of a constant radioactive source. Then Eq. 17 becomes

$$D \text{ (the number of radioactive nuclei formed)} = (R/\lambda_D)(1 - e^{-\lambda_D t}) \tag{19}$$

If for λ_D there is substituted its value in terms of the half-life, $\tau_{1/2}$, there follows

$$D = \frac{R\tau_{1/2}}{0.693}(1 - 2^{-t/\tau_{1/2}}) \tag{19a}$$

It is seen that in a bombardment time equal to one half-life of the radioactive product ($t = \tau_{1/2}$), 50% of the saturation or maximum yield is obtained. The most that can be obtained is given when t is so large that the expression in parentheses approaches its maximum value of unity. Thus the maximum yield is given by $R\tau_{1/2}/0.693$. Inspection of Eq. 19a shows that approach to this value is asymptotic, 75% being attained in two half-lives, 87.5% in three half-lives, and so on. It is obvious that continuation of bombardment beyond three or four half-lives results in very little additional yield.

The generalization of the equations of radioactive transformation to chains with three or more components is available in the literature.[6] However, such chains are encountered rarely in biological applications of the tracer method, so that further discussion can be omitted.

E. Branch Disintegrations

The possibility exists that an isotope can decay in several ways so that its disintegration constant is composed of several partial disintegration constants. As an example, the isotope of thorium, ThC ($_{83}Bi^{212}$), emits α particles with two different energies. One of these α particles is associated with the nucleus, ThC′ ($_{84}Po^{212}$), and has a calculated half-life of approximately 10^{-11} sec. The other comes from ThC. Gamma and beta radiations are also given off by ThC. The relations involved are shown schematically as follows:

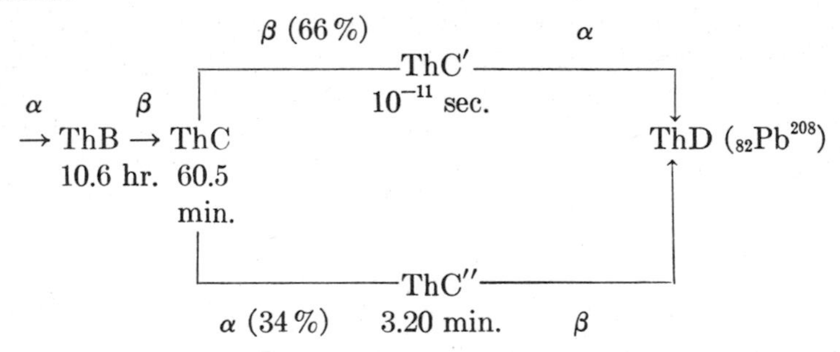

It is seen that in 66% of the disintegrations ThC emits a β particle. The remainder of the disintegrations go by α-particle emission. Similar branching reactions occur in the natural radioactive series for RaC and AcC.

In such a branching reaction, the total disintegration constant is equal to the sum of all the disintegration constants for each of the branching processes, i.e.,

$$\lambda = \lambda_1 + \lambda_2 + \cdots$$

[6] Bateman, H., *Proc. Cambridge Phil. Soc.* **15**, 423 (1910). See also Rutherford, E., Chadwick, J., and Ellis, C. D., "Radiations from Radioactive Substances," Chapter I. Cambridge U. P., 1930; Hull, D. E., *J. Phys. Chem.* **45**, 1305 (1941).

Thus, in ThC, $\lambda_1 = 0.34\lambda$ and $\lambda_2 = 0.66\lambda$. The total decay will follow the usual $e^{-\lambda t}$ law, where λ is the total disintegration constant, because the decay is due to loss of the stock of nuclei which have not disintegrated. To determine partial disintegration constants one may determine the tot l decay constant and the "branching ratio" which is the fraction of disintegrations following a given branch. Among the artificially radioactive isotopes, branching occurs mainly with alternative positron or negative electron emission. Because K capture is also possible with positron emitters, a three-way branch is likely. Thus ${}_{29}Cu^{64}$, which lies between the isobars ${}_{28}Ni^{64}$ and ${}_{30}Zn^{64}$, can decay in three modes, i.e.,

$${}_{28}Ni^{64} + \beta^+ \leftarrow {}_{29}Cu^{64} \rightarrow \beta^- + {}_{30}Zn^{64}$$

and

$${}_{29}Cu^{64} + e_k^- \rightarrow {}_{28}Ni^{64}$$

The K-capture process results in emission of the characteristic x-rays of the daughter Ni because electrons from the outer shells drop into the hole left in the K shell by loss of the K electron. The x-ray emission is found to have the same half-life ($\tau_{1/2} = 12.8$ hr.) as the positrons and negative electron emission.

F. Radioactivity Units

The original unit of radioactivity, the curie, was the quantity of radon in equilibrium with 1 g. of its parent, radium. (This is 0.66 mm.^{3} of radon gas at 0°C. and 760 mm. Hg pressure.) The restricted nature of such a unit led to the proposal that the curie unit be extended to include the equilibrium amount of any decay product of radium. The absolute radioactivity of 1 g. of radium has been measured many times, with the average value clustering around 3.7×10^{10} disintegrations/sec. This figure was chosen arbitrarily by the International Radium Standard Commission in 1931 as the curie unit of radioactivity with a decision as to the third figure deferred until general agreement could be reached. It was never intended that the curie unit be applied to substances not in the radium family. However, the advent of the artificial radioactive elements resulted in the curie units being taken over for use with any radioactive substance. The curie has now come to mean that amount of radioactive isotope required to give 3.7×10^{10} disintegrating nuclei per second.

The determination of radioactivity in terms of the curie requires a measurement of the absolute rate of disintegration. This is a most difficult measurement in general, because a given disintegration may involve emission of a number of types of radiations with varying energies and varying detection efficiencies. It is necessary to know the disintegration scheme of the radioactive isotope accurately. Relatively few such schemes have been worked out. It is obvious that a curie of an isotope with pure β emission will give radically different response in a given detection device from that exhibited by one decaying mainly by K capture.

More confusion results, however, from the use of the curie unit as a measure of intensity of γ radiation. This unfortunate practice, arising mainly from radiological use of radon and radium, involves comparison of γ intensity of the radiation source with that from a unit amount of radium in equilibrium with its decay products with some absorber placed between sample and detector in the two measurements. One then speaks of "curie-equivalents." Thus some investigators use the curie to describe any radioactive source whose γ rays produce the same effect in a detector as a curie of radon. Inasmuch as the response of the detector varies profoundly with the type of radiation and with the energy of any given radiation, it is obvious that no meaning in terms of disintegration rates can be attached to the curie-equivalent. The irrationality of this practice reaches its ultimate absurdity with isotopes which emit practically no γ radiation.

Efforts have been made to bring order into radioactivity standards by proposing an arbitrary unit which does not depend on measurement of the absolute disintegration rate of radium but instead is defined arbitrarily. Condon and Curtiss, for instance, have proposed the rutherford (rd.) as the basic radioactivity unit. This unit would equal 10^6 disintegrations/sec. They also propose a new unit for γ-ray intensity, which requires description of the unit for quantity of γ radiation called the roentgen (r.). This unit has been defined as the quantity of x- or γ radiation which, in passage through a volume of air weighing 0.001293 g., produces ions of either sign equivalent to 1 electrostatic unit (esu.) of electricity. The mass of air referred to is 1 cm.^{3} of air at standard conditions (0°C., 760 mm. Hg). This complicated definition will be elaborated later (see Chapter IV). For this discussion it is necessary to remark merely that the roentgen is a unit of energy dissipation and not of γ-ray quality or intensity. The quantity defined exactly by the roentgen is the quantity of ionization produced by excited electrons in the wake of the γ rays. Nor does this unit depend on the time required to produce the ionization. Hence rate of γ-ray dosage requires a unit such as roentgens per unit time. Gamma-ray intensity cannot properly be referred to in terms of roentgens per second because such a unit is a unit of ionization intensity and not a unit describing the number of γ photons flowing through any defined point or volume. Condon and Curtiss have proposed an unambiguous physical unit for radioactive source strength independent of measurements of absolute disintegration rate. They have recommended the *roentgen per hour at 1 meter* (*rhm.*). In accordance with this proposal, 1 rhm. of I^{131} is that amount whose unshielded γ radiation produces 1 r./hr. of ionization in air 1 meter from the source. The γ

radiation from 1 g. of radium (curie-equivalent) in equilibrium with its decay products gives an effect equal to 0.969 rhm.

In referring to quantities of radioactivity in most tracer applications it is convenient to use units smaller than either the curie or the rutherford. The usual prefixes milli- and micro- (meaning 1/1000 and 1/1,000,000, respectively) are employed. Thus 1 millicurie (mc.) is 0.001 curie. Various conversion factors and associated information are included in Appendix 1.

5. NUCLEAR REACTIONS

Processes for radioactive tracer production are based on the interactions between a small number of fundamental particles and the stable nuclei of the elements. These interactions or nuclear reactions resemble ordinary chemical reactions in that they have a heat of reaction (mass change) and an energy of activation, and also exhibit varying reaction rates depending on the experimental conditions. A typical reaction for the sodium and helium nucleus may be written as follows:

$$ {}_{11}Na^{23} + {}_{2}He^{4} \rightarrow {}_{1}H^{1} + {}_{12}Mg^{26} \qquad (20) $$

It is convenient in writing such reactions to use an abbreviated form in which the target nucleus is indicated first, followed by parentheses containing, seriatim, the reacting particle, the emitted particle, and the product nucleus. Reaction 20 would be written as $Na^{23}(\alpha,p)Mg^{26}$, according to the notation for fundamental particles as presented earlier. It is important to note that the analogy between chemical and nuclear reactions breaks down in the following respects:

1. The atomic numbers of the nuclei change.
2. Mass number is conserved but total mass is not.
3. Heats of reaction and activation are five or six orders of magnitude greater for nuclear reactions than for chemical reactions.
4. Nuclear reactions are concerned with individual nuclei so that the statistical approach used describing ordinary chemical reactions, i.e., mass law, is not valid.

The nature of nuclear reactions has been elucidated to a large extent by Niels Bohr.[7] The present concept pictures a two-step process in which there is an amalgamation of particle and nucleus followed by breakup of the intermediate nucleus so formed. The target nucleus constitutes a system not unlike a drop of liquid, since it contains many particles (protons and neutrons) among which the total energy is distributed more or less equally and among which there is a constant interaction. The impinging particle, on entering such a system, distributes its kinetic energy and energy of binding equally among all the nuclear particles so that the compound

[7] Bohr, N., *Nature* **137**, 344 (1936).

nucleus is raised in energy content. Thus, if a target nucleus absorbs a proton with a kinetic energy of 5 Mev., it can be shown that, because the average binding energy per particle for most nuclei is about 8 Mev., the compound nucleus (target + proton) may have its energy content raised to a level 13 Mev. above the ground state. This process is assumed to occur exceedingly rapidly, i.e., in the time it would take for a 5-Mev. proton to traverse typical nuclear distances which are 2 to 8 $\times 10^{-13}$ cm. This time is of the order 10^{-20} to 10^{-21} sec. Once the compound nucleus is formed, the excitation energy is divided among all the nuclear particles so that on the average no one particle has sufficient energy to escape from the nucleus. Consequently, a long time may elapse before a particle breaks away. The disintegration of the compound nucleus is, therefore, a process quite independent of the initial amalgamation. The nucleus does not "remember" which of its constituent particles was the impinging particle.

The time required for the emergence of a nuclear particle depends on the chance that sufficient energy will finally concentrate on one particle. This may be a time 10^6 to 10^{10} times as large as the initial time required for nuclear amalgamation. Furthermore, the chance is vanishingly small that the escaping particle will take off all the initial excitation energy on leaving, so that the residual nucleus may still be left in an excited state from which it can drop ultimately to the lowest energy state with a very long characteristic time. The delay in dissipating this residual energy is brought about by the evocation of the processes involved in β and γ emission, which are the only mechanisms available for the loss of the relatively small energy left after the departure of the heavy nuclear particles.

The Bohr theory is essentially statistical in nature and can be expected to hold best for nuclei containing a large number of nuclear particles. It may be expected to undergo much modification as experimental data become more elaborate.

To initiate a nuclear reaction it is necessary, of course, to bring the reactants together. Neutral atoms cannot be used because the closest the two nuclei can approach is the total atomic radius of some 10^{-8} cm. As already discussed (see pp. 2, 8, 9), the nuclear forces do not extend appreciably beyond 10^{-12} cm. To bring two nuclei within reacting distance, one of the nuclei, preferably the light reactant, is stripped of its negative extranuclear electrons and accelerated by electrical means to a high kinetic energy, so that it can penetrate to the nucleus of the target atom. Because such a projectile is positively charged, as is the target nucleus, electrostatic (Coulomb) repulsive forces act to prevent the two nuclei from achieving contact. As the projectile approaches the nucleus, the potential energy of the system (nucleus + projectile) rises so that work must be done on the nuclei to decrease the distance of separation. If the potential energy, $V(r)$, is plotted against

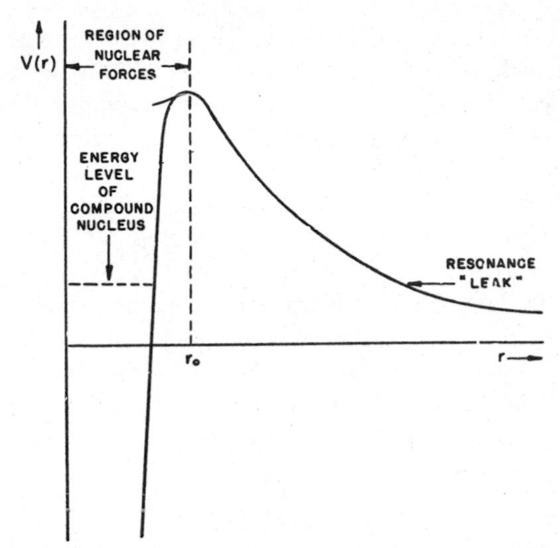

FIG. 5. Nuclear potential energy as a function of distance of separation of a charged bombarding particle from the target nucleus.

the distance of separation, (r), a curve of the type shown in Fig. 5 is obtained. The energy rises to the point r_0 corresponding on the abscissa to the distance at which the Coulomb forces are counterbalanced and finally overcome by the attractive forces which act to hold nuclear particles together. In the region $r < r_0$, the potential energy decreases abruptly to a large negative value representing the binding energy of the intermediate nucleus. Thus the energy curve passes through a maximum for a value of r which may be considered roughly as the "nuclear radius," namely, that distance at which nuclear attractive forces become operative.

The bombarding particle must climb a "hill," so to speak, before it can enter the nucleus.[8] The height of this hill or "potential barrier" is analogous to an activation energy. The analogy to a hill is imperfect, however, in that it is not necessary that a particle surmount the hill to enter the nucleus. On a quantum mechanical basis an incident particle at a point $r > r_0$ (see Fig. 5) has a finite probability of also appearing at a point $r < r_0$ for a given incident energy. Gamow has shown that this probability decreases exponentially with decreasing energy.[9] Even at relatively large distances and for relatively low energies there is some "leakage" of particles through

[8] It is from measurements of the energy required to obtain deviations from the kind of particle-scattering expected from the Coulomb force that the values of nuclear radii are calculated. On the basis of these calculations the empirical relation given in Eq. 7 is obtained.

[9] Gamow, G., *Z. Physik* **51**, 204 (1928); see also Gurney, R. W., and Condon, E. U. *Nature* **122**, 439 (1928).

the hill. This phenomenon is enhanced if there exists an energy configuration of the intermediate nucleus (dotted line in Fig. 5) which has a potential energy content equal to the incident energy of the bombarding particle. In this case "resonance" absorption can occur with concomitant increase in probability for capture.

The probability that a nuclear reaction takes place once the reactants are brought together is conveniently represented by use of the concept "cross section" borrowed from kinetic theory. The cross section, σ, of a nuclear process is the portion of beam area of bombarding particles which is removed by the process in question when the beam is incident on a *single* target nucleus. Suppose a very thin element of target, dx centimeters thick, is exposed to a beam of particles with intensity I, where I is understood to be the number of particles per unit area of beam. The diminution in beam intensity, dI, brought about by the process with cross section σ is given by

$$dI = \sigma \cdot N \cdot I \cdot dx \tag{21}$$

where N is the number of target nuclei per unit volume in cubic centimeters and σ is measured in square centimeters. Since σ is dependent on the kinetic energy of the particles, this expression is valid only for targets so thin that negligible energy loss as well as beam diminution occurs in transit through the target material. Under these conditions, σ is referred to as the "differential" cross section. In a target so thick that the beam is completely absorbed, the total number of nuclear events can be calculated by integration of Eq. 21, provided that the variation of σ with energy is known. Much of the data of nuclear physics is concerned with the dependence of σ on various factors in nuclear reactions, including not only the energy and type of reacting particles but also such factors as symmetry properties of the interacting nuclei, changes in angular momentum for the reactants and products, and spin interactions. A detailed discussion of these matters is not relevant here. It is appropriate, however, to present some general considerations relating to the use of those particles of major importance for tracer production, namely, the neutron and the deuteron.

6. NEUTRON-INDUCED TRANSMUTATIONS

A. General Remarks

The neutron is uncharged and hence experiences no Coulomb repulsion on approaching a target nucleus. The probability of neutron capture is determined mainly by the time it lingers in the vicinity of the nucleus. Since this depends inversely on the velocity of the neutron, it is obvious that the capture probability, and hence the value of σ, for neutron capture will depend to a first approximation on the inverse of its velocity. It is advantageous, therefore, to lower neutron velocities in order to achieve maximal transmutation effects.

Once the neutron is captured, the intermediate nucleus is raised some 8 Mev. above the ground state of the original target nucleus. In this energy region there exist nuclear energy levels spaced rather closely (1 to 100 ev.). Emission of heavy particles such as neutrons, protons, and α particles is quite unlikely, since little or no chance exists, after the distribution of the original 8 Mev. of binding energy among all the nuclear particles, for sufficient energy to concentrate on a single particle again and permit escape. This is all the more true of charged particles, such as protons, which must overcome the resistance of the potential barrier to escape, just as they must overcome the same barrier to penetrate a nucleus in the reverse process of capture. Hence the excess energy left, after capture of a slow neutron and subsequent usual γ emission as the compound nucleus drops to the ground state, is most likely to be dissipated by emission of light particles, such as electrons, resulting from redistribution in number of neutrons and protons to form a more stable nucleus, or by emission of γ rays if the compound nucleus is relatively stable and requires no redistribution of particles. Therefore, slow neutron capture reactions will have relatively high cross sections and will, in general, lead to emission of γ rays and formation of stable nuclei, or emission of β and γ rays if unstable nuclei are formed. Since resonance levels also exist, there may be remarkably high capture cross sections for slow neutrons because of resonance interaction. Thus, capture cross sections 10^3 to 10^4 times the geometrical nuclear cross section, derived from charged particle scattering experiments, are found for certain elements, such as boron, cadmium, and some rare earths, i.e., dysprosium and gadolinium.

As the neutron energy is increased, the cross section for capture decreases, following the inverse velocity relationship. With further increase in energy the cross section begins to rise again because probability of capture followed by heavy particle emission increases with increasing energy. However, the capture probability rarely reaches that found at low energies. At high neutron energies ($\sim$5 to 10 Mev.) the activation energy is sufficient so that heavy particle emission becomes much more probable because sufficient energy is available to concentrate in a relatively short time on some one particle. The re-emission of a neutron is favored because no potential barrier exists against its escape. The proton, being singly charged, should escape more easily than the doubly charged α particle which has twice as high a barrier to penetrate. However, these general predictions are subject to many uncertainties introduced by other factors, such as those mentioned above (symmetry considerations, changes in angular momentum, etc.).

It should be noted that during re-emission of a neutron the emergent particle will in all probability possess an energy relatively lower than that with which the incident neutron entered. This follows from the low probability that *all* the energy of the incident particle will be concentrated on a single particle.

Production of tracers involving (n,γ) reactions is favored at low energies, whereas (n,p) and (n,α) reactions have higher cross sections at high neutron energies. Some interesting exceptions to this statement are important for the production of the long-lived isotopes of carbon and sulfur, as will be seen when these elements are considered (see Chapters X and XI). In consummating the (n,γ) reaction, it is necessary to slow the neutron to low velocities because neutrons are produced as energetic secondary particles in nuclear reactions. Since neutrons are uncharged. energy loss will not take place by mechanisms involving electrical interaction with the electrons of the matter through which the neutrons move. A collision mechanism is needed. In particular, slowing of neutrons is most efficiently accomplished if they are allowed to collide with particles of equal mass, such as protons. The energy loss at each collision with a proton is roughly 50% on the average, so that a small number of such collisions will suffice to reduce a neutron energy of several million electron volts to that corresponding to thermal energies (~0.03 ev.). Heavy nuclei, such as lead, will hardly affect neutron velocities on scattering, since little energy is lost by the light neutron colliding with a very heavy nucleus, just as a tennis ball loses little energy in colliding with a wall. From the practical standpoint, a bombardment involving the (n,γ) process should be carried out with the target embedded in some hydrogenous material such as water or paraffin. Furthermore, no highly absorbing material, such as boron, cadmium, dysprosium, or gadolinium, should be present. Soft glass rather than Pyrex should be used because Pyrex contains considerable quantities of boron.

B. Neutron Sources

Two types of installation are of major importance in the production of neutrons—the uranium pile reactor[10] and the cyclotron.[11] It can be assumed that the reader is familiar with these sources. The reactions employed in neutron production may be listed as follows:

1. The Uranium Fission Reaction. In this reaction the rare isotope of uranium, $_{92}U^{235}$, is split, on absorption of a slow neutron, into two nuclei of approximate atomic numbers 30 to 49 and 50 to 63. During each fission a number of neutrons is emitted, some of which may be used to build up and maintain the fission reaction chain. The rest escape from the reacting mass and become available, therefore, for activation of materials. The neutron yield from this process is so high that macroscopic quantities of new

[10] Friedlander, G., and Kennedy, J. W., "Nuclear and Radiochemistry," Chapter 12. Wiley, New York, 1955.

[11] Livingston, M. S., *J. Appl. Phys.* **15**, 2 (1944); **15**, 128 (1944).

isotopes can be prepared. Although most of the neutrons produced in this process are fast, the moderator necessary for successful operation of the chain reaction makes most of the neutrons available at thermal energies.

Because nuclei of medium atomic weight (fission products) have greater binding energies than do heavy elements, a great release of energy, amounting to almost 200 Mev., occurs during fission. The importance of the fission process is largely due to this fact in addition to the realization of a chain reaction. Nuclei other than U^{235}, like Pu^{239} and other transuranic elements, also can be used as target nuclei. A discussion of the vast mass of data on yields of fission products is available in some of the references listed at the end of this chapter.

2. The Beryllium-Deuteron Reaction. This may be written

$$_4Be^9 + {}_1H^2 \rightarrow {}_5B^{10} + {}_0n^1 \tag{22}$$

The reaction is exothermic, with a heat of reaction approximately equal to 4 Mev. The energy of the neutrons depends on the deuteron energy. The maximum energy attainable is the sum of the heat of reaction and the kinetic energy of the deuteron so that a 12-Mev. deuteron will give neutrons with a maximum energy around 16 Mev. The neutron energies can be controlled by varying the angle of incidence of the initial deuteron beam, as is to be expected from the collision mechanics of the process. Furthermore, a neutron energy spectrum, with groups at various energies lower than the maximum, is found because there are numerous energy levels in ${}_5B^{10}$, the product nucleus, so that the emergent neutron can escape with different discrete energies depending on the state of excitation in which ${}_5B^{10}$ is left. This reaction is most commonly employed because it has a cross section second only to the reactions with lithium and with deuterium, described below, and because beryllium targets can be prepared to withstand high deuteron beam currents, whereas this is not the case with lithium and deuterium.

3. The Lithium-Deuteron Reaction. This reaction may be written

$$_3Li^7 + {}_1H^2 \rightarrow {}_4Be^8 + {}_0n^1 \tag{23}$$

It is more exothermic than the reaction with beryllium and possesses a higher cross section as well. It is used if very high-energy neutrons are required. It is difficult, however, to prepare lithium targets which can dissipate the heat generated by even moderate deuteron beam currents (10 to 50 μa.).

4. The Deuterium-Deuteron Reaction. This reaction, written as

$$_1H^2 + {}_1H^2 \rightarrow {}_2He^3 + {}_0n^1 \tag{24}$$

has the highest cross section at moderate deuteron energies of any deuteron reaction known. This is owing to the very small potential barrier of the light singly charged deuterium nucleus. However, targets containing deuterium which can withstand large beam currents are not available. This reaction is used rarely for tracer production.

5. Alpha-Particle Reactions. Alpha-particle reactions, such as

$$_4Be^9 + {}_2He^4 \rightarrow {}_6C^{12} + {}_0n^1 \tag{25}$$

are used only if naturally radioactive sources such as radium, polonium, or thorium are available.

C. NEUTRON REACTIONS

1. Slow Neutron Capture (n,γ). The capture of a slow neutron results in γ-ray emission (with exceptions noted later in this section). The general equation for the process is

$$_ZT^A + {}_0n^1 \rightarrow {}_ZP^{A+1} + \gamma + \Delta E \tag{26}$$

where T is the target nucleus with mass A and charge Z, P is the product nucleus, and ΔE is the heat of the reaction which appears mainly as γ-ray energy. A typical reaction is

$$_{17}Cl^{37} + {}_0n^1 \rightarrow {}_{17}Cl^{38}\ (\tau_{\frac{1}{2}} = 37 \text{ min.}) + \gamma \tag{27}$$

Over one hundred such reactions are known which result in a radioactive nucleus. Since the ratio of neutrons to protons is increased in this reaction, one may expect that in nearly all cases the resultant nucleus will be a negative β-ray emitter. All slow neutron capture reactions of this type are exothermic, the heat of reaction being largely the binding energy of the captured neutron ($\sim$8 Mev.). As explained above, the yields are greatest for low neutron energies and particularly for energies associated with resonance processes.

2. Neutron Capture with Proton Emission (n,p). The general equation for this reaction may be written

$$_ZT^A + {}_0n^1 \rightarrow {}_{Z-1}P^A + {}_1H^1 + \Delta E \tag{28}$$

The symbols have the usual significance. Because a neutron is added and a proton subtracted, a negative electron emitter can be expected for the product nucleus. The product nucleus will, therefore, revert to the target nucleus in the disintegration following the (n,p) reaction. A typical cycle is represented by the following reactions:

$$_7N^{14} + {}_0n^1 \rightarrow {}_6C^{14} + {}_1H^1 \tag{29}$$

$$_6C^{14} \rightarrow {}_7N^{14} + \beta^- \tag{29a}$$

P, the product nucleus, must be heavier than T, the target nucleus, in order that it emit negative electrons in the decay process. The difference in mass between P and T gives the maximum energy of the β rays emitted. This energy, subtracted from the mass difference between neutron and proton, gives ΔE, the heat of the reaction. If the maximum β-ray energy of P is less than the mass difference $(n - p)$, then the (n,p) reaction is exothermic and can be initiated by neutrons with zero kinetic energy.

However, the potential barrier always present for charged particles interferes with proton emission. Only when the atomic number, and consequently the potential barrier, is low can appreciable proton emission occur at low neutron energies. The yield may be considerably enhanced, however, if a resonance level is involved. This seems to be true for two important instances, namely, N^{14} (n,p) C^{14} and Cl^{35} (n,p) S^{35}. These two reactions are of major importance in the production of the important long-lived radioactive carbon and sulfur isotopes. The barrier height against proton emission is 2 Mev. for nitrogen and 5 Mev. for chlorine. Nevertheless, high yields of C^{14} and S^{35} are obtained with slow neutrons. These two reactions constitute important exceptions to the general rule that slow neutron capture invariably results in γ-ray emission.

As the neutron energy is increased, the (n,p) reaction becomes more probable because sufficient energy to nullify the barrier is supplied. Over sixty (n,p) reactions have been studied, and the radioactive products have been found to be negative β emitters for the most part. In the region of higher atomic numbers ($Z > 35$), it is possible for stabilization to occur either by negative or positive β-ray emission because stable isobars occur with some frequency at higher atomic numbers. For this reason some positron emitters are also noted.

3. Neutron Capture with α-Particle Emission (n,α). This reaction occurs practically only with high-energy neutrons because of the high barrier to α-particle penetration. The general equation for the reaction in the usual nuclear terminology is

$$_{Z}T^{A} + {}_{0}n^{1} \rightarrow {}_{Z-2}P^{A-3} + {}_{2}He^{4} \tag{30}$$

Most of the reactions studied have been in the region of atomic number ≤ 30. The reaction is quite rare at high atomic numbers because of the great barrier height. In general, competition from neutron or proton emission reactions lowers the yield of (n,α) considerably so that it compares unfavorably with other high-energy reactions.

4. Neutron Capture with Neutron Emission (n,n). This reaction has a large probability at high energies because of the absence of a barrier to neutron emission, as discussed before. The emergent neutron will have considerably less energy than the impinging neutron, so that the whole process may be considered as an inelastic collision of target nucleus and neutron. The energy remaining in the target after neutron emission can be dissipated as γ radiation. Since no change occurs in n–p ratio in this reaction, the product nucleus is merely an excited form of a stable nucleus. Thus, in the case of In^{115},

$$In^{115} + {}_0n^1 \rightarrow In^{115*} + {}_0n^1 \tag{31}$$

an isomer of In^{115} is formed which decays only by γ emission with a 4-hr. half-life.

The reaction is of some interest because the limiting cross section at high energies approaches the geometrical cross section of the nucleus.

5. Multiple Ejection of Nuclear Neutrons $(n,2n)$. The general reaction may be written

$$_ZT^A + {}_0n^1 \rightarrow {}_ZP^{A-1} + 2{}_0n^1 \tag{32}$$

The ejection of two neutrons from a nucleus after capture of one neutron requires at least the energy corresponding to the binding energy of the extra neutron released. This energy is, on the average, about 5 to 8 Mev. so that this reaction is usually endothermic by approximately 5 to 8 Mev.

The energy can also be supplied by using not a neutron but a high-energy γ ray or deuteron, i.e., (γ,n) or $(d,2n)$. In general, multiple emission at high energies is usually associated with emission of neutrons because the neutron is favored over other heavy particles for reasons already stated. As in the (n,n) reaction, the $(n,2n)$ reaction also has a cross section approaching in magnitude the geometrical cross section as the energy is increased. Many cases have been studied in which radioactive isotopes are formed. In these cases positron emitters are usually formed because the n–p ratio is lowered.

6. Concluding Remarks. Of the reactions described above, only the (n,γ) and (n,p) reactions are of major importance in tracer production. This follows from the high cross sections usually encountered in these reactions. Low energies are most favorable for the (n,γ) and some important (n,p) reactions, so that some arrangement to increase the ratio of slow to fast neutrons in neutron sources is required to attain maximal yield. Material which is effective in slowing neutrons must not contain substances which themselves exhibit large neutron capture cross sections.

7. DEUTERON-INDUCED TRANSMUTATIONS

A. General Remarks

Deuteron-induced reactions differ radically from those induced by neutrons. The deuteron is a singly charged complex of proton and neutron. Hence, a potential barrier exists between bombarding deuterons and target nuclei. Increasing capture cross sections are obtained with increasing rather than decreasing energy. The deuteron is favored over other charged particles for transmutation purposes mainly because it is a singularly loose combination of neutron and proton. The binding energy of the deuteron is only 2.18 Mev., or about 1 Mev. per particle. The deuteron may "polarize" partially in the nuclear field, the neutron component being turned toward the nucleus, the proton component being repelled. The neutron component can be captured by the target nucleus without complete capture of the deuteron as a whole.[12] Hence, a reaction which amounts formally to an (n,γ) reaction is initiated at energies much lower than would be required if complete capture of the deuteron were necessary. Because of the small binding energy of the deuteron, practically all deuteron reactions involving single particle emission—(d,n), (d,p), or (d,α)—are exothermic. As a consequence of these factors, deuteron reactions usually give higher yields than proton or α-particle reactions. Furthermore, in the cyclotron, a deuteron beam of a given energy is easier to obtain than a proton beam of the same energy. Higher intensities are obtainable with singly charged particles such as deuterons and protons than are obtainable with the doubly charged α particles. All these considerations point to the use of deuterons rather than any other charged particle at present for tracer production.

The energy dependence of the capture cross section for various deuteron reactions can be predicted in a rather broad way, although specific features require an intimate knowledge of nuclear forces which is not as yet available. The following facts emerge from experimental studies:

1. At low energies (i.e., kinetic energies less than the barrier height), reaction yields increase exponentially with energy, following the Gamow penetration probability which is also an exponential function of energy. Resonances are observed in the very low-energy region (0.1 to 0.5 Mev.), as is to be expected from the energy levels of nuclei disclosed by work with slow neutrons. However, as far as yields are concerned, the exponential increase with energy is the major factor, becoming unimportant only after the energy exceeds the barrier height. This height is approximately 8 Mev. at atomic number 50.

2. Once the barrier is surmounted, the total capture probability is essentially constant with increasing energy. Thus the differential capture cross-section value is constant and independent of energy. Increase in yield is due only to increased penetration of target material. This increase in range is proportional to the 3/2 power of the energy. If the differential cross section for a particular transmutation is plotted against the deuteron energy, there will be an initial exponential rise followed by a flat portion where the cross section becomes constant (see Fig. 6). At extreme energies,

[12] Oppenheimer, R., and Phillips, M., *Phys. Rev.* **48,** 500 (1935).

competition with new reactions rendered more probable at high energies can lower the cross section from its maximum constant value.

3. As pointed out in the discussion of neutron reactions, charged particle emission is less probable than neutron emission; i.e., the (d,n) reaction should be more probable than the (d,p) reaction. Actually, this is not true unless complete capture of the deuteron is involved. The polarization phenomenon mentioned above allows a (d,p) reaction to occur without total capture, so that the yield from (d,p) reactions at low energies (2 to 5 Mev.) for low and medium atomic numbers may be considerably higher than for (d,n) reactions. The (d,n) reaction cross section rises above the (d,p) cross section once sufficient energy is available to allow the deuteron to be absorbed completely. This effect is especially noticeable when sufficient energy is available to initiate the $(d,2n)$ reaction. At these energies the $(d,2n)$ cross section rises rapidly above the (d,n) and (d,p) cross sections.

4. Those deuteron reactions which are endothermic show no yield below a certain threshold energy which is equal to the negative heat of reaction. Once this energy is exceeded, the cross section will increase exponentially to some limiting value in a manner similar to that observed for exothermic reactions.

B. Deuteron Reactions

1. Deuteron Capture with Proton Emission (d,p). The general equation for this process may be written

$$_{Z}T^{A} + {}_{1}\mathrm{H}^{2} \rightarrow {}_{1}\mathrm{H}^{1} + {}_{Z}P^{A+1} \tag{33}$$

This reaction can occur either with partial capture or with total capture, as discussed above. The reaction is formally equivalent to (n,γ), so that the nature of the products is the same as for the (n,γ) reaction. Thus, mainly negative β-particle emitters are formed. Although all elements can be transmuted in the (n,γ) process by any neutron, it requires $\sim$16-Mev. deuterons to accomplish the (d,p) reaction on all elements in good yields.

2. Deuteron Capture with Neutron Emission (d,n). As discussed previously, this reaction has the highest cross section in the energy range in which deuteron reactions proceed entirely by total capture. The general equation is

$$_{Z}T^{A} + {}_{1}\mathrm{H}^{2} \rightarrow {}_{0}n^{1} + {}_{Z+1}P^{A+1} \tag{34}$$

3. Deuteron Capture with α-Particle Emission (d,α). Much the same remarks as were made for (n,α) hold for this type of reaction. The (d,α) reaction has some importance in the single case of production of long-lived radioactive sodium (Na^{22}). The general reaction is

$$_{Z}T^{A} + {}_{1}\mathrm{H}^{2} \rightarrow {}_{Z-1}P^{A-2} + {}_{2}\mathrm{He}^{4} \tag{35}$$

4. Multiple Neutron Emission after Deuteron Capture $(d,2n)$. This reaction has the highest cross section at high deuteron energies (12 to 16 mev.). It is formally equivalent to (p,n), so that if it is desired to make an isotope which requires substitution of a proton with a neutron, as in most positron emitters, this reaction can be used with good efficiency despite the usually high threshold energy resulting from its endothermic nature. The general reaction is

$$_{Z}T^{A} + {}_{1}H^{2} \rightarrow {}_{Z+1}P^{A} + 2{}_{0}n^{1} \tag{36}$$

5. Tritium Emission (d,H^3). This interesting reaction has been observed in a very few instances, the most important being that in beryllium, $Be^9(d,H^3)Be^8$. It occurs with an extremely high cross section in the deuterium-deuteron reaction:

$$_{1}H^{2} + {}_{1}H^{2} \rightarrow {}_{1}H^{3} + {}_{1}H^{1} \tag{37}$$

6. Competition and Yields of Deuteron Reactions. It can be seen that, once a deuteron at high energy is absorbed by a nucleus, the intermediate nucleus formed may disintegrate in a variety of ways. In fact all reactions that are energetically possible will occur to a greater or lesser extent. A case often cited is that of copper. Some possible reactions are presented in the following scheme:

$$_{29}Cu^{63} + H^{2} \rightarrow [{}_{30}Zn^{65}] \begin{cases} \rightarrow {}_{30}Zn^{64} + {}_{0}n^{1} & (a) \\ \rightarrow {}_{30}Zn^{63} + 2{}_{0}n^{1} & (b) \\ \rightarrow {}_{28}Ni^{61} + {}_{2}He^{4} & (c) \\ \rightarrow {}_{29}Cu^{64} + {}_{1}H^{1} & (d) \\ \rightarrow {}_{29}Cu^{62} + {}_{1}H^{3} & (e) \end{cases} \tag{38}$$

Reaction *a* might be expected to have the highest yield starting at neutron energies of 4 to 5 Mev. However, because of the partial-capture phenomenon, reaction *d* begins first at an energy of some 2 to 3 Mev. falling off in yield relative to reaction *a* only when bombarding energies of some 10 Mev. are reached. The (d,α) reaction (reaction *c*) begins approximately at the same voltage as the (d,p) reaction (reaction *d*), but with much smaller yields, never equaling at any time either (d,n) or (d,p) yields throughout the energy range 3 to 16 Mev. The (d,H^3) reaction has the lowest cross section of all the reactions possible and shows appreciable yield only at relatively high bombarding energy. The $(d,2n)$ reaction has a threshold energy of some 6 Mev. and so does not compete seriously with (d,n) and (d,p) until this energy is exceeded appreciably. At 13 to 16 Mev., however, the cross section of $(d,2n)$ rises so high that the total yield from this reaction is comparable with both (d,n) and (d,p).

Experimental curves[13] for the processes $Na^{23}(d,p)Na^{24}$, $Br^{81}(d,p)Br^{82}$, and

[13] Clarke, E. T., and Irvine, J. W., Jr., *Phys. Rev.* **66**, 231 (1944); **70**, 893 (1946).

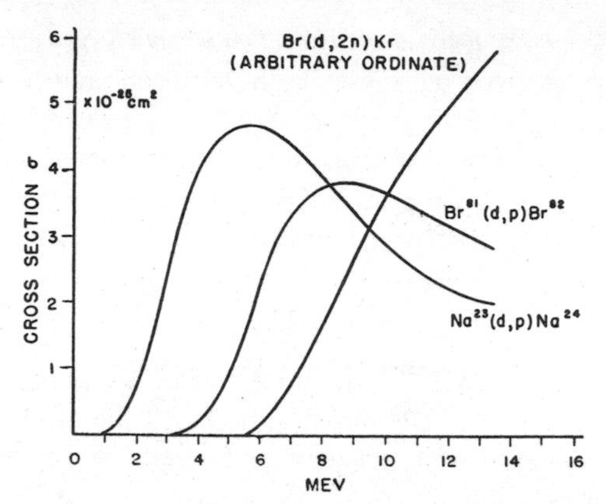

FIG. 6. Cross-section-energy relations for various deuteron-induced reactions in sodium bromide. (After Clarke and Irvine.[13])

$Br^{79,81}$ $(d,2n)$ $Kr^{79,81}$ are shown in Fig. 6. Here the differential cross section, σ, is plotted against the deuteron energy, E. It is seen that for the (d,p) reaction on sodium the cross section is appreciable only at 1 to 2 Mev., rising rapidly to a maximum value of 4.7×10^{-25} cm.^{2} at 5.5 Mev. Beyond this bombarding energy competitive reactions lower the (d,p) cross section until, at 14 Mev., the cross section is less than half its maximum value at 5.5 Mev. The total yield of Na^{24} is obtained by integration of the curve for σ vs. E of Fig. 6. The so-called thick target yield or integral curve of yield vs. energy is given in Fig. 7. It will be appreciated that the yield obtained is the area under the curve in Fig. 6 relating the differential cross section, σ, to the energy. Thus, although maximum σ is obtained at 5.5 Mev., most of the area is contributed beyond this point. Thus the yield at 14 Mev. is some six times that at 5.5 Mev. The increased penetration of the deuterons at high energies compensates for the lowering of cross section so that it is more economical to bombard at high energies, provided there is approximately the same beam intensity available.

It should be noted that at 5.5 Mev. a yield of approximately 0.5 mc./μa.-hr. of deuterons is obtained. At 14 Mev. the yield is 3.0 mc./μa.-hr. Thus six times the beam is required at 5.5 Mev. as at 14 Mev. to get the same yield in the same time.

With respect to the (d,p) and $(d,2n)$ reactions on bromine shown in Fig. 6, the effect of the competition of the $(d,2n)$ process with the (d,p) process is quite clearly apparent. Undoubtedly the (d,n) process (not shown) is also responsible for decline of the (d,p) reaction. The curve for the (d,n) reaction would be intermediate between the (d,p) and $(d,2n)$ curves. The corresponding integral curves for total yield are given in Fig. 7.

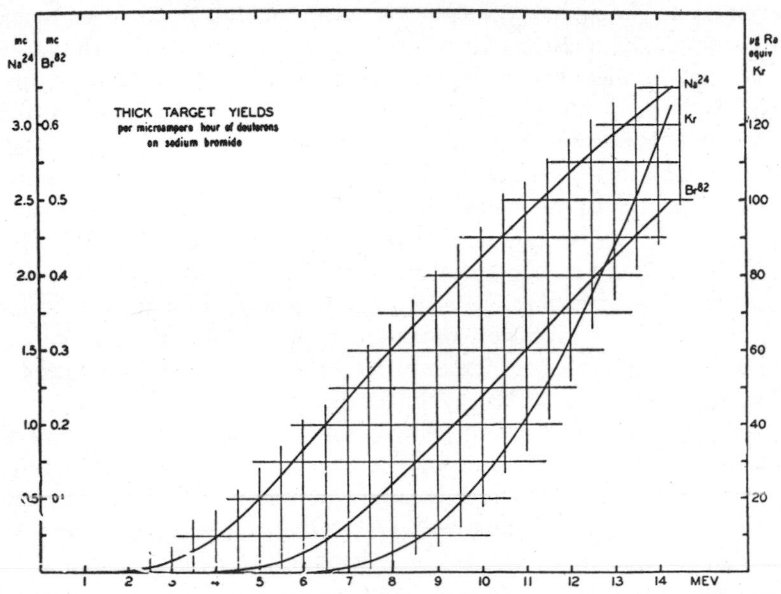

FIG. 7. Thick target yield curves for processes shown in Fig. 6. (After Clarke and Irvine.[13])

8. TARGET TECHNIQUES AND RADIOCHEMISTRY

A. TARGET CHEMISTRY

All tracer experiments begin with target manipulations which may be carried out both during and after bombardment. The vicissitudes undergone by the target may modify and determine to a large extent procedures for the later stages of any tracer experiment. The main factor in all target preparations is the large heat input from the bombarding beam of nuclear particles. Only a small fraction of the deuteron beam in a cyclotron is utilized in nuclear processes. Even at bombarding energies as high as 16 Mev. there is capture of no more than 0.1 to 0.2% of the incident deuterons by a target of low atomic number such as beryllium. The deuterons for the most part are slowed by electrical interaction with the electrons in the target material, eventually coming to rest as neutral atoms or, what is more probable, reacting to form a deuteride. If a deuteron beam with an intensity of 100 μa. and an initial energy of 10 Mev. is absorbed in a target, 1 kw. of power or 250 cal. of heat input per second must be dissipated.

Neutron bombardments with the cyclotron pose no particular heat dissipation problem, since only the fraction of the primary accelerated particles (deuterons) captured can give rise to the secondary neutrons, and this fraction is no more than 0.1 to 0.2%, even at high bombarding energies. Furthermore, the neutrons are emitted in all directions, and the radiation flux is much smaller than in the initial deuteron beam. In the uranium pile, however, the neutron flux approaches that obtained in the cyclotron for the primary deuteron beam. This can be seen from the published data on neutron activation[14] in which yields attained with the pile are given which appear to be 10^4 to 10^5 times the comparable neutron activation yields obtained with similar target placement in the cyclotron source. Hence, power dissipation for neutron activation in the pile presents problems similar to those encountered with deuteron activation in the cyclotron. Since few details are available on the manner in which these problems are handled in the pile, this discussion will be concerned only with target techniques as practiced with the cyclotron.

Two types of targets are in general use in cyclotrons. In one arrangement, the cyclotron beam is brought out of the vacuum chamber through a thin metal foil "window." The target material is placed in a chamber external to the vacuum chamber and irradiated by the emergent beam. This arrangement is known as the "external" or "bell-jar" target[15] and has many advantages. Chemically active substances, such as phosphorus, can be bombarded by using an atmosphere of inert gas, such as helium, which produces no contaminating activities and which aids in cooling the phosphorus by heat conduction. The bombardment of materials such as phosphorus cannot be carried out inside the cyclotron because the resulting vaporization of the phosphorus would destroy the vacuum and produce failure in cyclotron operation.

The cyclotron beam emerges spread out over an area which is considerable when compared to its area inside the vacuum chamber. Hence, it is a relatively simple matter to water-cool external targets because the beam intensity, i.e., heat input per unit area, is lowered. As against these advantages, there are the disadvantages that the available external beam is rarely more than 20% of the internal beam and that a large amount of target material must be used to absorb the beam. A lowering in "specific activity" (see p. 10) results if the target reaction is one in which the target nucleus and product nucleus are identical chemically. Furthermore, the beam transmitted to the external target is limited by the window material which melts despite vigorous air cooling at beam intensities of 100 to 200 μa. For a detailed discussion concerning the mechanical details of bell-jar target chambers, the reader should consult the literature.[15]

When the target material can be bombarded with safety inside the cyclo-

[14] *Science* **103**, 697 (1946).

[15] Kurie, F. N. D., *Rev. Sci. Instr.* **10**, 199 (1939). Also see reference 11 for a general discussion of target techniques.

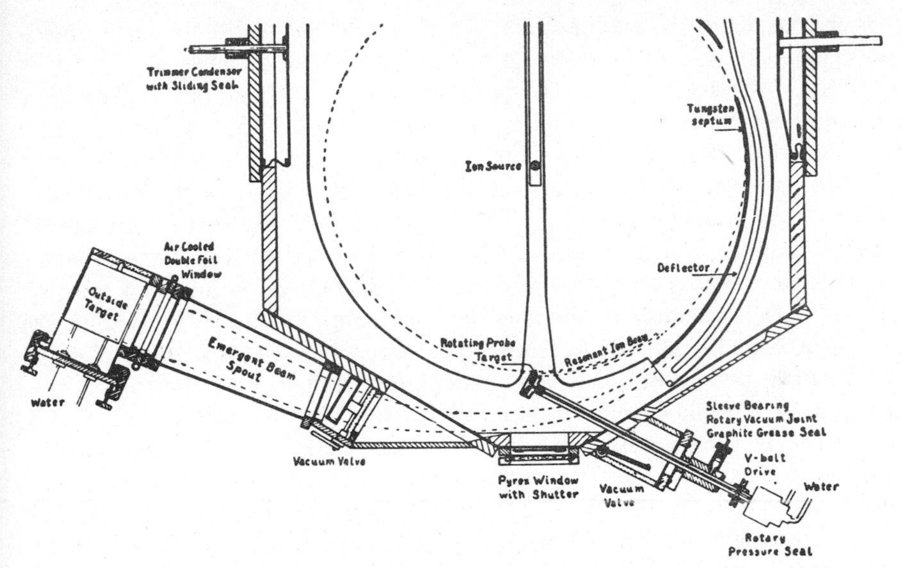

FIG. 8. Target arrangement in the M.I.T. cyclotron. (After Livingston.)

tron vacuum chamber, much larger beams can be employed and these beams can be concentrated over a very small area. The target in this case is referred to as an "internal" or "probe" target.[16] Beams as high as 1 ma. can be obtained over areas of less than 1 sq. in. This amount of heat input per unit area cannot be dissipated merely by water-cooling, and hence devices for rotating and rocking the probe are used. The beam intensity is limited to a power dissipation in the neighborhood of 30 kw./sq. in. Target materials are in most cases limited to metals or metallic compounds which possess high heat conductivity and which can be soldered or plated to copper backing surfaces which, in turn, are water-cooled. In Fig. 8 the target arrangement used in the cyclotron at the Massachusetts Institute of Technology is shown. Both types of targets have been used simultaneously.

It is desirable that a maximum portion of any target be made up of nuclei which are the reactant in the nuclear process to be used. Thus, if radioactive phosphorus (P^{32}) is to be made by the $P^{31}(d,p)P^{32}$ reaction, the best target is one made of pure phosphorus. However, it may be necessary to bombard the target in a vacuum or under conditions in which comparatively efficient cooling is impossible. It will not be possible to use pure phosphorus in such a case because its poor heat conductivity leads to overheating and excessive volatilization. In this contingency some more durable material such as a metal phosphide is required. The phosphide which combines high phosphorus content and heat stability is the target indicated.

[16] Wilson, R. R., and Kamen, M. D., *Phys. Rev.* **54**, 1031 (1938).

Other considerations, such as ease of bonding to the metal-cooling surface and gas content, may be decisive in the final choice of target.

Because the necessity for cooling exists and because this implies bonding to metal surfaces such as copper, iron, silver, or platinum, it follows that a chemical procedure for final purification of the target material must be elaborated. Each of the contaminating materials will give rise to its characteristic radioactive products. These, in general, will cover a range of atomic number from two less to one more than that of the contaminant, if deuterons are the bombarding particles. In the bombardment of ferrous phosphide with deuterons, the chemical procedure must purify phosphorus and its attendant radioactive isotope P^{32} from activities associated not only with sulfur, but also iron, cobalt, manganese, copper, zinc, and nickel activities. If the phosphide contains impurities such as sodium, chlorine, cobalt, nickel, and arsenic, the procedure must take these into account as well as elements like selenium, germanium, gallium, chromium, and vanadium. In most cases the procedure may be simplified by the occurrence of only short-lived impurities which disappear rapidly enough to constitute no contamination problem.

It is quite apparent that the production of any radioactive isotope is a unique chemical problem. Few remarks of a far-reaching validity are possible. It may be said that ease of chemical purification quite often overrides considerations of target efficiency based on isotope abundance. These matters, as well as others mentioned in the following paragraphs, will be discussed in more detail under the special sections devoted to the preparation of particular isotopes.

B. Separation of Isotopes by the Szilard-Chalmers Process

Under certain conditions it is possible to isolate radioactive isotopes from inactive isotopes of the same element during neutron bombardment by taking advantage of the recoil energy imparted to the product nucleus during capture of the neutron. This type of process is known generally as the *Szilard-Chalmers reaction*, named after the investigators who first demonstrated its existence.[17] The initial studies showed that when an alkyl halide was bombarded with slow neutrons most of the radioactivity associated with the halogen atoms could be extracted with water. It was deduced that the radioactive halogen was present mainly in some inorganic form such as halide ion or atom.

In ethyl iodide, capture of a neutron by the iodine atom leads to rupture of the C-1 bond, the active iodine then recoiling into the body of the liquid where it may exist as a neutral atom or ion. On addition of water, most of this iodine emerges as iodide and enters the water phase in which it is more

[17] Szilard, L., and Chalmers, T. A., *Nature* **134**, 462 (1934).

soluble. The activity is associated with a very minute fraction of the original iodine so that a high specific activity is obtainable.

$$C_2H_5I^{127} + {}_0n^1 \rightarrow C_2H_5 + I^{128*} \tag{39}$$

If the active atom formed enters into an exchange reaction with the inactive isotope present in the parent material, the yield of the process is lowered, thus:

$$C_2H_5I^{127} + I^{128*} \rightarrow C_2H_5I^{128*} + I^{127} \tag{40}$$

Such an occurrence would vitiate the advantage gained in using this kind of reaction.

In general, the total yield obtained in this type of process represents a balance between the initial capture and subsequent exchange reactions. It is possible to arrange conditions of irradiation so that exchange reactions are minimized and maximal yields obtained. Details of these procedures are best presented in connection with discussion of production processes related to particular isotopes.[18] The general conditions for efficient utilization of the Szilard-Chalmers method are:

1. Sufficient recoil energy must be available to the nucleus on capture of the neutron so that the chemical bond can be ruptured.
2. There must be no exchange of activated atoms with inactive atoms.
3. The radioactive atoms must be chemically separable.

In clarification of the first requirement there follows a calculation of the energy involved in the Szilard-Chalmers process. One begins by assuming conservation of momentum between recoiling nucleus and emitted γ ray, immediately after capture of the neutron. Because the neutron is virtually at rest when captured, the only energy available is the energy of binding of the neutron to the nucleus. The momentum of the γ ray, p, is given by the relation

$$p = E_\gamma/c \tag{41}$$

where E_γ is the energy of the γ ray and c is the velocity of light. The recoiling nucleus will possess an equal and opposite momentum. If its mass is M and the recoil energy is denoted by E, then

$$E = p^2/2M \tag{42}$$

Substituting for momentum in terms of the γ-ray energy, it is seen that

$$E = (E_\gamma/c)^2 \cdot \tfrac{1}{2}M = E_\gamma^2/2Mc^2 \tag{43}$$

[18] See Friedlander, G., and Kennedy, J. W., "Nuclear and Radiochemistry," Wiley, New York, 1955, for a general discussion of chemical phenomena attending nuclear recoil as well as for a general review of the applications of radioactive tracers to chemical problems.

It is convenient to express E in units of millions of electron volts. Since 1 m.u. (M_Hc^2) is equivalent to 931 Mev., E is given as $E^2/1862M$. When this expression is multiplied by 10^6, E is given in terms of electron volts as $536(E^2/M)$. This formulation neglects relativistic effects due to variation of electron mass with velocity. If a 5-Mev. γ ray is emitted by a nucleus of mass 50, the recoiling nucleus will have an energy equal to $(536 \times 25)/50$, or 268 ev. A heavy element, such as lead with a weight close to 200, will have a recoil energy of only some 65 ev. In most circumstances, therefore, neutron capture will lead to rupture of the chemical bond. Hence the first requirement is not stringent. It should be remarked, however, that Eq. 43 gives a maximal value for the nuclear recoil energy on the assumption that only one γ ray is emitted. Edwards and Davies[19] have discussed the possibility that more than one quantum may be involved, in which case the nuclear recoil energy can be much less than that calculated by Eq. 43. If the two quanta are given off in opposite directions, the nuclear recoil energy can become zero.

The second requirement with regard to exchange reactions is rather severe. The recoil atoms are effectively at very high temperatures because their energies are ten to one hundred times ordinary chemical energies. It is to be expected, therefore, that such recoil atoms will readily undergo reaction under circumstances in which ordinary atoms would not react. An activated bromine atom could displace a chlorine atom from a molecule such as C_2H_5Cl. Irradiation of a molecule such as C_2H_5Br will lead to formation of appreciable amounts of active $C_2H_4Br_2$. The phenomenon of exchange is important in radiochemistry, and the study of exchange reactions between different chemical forms of the same elements can be utilized to give information on strength of chemical bonds as a function of solvent, temperature, etc. From the biological viewpoint, the existence of exchange reactions is important in interpretation of tracer experiments, because molecules once labeled must retain the label against all processes except those involved in the actual metabolism of such a molecule. This matter will be considered in detail in Chapter VI. For the present discussion certain generalizations, admittedly imprecise, should be made. If a given element is present in different states of combination in a homogeneous system, an exchange reaction will not proceed at a measurable rate unless reversible equilibria can be found through which such exchanges can take place. This would imply little exchange between two forms of chlorine such as chloride ion and chloroform. However, an exchange between chloride and chlorate is quite probable because there exists an oxidation-reduction equilibrium between these two forms. Another good example of an exchange equilibrium brought about through an intermediate equilibrium is the exchange

[19] Edwards, R. R., and Davies, T. H., *Nucleonics* **2**, No. 6, 44 (1948).

between iodine and iodide ion. Here the formation of the symmetric I_3^- ion is involved. Such exchanges, in general, involve some collision mechanism with formation of intermediates. It is also possible to effect exchange reactions by simple electron transfer mechanisms. Thus an exchange is observed to occur between two such complexes as MnO_4^{-2} and MnO_4^-. In this case there exists only a difference in electrical charge so that transfer of an electron brings about the same result as though there had been actual exchange of the central atom.

Mention of an interesting application of the Szilard-Chalmers process may be interpolated here. Anderson and Delabarre[20] have found it possible to activate cobalt in vitamin B_{12} by the $Co^{59}(n,\gamma)Co^{60}$ reaction with high Co^{60} retention and small loss of biological activity. Cobalt is known to be bound firmly in the vitamin, and it appears that, despite the large energy of recoil, a surprisingly large fraction of the cobalt remains in the parent molecule. Anderson and Delabarre sealed 8.8 mg. of crystalline vitamin B_{12} in a quartz ampoule *in vacuo*. The vitamin was then exposed to a thermal neutron flux of 1×10^{13} neutrons/cm.^{2}/sec. for 7 days at $\sim$80° C. After irradiation, the sample was allowed to stand for two months so that all short-lived activities disappeared. No visible alteration in the crystals was observed, nor was any water-insoluble material formed. The crystals were dissolved and purified by paper and column chromatography. One fraction which behaved in the same manner as authentic B_{12} was analyzed and found to be identical chemically and biologically with vitamin B_{12}. The purified material was shown to account for $\sim$80% of both the original radioactivity and bioactivity. Earlier claims for similar high retentions during activation of sulfur in cystine[21] have not been confirmed,[22] however. Thus, it appears that simple energetic considerations are not sufficient to predict degrees of retention of atoms in complex molecules.[23]

C. Survey of Radiochemistry

In developing chemical separation procedures, the nuclear chemist has available a large reservoir of facts from the field of natural radioactivity. The basic principles of radiochemistry have been well worked out for the naturally radioactive elements by such pioneers as Hevesy, Paneth, Fajans, Hahn, and others. In the rest of this section, a brief survey of radiochemistry will be given, with particular emphasis on chemical separation procedures involved in the preparation of tracer materials from activated targets.

Hahn has systematized the precipitation phenomena involved in the co-

[20] Anderson, R. C., and Delabarre, Y., *J. Am. Chem. Soc.* **73,** 4051 (1951).
[21] Ball, E. G., Solomon, A. K., and Cooper, O., *J. Biol. Chem.* **177,** 81 (1948).
[22] Lipp, M., and Weigil, H., *Naturwissenschaften* **30,** 189 (1952).
[23] See Willard, J. E., *Ann. Revs. Nuclear Sci.* **3,** 214 (1953).

precipitation of micro amounts of radioactive elements with bulk precipitates of other substances. These coprecipitation phenomena are classified conveniently as follows:

1. Isomorphous Substitution. If the radioactive element is truly isomorphic with the precipitating element, the radioactive element will be incorporated in the precipitate. The distribution of the radioactivity throughout the solid is continuous and not affected particularly by the precipitation procedure. Washing will not remove the radioactivity.

2. Superficial Adsorption. If the precipitate forms a surface layer which has surface-active portions bearing an electric charge opposite in sign to the radioactive ion, there is a tendency for coprecipitation at the surface to occur. This is especially true when the radioactive ion tends to form an insoluble compound with the surface ions of opposite charge. The condition of the surface is affected markedly in sign and effective area by conditions of precipitation. Thus, a freshly formed large surface (finely divided form) will tend to occlude the radioactive ion more than an aged coarse precipitate with relatively small surface area. It is possible to alter the amount occluded by washing procedures.

3. Anomalous Substitution. Sometimes it is found that a coprecipitation with isotropic distribution occurs, as in true isomorphous substitution, even though formation of solid solutions cannot occur with weighable amounts of the two components. It can be shown that $PbCl_2$ is an anhydrous rhombic crystal and $BaCl_2 \cdot 2H_2O$ is a hydrated monoclinic crystal. Nevertheless, a radioactive lead isotope such as RaD will distribute itself continuously in crystalline $BaCl_2 \cdot 2H_2O$. An explanation lies in the possibility that, at the low concentrations of lead isotope encountered when working with radioactive lead solutions, a limited solid solubility is possible.

In addition to these three classes of coprecipitation there are some ill-defined cases in which irregular internal adsorption takes place just as though portions of mother liquor had been trapped in the crystal. In chemical separation procedures involving artificial radioactivity, it is always possible to work with macroscopic amounts of the element in question simply by adding a quantity of inactive isotope. Under these conditions, dependence on coprecipitation phenomena is minimized. The necessity for a thorough study of such phenomena in natural radioactivity was largely derived from the fact that no stable isotopes of some of the radioactive elements involved existed. Nevertheless, a knowledge of conditions under which coprecipitation of radioactive elements with substances chemically dissimilar can occur is vital in properly designing separation procedures.

In devising chemical procedures involving artificial radioactive substances it is important to realize the amounts of such material involved. Suppose a target of phosphorus is bombarded with deuterons to produce radioactive phosphorus by the $P^{31}(d,p)P^{32}$ reaction. Let it be assumed that a chemical purification is carried out and there is obtained a sample of phosphorus which is found to exhibit a radioactivity corresponding to 1.2×10^{10} disintegrations/sec. in a total of 1 g. of phosphorus. From the fundamental radioactivity law (Eq. 9) the rate of decay is related to the total number of P^{32} atoms present as

$$dN/dt = -1.2 \times 10^{10} = -\lambda N$$

where λ is the characteristic disintegration constant; λ can be calculated, if the half-life ($\tau_{\frac{1}{2}}$) of P^{32} is known (Eq. 12). Thus

$$\lambda = 0.693/\tau_{\frac{1}{2}}$$

Since $\tau_{\frac{1}{2}}$ for P^{32} is 14.3 days or approximately 1×10^6 sec., λ is approximately 7×10^{-7} sec.$^{-1}$. Thus N will be $1.2 \times 10^{10}/7 \times 10^{-7}$ atom. This is $32 \times 1.7 \times 10^{16}/6 \times 10^{23}$ or 9×10^{-7} g. (somewhat less than 1 γ) as P^{32}. Thus the specific labeled content is 9×10^{-7}, since a total of 1 g. of P^{31} is present in this sample. This represents an extremely active sample of phosphorus.

It may be necessary to work with samples in which there are only 1.2 disintegrations/sec. Such a sample is weaker by a factor of 10^{10} than the one described and would correspond to 9×10^{-17} g. as P. If such a small amount of P^{32} were formed in a reaction such as $Cl^{35}(n,\alpha)P^{32}$, and if no P^{31} were present as contaminant, it is quite evident that the amount of phosphorus present would be far below that which could be relied upon to give characteristic preparation reactions of phosphorus. There would certainly be too little phosphorus present to form any of the characteristic phosphorus precipitates on addition of the proper reagents, since the solubility product would never be exceeded at such high dilutions. It is usually advisable to add small quantities (milligrams) of the element to be purified so that ordinary chemical manipulation is possible. The amount of such material added is minimal to keep the specific activity maximal. Such material is called "carrier." In any bombardment in which an element of mass A and charge Z is transmuted to other elements with charge $(Z + 2)$ to $(Z - 2)$, one adds carrier (1 to 10 mg.) of each element so formed, so that in the subsequent chemical separations each radioactive isotope follows its characteristic chemistry. The carrier, of course, must be equilibrated chemically with the radioactive isotope before a chemical separation is carried out.

In certain very special cases, carrier is not needed to effect a separation. Since no dilution occurs, extremely high specific activities can be attained. Thus, differential distribution between two immiscible solvents can be exploited to separate radioactive gallium (Ga^{68}) from zinc after bombardment of a zinc target with deuterons. The zinc target is dissolved in HCl, the acidity adjusted to 6 N, and the gallium extracted as $GaCl_3$ with diethyl ether. The removal of the radioactive gallium is as efficient as when it is present in macroscopic amounts. A similar technique is not applicable to the separation of radioactive iron where carrier is found necessary to ensure proper distribution.

Another type of separation of radioactive impurities can be carried out with precipitates such as MnO_2 and $Fe(OH)_3$. Thus, in the purification of a material such as radioactive calcium from the $Ca^{44}(d,p)Ca^{45}$ reaction, contaminating heavy metal activities can be removed by adding a small quantity of iron and precipitating $Fe(OH)_3$ with carbonate-free ammonia. All heavy metal activities can be removed in this fashion even though no specific carrier is added. This is in spite of the expectation that the isotopes concerned would not precipitate as the hydroxides because they are present in too low a concentration to exceed the solubility product constant for the metal hydroxides. This type of precipitate is referred to as a "scavenger," its action being based mainly on surface adsorption.

In addition to precipitation procedures, carrier-free separations can be carried out electrochemically. In the cases cited above involving radioactive copper and zinc, immersion of a piece of lead foil in the radioactive solution will result in the plating out of copper while zinc is left in solution. This method is applicable, in general, to all cases where a large difference exists in oxidation-reduction potential between radioactive components. Other methods based on volatility, distribution between immiscible solvents, adsorption, etc., have been employed.

A type of chemical separation based on nuclear isomerism is of special interest in tracer research. Suppose a radioactive nucleus exists in two energy states, the lesser of which is radioactive by β-ray emission. Decay may take place in two steps, the first being emission of γ radiation in the transition from the upper energy to the lower energy state, and the second being emission of a β ray as the nucleus decays to a stable isotope. During the initial γ-ray emission, the emitting nucleus can gain recoil energy in two ways. One mechanism results from the possibility of a process known as "internal conversion," in which the γ-ray energy appears in an extranuclear electron which may be emitted instead of the γ ray.

When internal conversion occurs, more energy is available than in the direct γ-ray recoil process. The two bodies concerned are the nucleus and the extranuclear electron. The momentum of the electron is most conveniently expressed in terms of the product of the magnetic field, H, and the radius of curvature, ρ, of circular path into which the electron is bent by the field H. The energy, E, available to the nucleus is given in electron volts by the expression

$$E = \frac{4.8 \times 10^{-5}(H\rho)^2}{M} \tag{44}$$

In this expression $H\rho$ is given in gauss-centimeters.

As an example of the isomer recoil process, one may consider the two isomeric states[24] of Te^{129}. In this nucleus an energy state 0.100 Mev. above the ground state exists. The upper energy state decays by emission of a 0.100-Mev. γ ray with a half-life of 32 days to the lower state which then decays by negative β-ray emission with a half-life of 72 min. If the γ ray transmits its energy directly to the nucleus, it can be calculated from Eq. 43 that $(536/M)\ (0.100)^2$, or about 0.04-ev. recoil energy is imparted to the nucleus. If the internal conversion process takes place, an electron in the inner K shell will be emitted with an energy which is the difference between the initial γ-ray energy (0.100 Mev.) and the energy required to to remove an electron from the K shell ($\sim$0.032 Mev.). Therefore, the electron energy will be 0.068 Mev. The $H\rho$ value for an electron with this velocity is approximately 910 gauss-cm. From Eq. 44 the energy available to the nucleus is

$$E = \frac{4.8 \times 10^{-5}(910)^2}{129} = 0.31 \text{ ev.}$$

Thus, the energy available in the internal conversion process is eight times as great as that from the direct γ recoil.

The available energy of 0.31 ev. is rather small compared to chemical binding energies (1 to 10 ev.), yet it is found that practically every internal conversion process results in a splitting of the chemical bond and ejection of the radioactive nucleus. If tellurium is synthesized into the compound $Te(C_2H_5)_2$, the tellurium atom is ejected during the 32-day isomeric transition in good yield, giving the pure 72-min. tellurium isomer, which collects on the walls of the containing vessel. It is interesting to note that mere γ-ray recoil is not sufficient to disrupt a chemical bond in some cases of nuclear isomerism. Thus Zn^{69}, which decays by an isomeric transition[25] with a half-life of 13.8 hr. to a negative β-ray-emitting isomer with a half-life of 57 min., has associated with the isomeric transition a γ ray of 0.440 Mev. energy which does not undergo internal conversion. The nuclear recoil energy in this case is approximately 1.3 ev. Despite this high energy, no detectable deposit of 57-min. zinc is found when the isomer is synthesized into zinc ethide, $Zn(C_2H_5)_2$. General experience leads to the conclusion that rupture of chemical bonds is found primarily in isomeric transitions in which internal conversion occurs. The reason for this is to be found in the excitation of the atom which follows ejection of the K electron.

The significance of all this to tracer methodology is that isotopes which display the phenomenon of nuclear isomerism coupled with internal conversion effects must be used with suitable precaution. Such tracers will continually escape from labeled positions in molecules by splitting chemical

[24] Seaborg, G. T., Livingood, J. J., and Kennedy, J. W., *Phys. Rev.* **57**, 363 (1940).
[25] Livingood, J. J., and Seaborg, G. T., *Phys. Rev.* **55**, 457 (1939).

bonds during the internal conversion process, but they are amenable to tracer application provided the decay of the tracer samples can be measured and the upper state activity distinguished from the lower state activity. Thus, in the case of the Br^{80} isomers, the upper state activity with a half-life of 4.5 hr. is sufficiently different from the lower state activity of 18 min. so that Br^{80} can be used as a tracer. It is required only that sufficient time be allowed to lapse so that all free 18-min. activity disappears, leaving only the 4.5-hr. activity in the samples measured. If the half-lives of the two states are nearly identical there is little hope of tracer application. It is highly inadvisable to use isomers as tracers if the disintegration relations between the isomers are not known in detail.

GENERAL REFERENCES

A. Texts

1. Blatt, J. M., and Weisskopf, V. F., "Theoretical Nuclear Physics." Wiley, New York, 1952.
2. Friedlander, G., and Kennedy, J. W., "Nuclear and Radiochemistry." Wiley, New York, 1955.
3. Halliday, D., "Introductory Nuclear Physics." Wiley, New York, 1950.
4. Hevesy, G., and Paneth, F. A., "A Manual of Radioactivity." Oxford U. P., New York, 1938.
5. Kaplan, I., "Nuclear Physics." Addison, Cambridge, Massachusetts, 1955.
6. Meyer, S., and Schweidler, E., "Radioaktivität." Teubner, Berlin, 1927.
7. Rutherford, E., Chadwick, J., and Ellis, C. D., "Radiations from Radioactive Substances." Cambridge U. P., New York, 1930.
8. Segrè, E., Ed., "Experimental Nuclear Physics." Wiley, New York, 1953.

B. Articles

1. Coryell, C. D., and Sugarman, N., "The fission products," National Nuclear Energy Series. Div. IV, Vol. 9. McGraw-Hill, New York, 1951.
2. Kohman, T. P., and Saito, N., Radioactivity in geology and cosmology. *Ann. Revs. Nuclear Sci.* **4,** 401 (1954).
3. Morrison, P., Introduction to the theory of nuclear reactions. *Am. J. Phys.* **9,** 135 (1941).
4. Whitehouse, W. J., Nuclear fission. *Progr. Nuclear Phys.* **2,** 120 (1953).

These references may be supplemented by consulting the above texts. Timely articles will also be found in *Annual Reviews of Nuclear Science*, published by Annual Reviews, Inc., Stanford, California, and in *Nucleonics*, published by McGraw-Hill Publishing Co., Inc., New York.

CHAPTER II

RADIATION CHARACTERISTICS OF TRACER ATOMS

1. INTRODUCTION

As we have seen in Chapter I, radioactive atoms emit three types of radiation, α, β, and γ rays, which are radically different in physical characteristics. The α rays are streams of doubly positively charged helium nuclei moving with relatively slow velocities (1 to 2×10^9 cm./sec.). The β rays are composed of electrons singly positive or negative in charge, moving with much higher velocities which approach, at high energies, the speed of light (3×10^{10} cm./sec.). The γ rays consist of electromagnetic quanta (*photons*) which are uncharged and move at the speed of light.

The paths of these particles as they move through matter differ as a result of these properties. The way particles emitted by radioactive bodies dissipate energy depends on whether they are charged or uncharged (see Chapter III). The charged particles (α and β) interact electrostatically with the electrons (and to a smaller extent with the nuclei) of the atoms in the matter through which they pass, leaving a wake of excited and ionized molecules.

The α particle, being rather sluggish and heavy, rarely deviates from its straight path. It is deflected only when it collides with a nucleus. Because its electrostatic interaction with the electrons of the absorbing material is intense, owing to its high charge and low velocity, the α particle moves only a short distance before coming to a stop. As it slows down, it captures electrons, so that by the time it stops it has become a neutral helium atom. The average energy loss along the path of the α particle fluctuates only slightly, and the distance traversed (range) in any given medium is quite sharply defined, depending on the energy with which the α particle was emitted. Because the fluctuations in energy transfer at each atomic encounter are random in nature, the ranges of α particles cluster with a Gaussian type of probability distribution around the ideal sharply defined range which they would all possess were there no fluctuations. This fluctuation in range is called "straggling" and introduces a small uncertainty in the α-particle range.

Much larger uncertainties in range are found with β particles, which move much faster, are lighter in mass by a factor of 7400 and are singly charged. As a consequence, the distances traveled by β particles are much greater for a given kinetic energy, and their paths more erratic. Many interactions or collisions with other electrons occur in which the β particle is deflected at a large angle to its initial direction. The particles follow a tortuous path in the absorbing medium, and straggling is their dominant feature. For a given energy of emission β particles may possess all ranges up to a maximum which is ill-defined in any absorbing material.

Gamma radiations (photons) are uncharged and massless relative to α and β particles. The photons can lose energy only by collision processes for which the probability is smaller, in general, than for those interactions whereby charged particles dissipate energy so that they possess the greatest penetrating power of all three radiations for a given energy. The path of the α particle in a medium such as air can be visualized as a straight line 1 to 10 cm. in length for the energies ordinarily encountered. Beta particles of similar initial energy can be imagined to wander in a rather random fashion with a total path length of meters.

Tracer elements which emit α particles are not of major importance in biology. Hence, in this chapter only the properties of β and γ radiation will be considered. The mechanism whereby radioactive bodies give rise to these radiations is not germane to tracer methodology.[1] Only those phenomena attendant on the interaction of these particles after leaving nuclei will be considered.

2. BETA RADIATIONS

A. The Nature of Beta Radiations

As noted in Chapter I, these radiations may consist of either positive or negative electrons. The positive electrons have an ephemeral existence only. On coming to rest or being slowed in flight, a positive electron combines with a negative electron, and both particles disappear—they are said to be "annihilated." These photons depart from the scene of the annihilation in exactly opposite directions because momentum must be conserved. The rest mass (mass at zero velocity) of both particles appears as two γ-ray photons, each with a kinetic energy equivalent to the mass of an electron (0.5 Mev.). Annihilation radiation is always associated with positron emission. Except for this annihilation effect and the difference in sign of electrical charge, the positive and negative electrons have identical properties. Thus, the absorption and scattering characteristics of negative electrons may be considered as identical with those of positive electrons. In this discussion electrons emitted from nuclei will hereafter be referred to as β particles.

The β particle has an electric charge of 4.803×10^{-10} absolute electrostatic unit and a rest mass of 9.107×10^{-28} g. This mass is approximately

[1] Consult Konopinski, E. J., *Revs. Mod. Phys.* **15**, 209 (1943), for a general review of β-decay processes.

$\frac{1}{1800}$ that of the proton. Tracer atoms emit β particles at high velocities which in some cases approach the velocity of light to within a few tenths of 1%. The mass of a β particle varies with its velocity in accordance with the special relativity theory. Because of the very small mass of the β particle, it is easily deflected in passing through an absorbing medium. Scattering effects are predominant in the interactions between β particles and matter. Therefore, it is difficult to estimate path length of a β ray which has traversed a given thickness of absorber.

The major consequence of the interaction of β particles, as well as other charged particles, with matter is the appearance of pairs of ions in the medium through which they pass. In air, approximately 30 ev. is required, on the average, for the dissociation of a single air molecule into a heavy positive ion and an electron. A 300,000-ev. β particle, therefore, can produce approximately 10,000 ion pairs in air before coming to rest. The efficiency of this process depends inversely on the square of the velocity of the β particle over most of the energy range encountered in tracers (0.01 to 2 Mev.). The ionization in a unit of path is quite low at high energies; most of the ionization effects occur after the β particles have been slowed considerably. For maximum efficiency in detection, the volume available for the ionization process must be large enough to include the whole range of the particle. An arrangement which catches the β particles at the end of their range is more efficient than one which includes their path only in its initial stages.

There is always a small amount of extranuclear electromagnetic x-radiation associated with the absorption of β particles. This is made up from several sources. In some cases the β particle changes its velocity in the field of the atomic nuclei and, as a consequence, emits electromagnetic radiation. This radiation is referred to as "bremsstrahlung" (p. 12) and introduces some complications in nuclear studies designed to determine the absorption characteristics of the β rays. In addition to this type of radiation, the β particle may excite atoms without dissociation. The disturbance of the atomic electrons results in the emission of the characteristic x-rays of the atom.

Tracer bodies emitting only β radiation will always show a small apparent γ- and x-ray emission because of these factors. These effects are not very appreciable, however. Perhaps only one to five such quanta accompany the emission of every 100 β particles.

B. The Absorption of Beta Particles

The nature of β-particle absorption may be understood best against the background of the phenomena of energy dissipation discussed in the previous sections. If a series of thin layers of some absorbing material is placed

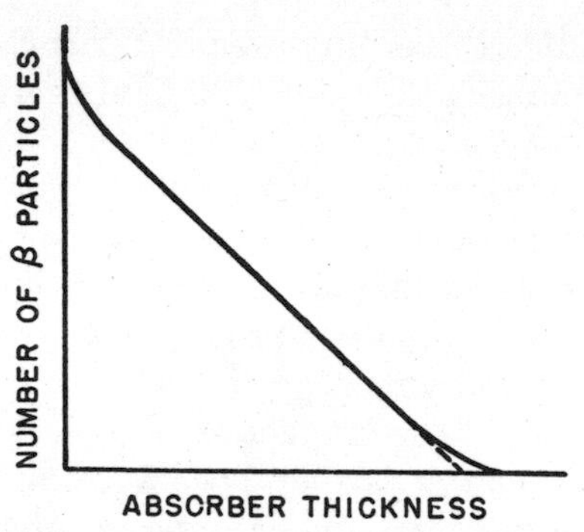

FIG. 9. Absorption curve for homogeneous electron radiation.

between a suitable detection device and a source of β particles homogeneous in energy, it can be expected that the number ot β particles detected will decrease as the number of absorbing layers increases because these β particles are easily scattered at large angles to their initial direction and so are lost from the beam of particles defined by the source and detector. In addition, β particles will lose energy by electrostatic interactions with the electrons of the absorbing matter. Consequently, both the number and energy of the β particles will diminish until, at an absorbing thickness depending on the energy of the particle, practically none will be found to emerge from the absorber. Because of the extreme straggling, as well as scattering and formation of bremsstrahlung, it is not easy to determine the exact absorption thickness at which complete absorption takes place. In Fig. 9 a typical absorption curve for β particles homogeneous in energy is shown. The number of β particles, N, is plotted as a function of the absorber thickness

TABLE 1

RANGE-ENERGY DATA FOR ELECTRON AND β RADIATIONS

Energy (Mev.)	Range R (extrapolated) (mg./cm.^{2} Al)
0.053	6
0.13	20
0.146	30
0.22	48
0.31	81
0.400	122
0.600	213
1.022	426
1.50	741
1.80	812

expressed in units of weight per unit area (mg./cm.^{2}). It is seen that over most of the range the curve is nearly linear. The end point or absorber thickness for complete absorption is estimated by extrapolation from the linear portion of the curve (dotted line).

The shape of such a curve depends very markedly on the manner in which the source is placed with respect to the detector. In particular, spurious end points resulting from excessive scattering around the edges of the absorber or originating from material surrounding the source or detector must be eliminated. The extrapolated ranges for homogeneous β radiations have been determined by a number of investigators.[2] Data selected from these and other studies relating range, R, in grams per square centimeter of aluminum to β-ray energy are exhibited in Table 1.[3]

The nature of the absorbing material does not markedly affect the range-energy relations when the range is expressed in units of weight per unit area, unless the absorber is a very heavy element. This fact is not surprising because the β particles are slowed down almost entirely by interaction with atomic electrons in the matter through which they pass, and the number of electrons per unit volume is very nearly proportional to the mass for the lighter elements in the periodic system.

Data like those in Table 1 are applicable both to homogeneous electrons and to continuous β radiations.[3] The end points of the range-energy curves appear to be the same for both kinds of radiation as long as the maximum energy of the continuous β spectrum is identical with the energy of the homogeneous electron radiation. As shown in Fig. 1, Chapter I, the β radiations from radioactive sources show a continuum of energies which results in a composite curve, the shape of which depends on the particular isotope studied.

The absorption curve for β radiations would be expected to be made up of a composite curve determined by summation of numerous curves of the homogeneous type shown in Fig. 9. Each energy would contribute a component determined by the energy distribution of the β-ray spectrum. It is apparent that great complexity attends the construction of an absorption curve for the nonhomogeneous β radiation, which is typical of β emitters, from known absorption curves for the homogeneous components present in varying degree. It so happens, however, that the summation effect results in an absorption curve which can be represented closely over most of its range by an exponential function. Thus, when the logarithm of the number

[2] Schonland, B. F. J., *Proc. Roy. Soc.* **A108,** 187 (1925); Madgwick, E., *Proc. Cambridge Phil. Soc.* **23,** 970 (1927); Marshall, J. S., and Ward, A. G., *Can. J. Research* **A15,** 39 (1937).

[3] Data presented are from a compilation by Katz, L., and Penfold, A. S., *Revs. Mod. Phys.* **24,** 28 (1952). This article provides an excellent discussion of all the available material on determination of β-ray energies by absorption procedures.

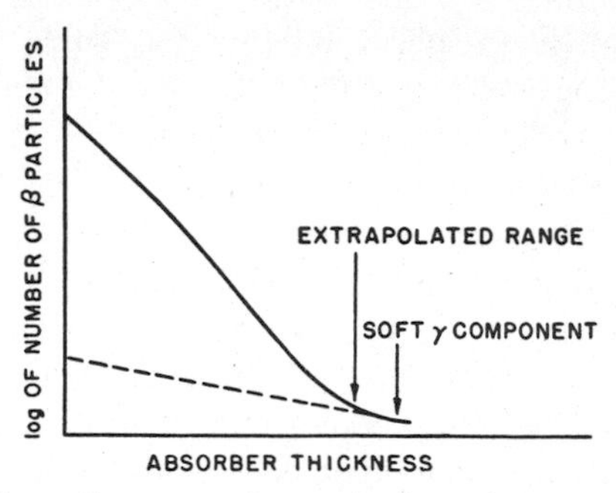

FIG. 10. Absorption curve for typical continuous β spectrum.

of β particles penetrating an absorber is plotted against absorber thickness, something very close to a straight line is obtained (see Fig. 10). Although this may be regarded as fortuitous, it is nevertheless fortunate, because it follows that over most of the β-particle spectrum the usual logarithmic expressions can be employed in a manner quite analogous to the treatment of radioactive decay discussed in Chapter I. For example, if S_0 represents the initial strength (number of β particles emitted in unit time) of a given β emitter, then the strength S, after passage through a thickness, t, of absorber is given by

$$S = S_0 e^{-\mu t} \tag{1}$$

where μ is the *absorption coefficient.* In a manner strictly analogous to Eq. 12 of Chapter I, μ may be related to the value of the thickness, $t_{1/2}$, required to halve the initial strength (half-absorption value) by the relation

$$t_{1/2} = 0.693/\mu \tag{2}$$

In Fig. 10, some departure from linearity is obvious. This will occur even in pure β emitters because of production of extranuclear γ radiations (bremsstrahlung) which are more penetrating in general than the β radiation with which they are associated. In most cases, however, deviation from linearity is noted even before the region of bremsstrahlung is reached because nuclear as well as extranuclear γ radiations are often associated with β radiation. In such cases, the curve is extended until sufficient range is available to estimate by extrapolation the γ-ray component. This can then be subtracted from the total absorption curve to give the true β-ray absorption. Complications arise if the γ-ray spectrum is complex and if some of the γ-ray components are soft enough to have absorption characteristics similar to the β radiations.

It can be appreciated that in most tracer applications the "true" β absorption is of little practical interest. It is the *apparent* absorption of the lumped radiation as influenced by sample orientation and peculiarities of the detection method which is the important phenomenon. Hence, procedures based on empirical determinations of absorption corrections are adequate in tracer experiments, and it is unnecessary to apply corrections based on calculations from the known disintegration schemes. Details of the manner in which absorption corrections are handled vary with each radioactive isotope and so are best deferred to discussion of specific isotopes.

In designing tracer procedures it is often helpful to have available data relating the maximum β energy of the tracer and its β-particle range. A number of empirical relations have been proposed,[4] any of which is equally useful.

Recently, Katz and Penfold[3] have digested and evaluated the great mass of data on ranges and energies of β radiations and emerged with the following relation between the range, R, in mg./cm.^{2}, and the end point energy, E_0 in Mev.:

$$R = 412E_0^n, (n = 1.265 - 0.0954 \ln E_0) \quad (3)$$

This formula is valid up to about 3 Mev. The agreement between ranges calculated according to this formula and those actually observed falls within a few per cent in nearly all cases. It is recommended that Eq. 3 be used up to 2.5 Mev., and that above this value another relation, as given in the following equation, be employed:

$$R = 530E - 106 \qquad (2.5 < E < 20 \text{ Mev.}) \quad (4)$$

In Fig. 11 the curves corresponding to Eqs. 3 and 4 are shown. In using such curves it should be remembered that E_{max} is approximately $3E_{av}$ (three times the average β energy), so that it is a simple matter to estimate absorbing thickness which it is not desirable to exceed for any particular isotope. Thus, for P^{32}, which has a value for E_{max} of 1.7 Mev., the average energy, E_{av}, is approximately 0.6 Mev. From Fig. 11, this corresponds to an average range of about 200 mg./cm.^{2} Al. It is advisable, therefore, to use absorption thicknesses in samples and detectors of less than 200 mg./cm.^{2} Al.

It has been customary for some workers to assume, as in Eq. 1, that over a limited region the intensity of the β radiation varies in an exponential manner with absorber thickness. The expressions derived empirically can all be related to Eqs. 1 and 4 as special cases. One such relation, which may be easier to use than Eq. 3, is

$$\mu/d = 22/E^{1.33} \quad (5)$$

[4] Feather, N., *Phys. Rev.* **35,** 1559 (1930); Widdowson, E. E., and Champion, F. C., *Proc. Phys. Soc. (London)* **A50,** 185 (1938).

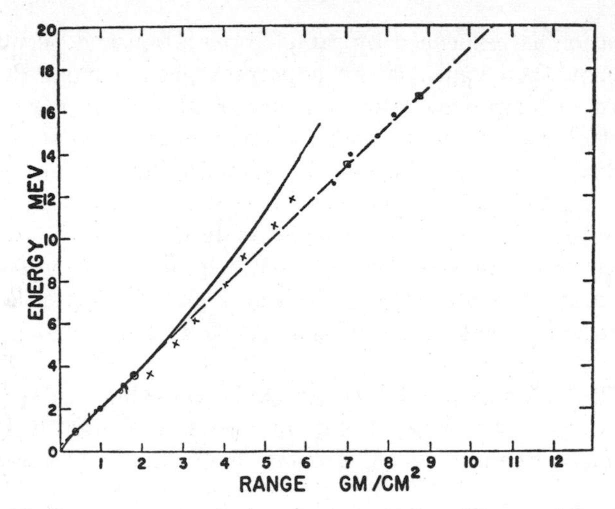

FIG. 11. Range-energy relations for β particles. (Katz and Penfold.[3])
—— Plot of $R = 412E_0^n$, $n = 1.265 - 0.0954 \ln E_0$
— — Plot of $R = 530E_0 - 106$, $E_0 > 2.5$

for energies between 0.1 and 3.0 Mev.[5,6] (μ/d = mass absorption coefficient; see p. 59).

C. Remarks on Scattering of Beta Particles

For tracer procedures it is important to emphasize that particles are easily scattered in passing through matter and that, as a consequence, detection procedures involving β-ray emitters must be standardized with respect to such factors as sample orientation and material for sample holders. If a β-ray source is placed at a considerable distance from a detector and an absorbing metal foil is interposed, the radioactivity measured is usually greater when the foil is placed near the detector than when it is placed near the source. This follows because β particles scattered from the foil have a greater probability of entering the detector when the foil is close to the detector.

The scattering produced by various materials is usually investigated in the following way. A given β-particle source is fixed in a standard position near the detector. The source is mounted on a very thin film so that few particles emergent in a direction away from the detector are scattered into it. A thickness of absorbing material is then placed on the side of the source away from the detector instead of between the source and detector. It is

[5] Evans, R. D., *Advances in Biol. and Med. Phys.* **1**, 163 (1948).

[6] Siri, W. E., "Isotopic Tracers and Nuclear Radiations," p. 58. McGraw-Hill, New York, 1949.

then determined whether the radioactivity measured is increased as a result of scattering back from the absorber. If I_0 is activity measured without the back-scattering and I is the activity with back-scattering, then the percentage reflection, which is given by $[(I - I_0)/I_0 \times 100]$, is found to vary with the material of the absorber, increasing with the atomic weight. The reflection increases also with absorber thickness until the thickness is so great that the reflected particles begin to be absorbed completely. For these reasons it is possible to introduce serious error into measurements of β radiation by failing to control the thickness or nature of material on which the source material is mounted.

Specific data on variation in detection efficiency brought about by changes in geometry will be given in Chapter III.

3. GAMMA RADIATIONS

A. Nature of Gamma Radiation

Gamma radiation is the nuclear analog of the visible and ultraviolet radiation emitted by atoms during electronic transitions. The atomic nucleus is supposed to exist in quantized energy states between which transitions occur. The rates of these transitions are governed by probabilities which are expressed in the form of "selection rules." When the transition occurs from a given energy state to states lower in energy, electromagnetic radiations in the form of photons are emitted. These radiations are the nuclear γ rays.

Gamma-ray photons may be described in terms of both wave and particle properties. For a given kinetic energy, E, a photon can be said to possess a frequency, ν, defined by the familiar relation $E = h\nu$, where h is Planck's constant. A wavelength, λ, can be associated with this frequency; $\nu = c/\lambda$, where c is the velocity of the photons (velocity of light, or of general electromagnetic radiation, *in vacuo*). The wavelength, λ, is often expressed in terms of the unit length 10^{-11} cm. (X.U.). The relation between E and λ is given by

$$\lambda \text{ (in X.U.)} = 12.38/E \text{ (in Mev.)} \tag{6}$$

Frequently the emission of γ radiation is consequent on the emission of α or β particles because in such cases the residual nucleus is often left in an excited state from which it may decay to the lowest energy state by emission of one or more photons. It has also been pointed out (Chapter I) that neutron capture leads to emission of γ radiation representing the binding energy of the new nucleus formed. The appearance of annihilation radiation in positron emitters has already been discussed as has the phenomenon of internal conversion (see pp. 12, 14, 46, 47, 50).

All energies up to about 3 Mev. are encountered in artificial radioactivity.

In addition, soft components such as x-rays and bremsstrahlung are found. Obviously, in all cases involving isomeric transitions of stable nuclei only γ radiation is emitted because the nuclear charge does not change. It is possible to cite at least one case (Fe^{55}) in which practically all the detectable radiation appears to be of x-ray nature.

B. Interaction of Gamma Radiation with Matter

1. Brief Survey. The interactions of γ radiation with matter are more easily defined theoretically than experimentally. There are five different types of interactions, three (1, 2, 3 as listed) of which have to do with the extranuclear electrons and two (4 and 5 as listed) with the nuclei of the absorbing matter. These are (1) Compton scattering, (2) coherent scattering, (3) photoelectric absorption, (4) pair production, and (5) photodisintegration.

Compton scattering is a process in which the γ-ray photon collides with an electron which may be considered to be free. The electron is bound to the atom with an energy which is negligible in comparison with the kinetic energy of the photon. After collision the photon with a lower energy recoils in one direction and the electron is ejected in another direction. In this process both energy and momentum are conserved. Because the photon has a lower energy after the collision, it follows from Eq. 6 that it has a longer wavelength. If φ is the angle between the initial and final direction of the photon, then the change in wavelength is given by

$$\Delta\lambda \text{ (in X.U.)} = 24.2(1 - \cos \varphi) \tag{7}$$

This process is dominant in elements of low and medium atomic number and for moderate γ-ray energies. Because the interaction is entirely concerned with extranuclear electrons, the Compton scattering effect per atom is proportional to the atomic number Z, just as is the scattering of β particles. For energies greater than 0.5 Mev., Compton scattering is inversely proportional to the γ-ray energy.

Coherent scattering results when the electron remains bound to the nucleus. The incident photon is scattered without change in wavelength. Obviously this process is encountered mainly with very low-energy γ rays. Such scattering results in diffraction phenomena if the atomic scattering centers form a regular pattern.

Photoelectric absorption is identical with x-ray photoelectric effects. In this process the photon loses all its energy and an extranuclear electron is ejected with a kinetic energy equal to the difference between the original photon energy and the energy with which the electron is bound to the atom. The process is favored by low γ-ray energies and even more so by high atomic numbers. The probability for photoelectric absorption per atom is proportional to Z^4/E^n, where n is a number ranging between 1 and 3 depending on what wavelength region is involved (x-rays or γ rays).

Pair production, as discussed in Chapter I, is a process formally the reverse of annihilation. If the photon has an energy in excess of that required to equal twice the rest mass of the electron ($\sim$1 Mev.), then it is energetically possible for the photon to be absorbed and for an electron pair, consisting of a positron and an electron, to appear. This phenomenon is concerned entirely with the nucleus of the atom with which the photon interacts. The probability of this type of interaction increases with energy and is markedly dependent on atomic number, increasing as Z^2 per atom.

Photo disintegration is also a nuclear interaction involving nuclear transformation resulting from absorption of a high-energy photon. The energy required for the process depends on the nuclear reaction initiated. This interaction is quite unimportant as a factor in γ-ray absorption and scattering.

2. Scattering and Absorption of γ Radiation. The dissipation of photon energy takes place mainly by Compton scattering, photoelectric absorption, and pair production. The detection of γ rays involves the detection of the secondary electronic radiations arising from these processes. It is found that a beam of γ rays, suitably collimated and of homogeneous initial energy, is attenuated in energy in passing through absorbing material. This attenuation occurs in an exponential manner quite identical with the energy loss experienced by ordinary light photons according to a relation known as Lambert's law. The energy loss per unit of absorbing thickness, dE/dx, occurring in passage through an element of thickness dx is a constant fraction, μ, of the incident energy, E.

Thus,

$$dE/dx = -\mu E \qquad \text{and} \qquad E = E_0 e^{-\mu x}$$

where E_0 is the initial energy, x is the thickness of absorber, and μ is known as the linear absorption coefficient, made up of various parts contributed by the three processes described above. As in the case of β-ray absorption, a quantity known as the *half-thickness value* can be used to denote that thickness of absorber required to diminish the initial energy of the photons by a factor of 2. The *half-thickness value*, $t_{1/2}$, is related to μ by the familiar relation $t_{1/2} = 0.693/\mu$. Other convenient expressions in use are:

1. The *mass absorption coefficient*, μ/d, which is the linear coefficient divided by the density. This coefficient is independent of the physical state of the absorber.

2. The *atomic absorption coefficient*, μ_A, which refers to the absorption per atom.

3. The *electronic absorption coefficient*, μ_e, which refers to the absorption

per electron. This coefficient is most useful when Compton scattering is dominant.

The contribution of the various absorption processes to the total absorption is shown in Figs. 12 and 13, in which the absorption coefficients are plotted as a function of photon energy for aluminum and lead. From these

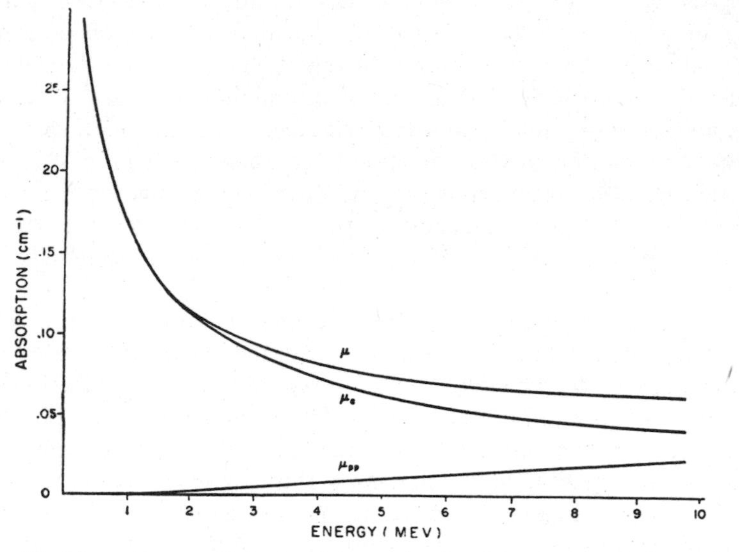

FIG. 12. Absorption coefficients of γ rays in aluminum; μ, total absorption; μ_{pp} , pair production; μ_C , Compton absorption.

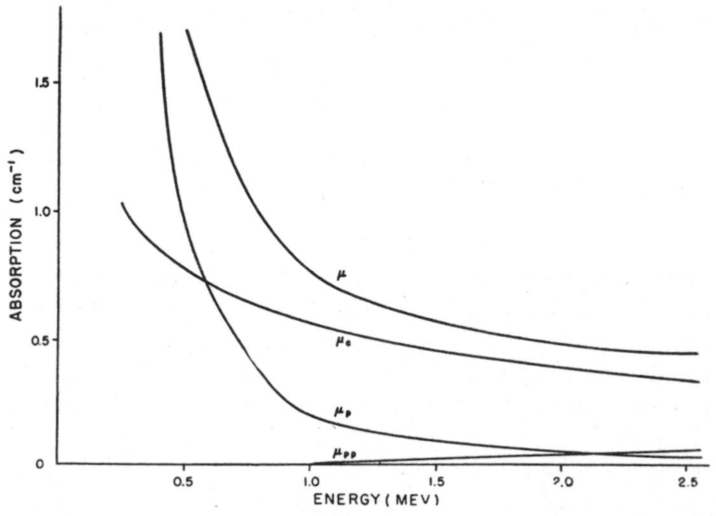

FIG. 13. Absorption coefficients of γ rays in lead; μ, total absorption; μ_C , Compton absorption; μ_p photoelectric absorption; μ_{pp} , pair production.

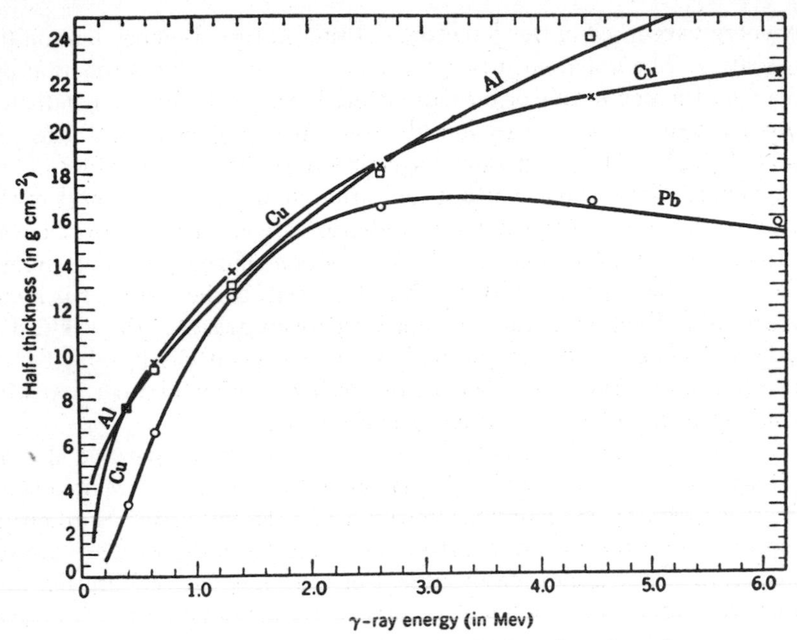

FIG. 14. Half-thickness values for Al, Cu, and Pb as function of γ-ray energy.[7]

curves it can be seen that the Compton process dominates energy dissipation in the energy range approximately 1 to 3 Mev. in lead and is almost entirely responsible for absorption in aluminum over most of the energy range. For greater clarity in visualizing the comparative absorption of two diverse elements such as aluminum and lead over an extended energy range, the relation between half-thickness value and photon energy is plotted in Fig. 14.

In most tracer radiations the emitted photons are nonhomogeneous in energy. Fortunately, it is only necessary to determine a mean absorption coefficient for the radiation as a whole. From the preceding discussion it can be inferred that close control of both the geometry and the nature of the absorbing material is important in achieving reproducible results. The absorption coefficient, μ, may be imagined in general as made up of a sum, Σ, of partial coefficients, μ_E, associated with photons of energy E. If the intensity of the γ radiations for each energy is symbolized as I_E, then μ may be written as $\Sigma\mu_E I_E/\Sigma I_E$, if it is assumed that the absorber is so thin that attenuation does not disturb the intensity ratio of photons with different energies. The efficiency of detection depends on the fraction of the

[7] As drawn by Friedlander, G., and Kennedy, J. W., "Nuclear and Radiochemistry," Wiley, New York, 1955; based on calculations from data by Davisson, C. M., and Evans, R. D., *Revs. Mod. Phys.* **24,** 79 (1952), and some experimental values given by Colgate, S. A., *Phys. Rev.* **87,** 592 (1952).

γ-ray energy absorbed in the detector, and this, in turn, is dependent on the γ-ray energy. The apparent absorption, μ_{app}, is an empirical function dependent on the conditions of measurement. If these conditions result in a certain efficiency, K_E, for each radiation, then μ_{app} can be written as $\Sigma K_E \mu_E I_E / \Sigma K_E I_E$. The absorption correction depends on the values of the K's. When conditions are such that ionization volume is small, most of the electrons to be detected as secondary radiation are contributed by the wall material so that the values of K and the detection efficiency are determined by the absorption coefficients of the wall material. If the chamber is large, an appreciable fraction of the secondary electrons arises in the gas of the ionization volume. To obtain the highest efficiencies it is advisable to use large volumes preferably filled with gases of high atomic number and with wall materials giving a strong photoelectric effect.

It should be emphasized that photon absorption, unlike particle absorption, is appreciably influenced by the nature of the absorber. It is important in all tracer researches involving detection of γ radiation to insure either that sample preparations are controlled so that the radiations are always emitted in the same atomic environment or that adequate calibration procedures are employed in applying corrections for changes in sample nature.

GENERAL REFERENCES

A. Texts

1. Compton, A. H., and Allison, S. K., "X-Rays in Theory and Experiment." Van Nostrand, New York, 1935.
2. Friedlander, G., and Kennedy, J. W., "Nuclear and Radiochemistry." Wiley, New York, 1955.
3. Rasetti, F., "Elements of Nuclear Physics." Prentice-Hall, New York, 1936.
4. Siri, W. E., "Isotopic Tracers and Nuclear Radiations." McGraw-Hill, New York, 1949.

B. Articles

1. Bethe, H. A., and Ashkin, J., Passage of radiations through matter. *In* "Experimental Nuclear Physics" (E. Segrè, ed.), Vol. I, p. 166. Wiley, New York, 1953.
2. Davisson, C. M., and Evans, R. D., Gamma ray absorption coefficients. *Revs. Mod. Phys.* **24,** 79 (1952).
3. Evans, R. D., Interaction of radiation with matter. *Advances in Biol. and Med. Phys.* **1,** 151 (1948).
4. Feather, N., Concerning the absorption method of investigating β-particles of high energy. *Phys. Rev.* **35,** 1559 (1930).
5. Katz, L., and Penfold, A. S., Range-energy relations for electrons and the determination of beta-ray end-point energies by absorption. *Revs. Mod. Phys.* **24,** 28 (1952).
6. Widdowson, E. E., and Champion, F. C., Application of the absorption method to the determination of the upper limits of continuous β-ray spectra. *Proc. Phys. Soc.* (*London*) **50,** 185 (1938).

A good compilation of data will also be found in Circular No. 499, Bureau of Standards, U. S. Dept. of Commerce.

Chapter III

Isotopic Assay

1. INTRODUCTION

The assay of radioactive tracer material differs fundamentally from that of stable tracer material. The principles of the former will be considered first. The quantitative determination of radioactivity is based on the ionization or excitation of matter by the radiations emitted by radioactive bodies. Until 1950, nearly all assay equipment was designed for observation of ionization in gases. The classic example of this type was the Geiger-Müller counter. Others were the proportional counter and various kinds of ionization chambers (electroscopes and electrometers). Since 1950, a new detector, the scintillation counter, which measures the interaction of radiation with either solid or liquid media, has been developed. This detector challenges the dominant position of apparatus dependent on collection or detection of ions in gases. In this chapter, the basic principles underlying the operation of both will be presented.

2. ASSAY OF RADIOACTIVITY

A. Basic Phenomena

Passage of charged particles through a gas results in the formation of ion pairs. Each pair consists of a heavy positive ion and a negative electron. This ionization results from electrostatic interaction between the electric charge on the moving particle and the atomic electrons of the gas molecules. The magnitude of the ionization in a given length of particle path depends mainly on two factors. The first is the velocity of the ionizing particle, which determines the duration of the interaction. The second is the charge of the ionizing particle, which determines the magnitude of the forces operative during interaction. The ionization produced is less, the higher the velocity and the lower the charge. Roughly, the ionization per unit path (specific ionization) varies directly as e^2 and inversely as v^2, where e is the charge and v the velocity of the charged particle. For a given energy, therefore, α particles, because they are doubly charged and slow moving, give a much higher specific ionization than β particles.

Uncharged radiation (γ rays, x-rays, etc.) can ionize only through collision mechanisms whereby secondary charged particles are set in motion. For this reason, detection efficiency for γ rays is usually less than that for α rays or β rays.

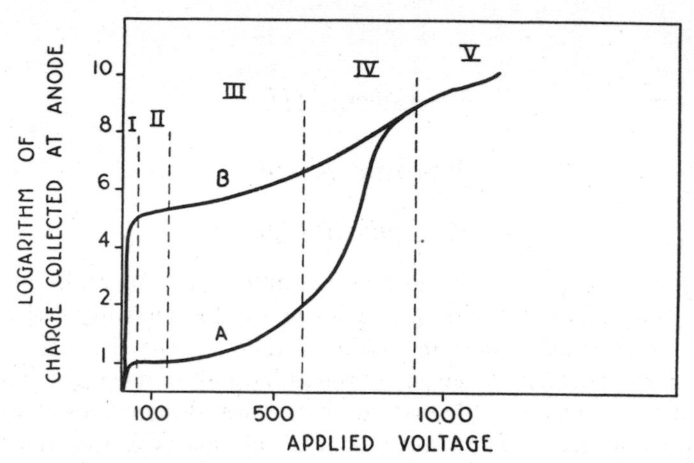

FIG. 15. Ionization-voltage relations. (After Montgomery and Montgomery.[1])

If a gas in which ionization is taking place is contained between electrodes on which a voltage is impressed, its constituent positive ions and electrons will move toward the electrodes. The positive ions will be collected at the cathode and the electrons at the anode. The magnitude of the charge collected depends on the impressed voltage, as shown in Fig. 15, in which the charge collected at the anode is plotted as ordinate against the applied voltage as abscissa. For convenience in representing the whole ionization range, the collected charge is plotted on a logarithmic scale. Curve *A* is drawn for the condition obtaining when the initial charge produced consists of only a few ion pairs (as for a β ray or a γ ray). Curve *B* covers the same range when 10^4 to 10^5 ion pairs are present initially (as for an α-ray ionization). In region I, i.e., at low potentials (0 to 100 volts), a fraction of the electrons produced reach the electrode, the rest being lost by recombination. As the voltage rises, electrons are swept more rapidly into the anode and fewer are lost by recombination until finally in region II practically all electrons formed in the gas are collected. Region II, therefore, represents a region of saturation charge.

To achieve a higher charge from a given initial ionization, some multiplicative process must be involved. In fact, as the voltage increases beyond region II into region III (100 to 500 volts), there is an increase in ionization due to production of new ion pairs by collision processes. The amplification attained depends on the voltage. As the voltage increases (region IV), production of ternary and quaternary radiations initiate a complicated discharge mechanism until finally (region V), at sufficiently high voltages, the total charge collected becomes independent of both the applied voltage and the initial ionization (curves *A* and *B* coincide).

[1] Montgomery, C. G., and Montgomery, D. D., *J. Franklin Inst.* **231,** 449 (1941).

The detection of the charges produced in this fashion is the fundamental problem in radioactive assay. The type of instrument employed depends on which one of the various regions shown in Fig. 15 the observer elects to use. All the detection devices employed can be considered as variations on a few basic instruments. These, in turn, are essentially a variant of the fundamental apparatus—a gas chamber contained within electrodes which is usually called an "ionization chamber."

B. Basic Instruments

1. General Remarks. In elaboration of the remarks in Section 2-A, the detection problem posed by the physical situation presented in Fig. 15 will be discussed at more length. The production of ion pairs in a system composed of two electrodes across which a voltage is impressed diminishes the applied potential in a manner governed by the familiar laws of electrostatics. Thus an ionization chamber is equivalent to an electrostatic condenser with a characteristic capacity, C, which is related to the charge, Q, and applied voltage, V, by Eq. 1:

$$Q = CV \tag{1}$$

One electronic charge, e, on a capacity of C (centimeters) produces a potential change given by $300e/C$ in volts. Thus, when the appropriate value is substituted for e (4.8×10^{-10} esu.), the potential change in microvolts (1μv. $= 10^{-6}$ volt) is found to be given by $0.144/C$. It is obvious that, when 1 to 500 ion pairs are to be collected and measured, the system must be so constructed that the capacity, C, has a minimal value. For 10-cm. capacity, a charge of 100 e will produce a change in voltage of 1.4 μv., whereas a capacity ten times as great will give a voltage change of only 0.14 μv. Such small potential changes can be detected only if they take place suddenly. In the construction of detection apparatus for such minute voltage changes, every effort is made to minimize capacity. It should be noted that in any capacitative system governed by Eq. 1 a general relation exists between the variation in Q and in C and V given by differentiating Eq. 1. Thus,

$$dQ = CdV + VdC \tag{2}$$

Any change, dQ, in the total charge is related not only to a change in voltage, dV, but also to a change in capacity, dC. The term VdC is operative only when the mechanical elements involved in the detection of the voltage change are in motion. In the various forms of electrometer (string, Compton, quartz fiber, etc.) voltage changes are observed by noting the deflection of a moving electrode subjected to a combination of electrostatic forces and mechanical restoring forces. If such electrometers are connected to the collecting electrode of an ionization chamber, an added capacity is introduced which changes as the electrometer electrode moves across the electrometer scale, thus introducing the term VdC of Eq. 2. In such apparatus the quantity, dQ, which it is desired to measure cannot be determined solely from the more conveniently measured change in voltage.

There are two types of electrometer, however, which do not depend on a mechanical motion of electrodes and for which the correction term VdC is unnecessary. These are the vacuum tube electrometer[1] and the quartz piezoelectric crystal electrometer.[2] There are also some moving-electrode electrometers so designed that there is a minimal or negligible change in capacity. For these electrometers the current resulting from ionization, dQ/dt, is related linearly to the rate of change of voltage, dV/dt. In all these instruments operating as current detectors, the experimental conditions relate mainly to curve B (regions I and II of Fig. 15), the region of moderate ionization currents and low applied voltage.

In most radioactive tracer determinations, however, it is desirable to count with surety single particles. This requires working with initial ionizations of as little as one ion pair, involving the ionization-voltage relations given in curve A. In regions I and II, the detection apparatus must include an extremely sensitive amplifier. It is also advantageous, in order to obtain the maximum pulse (dV/dt), to have a system with very low capacity. In region III, some amplification is achieved directly in the ionization chamber, so that requirements on over-all amplification are somewhat relaxed. In this region it should be noted that the current collected at a given voltage is proportional to the initial ionization. In other words, curves A and B become parallel. Thus, by using an amplifier which responds only to initial pulses corresponding to the upper curve B but not to the lower curve A, it is possible to distinguish whether the pulses are due to heavily ionizing particles (α rays) or to low initial ionization (β and γ rays). An ionization chamber operated in this way is called a "proportional" counter. In region V, the ionization chamber introduces, through its discharge mechanism, a considerable amplification which can be used to feed an easily detectable pulse to an amplifier which, in turn, can actuate a mechanical recorder. The size of the pulse does not depend on either the voltage or the initial ionization. The ionization chamber is "triggered" so that any ionizing particle entering the sensitive volume will initiate a discharge which can be detected. The most familiar example of this type of discharge mechanism is encountered in the Geiger-Müller tube counter, which is a low-capacity ionization chamber named for the two scientists first prominent in the design and application of such counters.[3] Consequently region V is termed the "Geiger-Müller" region. In further discussions this term will be indicated by the abbreviation "G-M."

[2] Curie, M.."Radioactivité," p. 15. Hermann, Paris, 1935.

[3] Geiger, H., and Müller, W., *Physik. Z.* **29**, 839 (1928); **30**, 489 (1929).

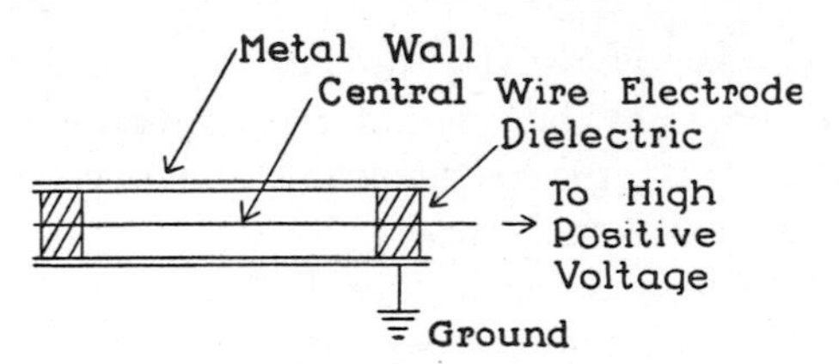

FIG. 16. Typical G-M tube geometry.

In the last few years great progress has been made in the development of electronic amplifiers and specialized ionization chambers. A discussion of amplifiers is not within the scope of this book. Some references are included at the end of this chapter for the reader who wishes to probe into details of amplifier design.

2. The Geiger-Müller Tube. In this section a short review of the various phases of the G-M discharge phenomenon will be presented. A typical geometry for the G-M tube is shown in Fig. 16. The central wire is coaxial with the outer cylinder and is of minimal diameter to reduce capacity. In use, the wire is usually maintained at a high positive potential with respect to the outer cylinder, which is at ground potential.

When an ionizing particle enters the tube a few ion pairs are formed in the gas space. Under the action of the impressed electric field, the electrons are accelerated toward the wire while the positive ions move toward the cylinder. The electrons move extremely rapidly compared with the much heavier positive ions. Hence, as a first approximation, the motion of the positive ions can be neglected. As the electrons come close to the wire they are accelerated at an increasing rate because the voltage gradient is steepest near the wire. When sufficient energy is gained, further ionization of gas by collisions between the accelerated electrons and gas molecules liberates more electrons which, in turn, are accelerated toward the wire. This leads to further ionization so that, through repetition of this process, one electron gives rise to many hundreds more. Such a process is called an electron "avalanche." While this process is going on, a stationary positive space charge due to the sluggish positive ions is created in the wake of the electrons. In addition, electromagnetic radiation appears because of the interaction of the electrons with the gas molecules. These radiations (photons) spread throughout the tube and may initiate further electron avalanches by ejecting photoelectrons from the walls of the G-M counter. It can be seen that a very large number of electrons is created from an initial few.

The discharge ceases eventually because the positive space charge diminishes the effective voltage gradient at the anode. The duration of the pulse to this point is very short ($\sim 10^{-7}$ sec.). Before the G-M tube can be used to initiate another pulse, the positive ions must diffuse sufficiently toward the cathode so that the original voltage gradient at the wire is restored. This portion of the discharge cycle is the longest, lasting from 10^{-3} to 10^{-4} sec., and represents the "dead time" of the tube counter.

Although the positive ions can initiate a spurious discharge and prolong the dead time by producing additional photons on striking the cathode, an organic vapor such as ethyl alcohol obviates this possibility. The role of the organic vapor as a "quenching" gas appears to derive from the large probability of predissociation from excited electronic states exhibited by polyatomic molecules.[4] In the usual gas mixture, a monatomic constituent, usually argon, is mixed with a small quantity of polyatomic vapor, usually ethyl alcohol. The positive argon ions in moving toward the cylinder wall make numerous collisions with the alcohol molecules and transfer charge with the result that the ion sheath which finally arrives at the cathode wall is composed almost entirely of positive alcohol ions. The high positive field of these ions draws electrons from the cathode so that excited neutral molecules which dissociate before they can strike the wall and liberate secondary electrons are created. The polyatomic gas also eliminates photoeffects at the cathode by absorbing ultraviolet photons formed in the avalanche process. A G-M tube filled with such a mixture is called a "fast" counter and, as such, needs no external electronic circuits to terminate the G-M discharge.

The pressure employed is commonly 10 cm. Hg, the argon comprising 90 to 95% of the total. In addition to ethyl alcohol, numerous quenching gases can be used, i.e., xylene, toluene, methylene bromide, propylene bromide, propylene chloride, nitromethane, and nitroethane. Helium is sometimes used in place of argon. A pure noble gas is unsuitable for fast G-M tubes because metastable ions are formed which diffuse slowly to the cathode wall and initiate spurious discharges by secondary electron emission.

The ionization-voltage characteristics of the G-M tube inferred in the above discussion from curve *A* of Fig. 15 may be presented in the form shown in Fig. 17, where pulse rate in counts per minute is plotted against the voltage. At low voltages, the charge collected during each pulse is too small to actuate the electronic amplifier used with the G-M tube. Hence no pulses are observed. As the voltage is increased, a value is reached at which the charge collected per discharge is just large enough to be detected. This is called the "starting potential" (symbolized V_0). Increasing the voltage further causes larger pulses which are detected with increasing efficiency until finally a voltage is reached at which practically all pulses are detected. A continued rise in voltage causes no further increase in pulse

[4] Korff, S. A., and Present, R. D., *Phys. Rev.* **65**, 274 (1944).

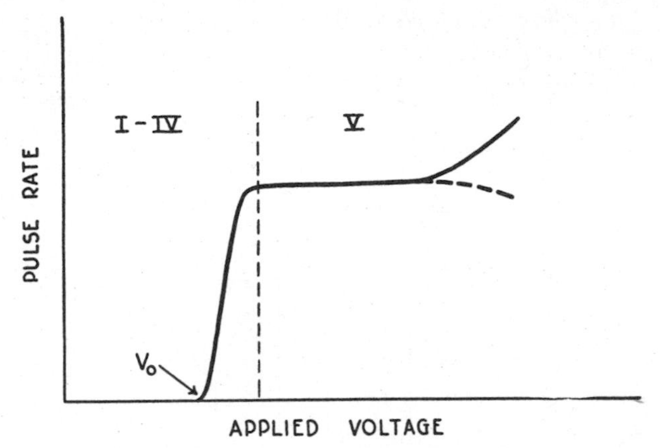

FIG. 17. G-M tube counting characteristic curve.

rate. It is in this voltage range known as the "plateau" that the G-M tube is operated. As the voltage is increased to higher and higher values, a continuous discharge sets in. It is heralded by a rapid rise in pulse rate, even though the source of radiation remains at the same intensity. It is also possible that the opposite behavior (shown by the dotted curve) is observed; if this happens, the characteristic curve falls. This results from an increase in dead time brought about by initiation of long-lived discharges usually associated with too small a leakage resistance in the output of the G-M tube. The plateau is limited, then, by the sensitivity of the amplifier at low voltages and the increasing probability of spurious discharges at high voltages. At very high voltages the G-M tube becomes "paralyzed," being in a condition of continuous discharge.

Use of a light gas such as helium results in a lower starting potential, V_0, than that observed with a heavy gas such as argon. The starting potential can also be lowered by lowering the gas pressure. Over a limited range the starting potential is linearly dependent on the pressure. If, for a typical G-M tube operating in air at 4 cm. Hg pressure, it is found that the starting potential is approximately 1000 volts, it will be found that the values for the starting potential at higher pressures fall on a straight line, the value at 8 cm. Hg being approximately 1300 volts. By decreasing the radius of the central wire it is possible to get more effective (steeper) voltage gradients for a given pressure; a G-M tube with a wire 0.003 inch in diameter will show a starting potential of approximately 1000 volts, whereas a G-M tube with a wire 0.010 inch in diameter may not become operative until 1200 volts is reached. This effect depends markedly on the pressure, being more accentuated at high than at low pressures.[5]

[5] Montgomery, C. G., and Montgomery, D. D., *J. Franklin Inst.* **231**, 463 (1941)

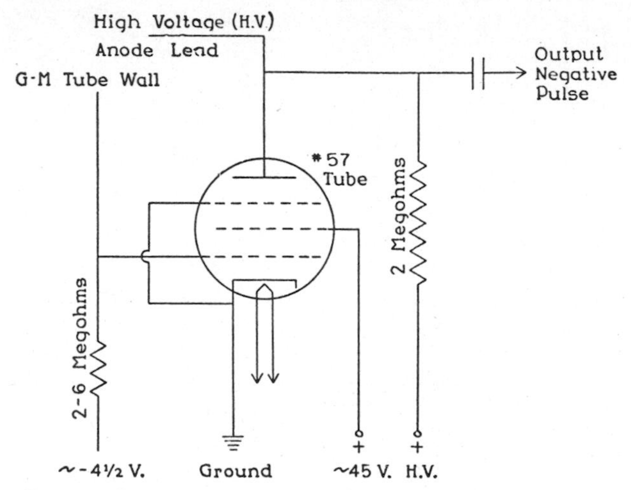

FIG. 18. Quenching circuit. (After Neher and Harper.[6])

It may be inferred from what has been said about quenching that there are types of gas fillings which result in "slow" G-M tube counters. Such tubes invariably result if the gas filling consists of a single component such as hydrogen or helium, or mixtures such as argon and oxygen, or argon and hydrogen, in which there are no polyatomic constituents. In this case it is advantageous to use one of a number of vacuum tube quenching circuits. In a typical circuit,[6] shown in Fig. 18, negative voltage is applied to a grid of a pentode (#57 tube) sufficient just to prevent flow of current to the plate from the filament. A discharge in the G-M tube causes positive charge to collect on the G-M tube wall. This charge, in turn, is impressed on the grid electrode, causing the vacuum tube to become conducting. The potential on the plate is thereby reduced, and, consequently, so is the potential on the central wire of the G-M tube. The discharge thereby ceases because there is not enough potential to maintain the discharge.

3. Construction and Operation of G-M Tube Counters. Most of the literature on this subject describes procedures for highly specialized research in which extreme demands as regards sensitivity and stability are made. Tracer requirements can usually be met with G-M tubes which are not particularly satisfactory from the standpoint of the cosmic-ray physicist. Nowadays, G-M tube counters for every kind of tracer assay can be obtained commercially at a cost which is low enough to make home-built counters uneconomical.

The materials of construction are extremely varied. The cathode cylinder can be made of almost any common metal. Brass, copper, iron, aluminum,

[6] Neher, H. V., and Harper, W. W., *Phys. Rev.* **49**, 940 (1936).

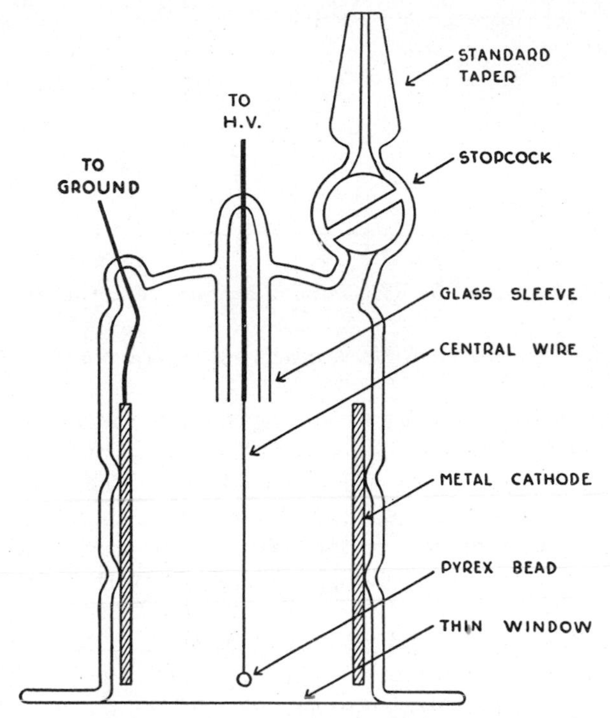

FIG. 19. Diagram of bell-jar G-M tube.

silver, or nickel is most frequently used. In some cases the envelope consists of glass, the inside of which is plated or sputtered with a metal such as silver. Special cathode materials are sometimes used to enhance sensitivity to γ radiation. Each type of cathode material requires special processing to obtain the most satisfactory results. The anode wire is most often iron or tungsten. Almost every conceivable geometry can and has been employed.

Thick-walled tubes, as shown in Fig. 16, are not suitable for the assay of isotopes emitting low-energy β radiation. Hence it is necessary to provide another type of counting tube with a thin window. A simple apparatus is shown in Fig. 19. A metal cylinder, usually brass or copper, is contained in a glass envelope, the electrical lead being brought out through a tungsten-glass seal. The glass envelope has a flange-ground end polished to a good flat, smooth surface.

The central wire of 4-mil tungsten is welded to a tungsten rod (about 10 mils) which is brought through a glass seal. A small Pyrex bead is fused to the free end of the wire to give the dielectric necessary to define the ionization volume as the space between the wire and the cylinder. This is essential because otherwise the ionization volume would be mainly that between

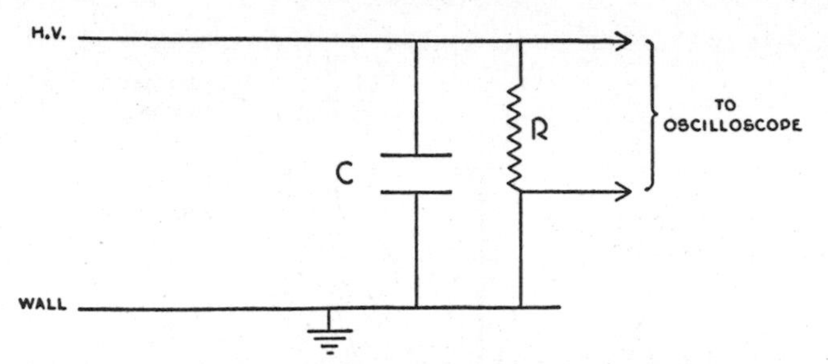

FIG. 20. Circuit for connection of oscilloscope to G-M tube.

the point and the adjacent wall—a much less sensitive and stable arrangement.

The bell-jar tube usually has a counting characteristic curve closely similar to the cylinder tube described above. The background counting rate to be expected with these tubes varies to some extent. The cosmic radiation which is responsible for the background count is not uniform, and so background will depend somewhat on the orientation of the counting tube. The magnitude of this rate for a bell-jar counter placed with its long axis vertical is usually 6 to 10 ct./min./cm.^{2} of window area. This may be minimized by enclosing the tube in a lead shield with a wall thickness of 2 to 3 inches. Such a shield is also helpful if the G-M tube is sensitive to light.

To test the counting tube, an oscilloscope is essential. For this purpose the oscilloscope (du Mond, RCA, or any commercial make) is connected to the tube, as shown in Fig. 20. A weak source of radiation is brought near the tube. The voltage is increased until the plateau of the counting curve is reached. In the plateau region the oscilloscope kicks should be sharp and of uniform height. On the sweep timing range of 200 to 1000 cycles the pulses should appear somewhat as shown in Fig. 21—that is, the pulses should show a sharp rise followed by an exponential fall which occurs in an interval of 10^{-3} to 10^{-4} sec. A convenient method for deciding what height of pulse

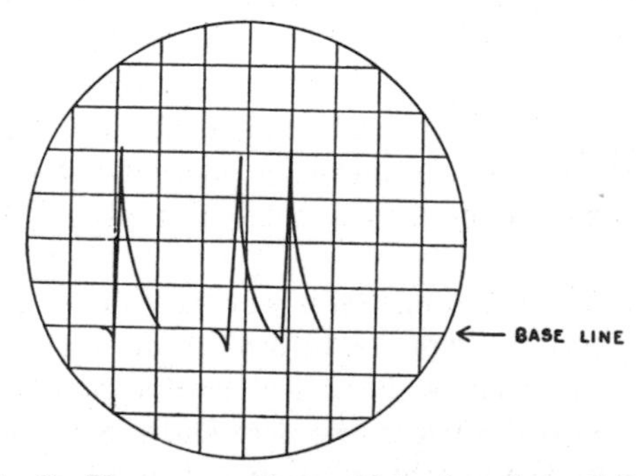

FIG. 21. Oscilloscope pattern of impulses from G-M tube.

to expect for a given amplification in the oscilloscope is to examine the pulses from a tube known to be operating satisfactorily. Unsatisfactory pulses which vary in height and breadth can usually be correlated with other symptoms such as inordinately high background, small plateau, and nonstatistical distribution of pulses. Another indication that the G-M tube is not operating adequately is failure of the counting rate to drop immediately after removal of a radioactive source.

Most troubles can be traced to vacuum leaks or electrical breakdown. A vacuum leak manifests itself by a steady increase in threshold voltage and shortening of the plateau region. In a tube containing 10 cm. Hg pressure of the argon-alcohol mixture, introduction of air raising the pressure by 1 to 2 cm. Hg can completely abolish the plateau region.

G-M tubes designed for internal counting of solids are available. In these the sample is mounted on a cylinder which may be slid back and forth over the sensitive volume so that background rate can be determined without dismantling the tube.

Inside counting procedures can be used which are adequate with regard to stability, ease of sample introduction, and length of time required for sample introduction. In one type of procedure, a large number of samples is mounted on the periphery of circular wheels or plates. This assembly is placed in an airtight container, usually made from vacuum desiccator parts into which is incorporated a G-M tube counter. A typical arrangement[7] is shown in Fig. 22 in which the container is constructed with two vacuum desiccator lids. A central shaft which can be rotated by twisting the lower ground joint drives each of two gear wheels on which the samples are placed. The counter tube can be positioned over either set of samples by shifting the top lid. The proper gas mixture is introduced through the bottom stopcock. Automatic sample changers are available commercially.

It should be remarked, finally, that all G-M tubes using dissociable vapors have a finite counting life dependent on the amount of such vapor lost during each pulse. Typical lifetimes range from 10^8 to 10^9 total counts. Thus a tube counting a sample with an activity of 10,000 ct./min. will last 10^4 to 10^5 min. at most before a refilling is required, provided some other occurrence does not bring the tube to grief.

4. Proportional Counters. The G-M tube counter is a versatile and nearly universal instrument for general use in β assay, but it is being displaced to some extent by the proportional counter. The proportional counter is so called because it operates in region III (Fig. 15) where the charge collected is proportional to the initial ionization. The proportional counter has a number of advantages over the G-M counter. It can be operated at reduced voltages, with consequent greater stability and reproducibility. The or-

[7] Labaw, L. W., *Rev. Sci. Instr.* **19**, 390 (1948).

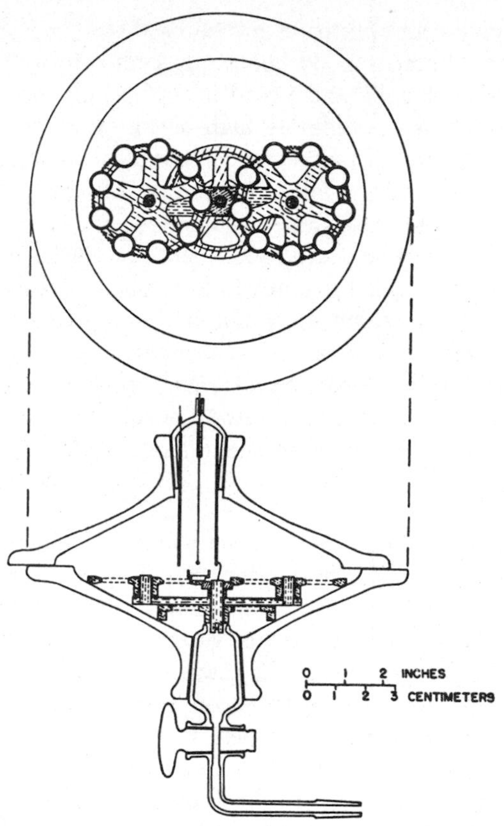

FIG. 22. Schematic diagram of inside G-M counter showing sample-holder assembly in bottom desiccator lid and cross-sectional view of assembled counter with sample in place under cathode. (After Labaw.[7])

ganic quenching gas is not consumed as rapidly as in the G-M tube. There is practically no dead time during each pulse, so that very fast counting rates are possible. The fact that the reduced voltage results in smaller pulses which require external amplification is not a serious drawback because inexpensive commercial vacuum-tube amplifiers are available and are perfectly adequate to ensure proper operation.

The most popular type of proportional counter is the "flow" counter. In this instrument, the counting gas mixture flows through the ionization chamber or cylinder at a pressure slightly higher than atmospheric. The samples are introduced through an airlock by means of a sliding shelf arrangement. Samples may also be assayed outside the counter, which may be equipped with a thin window.

By collecting all the ionization produced by a given radiation it is possible to measure the energy of the radiation because, when this is done, the pulse height is proportional to the energy. In this way, the proportional counter can be used as a spectrometer. By suitable analysis of the pulses arriving at the input of the auxiliary amplifiers (see p. 87) it is possible to discriminate between the various radiations and thus to count one kind of radiation in the presence of another. This is an important advantage in many tracer assays. In particular it enables investigators to reduce undesirable background response by rejecting pulses outside the energy range which is characteristic of the radiation being measured.

5. Electroscopes and Electrometers. Up to this point, discussion has centered around mechanisms basic to pulse-discharge counters in which primary ionization is amplified mainly within the ionization chamber. Alternatively, the primary ionization is detected by means of amplification external to the ionization chamber. It is advantageous to increase the ionization volume as well as the pressure so that there is maximal ion pair production per single ionizing particle. By use of an electrometer vacuum tube, such as the FP-54 (General Electric) or D-96475 (Western Electric), operating as a direct-current amplifier, the ionization resulting from irradiation of the sensitive volume of the chamber may be detected. A sketch of a typical ionization chamber is shown in Fig. 23. The radiations enter through a thin window, usually aluminum or mica. Because the chamber is operated with air at atmospheric pressure this window can be made as thin as warranted by the energy of the radiations studied.

The central electrode is maintained at ground potential; the wire cage is charged negatively 200 to 300 volts. With respect to the cage, the central electrode is therefore the anode. The negative charge on the wire cage acts also as a deterrent to negative electrons produced by α-particle contamination from the chamber wall.

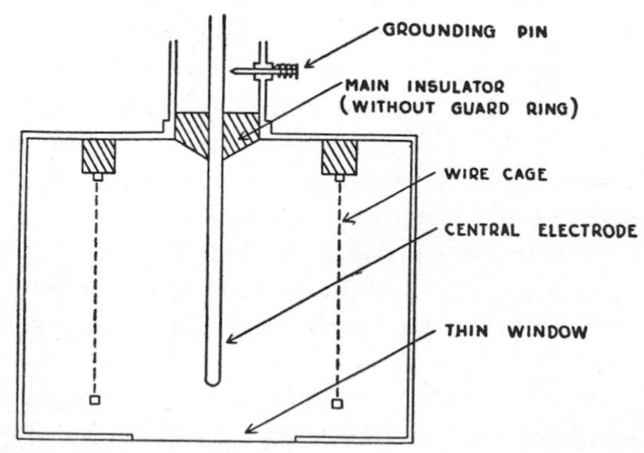

FIG. 23. Schematic representation of a typical ionization chamber.

The central collecting electrode is connected to the grid of the electrometer tube with as short a shielded lead as possible. Usually the electrometer is housed in a chamber which can be evacuated and mounted directly over the ionization chamber.

Vacuum tube electrometer circuits for use with such an ionization chamber are described in the literature. In practice, the apparatus is operated in two ways. In the first, the current from the ionization chamber flows through a high-resistance grid leak, the resulting drop in grid potential being indicated by a change in plate current which is detected with a galvanometer. High resistances in steps of 10 from 10^7 to 10^{12} ohms are provided as grid leaks. The sensitivity depends on the magnitude of grid leak used. Thus, for a resistance of 10^{11} ohms, a change of 1 mv. in grid potential is equivalent to a grid current of 10^{-14} amp., which is the magnitude of current to be expected from β-ray sources of moderate energy undergoing approximately 5000 disintegrations/min.

The second mode of operation feeds the ionization current from the collecting electrode directly to the grid of the vacuum tube with no other resistance involved. The grid "floats" at a potential which changes as the current flows to it from the chamber. This results in a steady drift, the velocity of which depends on the magnitude of the ionization current. Usually there is a natural drift due to the inherent tube leakage as well as ionization background. It is possible to annul this to a large extent by proper construction of the chamber. Also, if the drift contributed by the tube is in a direction opposite to that induced by ionization current from the collecting electrode, it can be compensated for by bringing a radioactive source near the chamber.

Special precautions with regard to contacts, insulation, and shielding of leads from stray electrical fields are required to give satisfactory operation. The important insulating bushing is that between the central electrode and the chamber wall or guard ring. Amber has been much used in the past, but recently special polystyrene insulation has been found more satisfactory. It is usually advisable to scrape the insulating surface with a clean knife free of oil or grease so that a fresh surface is formed before installation. A more uniform result can be achieved by mounting the insulator on a lathe and taking a small cut off the surface; this can be done with a fresh, oil-free cutting tool. To restore the collecting electrode to ground potential after a measurement, a special grounding pin is inserted in the pipe leading from the electrode. The best geometry for the grounding pin is a sharp point held near the collecting electrode lead so that a very small motion and tiny contact area are involved. Without this precaution, the device will react violently, with the galvanometer spot shooting off scale every time the chamber is grounded.

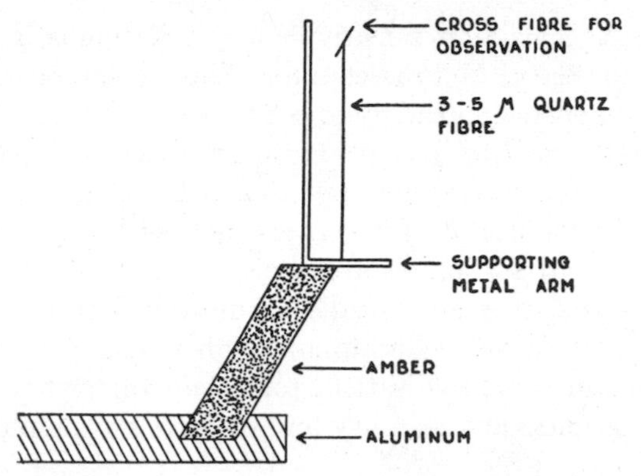

FIG. 24. Diagram of sensitive element in the Lauritsen electroscope.

In general, this type of instrument is not at present capable of the precision with weak sources attainable by use of the G-M tube counter.

A very useful instrument of the integrating type applicable to the assay of samples of moderate intensity is the Lauritsen quartz-fiber electroscope.[8] This simple device is essentially a refinement of the familiar "goldleaf" electroscope. Its moving electrode is a gold-covered quartz fiber 3 to 5 μ in diameter. As sketched in Fig. 24 this fiber is supported transversely on a copper arm which in turn is embedded in a good nonconductor such as amber. A charge of 100 to 200 volts applied to the copper arm causes the fiber to diverge. Its motion is observed with a telescope, a graduated scale being focused on the objective for measuring rate of drift as ionization takes place. The electroscope elements are enclosed in an aluminum can. Background drifts are usually $\sim$0.002 to 0.005 div./sec. The motion is linear over much of the scale. Response to ionization is linear up to 3 to 5 div./sec. The instrument as supplied by the manufacturer[9] comes equipped either with a thick-wall chamber or with a thin window. If no thin window is supplied and it is desired to detect soft radiation, the can should be carefully removed and a portion of the wall cut away. While this operation is proceeding, the electroscope should be kept covered in a dust-free atmosphere because the sensitive fiber element easily catches dust particles with a consequent reduction in sensitivity. A thin window conveniently made from thin aluminum foil ($\sim$0.1 to 0.5 mil) is cemented to the open portion of the can with shellac. The can is then replaced carefully. The necessity for care in moving the can off and on is dictated by the fact that small air movements can disturb the delicate fiber seriously. Careless manipulation can break it.

Although the instrument shows great sensitivity to α radiation because

[8] Lauritsen, C. C., and Lauritsen, T., *Rev. Sci. Instr.* **8**, 438 (1937).

[9] These electroscopes can be obtained from F. C. Henson Co., 3311 East Colorado Street, Pasadena, California.

of the high specific ionization, sensitivity to β radiation is, at best, considerably less than for the G-M tube counter. Thus, a sample emitting β rays with a continuous energy spread up to $\sim$2.0 Mev. at the rate of 100 ct./min. will give an effect equal to background on the electroscope, whereas such a sample is some ten to one hundred times the background of the G-M tube counter. By the use of heavy gases such as argon, methyl bromide, or freon, the sensitivity of electroscopes (as well as other atmospheric ionization chambers) to γ and x-radiation may be enhanced considerably. For work with samples of moderate and high intensity there is no instrument available which is more satisfactory than this type of electroscope. It possesses ruggedness and stability and requires no complicated amplifier mechanism.

It has been shown[10] that for low-energy β sources the Lauritsen electroscope can be modified so that sensitivity compares very favorably with the bell-type thin-window G-M tube counter. When the path length of the β radiations is only a few centimeters in all, most of the ionization can be collected inside the electroscope chamber, so that if a radioactive sample of an isotope such as C^{14} or S^{35} is placed directly inside the electroscope, the response compared to background is about the same as for the bell-jar G-M tube counter with sample adjacent to the window of the counter tube. In Fig. 25 the schematic arrangement for the modified electroscope can is shown.

In somewhat similar fashion recent developments have demonstrated that ionization chambers can be designed which can attain sensitivities equaling the performance of the best G-M tube counters. It may be remarked that this type of instrument has many advantages. It can be operated with air at atmospheric pressure. It can be made rugged and more flexible for adaptation to different forms of radioactive samples. The electronic equipment may be considerably simplified over that required for G-M tube operation. A development along these lines is the device known as a "vibrating-reed electrometer" in which a mechanical motion (movement of a diaphragm or metal reed) is employed to move an impressed charge in the electrostatic field of a condenser.[11] The movement of the diaphragm results in production of an approximately sinusoidal alternating voltage. The alternating-current signal is amplified by means of a conventional alternating-current amplifier. The stability of operation is unusually good when the vibrating element is properly fabricated. This instrument is now obtainable commercially.[12]

[10] Henriques, F. C., Jr., Kistiakowsky, G. B., Margnetti, C., and Schneider, W. G., *Ind. Eng. Chem., Anal. Ed.* **18,** 349 (1946).

[11] Palevsky, H., Swank, P. K., and Grenchik, R., *Bull. Am. Phys. Soc.* **21,** No. 3, 23 (1946).

[12] Applied Physics Corp., 362 W. Colorado Street, Pasadena 1, California.

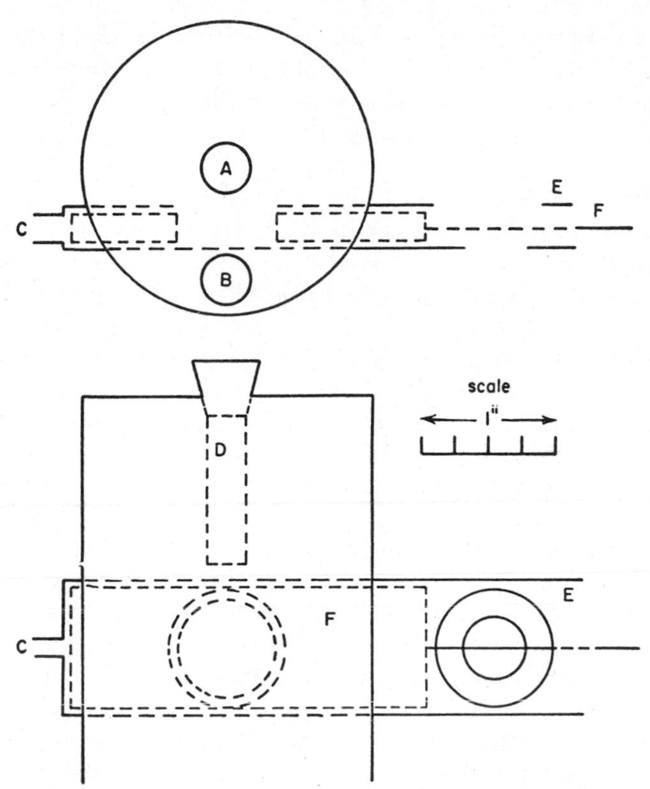

FIG. 25. Modified Lauritsen electroscope case. (After Henriques *et al.*[10]) *A*, window. *B*, opening for boat. *C*, to drying tube (magnesium perchlorate). *D*, boat for drying agent. *E*, sleeve for sliding bar. *F*, sliding bar for introducing sample disks.

A readily available type of electrometer which is proving quite helpful when used in conjunction with a properly designed ionization chamber is the Lindemann electrometer.[13a] This portable instrument is a modification of the conventional quadrant electrometer and is small, very rugged, and insensitive to tilt. Its sensitivity is almost the same as the vibrating-reed type, and it is relatively inexpensive.

C. Assay in Liquid and Solid Media—Scintillation Counters

Assay apparatus which measures the collection of ions in gases has been considered in the preceding sections. Historically, however, other methods were exploited first. A classic example is the fogging of photographic emulsions by radiation, which was used as early as 1895 to demonstrate the existence of radioactivity. Many other phenomena which result from the

[13a] The Lindemann electrometer can be obtained from the Cambridge Scientific Instrument Co., Ltd., Cambridge, England.

passage of radiation through liquids and solids can be used to detect radiation. These include deposition of colloids, coloring of crystals, and excitation of fluorescence or phosphorescence in appropriate media. Assay instrumentation based on the last-mentioned phenomenon is the topic of this section.

Sir William Crookes discovered in 1903 that a zinc sulfide screen exposed to α radiation in a dark room emits light. His apparatus consisted of the fluorescent screen, which was placed behind a small source of radium salt, and a microscope. He called this arrangement a "spinthariscope" and demonstrated with it that the luminosity of the screen was made up of single flashes of light occurring very rapidly. A few years later, Regener devised methods for counting the light flashes, which called for simultaneous observations by two persons. Each observer independently recorded the flashes he observed on a moving tape; whenever coincidences occurred, they were counted. This was the classical procedure used by Sir Ernest Rutherford and his school in their epoch-making researches on radioactivity and atomic transmutation early in the century. Because of the limitations inherent in this method, however, it was superseded by the methods based on ion collection in gaseous media, which culminated in the G-M counter and the proportional counter, which have already been discussed.

It is important to note that there is a fundamental difference between scintillation detectors and G-M tubes or proportional counters, although both depend on interaction between the radiation emitted and the electrons of the matter traversed. This difference lies in the mechanism of energy storage used in the instrument. Detectors of the G-M or proportional type depend on the removal of electrons from atoms or molecules in the gaseous state. The scintillation detector depends on the formation of excited states in which the electrons are retained in the atoms or molecules. When the excited atoms return to the unexcited ("ground") state, radiation is emitted in the form of quanta of light. In fluorescent or phosphorescent material, the energy of these light quanta is in the ultraviolet or visible range (0.1 to 10 ev.). Any material, liquid or solid, which can be brought to excited states by interaction with radiation can be made phosphorescent or fluorescent.

Since 1945, great improvements in ways of converting light into electric pulses have removed the limitations on scintillation methods formerly imposed by the human eye. Gains in sensitivity of 10 million or even 100 million in the closeness of the pulses which can be separated are now achieved routinely. Moreover, the work of Kallman and others has shown that the scintillation counter can be used proportionally because the pulse height is related linearly to the radiation energy dissipated in the phosphor. As a result, the scintillation method is being employed in an ever-increasing number of applications for the detection of radioactivity in tracer research.

The components of a typical scintillation detector consist of the phosphor, the photomultiplier tube, and the connection between the phosphor and photomultiplier. The "phosphor" is the crystal (or liquid) which transforms the radiation emitted by the radioactive body into ultraviolet or visible light. There are many kinds of phosphors. All phosphors used as scintillators must meet one basic requirement: they must be transparent to their own radiation. Hence, there is a premium on those materials which can be grown as large, clear crystals.

Inorganic crystals like zinc sulfide or sodium iodide are the most commonly used phosphors. Usually, phosphorescence yields are greatly increased by incorporating a little of a special kind of impurity into such crystals. For instance, the zinc sulfide phosphors may contain a little manganese, and the sodium iodide crystals may contain a little thallium.

The function of the impurity can be understood on the following basis. Excitation of an atom in a crystal lattice results in displacement of one of the atomic electrons and production of a residual positive region or "hole." In perfect crystals the normal consequence is recombination of the electron and the hole. During recombination, the energy originally used in creating the electron and hole is released either as radiant electromagnetic energy or as heat. The amount of fluorescence produced in the phosphor depends on the relative proportion of the energy in these two modes of dissipation. In a "perfect" crystal, the excited atomic system couples rapidly with other atoms in the lattice and the excitation energy of the former is used up in exciting the latter. In this process the energy is degraded into various modes of vibration and appears as heat.

If there is a means of shielding the excited atom from its neighbors or preventing energy transfer, more electromagnetic energy will be emitted and less produced. This result can be produced by introduction of the proper imperfection in the crystal. Most often this can be done in a controlled way by adding a specific impurity. However, even in pure materials there can be active centers for fluorescence which occur because of occasional dislocations resulting from the manner in which the crystal is grown.

Some pure substances are efficient phosphors. An example is naturally occurring calcium tungstate (scheelite). Crystals which contain heavy elements like iodine and tungsten are particularly suitable for detecting γ rays because elements of high atomic number are most efficient in converting high-energy electromagnetic radiation into excitation of the atoms in the phosphor (see Section 3-B, Chapter II).

Organic phosphors are somewhat different in behavior from the inorganic types. All the organic phosphors are transparent to their own radiation and seem to emit shorter flashes of light. Resolution times as small as 10^{-10} sec. are reported, as against resolution times one hundred times as long, or more, in the inorganic phosphors. Whichever kind of crystal is used, however,

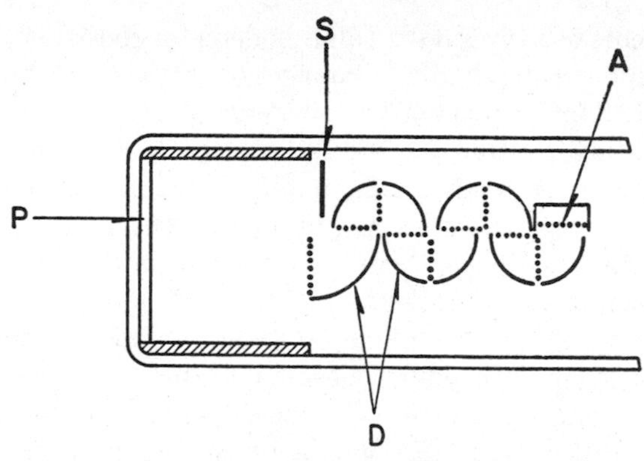

FIG. 26. Schematic representation of RCA photomultiplier tube. (Redrawn from article by Linden.[13b])

scintillation counters are very fast even by comparison with proportional counters. The recent development of liquid organic phosphors has made possible almost unlimited increases in phosphor size.

Similarly, solid phosphors can now be machined to any desired shape by procedures in which an organic phosphor is dissolved in a material, such as a styrene monomer, which is then polymerized into a plastic. An important advantage of this flexibility is that small sources of radioactivity (e.g., a small test tube containing a few milliliters of Fe^{59}-labeled plasma) can be placed directly inside the phosphor. In other cases, especially with low-energy β emitters, the source can be dissolved in a liquid phosphor to give a high detection efficiency.

The second component of the scintillation detector is the photomultiplier tube. It is a device for converting the light flash from the phosphor into an electric pulse. The photomultiplier is a vacuum tube in which numerous photosensitive electrodes are placed in such a way that, when light hits the first electrode, electrons produced by the photoelectric effect are drawn off and impinge on the second electrode, from which in turn the electrons go to another electrode, and so on through the series. At each electrode, the number of electrons multiplies, so that very large electrical pulses are finally obtained. The pulse is then led into an amplifying apparatus, and then to a register, as in the usual counting arrangement.

At present, two types of photomultiplier are in use. One, developed by the Radio Corporation of America, depends on a tricky arrangement of curved electrodes so placed that the electrons are multiplied at each electrode and also focused, by electrostatic means, onto the surface of the following electrode (see Fig. 26). In the tube shown, P is the transparent

[13b] Linden, B. R., *Nucleonics* **11,** No. 9, 30 (1953).

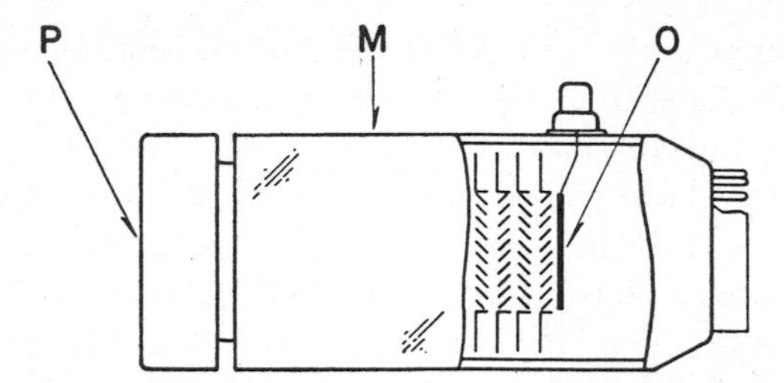

FIG. 27. "Venetian blind" type of photomultiplier tube.

photocathode, S is a shield on which potential can be varied to obtain optimal collection of photoelectrons, D stands for the dynode (two being indicated in the figure), and A is the anode. When the tube is operating, electrons from the photocathode, P, enter the first dynode, where their number is increased by secondary emission from the activated silver-magnesium alloy surface of the dynode. The first dynode is the largest so that photoelectron collection is enhanced. The electrons from the first dynode are then collected at the second dynode, where another increase in number of electrons takes place, and so on until the final electron beam is collected by the wire-mesh anode, A. All the dynodes (except the last box-shaped one surrounding the anode) have a mesh screen attached to them; each screen faces the dynode preceding it to enhance secondary electron collection.

Another type of photomultiplier tube, developed by E.M.I. Laboratories in England, has a linear geometry (see Fig. 27). The photomultiplier tube with photosensitive cathode, P, receives light from a phosphor and emits photoelectrons. The arrangement of the following electrodes shown is of the "Venetian blind" type. The photoelectrons from P are accelerated by an applied field in the tube envelope, M, to the first in the series of electrodes, where they produce a greater number of secondary electrons. These in turn are drawn by the applied field to the next electrode and the process repeats until collection at the last electrode, O. Each electrode is shielded from the preceding one by a fine-meshed grid to avoid the influence of the electric field of each preceding stage and thus facilitate escape of the secondary electrons.

The process continues until a large avalanche of secondary electrons arrives at the collecting electrode (anode), O. The number of electrons emitted for each incident electron at a given electrode is known as the "multiplication factor per stage." This usually varies between 3 and 5. Usually between 10 and 15 stages are present. The over-all multiplication will be the single stage factor to the power n, where n is the number of stages. Thanks to the development of the transistor, the design of proper external amplifiers which match the great resolution of the phosphor and phototubes is possible.

The third component to be considered is the connection between the phosphor and the tube. It is obvious that the efficiency of the counter depends critically on maximum transmission of light from the phosphor to the tube and on the exclusion of all other light. Unless the phosphor is placed directly against the tube in a light-shielded housing, it is necessary to construct a light guide. The principle of the light guide is well exemplified by the old experiment of Sir William Bragg, who showed that light could be efficiently transferred along a water jet under certain conditions. The underlying principle is that of total internal reflection. Light coming from inside a dense medium is reflected back if it encounters a less dense medium at an angle that varies with different media. A rod of plastic material like Lucite can be used to demonstrate this phenomenon. If light enters one end of the rod, it will pass along to the other with little loss so long as there are no bends so sharp that light cannot strike the limiting surface at less than the necessary ("critical") angle. If this requirement cannot be satisfied, the polished surfaces of the rod can be silvered to increase reflection.

It is important to remember that, because the photosensitivity and multiplication factor vary with each tube, a value for the multiplication factor for stable operation must be determined by trial and error.

The requisites for operation of scintillation counters can be summarized as follows. First, the radiation energy from the radiation source must be converted into energy of excitation of the constituent atoms of the phosphor with maximum efficiency. Second, de-excitation of the phosphor should result in the emission of radiation of a kind to which the phosphor is largely transparent. Further, the wavelengths of the phosphorescent radiation should match the wavelengths to which the photomultiplier is characteristically most sensitive. Third, the photosensitivity of the photomultiplier must be high.

The over-all efficiency of the scintillation device can be expressed as a product of the following factors: (1) the over-all gain; (2) the conversion efficiency of the energy of the radiation; (3) the transmission factor of the phosphor for its own radiation; (4) the geometry factor expressing the fraction of light collected by the tube; and (5) the photosensitivity of the tube, which is the number of electrons per unit photon energy. For the detection of α particles from polonium by a ZnS phosphor scintillator, the over-all quantity of electric charge per pulse is about 2×10^{-9}. Storage of this charge momentarily in a capacity of 20 $\mu\mu$f. will alter the voltage of the anode by 100 volts. This is a very large pulse and indicates the magnitude of sensitivity obtainable routinely.

The operation of a scintillation counter is completely different from that of a G-M tube or proportional counter. The scintillation counter has no "plateau" in the sense that this term is used in describing the counting characteristics of a G-M tube. The scintillation counter is operated with a discriminator which suppresses all pulses below a given energy. The height of the pulse supplied by the phototube increases with both the energy of the exciting radiation from the source and the voltage applied to the phototube. Consequently, a change in the applied voltage will change the height of the pulses fed to the discriminator which, in turn, transmits a changed number of pulses to the counting circuit. A continuous variation in counting rate can be obtained by changing either the discriminator setting or the tube voltage. The problem in scintillation counting is to find a combination of settings which is best from the standpoint both of sensitivity and of stability. This can be accomplished readily when the energy of the radiations is so great that most of the pulses resulting are higher than those due to the thermal background. In this case, the discriminator can be set to accept most of the pulses at a level slightly higher than that which cuts out the background. A small variation in the discriminator setting in the region just above background will not change the counting rate appreciably. Likewise, keeping the discriminator level constant and making a small change in the tube voltage will cause little change in the counting rate. In this way a kind of plateau is established. For radiation with energies too low to adopt this procedure it may be difficult to find a plateau. The best conditions for counting must be found by trial and error and represent a compromise between a high counting rate and a tolerable background.*

D. Auxiliary Instrumentation

The pulses produced by the various kinds of counters and ionization chambers usually require amplification before they can be recorded. Some discussion of such devices has already been presented (see p. 66). A typical circuit for detectors producing a steady current is shown in Fig. 28. This type of balanced direct-current amplifier requires careful stabilization and particular care in minimizing fluctuations in battery voltage. If the detector provides a pulse and not a steady current, an alternating-current amplifier can be used. A circuit commonly employed is shown schematically in Fig. 29. A cathode ray oscilloscope is useful in monitoring the performance of pulse counters (see p. 72). A "scaling" circuit for reducing the counting rate by a given factor is often included.

The circuitry required to complete any given assay apparatus varies with the nature of the detector. No amplifier is required for G-M tubes, nor are circuits with high scaling factors needed because of the long dead time. Ordinary voltage stabilization is sufficient because of the wide plateau. On the other hand, scintillation counters and proportional counters both require excellent voltage stabilization and stable amplifiers with gains between 10^2 and 10^5.

* For representative procedures involving use of liquid scintillation media in sample preparation and assay of H^3 and C^{14}, see Appendix 5.

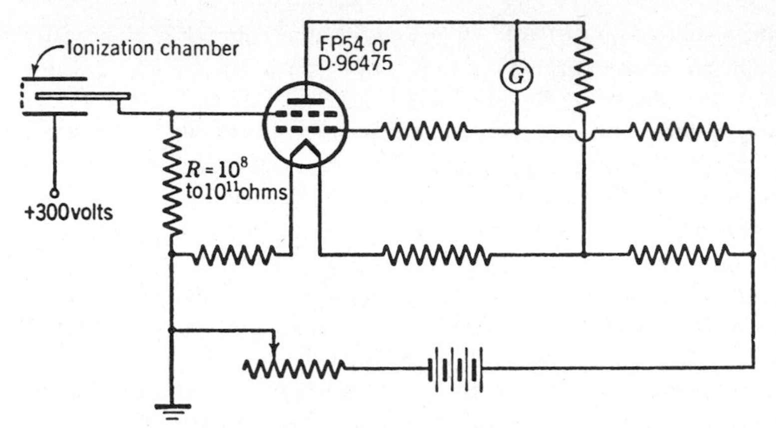

FIG. 28. Ionization chamber with balanced d-c amplifier. (After Friedlander and Kennedy.[14a])

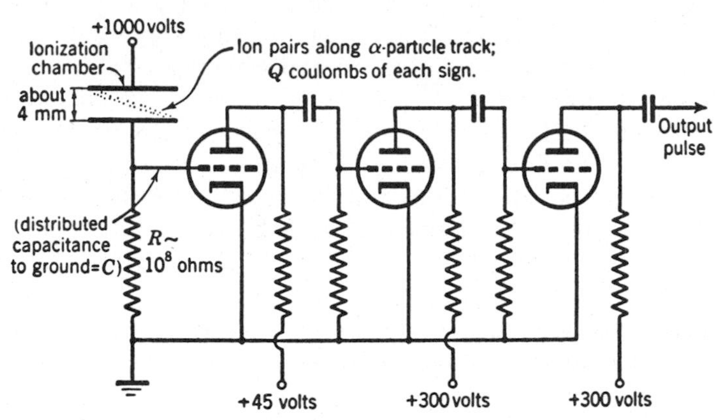

FIG. 29. Diagram of ionization chamber with linear pulse amplifier. (After Friedlander and Kennedy.[14c])

A convenient circuit for application in radioactive assay is the "counting-rate meter."[14b] in which pulses from the electronic amplifier are fed to a large condenser shunted by a resistance. The leakage current through the resistor is proportional to the pulse rate. In use, the pulses are equalized in the amplifier before being collected on the condenser. A direct reading of the leakage current on a microammeter is made, or, alternatively, the current is amplified by using a vacuum tube voltmeter with an output voltage pro-

[14a] Friedlander, G., and Kennedy, J. W., "Nuclear and Radiochemistry," p. 227. Wiley, New York, 1955.

[14b] Evans, R. D., and Meagher, R. E., *Rev. Sci. Instr.* **10,** 339 (1939); Evans, R. D., and Alder, R. L., *ibid.* **10,** 332 (1939); Gingrich, N. S., Evans, R. D., and Edgerton, H. E., *ibid.* **7,** 450 (1936).

[14c] Friedlander, G., and Kennedy, J. W., see p. 229 in reference 14a.

portional to the pulse rate. The output voltage may be recorded photographically, or mechanically with a pen recorder.

Often circumstances permit correlation of pulse height with the initial amount of energy dissipated in the detector, so that it is possible to set grid voltages on the amplifying tubes to values which can cut off pulses above or below a certain value. This sort of discrimination can be useful in determining energies of radiations and in other ways. A calibrated adjustable discriminator makes it possible to measure the counting rate as a function of discriminator settings and therefore to obtain a curve from which the pulse height distribution can be derived.

In practice, this can be better accomplished by means of a "single-channel analyzer." In this instrument, two discriminators are used. The circuits are so designed that only pulses of a height between the two discriminator settings can pass. The two settings may be varied independently, or they may be moved up and down together through the voltage range of interest. The distance between them is kept constant. Such an instrument makes possible great refinements in tracer assay because the investigator can set his detector for any given energy and thus achieve maximum sensitivity for a given isotope.

E. Corrections in Radioactive Assay

The results obtained with any of the detectors described require correction in varying degrees because one or all of the following factors must be taken into account: (1) resolving time, (2) background, (3) variations in efficiency for a given radiation, (4) absorption losses, and (5) sample geometry and back-scattering.

1. Resolving Time. Resolving time is the time required for a detector to recover from one pulse so that it can count another. During each pulse, there is a dead time when the detector cannot respond. The dead time in scintillators and in proportional counters is too short to be of any significance (10^{-6} to 10^{-10} sec.). In the G-M tube, however, the dead time is usually long enough (10^{-3} to 10^{-4} sec.) to constitute a serious limitation on the counting rate permitted. A simple method for determining the number of counts lost as a function of the sample strength is the following.

A solution of some radioactive isotope (a convenient one is P^{32} as phosphate), is prepared which contains approximately 50,000 ct./min./ml. This can be done by remembering that with G-M tubes of the bell-jar type with a thin window, 1 μc. of P^{32} gives about 1.5×10^5 ct./min. when measured as the solid directly under the window. Aliquots corresponding to various counting rates from 500 to 15,000 ct./min./ml. are made up accurately by dilution. Two or three 1-ml. samples of each dilution are then evaporated on small watch glasses; the sample should be contained

TABLE 2

TYPICAL DATA FOR COINCIDENCE CORRECTION

Aliquot	Actual Rate (ct./min.)	Background (ct./min.)	Actual rate Corrected for Background (ct./min.)	Real (Calc.) Rate (ct./min.)	Correction*
0.050	334	32	302	302	1.00
0.100	636	32	604	604	1.00
0.520	3107	32	3075	3140	1.02
0.800	4460	32	4428	4832	1.09
1.000	5182	32	5150	6040	1.17
1.500	7276	32	7244	9060	1.25

* It is not advisable to apply corrections higher than 15 to 20%.

within an area no larger than half that of the window diameter if a bell-jar counter is used, or half the cylinder diameter if a cylinder counter is used. These samples are counted and tabulated. The initial solution should contain no more than 5 to 10 mg. of phosphate per milliliter so that self-absorption can be neglected.

At low counting rates, the apparent counting rate will be equal to the true counting rate so that the latter is known from the actual rate observed. As the counting rate increases, the apparent rate begins to fall below that calculated from the dilution ratios. A sample data sheet is shown in Table 2. The statistical deviations are omitted for clarity of presentation. The ratio of the real rate to the apparent rate is the correction factor. This may be plotted as a function of the apparent counting rate for convenience in interpolating between the special values for which it has been determined by this means. To avoid any possibility of serious error the counting rate should not exceed that corresponding to 10% correction.

A more accurate procedure is that based on the analysis of the relation between the true counting rate, the apparent counting rate, and the resolving time. Suppose two sources give approximately equal counting rates, N_1 and N_2. Counted together they give a rate N_{12}. It can be shown[15] that T, the resolving time, is given by

$$T = 2(N_1 + N_2 - N_{12})/(N_1 + N_2)N_{12} \qquad (3)$$

Also T is related to the true counting rate, N_0, and to the apparent N by the equations[16]

[15] Beers, Y., *Rev. Sci. Instr.* **13,** 72 (1942).

[16] Skinner, S. M., *Phys. Rev.* **48,** 438 (1935); Ruark, A. E., and Brammer, F. E., *ibid.* **52,** 322 (1937); Volz, H. A., *Z. Physik* **93,** 539 (1935); Schiff, L. I., *Phys. Rev.* **50,** 88 (1936).

$$N = N_0 e^{-N_0 T} \tag{4}$$

$$N_0 = N(1 + NT) \tag{4a}$$

By expansion in series it is found that these equations are identical if the term $\frac{1}{2}N^2T^2 \ll 1$. T is thus determined by measurement of N_1, N_2, and N_{12}, and the rate N_0/N calculated by Eq. 4 or 4a. In most tracer work the first procedure is adequate because precision to better than 1% in the correction ratio can be attained.

2. Background Correction. The counting rate for any sample is the sum of the rate of radiation from the sample and the background rate of the counting tube. Obviously, the latter must be subtracted to find the activity of the sample. The influence of the background counting rate on the precision of assay is discussed in Section 2-F.

3. Variation in Efficiency. Any detector will show varying response to a given radiation over a period of time. The reasons for such variation are many. For instance, the counting voltage may not be accurately reproducible from day to day because an insensitive voltmeter may be used to determine its value. Consequently, a sample of radioactive phosphate may show 5000 ct./min. on one day, and *one half-life* later (14.3 days) it may give 2400 ct./min., instead of 2500. For this reason a long-lived standard source of radioactivity should be prepared to calibrate the efficiency of the counting tube from day to day. A convenient standard is uranium oxide (U_3O_8). A sample with a counting rate low enough so that no resolving-time correction need be applied is made up and mounted permanently. It is important that no change in apparent activity due to the absorption of moisture by the standard occur.

An adequate procedure (see Section 2-G) is to spread 15 to 20 mg. of precipitated and ignited U_3O_8 on an aluminum holder and secure it in place with Scotch tape as a cover. The sample can then be left in the open to age for a week while the tape gains the water it usually takes up. An alternative is to keep the standard always dry in a desiccator. It is advisable to prepare a control sample of the isotope being assayed to check that the sample is decaying properly.

4. Absorption Losses. Starting with a material of given specific activity, the source strength measured should increase linearly as the amount of material is increased. Actually, as the thickness increases, radiations from the lower layers begin to be lost by absorption in the sample material. Eventually the sample is so thick that only the top layers contribute to the assay. Thus, if source activity is plotted against source thickness, there results not a straight line but a curve which bends away from linearity and approaches a limiting value. Such a curve is shown for C^{14} in Fig. 30.[17] The

[17] Reid, A. F., *in* "Preparation and Measurement of Isotopic Tracers" (D. W. Wilson, A. O. Nier, and S. P. Reimann, eds.), p. 103. J. W. Edwards, Ann Arbor, Michigan, 1946.

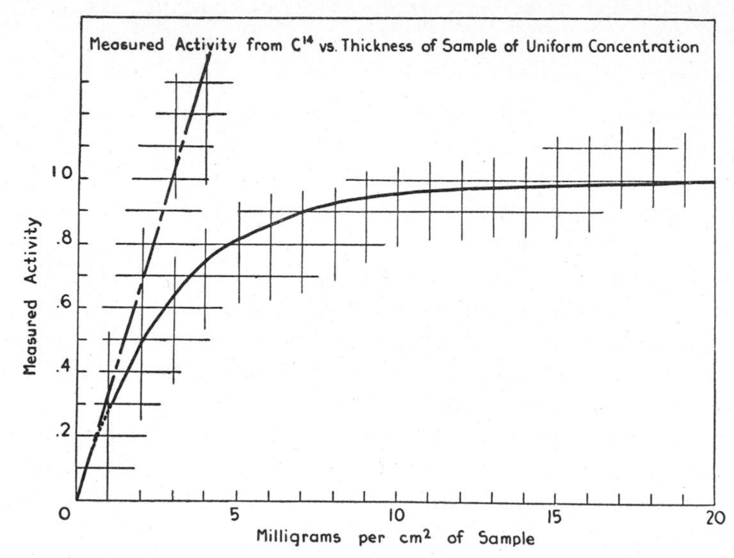

FIG. 30. Absorption curve for C^{14} radiation. (After Reid.[17])

thickness at which saturation is reached depends on the energy of the radiation and can be estimated roughly from the energy-range relations discussed in Chapter II. Saturation is reached at a thickness slightly greater than the range of β rays emitted. The limiting value for the specific activity at infinite thickness compared to that at zero thickness is about 10%. The absorption correction is made most conveniently by construction of a curve expressing the ratio of apparent activity to true activity as a function of thickness expressed in milligrams per square centimeters. In Fig. 30 it will be seen that the true rate is given by the straight line drawn with slope equal to the initial slope of the actual counting curve. The accuracy of the ratio at any thickness depends on the accuracy with which this initial slope is drawn. The counting curve deviates from linearity only slightly over a range such that this line can be drawn with sufficient accuracy for ratios extending up to 5 or 6 mg./cm.^{2} The ratio at 1 mg./cm.^{2} is 0.290/0.330, or 0.88; at 2 mg./cm.^{2} it is 0.495/0.665, or 0.75. By continuing in this fashion a table of ratios is obtained which may be plotted as in Fig. 31. The initial slope depends also on the window thickness and geometry of the sample so that a new calibration must be made if a window is replaced or the sample position changed.

To illustrate the use of the curve in Fig. 31, suppose two samples of $BaCO_3$ with equal specific activities, one weighing 10 mg., the other 20 mg., are counted on a disk with sample area 10 cm.^{2}. The first sample is found to have 630 ct./min., the second 1060 ct./min. Corrected for ab-

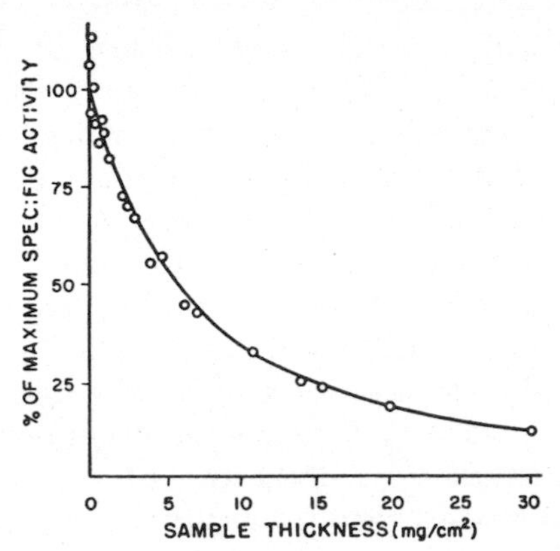

FIG. 31. Absorption correction curve for C^{14}. (After Yankwich *et al.*[18])

sorption, the true rates are 630/0.88 and 1060/0.75, or 715 and 1410, respectively. The second sample contains twice the carbon of the first, so that reduced to the same carbon content its activity is 1410/2, or 705. Thus, agreement within 2% is obtained.

Such precision is typical for a procedure of this nature and is quite adequate for tracer work. This procedure applies to all radioactive isotopes, with the proviso that the values for thickness of sample at which absorption becomes appreciable depend on the radiation energy of the isotope being studied. C^{14}, with its maximum β-ray energy of 150 kv., shows an appreciable correction at 1 mg./cm.^{2}. P^{32}, with a maximum energy some ten times as great, will not begin to be noticeably affected until thicknesses of 15 to 20 mg./cm.^{2} are reached.

It is apparent that the determination of activity is considerably simplified when it is possible to use "infinitely" thick samples because no absorption corrections need be made. The value of the ordinate at saturation thickness depends only on the specific activity of the sample. Thus in Fig. 32 there is plotted the dependence of activity in thickness for two C^{14} samples, A and B, identical in all respects except that sample A has twice the specific activity of sample B.

It is important that infinitely thick samples be identical with regard to mounting and surface area in order that the ratio of saturation activity be taken as the ratio of specific activity. If the surface areas are different, one must correct sample counts to constant area. As an example, suppose three thick samples, A, B, and C, are counted. Sample A shows 2000 ct./min. and is spread over an area of 2.5 cm.^{2}; B and C cover an area of 3.0 cm.^{2} and exhibit 2400 and 1800 ct./min. Sample B corrected to an area of 2.5

[18] Yankwich, P. E., Rollefson, G. K., and Norris, T. H., *J. Chem. Phys.* **14,** 131 (1946).

cm.[2] has 2400 (2.5/3.0) = 2000 ct./min.; hence B has the same specific activity as A. Sample C has 1800 (2.5/3.0) = 1500 ct./min., hence three-quarters the specific activity of A or B. If a contingency arises in which a sample must be counted in the region intermediate between infinite thickness and thickness, one may correct to infinite thickness by using the ratio of the ordinate at the abscissa involved to that at infinite thickness. Thus, if sample A were counted at the sample thickness indicated by the arrow in Fig. 32, the value found, 1600 ct./min., multiplied by the ratio 2000/1600 would give the infinite thickness value of 2000 ct./min.

It may also be remarked that in the self-absorption curves of the type shown in Fig. 31 it is sometimes observed that there is an initial increase followed by a decrease rather than a steadily decreasing function, as would be expected. This is owing to "self-focusing," a phenomenon which probably has its origin in a rather complicated scattering effort discussed below. For C^{14} such effects are confined to the region between 0 and 0.6 mg./cm.^{2}. To avoid uncertainties introduced by self-focusing, a valid procedure is to extrapolate to zero thickness, using the linear portion of the absorption curve beginning just beyond the region in which self-focusing occurs—e.g., from 1 mg./cm.^{2} on in Fig. 31.

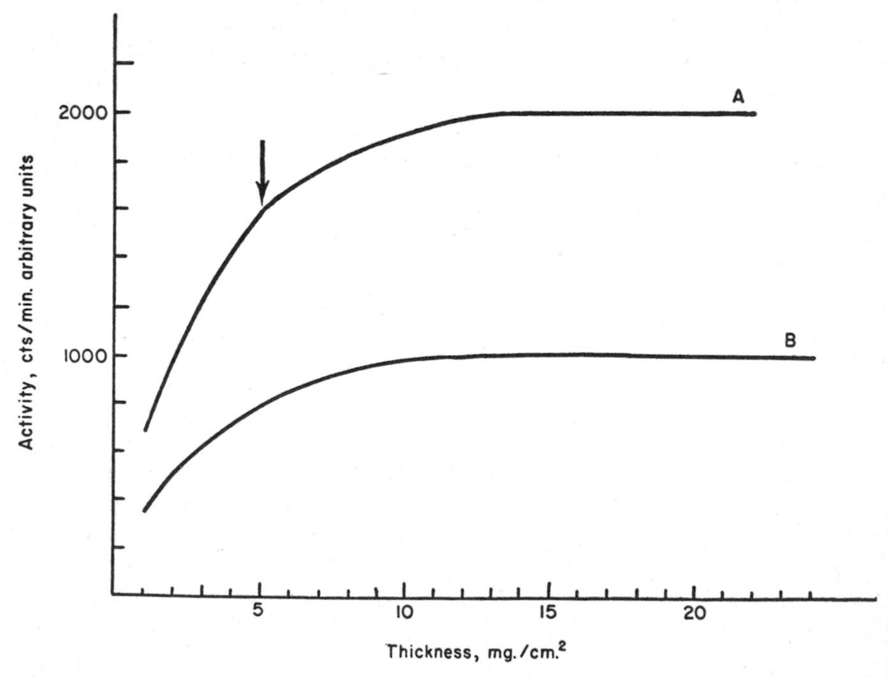

FIG. 32. Comparative activities of two C^{14} samples with specific activity of one sample (*A*) twice that of another (*B*) as function of sample thickness.

5. Sample Geometry. In the preceding discussion of absorption loss it has been assumed that the sample materials in any series of measurements are always spread homogeneously on identical backing material and counted in a fixed position with respect to the G-M tube. The extent to which these requirements are critical depends on the softness of the radiations concerned. With a low-energy β emitter like C^{14} or S^{35} a variation of 1 mm. in placement of samples can introduce an error of 5%. With a high-energy emitter such as C^{11} or P^{32} such displacement gives a vanishingly small error. It is wise to ascertain that all samples are spread over a fixed area well within the area of the tube window or cylinder. Variations in tube sensitivity sometimes are noted if radiations enter in different portions of the sensitive volume.

Sometimes it is necessary to use absorbers in assaying tracer samples. When this is done, care must be taken to ensure uniform geometry of absorber in relation to sample if large errors are to be avoided. A good example of the effect of absorber position on the counting rate can be given from a report by Johnson and Willard.[19] In one experiment a P^{32} sample (E_{max} of β = 1.69 Mev.) was used as a source placed 5.4 cm. from the G-M window. Between the source and the counter a 120.7-mg./cm.^{2} Al absorber was placed directly on top of the source, in which position 6368 ct./min. were recorded. When the absorber was moved upward toward the tube window, the counting rate gradually decreased to 4288 ct./min. with the absorber 0.9 cm. from the window and 4.5 cm. from the sample.

Similar results were obtained with a Co^{60} γ source which emits 1.1- and 1.3-Mev. γ rays. When the radiations were defined ("collimated") by passage through a hole 1.5 cm. in diameter and 15 cm. long inside a lead block and the experiment was repeated, the opposite effect was noted. As the absorber moved from the emergent radiation source to the tube window, the counting rate increased. Thus for the β source and with a 120-mg./cm.^{2} Al absorber the counting rate was 369 ct./min. with the absorber 5.8 cm. from the G-M tube window, and it increased to 1185 ct./min. as the absorber was placed 1.3 cm. from the G-M tube window.

These effects can be understood by reference to Figs. 33 and 34. In Fig. 33, picturing the case of the collimated beam, a fraction of the radiation, A, is transmitted directly to the counter regardless of where the absorber is placed. The remaining radiation is scattered out of the direct path of the beam with an angular distribution which is independent of absorber position, but the amount of this radiation which enters the window depends on the distance of the absorber from the window. Thus radiation indicated as B_1, scattered by the absorber in position 1 does not enter the counter, whereas the same radiation at position 2, denoted B_2, can still get into

[19] Johnson, F., and Willard, J. E., *Science* **109**, 11 (1949).

the counter. Thus the actual counting rate increases as the absorber is brought near the G-M tube when the initial radiation is collimated. When the radiation is uncollimated, the situation depicted in Fig. 34 results. The scattering effect still occurs, but a larger effect due to decrease in total radiation intercepted by the absorber as it is moved away from the source is superimposed. Thus radiation C_{IV} escaping the absorber and G-M tube when the absorber is in position IV has some chance of being scattered along direction C_{III} and entering the G-M tube when the absorber is in position III.

By varying the area and thickness of the absorber, the energy of the radiation, and the degree of collimation, it is possible to arrange matters so that a minimum in the counting rate is observed as the absorber is moved between sample and G-M tube. A special case of this type is the

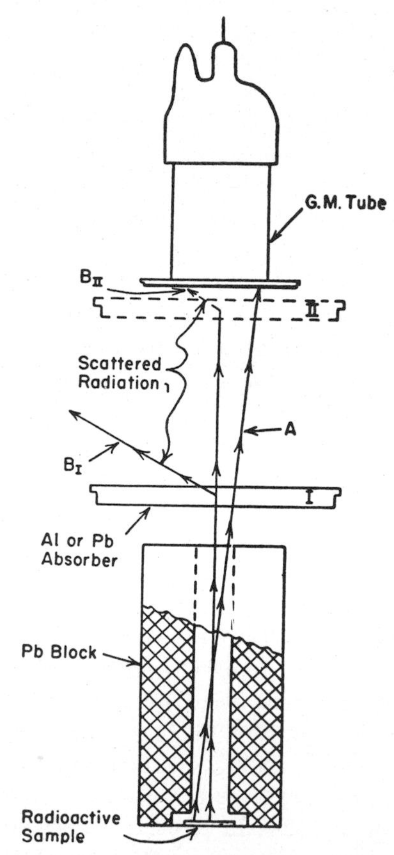

FIG. 33. Effect of position of absorber on counting rate of collimated radiation. (After Johnson and Willard.[19])

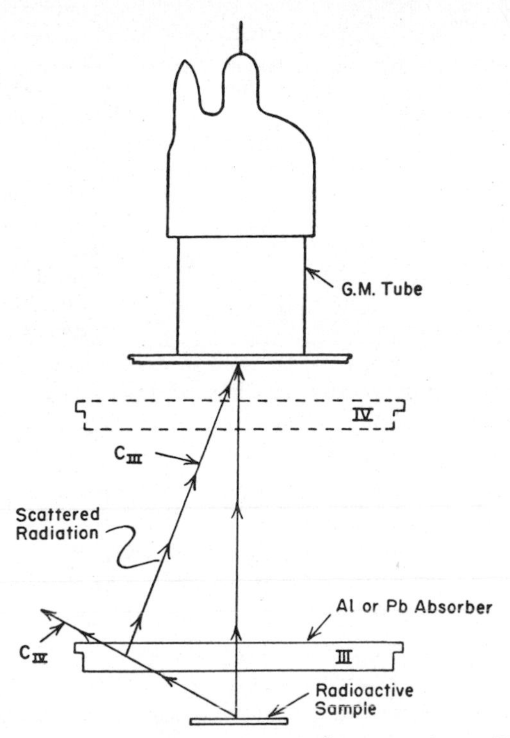

FIG. 34. Effect of position of absorber on counting rate of uncollimated radiation. (After Johnson and Willard.[19])

so-called "self-focusing" which is characterized by an initial increase in counting rate as a sample is diluted with inactive material.

The relation of back-scattering to self-absorption is well exemplified by the measurements of Yankwich and Weigl.[20] A sample of C^{14}-barium carbonate was mounted on an "infinitely thin" plastic film. The aluminum equivalent thickness of the G-M tube window and intervening air was 2.3 mg./cm.^{2} in one set of experiments. Thick layers of various materials were then placed behind the sample. The results are recorded in Table 3. Thus $BaC^{14}O_3$ counted on a backing of aluminum will exhibit back-scattering ratios varying from 1.16 to 1.35 as the thickness of $BaCO_3$ is increased. With C^{14}-wax the effect would be in the opposite direction, falling from 1.16 to 1.07.

The internal back-scattering results in self-absorption curves which vary in shape, depending on the nature of the active material. This effect is not too troublesome when the average atomic number of the elements in

[20] Yankwich, P. E., and Weigl, J. W., *Science* **107,** 631 (1948); see also Nervik, W. E., and Stevenson, P. C., *Nucleonics* **10,** No. 3, 18 (1952).

TABLE 3
BACK-SCATTERING OF C^{14} β PARTICLES (AFTER YANKWICH AND WEIGL[20])

Scatterer	Relative activity
Air	1.00
Platinum	1.51 ± 0.02
$BaCO_3$	1.35 ± 0.01
Glass	1.17 ± 0.01
Aluminum	1.16 ± 0.01
Paper	1.07 ± 0.015
Ceresin wax	1.07 ± 0.015

the materials compared is not greatly different (as in wax or paper) but may be considerable in comparing materials like $BaCO_3$ and wax. The effect to be expected is roughly the ratio of the scattering coefficients as given in Table 3. The material with the higher scattering coefficient will show less self-absorption. Thus, a self-absorption curve for $BaCO_3$ will lie above one for wax with ordinates differing by about 25%, i.e., $1.35/1.07 = 1.26$. This means that at, say, 10 mg./cm.^{2} a C^{14} sample measured as $BaCO_3$ will show a specific activity relative to zero thickness of 0.37, whereas one measured with wax will give a value of 0.28.

It is plain that in the comparison of tracer samples every effort should be made to ensure that backing materials are standardized and uniform and that the chemical composition of samples is maintained sufficiently constant so that effects of the type discussed above are minimized.

F. STATISTICAL ASPECTS OF RADIOACTIVE ASSAY

The ultimate accuracy of radioactivity measurements is limited by the statistical fluctuation inherent in random processes. It can be shown by analysis and confirmed experimentally that radioactive decay follows what is known as the *Poisson distribution law*. To see what this means it is necessary to recall briefly certain considerations from statistical theory.

Any given radioactive sample contains a large number of atoms, n. The problem is to determine the probability, $W(r)$, that exactly r of these atoms will disintegrate (be chosen) in any given time interval. The problem is formally related to the classical statistical question: given a very large set of objects, n in number, in which there are objects of type w with a probability of occurrence, p, what is the probability, $W(r)$, that exactly r objects are of type w. It is found that $W(r)$ is given by an expression formally descriptive of the well-known binomial distribution, namely,

$$W(r) = \frac{n!}{(n-r)!r!} p^r(1-p)^{n-r} \tag{5}$$

The derivation of this relation can be found in any text on statistics.[21, 22] This distribution law can be applied to the case of a batch of radioactive atoms in the following manner.

Let $W(m)$ be the probability that exactly m disintegrations occur in time t, starting with N_0 atoms at time t_0. Consider N_0 as the number n of objects in the random set being observed and m as the number r of type w (property of disintegrating in time t). The binomial expression becomes

$$W(m) = \frac{N_0!}{(N_0 - m)!m!} p^m(1 - p)^{N_0 - m} \tag{6}$$

The expression $(1 - p)$ denotes the probability that an atom does not decay in time t and is given by the ratio of the atoms surviving, N, to those initially present, N_0. This ratio by the fundamental decay law (Chapter I) is

$$N/N_0 = e^{-\lambda t} \tag{7}$$

The probability that an atom disintegrates, p, is therefore $1 - e^{-\lambda t}$. Substitution for p and $(1 - p)$ in Eq. 6 gives

$$W(m) = \frac{N_0!}{(N_0 - m)!m!} (1 - e^{-\lambda t})(e^{-\lambda t})^{N_0 - m} \tag{8}$$

In the usual situation encountered with radioactive samples, one observes a large number of disintegrations in a time very short compared to the half-life. This may be stated formally as a set of conditions, $\lambda t \ll 1$, $N_0 \gg 1$, and $m \ll N_0$. One may approximate $e^{\lambda t}$ as equal to $1 + \lambda t$ and use Sterling's relation, $x! = \sqrt{2\pi x}\, x^x e^{-x}$, to put the expression in a more convenient form. What emerges with these approximations and conditions is the Poisson distribution having the form

$$W(m) = M^m e^{-m}/m! \tag{9}$$

In this expression, M is the true average and may be obtained as the arithmetic mean of a large number of observations. Thus, the probability of obtaining any given number of counts, m, where the average is M, is given as $W(m)$ according to Eq. 9.

The fluctuations observed experimentally can be related conveniently to one fundamental parameter, the standard deviation of a single observation. This is denoted by σ' and defined as the square root of the average value of the square of the individual deviations. From the Poisson distribution law, this quantity is equal to the square root of the number of counts observed. The probable error, defined as the error which is as likely to be exceeded as not, is $0.6745\sigma'$. The chance that a single observation has associated with it an error larger than the probable error diminishes rapidly, being less than one in a thousand for an error five times the probable error.

For strong samples the standard deviation can be computed with fair accuracy simply by taking the square root of the number of counts. For weak samples the background rate contributes appreciably to the statistical error. The combined error for any sample is the square root of the sums of the squares of the individual errors. Suppose a sample is counted for 30 min. and found to give 3600 counts. The background determined for 50 min. gives 2500 counts. The background rate is 50 ct./min., and the standard deviation $(2500)^{1/2}/50$ or 1 ct./min. The sample has a rate of 3600 counts in 30 min. or 120 ct./min. The deviation is $\pm(3600)^{1/2}/30$ or ± 2 ct./min. The strength of the sample is $120 - 50 = 70$ ct./min. The total deviation is $\pm(1^2 + 2^2)^{1/2}$ or ± 2.2 ct./min. The sample strength is therefore 70 ± 2.2 ct./min.

Before calculations of the above nature can be applied it is necessary to ascertain that the counting data actually constitute an acceptable statistical aggregate. For this purpose a number of tests are available.[21–23] It will be sufficient to note here that adequate data will yield essentially the same standard deviation whether a long single count is made or whether this count is divided into many small counting intervals. In the first case the standard deviation is calculated as discussed above. In the second case the standard deviation of the average value is calculated from the square root of the sum of the "residuals" divided by the number of observations. The residuals are the differences between the actual counts found and the "true" value which is the arithmetic mean. The standard deviation of the mean in the normal probability distribution is given by the relation

$$\sigma = [\Sigma_m(A_m - \bar{A})^2/m(m - 1)]^{1/2} \tag{10}$$

Here the A_m is the individual counts, $\bar{A}$ the arithmetic mean, and m the number of times the counts are made. It will be noted that for large m Eq. 10 reduces to

$$\sigma = \frac{1}{m}\left[\sum_m (A_m - \bar{A})^2\right]^{1/2} \tag{10a}$$

As an illustration, suppose the sample discussed previously is counted not for 30 min. but six times at 5-min. intervals. The total count of 3600 is made up of a set of numbers such as 615, 577, 582, 617, 611, and 598,

[21] See Fisher, R. A., "Statistical Methods for Research Workers." Oliver and Boyd, London, 1936; Lindsay, R. G., and Margenau, H., "Foundations of Physics," Chapter IV. Wiley, New York, 1936.

[22] Rainwater, L. J., and Wu, C. S., *Nucleonics* **1,** No. 2, 60 (1947); **2,** No. 1, 42 (1948).

[23] Pearson, K., *Phil. Mag.* **50,** 157 (1940).

the average, $\bar{A}$, being 600. The residuals, $(A_m - \bar{A})$, are +15, −23, −18, +17, +11, and −2. It is seen that positive and negative values occur with nearly equal frequency. The arithmetic average of the residuals is 14.2. The residuals do not exceed twice the average in any case. This is inside the statistical expectation, because a result deviating by more than three times this average is not too unlikely. It is common practice to discard a result with a residual four times the average residual as arising from some systematic fluctuation not of statistical nature. The standard deviation based on Eq. 10 is calculated to be 7.0, so that the result of the assay is 600 ± 7.0 ct. in 5 min., or 120 ± 1.4 ct./min. Calculated from the total count of 3600 in 30 min., σ' is 2.0 ct./min., so that the assay yields 120 ± 2 ct./min. It can be seen that standard deviation calculated on the small sample basis (σ) is somewhat smaller than the deviation expected on the basis of the single observation (σ'), since, by using the small sample procedure, it has been possible to take into account the rather close clustering of the residuals around the average. Had there been several 5-min. observations when residuals as high as 60 occurred, then σ would have been larger than σ'.

In summary, it appears best to calculate σ by Eq. 10 when working with weak samples. Strong sample deviations can be represented adequately by the square root of the number of counts.

The main significance of these remarks on the statistics of counting is related to the determination of samples with an activity equal to or less than background. It can be calculated that with a normal background rate of 15 to 20 ct./min., an hour of counting is required to establish a result with a standard deviation no more than 5%. The practical limit for precision work is in this range. *It is not advisable to place reliance on a result which is 10% or less of background.*

It should be emphasized that the procedures discussed do not represent an adequate check on the statistical acceptability of counting data. A rigorous discussion of these matters is outside the scope of this book. In practice, inadequate tube counting is usually easy to ascertain without resorting to statistical analyses. For instance, in G-M tube counting, inspection of the oscilloscope pattern for ionization bursts and abnormal pulse patterns will usually reveal trouble long before it is suspected from the appearance of the data.

The considerations presented above relating to standard deviation, σ, have dealt with measurements involving discrete counts recorded by a counting device. It may be inquired what method is valid for estimation when instruments which integrate counting data are used. One may consider such instruments as made up of a resistance, R, and capacitance, C, combined in such a manner that the counting rate is averaged over an interval of time corresponding in magnitude to the time constant of the integrating circuit RC. It can be shown that, if the counting rate in counts per minute is denoted by A, the value for σ is given by the expression

$$\sigma = \sqrt{A/t} \tag{11}$$

where t is a time interval given by the expression $2RC/60$ in minutes, R being expressed in ohms and C in farads. Thus σ in terms of A and RC is given by

$$\sigma = \sqrt{30A/RC} \tag{12}$$

An approximate value for the time constant of any given integrating device can be estimated by noting the time required for the radioactivity rate as indicated by the deflection of the recording element to drop to $1/e$ of its steady value after removal of the sample.

An instrument such as the Lauritsen electroscope measures the rate of drift of a moving element. It is necessary in such a case to estimate the number, n, of ionizing particles which enter the instrument during the time of measurement. If the activity, A, is expressed in any units (divisions per second is usual), then the standard deviation is $A/\sqrt{n}$.

In conclusion, it should be remarked that no assay should be conducted for a time longer than is necessary to make the standard deviation less than errors of a systematic nonstatistical kind such as sampling uncertainty, biological fluctuations in source material, and uncontrollable chemical losses.

G. Standards in Radioactive Assay and Determination of Tracer Intensity

The determination of the intensity of absolute tracer activity is important because it is desirable to prevent radiation damage in organisms exposed to tracer radiations. To measure absolute disintegration rates, it is necessary, as discussed in Chapter I, to know the details of the disintegration process; that is, it must be known how many β and γ rays are involved in each transition, what the energy distribution of the emitted particles is, and what the efficiency of the detection apparatus is for each particle. Fortunately for most tracer experiments, only a rough estimate of the absolute radioactivity is required so that standardization procedures can be simplified somewhat.[24a]

The measurement of β-ray activity requires a standard which emits β rays, preferably identical in energy distribution. A popular standard is a radioactive daughter of uranium, UX_2. The parent uranium ($_{92}U^{238}$) decays

[24a] See Deutsch, M., Downing, J. R., Elliot, L. G., Irvine, J. W., Jr., and Roberts, A., *Phys. Rev.* **62,** 4 (1942), for a detailed presentation of some standardization methods.

by α-particle emission ($\tau_{1/2} = 4.6 \times 10^9$ yr.) to UX_1, which disintegrates in turn, emitting a low-energy β ray ($\tau_{1/2} = 24.5$ days), producing the energetic β-ray emitter UX_2 ($\tau_{1/2} = 1.14$ min.). A fraction of the UX_1 disintegrates by another mode to UZ, a nuclear isomer of UX_2, but this effect is too small to interfere appreciably with the use of the standard, assuming 100% disintegration in the normal way of UX_1 to UX_2. The β radiation from UX_2 has a rather high energy ($E_{max} \simeq 2.3$ Mev.), so that it is best used for assay of tracer materials emitting energetic radiation.

In recent years, the U. S. National Bureau of Standards, in collaboration with laboratories in Canada and England, has developed a series of β and γ standards, using P^{32}, I^{131}, Co^{60}, C^{14}, Tl^{204}, Sr^{90}, Y^{90}, and Au^{198}. Data on the various standards available can be found in a report by Seliger and Schwebel.[24b] Standards are also available from commercial firms.

For those who wish to prepare their own standards, a simple procedure is as follows. A weighed quantity of uranium oxide (the composition of which is known accurately) is spread homogeneously on a metal backing. This may be accomplished by using a suspension of finely powdered U_3O_8 in acetone to which a little Duco cement is added. The metal backing should be of low atomic number and not thicker than 1 mm. to avoid excessive back-scattering. Of course this thickness is not important if the unknown and the standard are mounted on identical backings. The UX_2 activity can be computed[25] from the weight of the uranium because the UX_2 is in radioactive equilibrium with the uranium. A microcurie of UX_2 radiation is contained in 3.5 g. of uranium oxide (U_3O_8). Since the UX_1 β radiation can be completely absorbed by 25 to 30 mg./cm.^{2} of aluminum, a thin aluminum foil some 3 to 4 mils thick is cemented over the oxide. The half-thickness of the UX_2 radiations is approximately 120 mg./cm.^{2} in aluminum, so that only a few per cent of the UX_2 radiation intensity is lost. Corrections for absorption in the window or walls of the detection

[24b] Seliger, H. H., and Schwebel, A., *Nucleonics* **12**, (1), 54 (1954).

[25] In the steady state (radioactive equilibrium) the number of uranium atoms disintegrating is equal to the number of UX_2 atoms disintegrating, so that

$$\lambda_U \cdot n_U = \lambda_{UX_2} \cdot n_{UX_2}$$

The weight in grams of the U_3O_8 molecule containing 3 atoms of uranium is $842.2/6.023 \times 10^{23}$, or 139.8×10^{-23} g. The weight of a moledule of U_3O_8 containing 1 atom of uranium is, therefore, $139.8 \times 10^{-23}/3 = 46.6 \times 10^{-23}$ g. Since 1 μc. of activity is 3.7×10^4 disintegrations/sec., the number of uranium atoms required to give an activity of 1 μc. is given by

$$3.7 \times 10^4/\lambda_U = 3.7 \times 10^4/4.9 \times 10^{-18} = 0.76 \times 10^{22}$$

The weight of U_3O_8 needed to supply an activity of 1 μc. is, therefore, $0.76 \times 10^{22} \times 46.6 \times 10^{-23} = 3.5$ g.

Similar calculations can be carried out for other uranium salts.

apparatus must be applied in the way described previously in this chapter (see Section 2-E). When a relatively insensitive instrument, such as an electroscope, is used, it is necessary to spread as much as 0.1 μc. on the backing to obtain a standard with a conveniently measurable radioactivity. On a surface of 10 cm.^{2} some 350 mg. is required. This introduces a certain amount of self-absorption corresponding to an effective thickness of 17 to 20 mg./cm.^{2} for which a correction must be made. With G-M tubes or sensitive ionization chambers, much thinner samples corresponding in activity to 10^{-2} or 10^{-3} μc. can be used so that no appreciable self-absorption occurs. An alternative standard is the β-emitting RaE ($\tau_{1/2}$ = 5 days) which is in equilibrium with RaD ($\tau_{1/2}$ = 22 yr.).

H. Radioautography

1. Introduction. The fogging of photographic emulsions, which has been mentioned as the oldest technique for radioactivity assay (see Section 2-C), can be exploited to visualize distribution of radioactivity. In general, this may be done in the following manner. The radioactive element is administered to the organism which distributes the element in some typical fashion. After a suitable interval, the tissue is washed free of any contaminating external radioactivity and dried. A tissue section is prepared by the usual techniques. This is pressed into close contact with photographic film. All regions of the organism containing radioactive element emit radiation which produces effects on the emulsion similar to visible light. On subsequent development of the film, all regions corresponding to localization of radioactive element are darkened so that a photographic image of the tracer distribution is obtained.

The image so produced is called a "radioautograph." Such radioautographs for investigation of the distribution of radioactive elements in animals were first made by Lacassagne and Lattes,[26] who investigated the deposition of polonium in renal rabbit tissue. With the advent of artificial radioactivity, a host of studies were made possible because radioactive isotopes of the important elements for biology became available.

Before proceeding to a description of experimental procedures, a few remarks of a general nature on the physical aspects of radioautography should be made.

The best resolution is obtained only if thin sections which can be placed in close contact with the film are prepared. It must be remembered that the radiations emerge in all directions from the localized spot, so that diffusion of the image can be minimized only by close contact of the film and the radioactive region. Hence a radioautograph cannot be made with good resolution if an intact animal carcass or plant is placed on top of the film.

[26] Lacassagne, A., and Lattes, J., *Compt. rend. soc. biol.* **90,** 352 (1924).

The exposure time depends on the amount of radioactive isotope concentrated in the section and is determined best by empirical means. The length of exposure also depends on the nature and energy of the radiations. Beta particles are much more effective than γ rays because, for a given initial energy, their range is much less. Beta-particle dosages are confined almost entirely in the film emulsion, whereas only a fraction of the γ-ray dose is dissipated in this fashion. The average energy of the β particles affects the total exposure to a certain extent. The dosage is not inversely proportional to the energy in a precise sense because the film does not have a linear response. Satisfactory blackening requires some threshold dosage. No general statements can be made as to total dosage required to produce a satisfactory image because what constitutes a satisfactory image depends on the investigator. Estimates vary from 1×10^6 to 1×10^8 β particles per square centimeter of film.

The response of autographs as regards resolution is best for α particles because of the homogeneous energy, short range, and high ionization intensity encountered with these radiations. Unfortunately none of the elements of major biological importance is an α emitter. However, many significant studies on localization of heavy elements such as lead, polonium, plutonium, and americium have involved as essential techniques radioautography with α emitters.[27]

2. Experimental Procedures. The problems involved in preparing radioautographs fall into two categories: (1) the method for handling tissues and (2) the processing of the emulsion.

The first category includes techniques for fixation, dehydration, embedding, and sectioning. Concerning fixation and dehydration, it may be said only that care must be taken to avoid leaching out or moving about material in the tissue. Specific procedures cannot be cited because each tissue is a problem in itself. As an example of the difficulty cited there will be recalled erroneous results obtained in studying bone deposition of plutonium and strontium when procedures for decalcification were employed.[28] The placement of dehydrated tissue in a suitable solid medium prior to sectioning may be accomplished in a variety of ways. Celloidin is often used, particularly for work with undecalcified bone. The tissue may also be frozen, a technique which is useful when it is desired to avoid contact with organic solvents. The usual technique involves impregnation with paraffin. Typical steps involved in this procedure, as described by Boyd,[29] follow.

After embedding in paraffin, sections 7 μ thick are cut and floated in water. The sections are lifted out on top of a clean microscope slide and

[27] Hamilton, J. G., *Revs. Mod. Phys.* **20,** 718 (1948).
[28] Axelrod, D. J., *Anat. Record* **98,** 19 (1947).
[29] Boyd, G. A., *J. Biol. Phot. Assoc.* **16,** 60 (1947).

then heated at 40°C. to smooth out wrinkles. The tissue is then refloated in water by carefully inserting the slide at an angle of about 45°, after which the section is transferred to a photographic plate in the dark. Excess water is removed using fine-grade filter paper as a blotter. The tissue section on the plate is dried in a light-tight box for a few hours, after which it is placed in the refrigerator in the dark for the rest of the exposure period. After exposure, the section in place on the emulsion is treated with xylene to remove paraffin, then with alcohol, and finally with water. After development of the image, the water-alcohol-xylene treatment is run in reverse rather than drying in air, which is slow. Paraffin sections are used most often when the section is mounted directly on the film.

Belanger and LeBlond have achieved improvements in resolution by eliminating the use of plates or film by a procedure in which the emulsion is poured over the stained section.[30] Endicott and Yagoda[31] have taken advantage of the increased sensitivity to nuclear particles exhibited by nuclear emulsions, i.e., thick, fine-grained emulsions developed for work with nuclear particles. Still another development involves the use of stripping emulsions.[32a]

It is important to remember that pressure, chemical reducing agents, alloys, and other agents can produce pseudophotographic effects. It is necessary to exclude obvious fogging agents such as formaldehyde and quinone. The reader is referred to the monographs by Yagoda[32b] and Boyd[32c] for further information on this as well as all other aspects of radioautography.

To particularize this discussion it is desirable to include a few examples of research employing radioautography. Arnon *et al.*[33] have described techniques for making radioautographs relating to the distribution of phosphorus in tomato fruits at various states of development. In one experiment, leaves and fruits were removed 36 hr. after introduction of labeled sodium phosphate (about 30 μc. of P^{32} per liter into nutrient solution bathing the roots of the plant. The leaves and fruits were cut into sections 2 to 4 mm. thick, and laid on pieces of thin paraffined paper which were then placed over x-ray film and wrapped in black paper. The samples were fixed in place by application of pressure, with a heavy glass plate as press.

Exposures of 1 hr. sufficed to give a clear picture of the P^{32} distribution.

[30] Belanger, L. F., and LeBlond, C. P., *Endocrinology* **39,** No. 1, 8 (1946).

[31] Endicott, K. M., and Yagoda, H., *Proc. Soc. Exptl. Biol. Med.* **64,** 170 (1947).

[32a] Pelc, S. R., *Nature* **160,** 749 (1947).

[32b] Yagoda, H., "Radioactive Measurements with Nuclear Emulsions." Wiley, New York, 1949.

[32c] Boyd, G. A., "Autoradiography in Biology and Medicine." Academic Press, New York, 1955.

[33] Arnon, D. I., Stout, P. R., and Sipos, F., *Am. J. Botany* **27,** 791 (1940).

Fruits from the same plant were compared at different stages of development by radioautography on the same plate, so that factors such as development time and exposure time were constant. The plant used was kept for 9 days in a nutrient solution containing 5 parts per million labeled phosphate with an activity of about 20 μc./l. of solution. In this case overnight exposure was used to obtain an autoradiograph which covered the whole range of uptake exhibited by the fruits. In other experiments the uptake of phosphate by seeds in the fruit was investigated, and it could be seen that the seeds of the fully ripened fruit were incapable of phosphate uptake, whereas in large green fruit the seeds assimilated phosphate readily. From the description of the procedure employed it will be appreciated that high resolution was not a requirement. Studies of this type, when carried out in conjunction with precison assay of the tissues by chemical fractionation, can be expected to be highly useful in clarifying many aspects of plant physiology.

In animal physiology, an interesting example of the application of radioautograph techniques is afforded by the work of Pecher,[34] who ascertained that strontium deposited almost entirely in bone, whereas phosphorus was distributed not only in bone but also more or less diffusely throughout the soft tissues. Two rats were treated, one with radioactive phosphorus as phosphate, and the other with radioactive strontium as lactate. Some days later the animals were killed and sections of the entire animals prepared. It was found that the strontium concentrated almost entirely in the skeletal structure, whereas the phosphorus was considerably more diffused. From what is known about phosphate retention it is probable that a radioautograph taken after a few weeks would have shown relatively more phosphorus accumulation in bone. In any case, the high concentration of strontium in bone suggests application to therapeutic bone irradiation. Such studies have been reported by Treadwell *et al.*,[35] who showed localization of strontium in areas of rapid bone formation and in particular in areas invaded by osteogenic sarcoma. In Fig. 35 is shown a roentgenogram of the leg section involved in a case of osteosarcoma with the corresponding radioautograph which illustrates the concentration of radioactive strontium.

An example of the radioautograph and its application to investigations at the clinical level may be presented from the work of Seidlin *et al.*[36] These

[34] Pecher, C., *Proc. Soc. Exptl. Biol. Med.* **46,** 86 (1941).

[35] Treadwell, A. deG., Low-Beer, B. V. A., Friedell, H. L., and Lawrence, J. H., *Am. J. Med. Sci.* **204,** 521 (1942).

[36] Seidlin, S. M., Marinelli, L. D., and Oshry, E., private communication. See also Seidlin, S. M., Marinelli, L. D., and Baumann, E. J., *J. Clin. Endocrinol.* **6,** 247 (1946). For earlier work, see Keston, A. S., Ball, R. P., Frantz, V. K., and Palmer, W. W., *Science* **95,** 362 (1942).

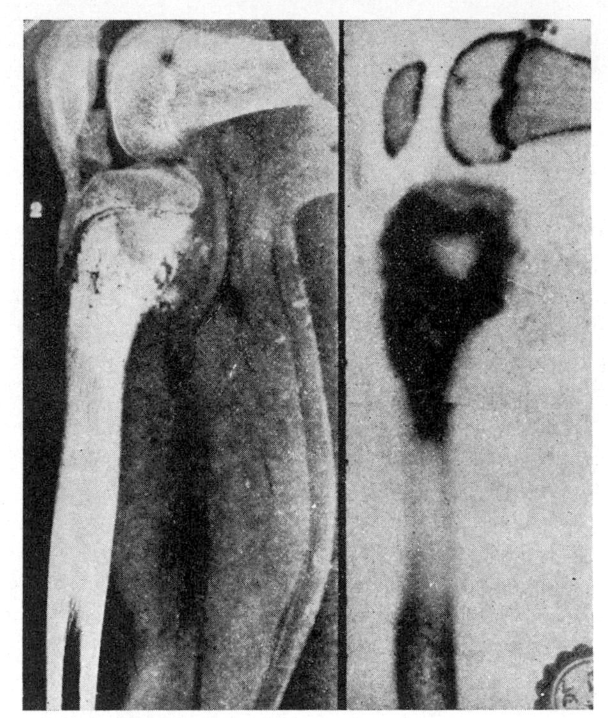

FIG. 35. *A*, roentgenogram of section of amputated leg showing osteogenic sarcoma involving the proximal epiphysis and metaphysis of tibia. *B*, radioautograph of same section showing concentration of Sr^{89} in tumor and in growing parts of normal femur. (After Treadwell *et al.*[35])

workers have studied the association of adenocarcinoma of the thyroid which occurred in connection with functioning distant metastases and hyperthyroidism. Hyperactivity of carcinomatous thyroid is not usually found. Such tissue, in fact, retains little iodine, as a general rule. Therefore, retention of iodine may be taken as a criterion for active thyroid function in the metastatic tissue.

In two cases hyperthyroidism was correlated with functioning of the metastatic thyroid tissue. In Fig. 36 a radioautograph of such tissue taken from the rib of a patient with metastatic carcinoma of the thyroid is shown. Comparison of the radioautograph with the photomicrograph of the same section shows definitely that tissue proved active by the staining test is also active with labeled iodine accumulation as the criterion.

3. Limitations and Prospects. The radioautograph technique can be used to demonstrate strikingly the movement and localization of tracer material. However, radioautography is not quantitative and is limited in ultimate resolution by three factors. First, the tracer radiations are emitted in all

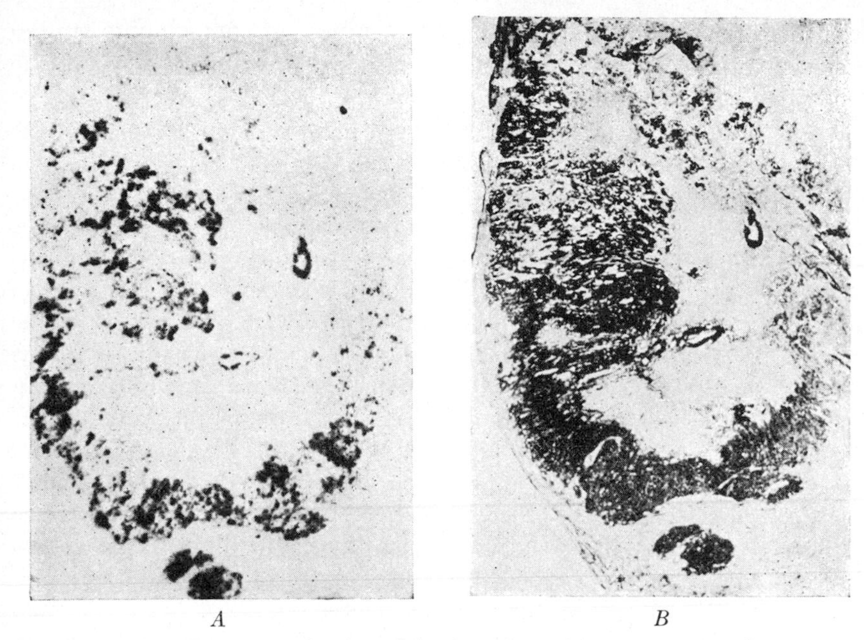

FIG. 36. *A*, radioautograph of unstained section of human metastatic tumor removed from rib. Distribution of I* in the viable tissue is indicated by dark areas. Magnification approximately 8×. *B*, microphotograph of section shown in *A* after staining with eosin-hematoxylin (same magnification). Note coincidence in I* distribution as found in *A* with viable regions shown in *B*. (After Seidlin *et al.*[36])

directions so that for each image point there corresponds a diffused area of blackening on the emulsion. Because these radiations penetrate the emulsion with varying range owing to the continuous distribution in β-particle energies, there is a limit to the definition of the image which can be obtained. Structure of tissue can be manifested only grossly even under the best conditions, owing to the lack of collimation of the radiation, especially if it is desired to keep the organism intact. Second, the grain of the photographic emulsion limits resolution to the range obtainable by conventional photomicrography. This factor is less important as a limitation than lack of radiation collimation. Third, the dosage to the tissue must be kept below the tolerance limits for radiation damage. Therefore, the tracer cannot be increased indefinitely simply to obtain increased intensity. The dosage limits depend on the energy of the radiations, the amount retained in the tissue, and the general distribution pattern of the tracer. These factors work most particularly against the extension of radioautography to the study of single cells. Bayley[37] has given a detailed discussion

[37] Bayley, S. T., *Nature* **160**, 193 (1947).

regarding the impracticability of conducting autoradiography studies designed to probe structural details of single cells with dimensions in the range 1 to 10 μ or less in diameter.

The interpretation of radioautographs cannot be undertaken without considerable knowledge derived from supporting chemical investigations. The radioautograph shows *where* the element goes but not *how* or in *what* form it gets there. For instance, in localization of iodine by thyroid, it is necessary to show that the iodine radiation producing the autograph is organically bound iodine (thyroxine, diiodotyrosine, etc.) rather than inorganic iodide. This requires biopsy and chemical fractionation of the tissue. Of course, once it is established that iodine entering the thyroid is retained exclusively as bound iodine, radioautographs can be used as a time-saving procedure for diagnosis of function in tissue. Similar considerations apply to all other elements investigated in this manner.

The extension of the technique to the range beyond ordinary micrography requires the solution of a formidable technical problem. This is the collimation and focusing of the tracer radiations. The difficulty lies in the existence of a continuum of β-ray energies. To attain a resolution comparable with that given by the electron microscope, the β particles must be rendered monoenergetic. Yet, when the particles emerge with all energies from zero to a maximum of several million electron volts, no simple electron microscope focusing procedure, like direct acceleration in a high electric field, is applicable. On the other hand, if only a portion of the β-ray spectrum is selected, then intensity is lost. These difficulties might be avoided by using tritium, because it is a tracer which has such a very low-energy maximum ($\sim$0.015 Mev.). Direct acceleration of the tritium β particles in an electron microscope with a focusing electric field of 100 kv. might produce β particles sufficiently monoenergetic to achieve a satisfactory radioautograph with resolution in the electron microscope range. Even in this case there would be the problem of tritium placement in the tissue to be solved. Tracer atoms which emit only γ radiation with a single conversion electron of high intensity would also be amenable to this type of extension in electron microscope technique.[38]

The extremely short range of the tritium β radiation has been exploited to visualize localization of tritium-labeled compounds in large cells like paramecia and yeast.[39] The organisms were grown in media containing tritium-labeled sodium acetate. Sections 1 μ thick were cut and autographed. Radioactivity was seen clearly in both endoplasm and ectoplasm. Many areas of concentration were found corresponding to food vacuoles,

[38] See Marton, L., and Abelson, P., *Science* **106,** 69 (1947).

[39] Eidinoff, M. L., Fitzgerald, P. J., Simmel, E. B., and Knoll, J. E., *Proc. Soc. Exptl. Biol. Med.* **77,** 225 (1951).

although a few appeared to resemble macronuclei. The mouth was outlined distinctly in some pictures. Beginning with levels of activity of approximately 10^7 disintegrations/min./mg., sufficient tritium was taken up to give adequate autographs after exposure of the sections to the film for 1 week. Since the average β particle from tritium is absorbed by 1 μ or less in nuclear emulsions, this nuclide is ideal for visualization of localization in photomicrographs and affords the greatest resolution possible at present.

3. ASSAY OF STABLE ISOTOPES

A. Introduction

The determination of the labeled content of tracers enriched with stable isotopes differs from the assay of radioactive tracers in that stable isotopes used as tracers are present in amounts which are appreciable on a percentage basis. The actual weight of the tracer element is significantly greater than that of the unlabeled element with its normal complement of isotopes. Hence, it is usual to determine the isotopic composition of stable tracer by measuring differences in weight. Because these methods are not so sensitive as those associated with measurements of radioactivity, tracer procedures involving stable isotopes are usually limited to experiments in which dilution of the labeling isotope is not so great as that permissible in experiments with radioactive tracers.

In a given element, the normal abundance of the labeling isotope will be A_0 per cent, that is, in every 100 atoms there will be A_0 atoms of the labeling isotope and $(100 - A_0)$ atoms of the other isotopes. If the element is enriched in the labeling isotope to the value of A per cent, and is used in a tracer experiment, it will be diluted with the normal mixture of isotopes, say X times. This means that for every 100 atoms of the labeling mixture, there will be $100X$ atoms containing A atoms of labeling isotope from the labeling mixture and $(X - 1)A_0$ atoms of labeling isotope from the diluting material. The total number of labeling isotope atoms is $A + A_0(X - 1)$, and the atom per cent of the labeling isotope is $(A + XA_0 - A_0)/X$ or $A_0 + (A - A_0)/X$. The difference between this value and the normal abundance, A_0, is defined as the *atom per cent excess*. This difference is $(A - A_0)/X$. But $(A - A_0)$ was the atom per cent excess of the original labeled element. In other words, if the concentration of the labeling isotope is expressed as atom per cent excess, and not as atom per cent, the ratio of its concentrations in the original and in the final material will be equal to the dilution factor, X, because

$$(A - A_0) \Big/ \frac{(A - A_0)}{X} = X$$

There are two principal methods which are applicable to the study of relative abundance of stable tracer isotopes. One is based on the use of combined electric and magnetic fields to analyze isotopic components when they are brought into a state of ionization in the gaseous state. The other depends on measurements of density differences between samples with different isotopic composition. Other methods also can be employed. Historically, methods based on isotopic shifts in line and band spectra were used first to demonstrate the existence of the important isotopes H^2, C^{13}, and N^{15}. In the special case of H^2 (deuterium) a microthermal conductivity method has been developed. For tracer work, however, the procedures based either on electromagnetic analysis of ionized isotopic material or direct density measurements of isotopic samples are the only ones of interest.

B. Density Methods

The determination of small differences in density arising from differences in isotopic composition is confined almost entirely to the assay of deuterium content in water, although it can also be applied to the assay of the heavy isotopes of oxygen in water. Various procedures are possible. The most common is called the "falling drop" method. In this method the time required for a standard drop of the water to be assayed to fall a given distance in an immiscible liquid of specific gravity slightly less than 1 is measured. The liquid is usually *o*-fluorotoluene, and it is contained in a long, thin, glass-stoppered tube mounted in a well-regulated water bath. Drops of uniform size are delivered from a specially designed micropipet.[40, 41] The rates of fall of a number of drops (8 to 10) are determined and averaged. The mean dropping time, T_0, of normal water is determined before and after each measurement of the dropping time, T, of the test sample. The reciprocals of these times are calculated, the differences taken ($1/T - 1/T_0$), and the corresponding deuterium content read off a calibration curve prepared from a series of accurately diluted samples of pure D_2O.

This method can be used to determine the deuterium content of water with an accuracy of 0.02%. Like other density methods, however, this procedure may need samples as large as 100 mg. to function best. Details of the apparatus required and procedures employed in combustion of samples will be found in a monograph by Glascock.[42a]

[40] Keston, A. S., Rittenberg, D., and Schoenheimer, R., *J. Biol. Chem.* **122,** 227 (1937).

[41] Popják, G., *Biochem. J.*, **46,** 558 (1950).

[42a] Glascock, R. F., "Isotopic Gas Analysis for Biochemists." Academic Press, New York, 1954.

If 2 to 3 ml. is available, recourse to what is called the "Cartesian diver" method is possible. This procedure calls for a glass or quartz bubble to be totally immersed in the liquid sample. The bubble will float at a fixed level depending on the temperature (at constant pressure) or the pressure (at constant temperature).[42b] The extremely small variation in liquid density brought about by changes in the temperature or pressure can be measured by noting the value for either temperature or pressure at which the float neither rises nor sinks. A difference of 0.0002°C. can be easily measured. This corresponds to a density difference of one part per ten million which is equivalent to a change in deuterium content of 0.0001%.

In an apparatus used by Schoenheimer and Rittenberg,[43] a glass float is constructed so that it maintains its equilibrium position at 0°C. under a pressure of 1 atmosphere. A reduction in pressure with the temperature remaining constant causes the float to sink because the water is more compressible. A linear relation exists between the density of the liquid and the pressure for which the float neither rises nor falls. The movement of the float is observed through a microscope fitted with two crosshairs in the eyepiece. The rate of fall is noted as the pressure is decreased. Likewise the rate of rise is measured as the pressure is increased. From these observations, a graph is constructed from which the pressure for no movement can be read off by finding the point where the linear plot intersects the ordinate for zero velocity. The float is calibrated with a solution of heavy water for which the density has been determined by the standard pycnometer method. (The pycnometer can also be used to measure densities with the same precision as in the falling drop method if several milliliters of water is available.) A change in pressure of 1 cm. Hg, which is a large effect, corresponds to a change in deuterium content of approximately 0.0004%. Obviously, scrupulous care must be taken to prepare the water free of dirt, air, etc. In practice the water is purified by distillation *in vacuo*. The volume of sample in this arrangement is about 2.5 ml., but this is no limitation in practice because of the high sensitivity of the method which permits dilution of most samples to the required volume.

In conclusion it may be noted that differences in index of refraction may be exploited for assay of differences in density of water.[44]

C. Assay by Electromagnetic Methods

The instruments employed in assay of isotopic composition by electromagnetic means are called "mass spectrometers" or "mass spectrographs." These two terms do not refer to different types of apparatus but only to

[42b] Richards, T. W., and Shipley, J. W., *J Am. Chem. Soc.* **34,** 599 (1912).
[43] Schoenheimer, R., and Rittenberg, D., *J. Biol. Chem.* **111,** 163 (1935).
[44] Crist, R. H., Murphy, G. M., and Urey, H. C., *J. Am. Chem. Soc.* **55,** 5060 (1933).

different ways of detecting and recording results. The principles on which these instruments are based have been widely described; the reader will find a modern account in the monograph by Inghram and Hayden.[45] In either the mass spectrometer or spectrograph four essential operations must be performed:

1. The material to be assayed must be vaporized at low pressure and then ionized. The positive ions produced from each isotope must be present in the same relative amount in the vapor as obtains for the isotopes in the original sample.
2. These ions must be rendered monoenergetic by electrostatic means.
3. After step 2, the ions must be collimated and focused by means of a magnetic field.
4. The ions focused at a collecting slit or on a photographic plate must be collected and the ion intensity must be measured.

The term *mass spectrometer* is restricted to machines wherein the ions are collected in ion chambers or by other electrical means. The term *mass spectrograph* refers to machines in which the detection is carried out by nonelectrical means—by photographic film or plate. The term *mass spectroscopy* is used loosely to indicate studies with both kinds of machines.

There are many ways in which the steps listed can be carried out. Hence, a great variety of mass spectrometers has been developed in recent years. One type, which serves as the basic model for many of the commercial machines available for tracer assay, is shown in Fig. 37.

Many precautions are required in using the mass spectrometer. In step 1, methods which do not fractionate the isotopes are necessary. This means that evaporation of solids or liquids cannot be used because the rate of evaporation depends on the atomic weight of isotopic components, the lightest isotope in any mixture being the most volatile. If the pressure in the ion source gets too high in this event, fractionation can occur because neutralization of the positive ions takes place at different rates for different isotopes.

It is best to introduce the sample as a gas, because it is possible to prepare gas samples and introduce them with minimal fractionation, as compared to direct volatilization in the ion source. Another source of error arises from the presence of hydrogen as a contaminant, because hydride formation may occur. Organic vapors present in the ion source can also cause errors by affecting the rate of volatilization of solids.

In step 2, the energy filter may be an electrical "lens" or a direct electrostatic field. In general, it is satisfactory to use direct acceleration, care being taken to keep the accelerating electric field constant. The voltage

[45] Inghram, M. G., and Hayden, R. J., *Natl. Research Council Nuclear Sci. Ser.* No. **14,** (1954).

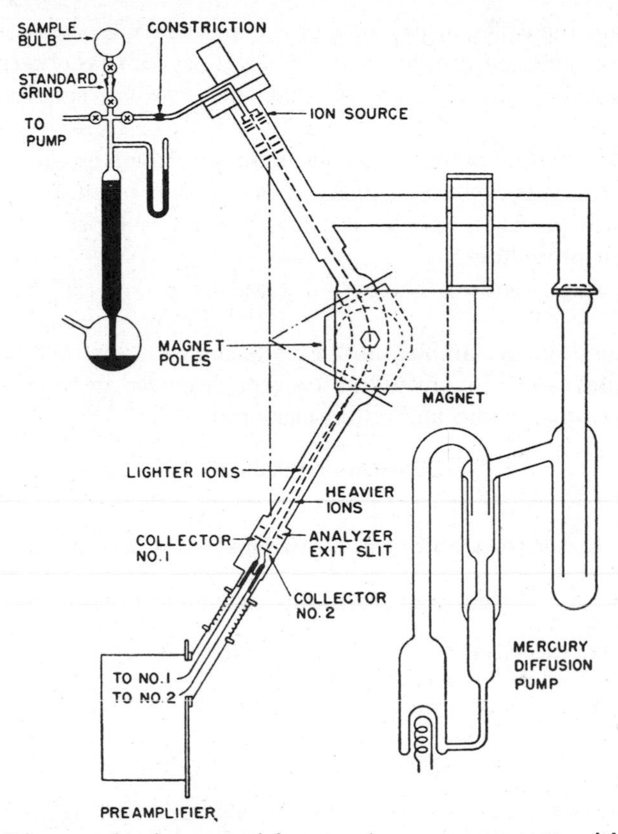

FIG. 37. Diagram showing essential parts of mass spectrometer with vacuum and gas handling system. (After Nier.[46])

must be high enough so that the initial spread in velocities introduced by the thermal conditions in the ion source is minimal.

The type of magnetic analyzer employed in step 3 is also optional. One type involves a homogeneous field which focuses the ion beams at an angle of 180° to the direction of the ion entry. Other types, of which that in Fig. 37 is one, utilize an inhomogeneous magnetic field with focusing at various angles, such as 127° and 90°. Once the machine is adjusted, the various mass components for a given charge can be focused at the collector by varying the accelerating voltage. If a voltage V_1 brings an ion beam of mass M_1 to the collector, then the voltage V_2 needed to collect an ion beam of mass M_2 will be given by M_1V_1/M_2.

If the machine is used with a single ion collector, as the various beams

[46] Nier, A. O., *in* "Symposium on the Use of Isotopes in Biology and Medicine" (H. T. Clarke, ed.), p. 89. Univ. Wisconsin Press, Madison, 1949.

are swept into the collector by varying the accelerating potential, fluctuations in ion production cannot be observed. They can be observed if two collection chambers are arranged so that each receives simultaneously a component beam.

The outstanding advantages of the mass spectrometer are:

1. The machine is applicable to the measurement of all isotopes.

2. Adequate precision can be attained for all isotopes without great alterations in procedures.

3. Small samples suffice. The actual quantity of material used is 1 to 2 $cm.^3$ of gas.

The preparation of samples and calculation of atom per cent excess depend on the isotope involved and are best discussed in those sections of this book devoted to the individual elements.

GENERAL REFERENCES

A. Texts

1. Bleuler, E., and Goldsmith, G. J., "Experimental Nucleonics." Rinehart, New York, 1952.
2. Boyd, G. A., "Autoradiography in Biology and Medicine." Academic Press, New York, 1955.
3. Calvin, M., Heidelberger, C., Reid, J. C., Tolbert, B. M., and Yankwich, P. E., "Isotopic Carbon." Wiley, New York, 1949.
4. Curran, S. C., and Craggs, J. D., "Counting Tubes." Academic Press, New York, 1949.
5. Curran, S. C., "Luminescence and the Scintillation Counter." Academic Press, New York, 1953.
6. Friedlander, G., and Kennedy, J. W., "Nuclear and Radiochemistry." Wiley New York, 1955.
7. Korff, S. A., "Electron and Nuclear Counters." Van Nostrand, New York, 1946.
8. Loeb, L. B., "Basic Processes of Gaseous Electronics." Univ. California Press, Berkeley, California, 1955.
9. Yagoda, H., "Radioactive Measurements with Nuclear Emulsions." Wiley, New York, 1949.

B. Articles

1. Curtiss, L., Measurements of radioactivity. *U. S. Dept. Commerce Circ.* 476 (1949).
2. Elmore, W. C., and Sands, M., "Electronics: Experimental Techniques," Div. V, Vol. 1, National Nuclear Energy Series. McGraw-Hill, New York, 1949.
3. Fitzgerald, P. J., Radioautography in cancer. *Cancer* **5,** 166 (1951).
4. Inghram, M. G., and Hayden, R. J., A handbook of mass spectroscopy. *Natl. Research Council Nuclear Sci. Ser. Rept.* No. **14,** (1954).
5. Norris, W. P. and Woodruff, L. A., Fundamentals of radioautography. *Ann. Revs. Nuclear Sci.* **5,** 297 (1955).
6. Odeblad, E., Quantitative autoradiography with P^{32}. *Acta Radiol.*, Suppl. 93, (1952).

7. Ramm, W. J., *in* "Radiation Dosimetry" (G. J. Hine and G. L. Brownell, eds.), p. 245. Academic Press, New York, 1956.
8. Rossi, B. B., and Staub, H. H., "Ionization Chambers and Counters." Div. V, Vol. 2, National Nuclear Energy Series. McGraw-Hill, New York, 1949.
9. Seliger, H. H., and Schwebel, A., Standardization of beta-emitting nuclides. *Nucleonics* **12,** No. 7, 54 (1954).
10. Staub, H. H., *in* "Experimental Nuclear Physics" (E. Segrè, ed.), Vol. 1, "Detection methods." Wiley, New York, 1953.
11. Swank, R. R., Linden, B. R., Harrison, F. B., Cowper, G., Kantz, A., Hofstadter, R., Cowan, C. L., Jr., Reeves, F., Koch, W., and Foote, R. S., Scintillation counting tubes. *Nucleonics* **12,** No. 3, 13 (1954).
12. Wilson, D. W., *in* "Preparation and Measurement of Isotopic Tracers" (D. W. Wilson, A. O. Nier, and S. P. Reiman, eds.) J. W. Edwards, Ann Arbor, Michigan, 1946.
13. Wilson, R. R., Corson, D. R., and Baker, C. P., Particle and quantum detectors. *Natl. Research Council Nuclear Sci. Ser. Rept.* No. 7, (1950).

Chapter IV

Radiation Hazards

1. GENERAL REMARKS

The passage of radiation through matter is characterized by events which involve the release of a great deal of chemical energy. Most frequently, atoms are ripped apart into pairs of ions or excited by the absorption of energy which, on a macroscopic scale, corresponds to molar heats of hundreds of kilocalories. Less frequently, an event may be catastrophic in nature, as when a neutron is captured by a nucleus with the consequent liberation of millions of electron volts which correspond, in a molar sense, to trillions of calories.

Disruption of atoms in biological tissue occurs at very much the same rate as in an equivalent quantity of air. Therefore, with a radiation dosage corresponding to 100 r. there will be about 10^{14} explosions on the atomic level per gram of tissue. The extent of the damage to the tissue depends on the nature of the radiation, particularly because the specific ionization (the amount of atomic disruption per unit of volume) varies with different kinds of radiation.

It is evident that radiation is not to be trifled with, regardless of the range of intensity involved. The mournful history of the early pioneers in x-ray and radium research underscores the need to emphasize the hazards inherent in working with radioactive substances. Radiation injury is insidious because, even though the energy from a given source may be great enough to injure tissue, it is not noticeable, and the damage it causes is often not apparent until long after exposure. On the other hand, reasonably simple precautions can exclude or render negligible the probability of harm from radioactive material.

As already discussed (see p. 22), a dose of 1 r., i.e., an "air" roentgen, means exposure of air to a field of radiation of such intensity that ions of either sign (+ or −) carrying a total of 1 electrostatic unit (esu.) of charge are formed per cubic centimeter of air. In 1 g. of air, $1/0.001293 = 773$ esu. of charge would be produced. This number of ion pairs requires the dissipation in air of 83 ergs of energy. On the assumption that the work required to pull an electron from an atom in tissue is the same as that in air, a tissue roentgen also requires dissipation of 83 ergs/g. of material. This condition is approximated for soft-body tissue but not for hard tissue such as bone.

It may now be seen how little sensible heat is required to cause radiation damage. A lethal dose for man or other mammals is approximately of the order 1000 r. At this level, each gram of tissue absorbs approximately 80,000 ergs. This corresponds to no more than 0.002 cal./g., or a rise in temperature at most of 0.002°C. This effect is associated with about 1.5×10^{15} ionizing events per gram of tissue. There are about 3×10^{22} atoms in this amount of tissue, so that the ratio of atoms disturbed to those unaffected is 5×10^{-8}. It is plain that only a few atoms in a hundred million need be altered to produce a lethal effect.

The actual damage which occurs at sublethal levels is manifest mainly in depression of activity or derangement of function in hematopoietic and epithelial tissue. However, a discussion of these effects is beyond the scope of this book. Similarly, it will be necessary to forego discussion of what little is known about mechanisms of radiation damage at the biochemical level.

The roentgen is a unit reserved for photon radiation. Another unit is required to describe dosage for particle radiation. It is now common practice to use the "roentgen-equivalent-physical" (rep.), which is defined as that amount of radiation which results in the dissipation of 83 ergs of energy per gram of tissue exposed. Other units which have been proposed to take into account variations in biological response to a given energy dissipation are discussed in Appendix 1.

It is important to include a few quantitative considerations about the role of specific ionization. A β particle with an initial energy of 2 Mev. traverses a total path length of about 1 cm. in tissue. The average specific ionization will be about 5×10^4 ion pairs per centimeter of tissue. An α particle of the same energy would travel only about 0.004 cm. in tissue while dissipating the same energy, and producing a similar quantity of ion-pairs, so that α particles are many hundreds of times as effective as β particles in regard to dosage and hence in regard to biological response. This factor is the basis for the common practice in medicine of limiting α-radiation dosage to only one-tenth or one-hundredth of the permitted β-ray dosage.

General experience indicates that continuous exposure of humans or animals up to about 0.03 r. (photon radiation) or 0.03 rep. (particle radiation) total-body dosage daily will result in few appreciable radiation effects. This is certainly true for exposure times of several years. It is not known, however, whether there may not be long-term effects at this level of dosage.

In most work with radioactive isotopes, general irradiation is not a hazard. The danger lies, rather, in exposure of limited portions of the anatomy. This restricted exposure may actually add a factor of safety to the daily permitted dose of 0.03 rep. because somewhat larger doses can be tolerated when they affect only a portion of the body. This is particularly true of the skin, which is not so radiosensitive, apparently, as other tissues.

All investigators subjected to radiation should familiarize themselves with procedures developed for radium and x-ray protection. In this connection a reference bibliography is given at the end of this chapter, but a brief discussion of protective techniques appears warranted here. In what follows, "strong" radioactive sources may be considered to be all γ-ray emitters in quantities equivalent to or greater than a few milligram equivalents of radium. The emphasis in this discussion will be on γ-ray emitters, because adequate shielding from β-ray emitters is usually obtained automatically by absorption of the radiation in the chemical glassware and clothing of personnel. If it is necessary to look into a beaker during a chemical procedure, goggles should be worn. The hands may be protected by rubber gloves and the use of long forceps. Personal cleanliness should be a fetish in all radiochemical laboratories because this habit helps to avoid contamination of tracer samples with extraneous or spurious radioactivities. It is essential that all manipulations involving strong radioactive samples be carried out in a special room far removed from the tracer research and detection laboratory. Special clothing consecrated to this purpose helps avoid the spread of contamination. All chemical operations on strong samples should be carried out in a hood with a strong draft so that activity introduced into the air by agitation and boiling of solutions is carried away and does not spread through the laboratory. This is especially important when gaseous radioactive isotopes are involved.

Protection against γ radiation is best afforded by working with remote control devices behind heavy lead shields at least 2 to 3 inches thick. The use of mirrors to observe the course of reactions behind the shield is strongly recommended. The temptation to reach in continually during a chemical procedure and remove the apparatus for a quick glance must be overcome. *Pipetting by mouth must be avoided.* Spillage of solutions should be minimized by use of rugged supports and stable structures for the chemical manipulations. Since there is always the danger of breakage, all glassware should be encased in close-fitting copper cans from which the solution can be recovered. These cans can be fitted with attachments for long-range handling. Numerous devices of this type are possible. A familiar simple design which can be employed is that encountered in Pyrex kitchenware wherein a tapered sliding union is used. Alternatively, operations may be conducted over a sturdy pan or tray.

In all manipulations, a major protective factor is minimizing of time of exposure. It may, for instance, be better to handle an intense γ-ray source with thin gloves rather than heavy gloves because the work can be done faster. However, distance is a more certain protective factor and whenever possible should be taken advantage of by means of remote control procedures.

2. HEALTH PHYSICS INSTRUMENTATION

The term "health physics" has been coined to describe procedures for protection of personnel involved in handling radioactive materials. Instrumentation for detection and measurement of radiation and for monitoring of radiation exposure is available from many commercial sources.

The most familiar radiation monitor is the pocket ionization chamber. This device is worn clipped to a pocket or lapel. The chamber is charged by means of an auxiliary apparatus (roentgen meter) which includes an electrometer and an illuminated scale for measuring applied charge. The scale is calibrated in units such as milliroentgens. At the end of the working day the chamber is inserted in the roentgen meter and the total charge received in the region where the chamber was worn is noted. Such a device does not indicate local exposure, i.e., to hands, eyes, or face, nor does it detect soft radiations which fail to penetrate the chamber wall.

Another monitoring procedure involves the use of photographic film which may or may not be specially treated to respond to various nuclear radiations. So-called "film badges" are made up in packets and kept covered with some type of opaque material. This covering material excludes light but also, unfortunately, soft radiation. Differentiation of hard β and γ dosage can be obtained by covering a portion of the film with a lead absorber. Film badges can be used to integrate dosages over rather long periods, although the most common practice is to wear them only a week or two. The use of film badges requires careful calibration and processing. If a number of laboratories are involved, it is advantageous to train one or two persons specifically for this job and for the responsibility of maintaining an adequate and reliable routine. Some commercial companies will process film badges at a nominal cost.

A number of portable G-M tube counters and direct-current amplifiers are available commercially for accurate monitoring and surveying of all kinds of radiation fields. Some models contain large ionization chambers with compact direct-current amplifiers operated from batteries, the total weight varying from 5 to 15 lb. Similar G-M tube counting sets are available. This type of instrument equipped with a thin-window G-M tube or ionization chamber is essential when monitoring soft radiation such as that encountered with C^{14} and S^{35}. Some survey meters have chambers with windows thin enough to detect α particles. Special devices which analyze airborne radioactivity either by collection of a gas sample or by aspiration of vapor through a retaining medium such as moist filter paper may also be obtained.

3. DOSAGE CALCULATIONS

One of the most practical aspects of health physics is the calculation or determination of dosage from ingestion of any given radioactive material. It is not a simple matter to measure or to calculate precisely either the actual dosage encountered for any particular set of conditions or the biological response to be expected. However, rough estimates of the dosage can be obtained from relatively simple considerations. In this section such estimates will be discussed.

An approach may be made to the general problem of dosage calculation by supposing that biological effects produced by radioactive material distributed uniformly in tissue are very much like the effects found with external radiation which dissipates an equivalent amount of energy in the same tissue. Actually, most radioactive materials have a distribution pattern which is not uniform either in space or time. This factor can be set aside in the initial approximations, however.

Most of the substances encountered in tracer work are β and γ emitters. Hard γ radiations ($E > 0.5$ Mev.) can usually be neglected because, for the most part, they are not absorbed as completely as β radiations. The correction factor arising from neglect of γ-ray effects is in most cases less than the uncertainty associated with calculation of the major dosage effect arising from β radiation. On the other hand, γ radiations of intermediate energy may require consideration, particularly when whole-body radiation must be minimized during localization of β radiation in therapy.

Consider the situation which prevails when β radiation assumes uniform distribution in tissue. First, it is necessary to know the average energy, E_{av}, of the β particles. If the energy spectrum of the β particles is known, the average energy per particle can be calculated. A good approximation is to assume the average energy to be one-third the maximum energy. Some such rule of thumb is necessary, because the data available usually refer to maximum energy rather than average energy.[1] Call C the concentration in rutherfords (rd.) per gram of tissue. In Appendix 1 it is shown that a roentgen-equivalent-physical of β radiation dissipates 5.24×10^{13} ev./g. of air. If this figure is assumed for tissue, the dosage rate, D, may be written in roentgen-equivalents-physical per minute:

$$D = \left[\frac{(1 \times 10^6 C)(10^6 E_{av})(60)}{(5.24 \times 10^{13})}\right] \tag{1}$$

and

$$D = 1.15 C E_{av}$$

[1] See Marinelli, L. D., Brinckerhoff, R. F., and Hine, G. J., *Revs. Mod. Phys.* **19**, 25 (1947).

where E_{av} is expressed in million electron volts. Converting to millicuries per gram gives $D = 42CE_{av}$. Thus, for 1 rd. of P^{32} per gram of tissue, where $E_{av} = 0.68$ Mev., D would be 0.78 rep./min. This estimate is more accurate than most based on such calculations because the average energy is known experimentally. When the $\frac{1}{3}E_{max}$ rule is used, an error as high as 20% may result simply from application of the rule. If the β spectrum is not simple, the contribution of each component must be estimated separately. If any conversion electrons are present, they will be homogeneous in energy and there will be no uncertainty regarding the energy to be used in calculations of dosage.

Some observers find it simple to remember Eq. 1 in the form which results when C is given in microcuries per gram or millicuries per kilogram and the time unit is the day:

$$D = 60CE_{av} \text{ rep./day} \tag{2}$$

Equations 1 and 2 assume that there is no appreciable decay during the period of exposure. The opposed case of complete decay during exposure may be calculated in the following way. The average life, T, of any radioactive atom is given by the relation (see Chapter I) $T = 1.44 \times t_{1/2}$, where $t_{1/2}$ is the half-life. The total dosage, D_t, will be the initial dose rate, D, multiplied by the total time each radioactive atom lives on the average ($1.44 \times t_{1/2}$), or

$$D_t = 1.44Dt_{1/2} = DT \tag{3}$$

The units of T must be chosen to correspond to those of D; i.e., if D is in rep./hr., T must be expressed in hours. D may be calculated by Eq. 1 or 2.

The intermediate case in which some decay occurs can be calculated by remembering that for any interval of time, t, the dosage, $D(t)$, will be proportional to the fraction of atoms disintegrated during time t. By the fundamental decay law

$$D(t) = D_t(1 - e^{-t/T}) \tag{4}$$

where D_t is the total dosage as given by Eq. 3.

No account has yet been taken of two factors of great importance which operate markedly in all tracer experiments on biological systems. First, the distribution of isotope is nonhomogeneous and changes with time. The most conservative estimate which can be made as regards this factor is to take the maximum concentration wherever found, assuming all tissues to be equally radiosensitive, and use this concentration for C. Alternatively, one may calculate the dosage rate for each individual tissue and take a grand average.

The second factor is elimination of radioactive material from any given tissue because of metabolic turnover. This may be considered as an added decay factor with its own characteristic half-life because such turnover gives rise to an excretion law which is exponential. This biological half-life must be compounded with the radioactive decay half-life to arrive at the proper half-life or average life to use in the formulas given above. The half-lives are added as reciprocals just as in the case of addition of radioactive half-lives. This follows from the consideration that, if λ_r is the constant for loss by radioactive decay and λ_B is the constant for loss by biological elimination, then the total constant is $\lambda = \lambda_r + \lambda_B$. Substitution in terms of average life, T, gives

$$\frac{1}{T} = \frac{1}{T_r} + \frac{1}{T_B} = \frac{T_B + T_R}{T_B T_r}$$

and

$$T = \frac{T_B T_r}{T_B + T_r}$$

Thus, if the average time for elimination of phosphorus from liver is 2 days and its average decay life is 20 days, then the over-all average life to be used in dosage calculation is

$$T = \frac{40}{2 + 20} = 1.8 \text{ days}$$

One may apply all the above formulas to α-particle emitters, assuming no enhancement of biological damage due to higher specific ionization, and remembering that α particles are not emitted in a continuous spectrum of energies but exhibit a discrete energy.

In applying these considerations to particular isotopes it is necessary to use wherever possible data available on element distribution, retention, and elimination. Biologically tolerable levels for a few nuclides are shown in Table 4.

4. SHIELDING

The most effective protection consists simply in making the distance between the observer and the source of radiation as great as possible, within practical limits. The law governing attenuation of radiation intensity with distance is the familiar inverse square law. A source emitting N radiations per second will exhibit a radiation flux at any point, P, given by $N_p = N/4\pi R^2$, where R is the distance from the source to point P, and N_p is expressed in radiation quantity per unit area. The ratio of intensity at any two points is equal to the inverse ratio of the squares of the distances. If N_1 is the intensity at P_1 which is R_1 units of length distant, and

TABLE 4

MAXIMUM PERMISSIBLE LEVELS FOR OCCUPATIONAL EXPOSURE[2]

Nuclide	Effective mean life (days)	In body (μc.)	In liquid (μc./ml.)	In air (μc./cm.^{3})
Ra^{226}	10^4	1.0	4×10^{-8}	8×10^{-12}
Sr^{89}	—	2.0	—	—
Sr^{90} ($+Y^{90}$)	5×10^3	1.0	8×10^{-7}	2×10^{-10}
Po^{210}	—	0.005	0.4	5×10^{-5}
C^{14} (as CO_2 in air)	—	—	—	1×10^{-4}
P^{32}	20	10	2×10^{-4}	—
Co^{60}	20	1	1×10^{-5}	—
H^3	10	1×10^4	0.4	5×10^{-5}
I^{131}	12	0.3 (0.18 in thyroid)	3×10^{-5}	3×10^{-8}

N_2 and R_2 are the corresponding quantities for point P_2, then

$$\frac{N_1}{N_2} = \frac{(R_2)^2}{(R_1)^2}$$

These considerations apply only to point sources emitting radiation uniformly in all directions. Morgan[3] has given the relation for γ-ray dosage rate, D, in milliroentgens (mr.) per hour at 1 foot from an unshielded source in the energy range 0.2 to 1.5 Mev. as approximately

$$D = 6CE \tag{5}$$

where C is the number of millicuries of γ-radiating isotope and E is the energy in million electron volts.

In attenuating γ radiation one may use absorbers of high atomic number. Calculations can be made as to the attenuation to be expected under ideal geometrical conditions (spherical shield) by remembering that γ-radiation energy is absorbed exponentially (Chapter II). The reader will find numerous tables of absorption coefficients for various energies of γ radiation and for various materials in the literature.[4] In practice it is best to use tabulated data only as a guide in designing shielding, leaving sufficient margin for extra thickness of material if survey meters indicate more material is needed. A safe tolerance rate to be used as a basis for γ-radiation shielding is 1 mr./hr.

[2] *Nucleonics* **8,** No. 2, 73 (1951).

[3] Morgan, G. W., U. S. Atomic Energy Comm. Isotopes Div. Circ. B-3, 1948.

[4] Parker, H. M., *Advances in Biol. and Med. Phys.* **1,** 276–277 (1948); see also Davisson, C. M., and Evans, R. D., *Revs. Mod. Phys.* **24,** 79 (1952).

As noted above, β radiation, at intensity levels usually met in tracer work, does not pose a particularly difficult shielding problem. Only the hardest β radiations can penetrate in significant amounts the usual materials encountered in chemical manipulations.

For further details, the reader's attention is invited to Appendix 2 in which a typical set of working rules for a radiochemical laboratory and a short bibliography of reference articles on radiation dosage and protection are presented.

GENERAL REFERENCES

A. Brochures (available from National Bureau of Standards, U. S. Department of Commerce)

1. Maximum Permissible Amounts of Radioisotopes in the Human Body and Maximum Permissible Concentrations in Air and Water, *Handbook 52.*
2. Recommendations for the Disposal of C-14 Wastes, *Handbook 53.*
3. Protection Against Radiations from Radium, Co-60, ands Cs 137, *Handbook 54.*
4. Safe Handling of Cadavers Containing Radioactive Isotopes, *Handbook 56.*
5. Photographic Dosimetry of X- and Gamma Rays, *Handbook 57.*
6. Radioactive Waste Disposal in the Ocean, *Handbook 58.*
7. Permissible Dose from External Sources of Radiation, *Handbook 59.*

B. Articles

1. Evans, R. D., Tissue dosage in radioisotope therapy. *Am. J. Roentgenol. Radium Therapy* **58,** 754 (1947).
2. Marinelli, L. D., Dosage determinations with radioactive isotopes. *Am. J. Roentgenol. Radium Therapy* **47,** 210 (1942).
3. Marinelli, L. D., Quimby, E. H., and Hine, G. J., Practical considerations in therapy and protection. *Am. J. Roentgenol. Radium Therapy* **59,** 260 (1948).
4. Morgan, G. W., Gamma and beta radiation shielding. U. S. Atomic Energy Comm. Isotopes Div. Circ. B-8, 1948.
5. Morgan, K. Z., Tolerance concentrations of radioactive substances. *J. Phys. & Colloid. Chem.* **51,** 984 (1947).

C. Texts

1. Spiers, F. W., *in* "Radiation Dosimetry" (G. J. Hine and G. L. Brownell, eds.), Chapter 1. Academic Press, New York, 1956.

Chapter V

PRACTICAL INTERLUDE

1. GENERAL REMARKS

All beginners who must bridge the gap between the booklore presented in previous chapters and the actual practice of tracer methodology invariably ask the same questions about the instrumentation required for a tracer laboratory, preparation of sample materials, manipulation of biological test systems with special regard to tracer requirements, and so on. It must be emphasized that tracer methodology is merely an adjunct to researches in biology and not an object in itself, even though it is a uniquely powerful tool.

In general, the choice of instrumentation and technique is determined by the kind of data required. At one extreme, use of only one isotope, easily assayed and readily available, may be needed. In such a case, the investigator needs only a single instrument, such as a simple electroscope; relatively crude manipulations in sample preparation and measurement will answer his purposes. At the other extreme, a complete radiochemical laboratory with all its complex physicochemical equipment may be required. In this chapter, general comments on techniques in isotopic assay will be presented. Amplification of these comments appropriate to the cases of particular nuclides will be given in later chapters.

2. SOME PROCEDURES ASSOCIATED WITH SAMPLE PREPARATION FOR RADIOACTIVITY ASSAY

Much of the art in tracer work lies in processing of material to bring it into a form suitable for assay. The procedures used are usually dictated by the choice of counting arrangement. In the case of α emitters this is determined by the short range of the radiation, which requires preparation of thin, even films, produced by evaporation or electrodeposition, and counting inside the sensitive volume of the detector. Nuclides that emit low-energy β rays, x-rays, and other soft radiations are often best assayed as gases. Compounds labeled with C^{14} fall into this category to some extent. The assay of H^3-labeled compounds invariably requires gas counting. This procedure makes good vacuum techniques and apparatus imperative.

Most nuclides which emit β rays and which are useful in tracer research can be prepared as thin, solid samples which can be assayed by means of thin-window G-M tubes or proportional counters. However, it is often difficult and tedious to obtain accurately reproducible assays because of such factors as self-absorption and uncontrolled geometrical placement. If such nuclides also emit γ rays, advantage should be taken of the instruments now available for γ counting, such as scintillation detectors. Absorption corrections for γ radiations are usually much less critical than they are in β assay. Scintillation counters are much more efficient than G-M tubes in detection of γ rays with energies greater than 50 kv. This fact can overbalance the circumstance that many more β rays are emitted than easily detectable γ rays (as in the case of P^{32}). Moreover, samples can be assayed with scintillation detectors in both liquid and solid states. This is an enormous advantage in biological work. Of course, for very low-energy β radiations where accompanying γ radiation is only "bremsstrahlung" and consequently of low intensity, the efficiency of the scintillator may be too small.

It has been remarked (p. 96) that every effort should be made to minimize effects arising from back-scattering and variation in chemical composition of samples. This means that some standard arrangement for source placement and a procedure for ensuring chemical homogeneity in sample comparisons are required. For instance, in the counting of C^{14}-labeled material, solid samples are prepared by combustion to CO_2, precipitation under standard conditions as $BaCO_3$, evaporation to thin homogeneous films, and placement in standard sample holders at standard positions relative to the counting tube. Standard sample holders can be constructed in the form of shelves contained in a rigid structure which supports the counting tube. The assembly is usually mounted inside a lead shield about 2 inches thick to lower the background produced by cosmic rays and other extraneous radiations.

It is sometimes helpful to have a set of standard absorbers which can be used to reduce activity, in the event a mistake has been made in a dilution factor. Absorbers for β radiation are usually of aluminum and can be cut to fit the sample shelves. A tab is left on one edge on which the thickness in milligrams per square centimeter can be marked. For γ-ray absorption, a similar set of absorbers made of lead can be employed. Another way of reducing counting rates is to use distance by placing samples on lower shelves. Factors between shelves can be determined with standard samples.

A good habit to cultivate is the constant monitoring of assay apparatus with standard samples. For this purpose, a set of standards with radiations very similar to or identical with those of the test samples should be used. Background rates, of course, must be checked every time an assay is made.

A variety of techniques may be employed in mounting solid samples.

If thin samples to be counted with the bell-jar G-M tube are required, some procedure must be devised to obtain films spread uniformly over an area which should not exceed three-fourths the window area. This requirement is especially important if the sample is placed close to the window because the sensitive volume of the counter changes markedly near the tube walls. The usual practice is to form a homogeneous suspension (slurry) of the radioactive material in a liquid in which it is insoluble or slightly soluble. The slurry is pipetted slowly onto a planchet; it must be stirred constantly during this time. Alternatively, the planchet can be placed on a rotating platform. The slurry is then transferred with a pipet or eyedropper, beginning by applying from the center outwards. An infrared lamp is usually used for evaporation. A stream of hot air from a hair dryer is also a convenient source of heat.

If the material to be assayed is entirely in solution or very small in amount, it can be pipetted onto thin cigarette paper placed in the planchet. The material will spread uniformly by capillary action. The paper is dried and its weight taken into account in calculating the absorption correction. Cigarette paper is uniform enough to assure reproducible results.

If enough material is available, filtration can be used. Most filtration procedures employ a commercially available Büchner funnel arrangement which can be dismantled readily for cleaning. The top section consists of a barrel which fits snugly over the filter stem which makes up the lower section. The lower section supports a filter paper cut to fit over the filter hole area. The upper section is pressed over the paper and lower section, making a tight closure. The whole assembly is placed in the flask, and the slurry is filtered. The filter is taken apart after filtration of the sample which has been dried in a preliminary way by passage of air through it. The sample can then be transferred to a mount and dried further before counting.

Other feasible procedures depend on centrifugation of slurried material. Usually, a detachable cap is placed at the base of a centrifuge tube. After centrifugation, the supernatant liquid is decanted off, the cap is removed, and the sample is dried in a stream of warm air.

Large amounts of solid are conveniently handled by making samples in briquet form. Dry powders are pressed into shallow, square-bottomed dishes by means of a spatula or piston and cylinder.

3. GAS COUNTING

Many of the difficulties inherent in the assay of solid or liquid samples can be avoided by going to the trouble of converting isotopic material to the gaseous state and assaying it in that form. Laboratories using hydrogen isotopes or the rare stable isotopes of nitrogen and oxygen will find it worth while to set up the best apparatus available for gas assay. For instance, once a vacuum line and its accompanying assay instrumentation are put together, it is no more difficult to burn C^{14}-labeled organic material to CO_2 and introduce this into a counter or ionization chamber than it is to make it into $BaCO_3$ and plate it for assay with a G-M tube.

Gas counting of a nuclide like C^{14} is twenty to fifty times as sensitive as solid counting at infinite thickness,[1a] when the bell-jar G-M tube with window thickness of 2 mg./cm.^{2} is used. In addition, better reproducibility can be obtained because the apparatus required for gas counting is usually better adapted to production of uniform samples than procedures in which organic material is plated directly or even converted to $BaCO_3$.

Proportional flow counters are almost as sensitive as gas-counting procedures. However, other advantages are inherent in gas counting. First, assay in the gas phase can reduce the amount of material needed from ten- to a hundredfold. Second, some combination of nuclides for double labeling experiments, such as H^2–H^3, C^{14}–C^{13}, etc., require gas assay for at least one of the nuclides so that it is more convenient to use the one procedure of gas-phase assay for both.

On the other hand, much more skill is needed to manipulate samples in gas counting, and the time consumed in preparation of samples can be very much greater than in procedures involving assay of solid samples. Few biologists possess the physical background necessary to acquire rapidly the skills needed in setting up vacuum lines and operating the equipment for gas-phase assay. Workers who must use hydrogen, nitrogen, and oxygen isotopes in tracer research, however, must learn the techniques necessary. It should be remarked that there are many machines in daily operation in biology, such as the Van Slyke apparatus, which make technical demands not much less in magnitude than those which arise in gas-phase assay.

4. SOME REMARKS ON VACUUM TECHNIQUE

An excellent monograph is available[1b] describing in detail the construction and manipulation of high vacuum equipment. Here only the general nature of the apparatus required will be discussed. There are always individual preferences in technical matters, and the recommendations to be given, it should be noted, reflect those of the writer.

First, the bench on which the vacuum line is set up should be lower than the usual laboratory table by a foot or so and equipped with scaffolding which can be anchored to a wall for stability. All the usual laboratory services (gas, water, air, etc.) should be supplied individually and not brought in by makeshift lines from other benches. A tank of oxygen can be secured to one end for on-the-spot glass blowing.

[1a] Glascock, R. F., *Biochem. J.* **52**, 699 (1952).

[1b] Glascock, R. F., "Isotopic Gas Analysis for Biochemists," Academic Press, New York, 1954.

Second, the vacuum system should be constructed out of standard tapers and joints for maximum convenience in assembling and disassembling the line. Anderson *et al.*[2] have described one such system in which the components of the combustion train can be dismantled for cleaning, repair, or replacement, and other units, such as monometers and traps, which require little servicing are permanently sealed. Glascock (see p. 12, reference 1b) has described another all-purpose vacuum system for the determination of H^3, C^{13}, and C^{14} in a 10-mg. sample.

Third, it is probably best to have available a good oil-diffusion pump backed by a fast mechanical pump. For most procedures the mechanical pump together with a Toepler pump should be adequate.

In all vacuum procedures, there is a considerable amount of art in obtaining accurate and reproducible results. Anyone wishing to set up vacuum apparatus such as is described in the literature for processing of isotopic samples must be prepared to spend time acquiring experience.

5. COMBUSTION OF LABELED MATERIALS

The advice about the necessity of acquiring experience with vacuum procedures in the previous section applies with even greater emphasis to combustion techniques because there are added requirements in working with isotopic material. For instance, as will be discussed later (see Chapter VI), fractionation of isotopes can occur if combustion is only partial. Hence, combustion must be quantitative. The various procedures available from the vast literature on procedures for combustion of organic compounds can be classified into two categories—wet and dry.

Two procedures are in general use. The first, which is a modification of the usual Van Slyke procedure, was designed originally for the determination of C^{14}. If there is a good vacuum line already available for processing gas samples, it is probable that there is no advantage in setting up the Van Slyke apparatus. However, biological laboratories which have a Van Slyke machine will find it convenient to adapt to a procedure based on it. Complete instructions for wet combustion with the Van Slyke apparatus can be found in papers by Van Slyke and his collaborators.[3-5] Lindenbaum *et al.*[6] have simplified the wet combustion method, using a gravimetric determination of carbon and avoiding manometric manipulations. Earlier, Claycomb *et al.*[7] described a similar wet combustion method. Wet com-

[2] Anderson, R. C., Delabarre, Y., and Bothner-By, A. A., *Anal. Chem.* **24,** 1298 (1952).

[3] Van Slyke, D. D., and Folch, J., *J. Biol. Chem.* **136,** 509 (1940).

[4] Van Slyke, D. D., Plazin, J., and Weisiger, J. B., *J. Biol. Chem.* **191,** 299 (1951).

[5] Van Slyke, D. D., Steele, R., and Plazin, J., *J. Biol. Chem.* **192,** 769 (1951).

[6] Lindenbaum, A., Schubert, J., and Armstrong, W. A., *Anal. Chem.* **20,** 1120 (1948).

[7] Claycomb, C. K., Hutchens, T. T., and Van Bruggen, J. T., *Nucleonics* **7,** No. 3, 38 (1950).

bustion techniques using permanent vacuum line installations are described by Glascock (see p. 88 *et seq.*, reference 1b).

The classical methods devised by Pregl are the basis for most of the dry combustion procedures. Naughton and Frodyma[8a] developed a method using a tube filling invented by Niederl and Niederl[8b] at temperatures of 650° to 700°C. This method was modified by other workers. One such modification has been described in the previous section. Payne *et al.*[9] describe a method for the combustion of relatively large amounts of material. The major innovation is the use of Inconel metal instead of quartz for the combustion tube. Glascock (see p. 116 *et seq.*, reference 1b) gives a detailed account of the use of wire-wound quartz tubes together with other procedures in connection with his permanent vacuum line installation for the assay of H^3, C^{13}, and C^{14}. A relatively simple combustion procedure is described by Keston *et al.*[10]

A major hazard in all combustions, especially those in which H^3 or C^{14} is assayed, is the occurrence of "memory" effects. By these are meant contamination of the line with radioactivity so that erratic contributions occur from one run to another, one sample contaminating the next. It is necessary to intersperse combustions of labeled samples with runs on unlabeled samples so that contaminants are swept out. Success in decontamination is signaled by the attainment of a satisfactory blank. No new assay should be attempted until proper blanks are obtained. With continued use, glassware tends to resist decontamination and eventually must be discarded.

An added complication arises when nitrogen is present in organic material. This is the production of nitrogen oxides which must be removed from products of combustion before collection of samples for isotopic assay. This problem has not been solved so that a routine prescription can be given. The method most often employed is reduction of the oxides of nitrogen with hot copper.[2] This method is time-consuming because it is necessary to freeze out all volatile products and to pump off all oxygen before passing the combustion products over the hot copper. A more serious difficulty is that the reduction is very temperature-sensitive. A variation of as little as 5°C. from the optimum temperature of about 455°C. can cause unsatisfactory results as regards carbon assay. This hazard may be eliminated by procedures based on the observation of Belcher and

[8a] Naughton, J. J., and Frodyma, M. M., *Anal. Chem.* **22,** 711 (1950).

[8b] Niederl, J. B., and Niederl, V., "Organic Quantitative Microanalysis." Wiley, New York, 1942.

[9] Payne, F. R., Campbell, I. G., and White, D. F., *Biochem. J.* **50,** 500 (1952).

[10] Keston, A. S., Rittenberg, D., and Schoenheimer, R., *J. Biol. Chem.* **122,** 227 (1937).

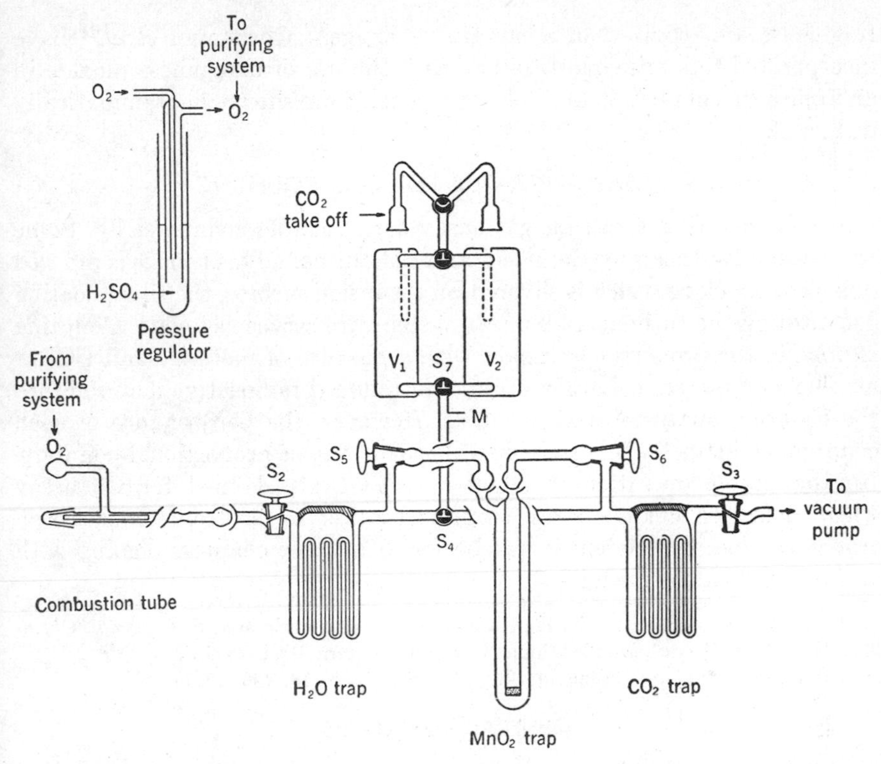

FIG. 38. Combustion line for preparation of gaseous isotopic samples. The sample to be burned is introduced in a platinum thimble and run into the high-temperature combustion furnace using an external magnet driven by an automatic advancing mechanism. S_2 and S_3 are stopcocks used to regulate flow of gas. S_4 and S_7 are stopcocks which communicate with the calibrated gas storage vessels V_1 and V_2. CO_2 samples can be removed from V_1 and V_2 by means of flasks attached to the "CO_2 take-off." If oxides of nitrogen must be removed, stopcocks S_5 and S_6 are manipulated to introduce the MnO_2 trap. Details of operation are given by Christman *et al.*[14]

Ingram, who have noted that manganese dioxide can be used as an absorbent for oxides of nitrogen.[11]

Another development, due to Kirsten,[12, 13] has been the introduction of combustion techniques in which the organic compound first is cracked thermally at high temperatures (930° to 970°C.), and then the organic

[11] Belcher, R., and Ingram, G., *Anal. Chim. Acta* **4,** 124, 401 (1950).
[12] Kirsten, W., *Anal. Chem.* **25,** 74 (1953).
[13] Kirsten, W., *Mikrochemie* **35,** 217 (1950).

fragments are oxidized in a stream of oxygen. Christman *et al.*[14] have incorporated this principle together with the use of manganese dioxide in an improved combustion line. The components are shown diagrammatically in Fig. 38.

6. ASSAY APPARATUS FOR GAS COUNTING

Special counting tubes for gas assay are available commercially. Some are based on a design by Bernstein and Ballantine.[15] This tube is composed of a glass envelope which is silvered on the inside surface, the silver coating constituting the cathode. A 2-mil tungsten wire serves as anode. Counting is done in the proportional region with a mixture of methane and CO_2 at atmospheric pressure. Earlier workers introduced radioactive material into the counting mixtures of G-M tubes. However, the performance of such counters is distinctly less reproducible than that of proportional counters. Routine stable operation and high sensitivity is claimed for C^{14} assay when ionization chambers are employed either as electrometers or electroscopes. Specific references will be found in later chapters dealing with assay of individual nuclides.

---

[14] Christman, D. R., Day, N. E., Hansell, P. R., and Anderson, R. C., *Anal. Chem.* **27,** 1935 (1955); Brookhaven National Laboratory Rept BNL-2309.

[15] Bernstein, W., and Ballentine, R., *Rev. Sci. Instr.* **18,** 496 (1947).

GENERAL REFERENCES

1. Comar, C. L., "Radioisotopes in Biology and Agriculture." McGraw-Hill, New York, 1955.
2. Francis, G. E., Mulligan, W., and Wormall, A., "Isotopic Tracers." Athlone Press, Univ. London, 1954.
3. Friedlander, G., and Kennedy, J. W., "Nuclear and Radiochemistry." Wiley, New York, 1955.
4. Glascock, R. F., "Isotopic Gas Analysis for Biochemists." Academic Press, New York, 1954.
   Niederl, J. B., and Niederl, V., "Organic Quantitative Microanalysis." Wiley, New York, 1942.

CHAPTER VI

SURVEY OF TRACER METHODOLOGY: BIOCHEMICAL ASPECTS, PART I

1. THE SIGNIFICANCE OF TRACER METHODS FOR BIOLOGY

The central feature in tracer methodology is the preparation of labeled samples of elements involved in biological processes. With such samples it becomes possible to distinguish and trace any molecule or atomic grouping the behavior of which is of interest in connection with biological function. It is apparent that the method makes contact with biology principally at the biochemical and physiological level.

At the biochemical level the biologist is interested in questions associated with the term "intermediary metabolism." The biochemist must solve such problems as the fate of a particular molecule in a given metabolic process, the manner in which component atomic groupings in such a molecule are mobilized as energy sources in the synthesis of organic material, and the importance of the molecule as a contributor to the structural elements of nuclear and cytoplasmic constituents of the living cell.

Such knowledge can be used by the physiologist who investigates the manner in which metabolic processes are integrated in the cell economy so that differentiation and growth proceed within their proper limits. Knowledge of metabolic patterns enables the radiobiologist to establish a rationale for investigations into the fundamental nature of radiation effects in cells. Clues for improvement of procedures in therapy are provided as a practical adjunct of such studies. It is plain that advances in knowledge of intermediary metabolism are fundamental to progress in all biology.

Biologists could not approach many problems in intermediary metabolism directly until the tracer method was developed because they could not follow the biological wanderings of a given atomic grouping once it had disappeared into the organism. For instance, the carbon skeleton of an ingested sugar molecule cannot be distinguished from cellular carbon, nor can any further distinctions be made as to which of the sugar carbons are incorporated into protein or glycogen and which into waste material. In principle such difficulties disappear when it is possible to label at will whichever carbon in the material fed is of interest. These considerations apply not only to carbon but also to all elements of biological interest.

Two methods for the investigation of intermediary metabolism exist using tracer atoms. The first, originated by Hevesy shortly after the discovery of isotopes by Soddy at the beginning of this century, depends on the use of the radioactive isotopes. The second, elaborated mainly by Schoenheimer and Rittenberg after the development of bulk separation procedures by Urey and his collaborators early in the 1930's, utilizes the rare stable isotopes of complex elements (elements with more than one isotope). In either case the methods are based on two facts which will be discussed at more length in a later section (see pp. 136–140). The first is that living organisms can distinguish one isotope from another with great difficulty, if at all. The second fact is that complex elements always exhibit a constant isotopic composition in the natural state. Thus it is necessary only to vary the natural isotopic composition of an element, either by changing the relative abundance of a rare stable isotope or by adding a radioactive isotope previously absent, to produce a labeled sample of an element. Such a tracer specimen will undergo the same chemical reactions as normal samples of the element with which it may be mixed, but it always can be distinguished from them because of its differences in physical properties such as mass or radioactivity.

The enormous possibilities of tracer methods cannot be summarized adequately in any text of reasonable size. Even a casual survey of the literature reveals a bibliography so extensive as to prohibit thorough coverage.[1]

It is proposed in the following discussion to survey the potentialities of tracer methods, illustrating with examples from published research as fully as space permits the various applications in biochemistry and physiology which are of major interest. An attempt will be made to indicate the pitfalls in the method, as well as the advantages. A few special topics which illustrate particularly well the unique contributions of tracer methods to biochemistry will be presented.

2. BIOCHEMICAL APPLICATIONS

A. General Tracer Requirements

The general procedure involves preparation of a labeled compound, introduction of the compound into the biological system, and later separation and determination of the labeled element in various biochemical fractions. Before results obtained can be interpreted correctly it is necessary to ensure that the label is adequate with regard to the following crite-

[1] See, for instance, "Isotopes—An Eight-Year Summary of Distribution and Utilization with Bibliography." U. S. Atomic Energy Comm., Washington, D. C., March, 1955.

ria:

1. The initial concentration of tracer must be sufficient to withstand dilution during metabolism. Suppose a sample of carbon dioxide (CO_2) labeled with C^{14} is employed in determining the fate of CO_2 during oxidation of glucose by an organism such as yeast. A sample of labeled CO_2 containing 10 mg. as carbon is to be administered to actively fermenting yeast cultures. During the course of the experiment 5000 mg. of carbon in the form of CO_2 originating from the fermentation of glucose is evolved. The original sample of $C^{14}O_2$ is diluted 500 times in the course of the experiment. The actual concentration of labeled CO_2 available at any time for metabolic participation is, therefore, reduced by a factor of anywhere up to 500.

If it is intended to measure the amount of CO_2 introduced into metabolic processes under these conditions, the original sample must be so concentrated that a dilution of 500 can occur and still allow the sample C^{14} to be detected with reasonable precision. Suppose that a total of 10 mg. as carbon can be assayed with the instrument in use. Hence if the sample contains initially 5000 ct./min./mg. of C, the CO_2 recovered may have an activity as little as 10 ct./min./mg. of C. Since 10 mg. can be used for measurement, a total effect of 100 ct./min. is expected. The background effect may be 50 ct./min. If a precision of 1% to 2% is needed to ascertain a smaller uptake of CO_2 than that assumed here (100%), a correspondingly stronger sample is needed. It appears advisable, therefore, to work at a level of specific activity a factor of 10 higher than that of the initial sample quoted. Dilution calculations, the nature of which depend on the particulars of the system studied, are a recurrent feature of all experimental tracer procedures. The simple considerations presented here will be elaborated and extended by reference to specific researches discussed later in this chapter.

2. Throughout metabolism the label must adhere to the particular molecule or portion of molecule with which it is originally incorporated. Stable tracer positions depend on the process studied and the isotope used and will be discussed in more detail elsewhere. In general, processes involving removal of tracer by direct "exchange" reactions must be minimized or obviated by control experiments. Thus tracer hydrogen attached to a carbon chain in groupings[2] such as —OH*, —NH_2*, and —C(=O)—C(H*)— is inadequate, in general, as a tracer for the molecule or atomic grouping to which it is so attached because the tracer hydrogen exchanges rapidly with hydrogen ions in the usual aqueous media independently of metabolic mechanisms. Tracers involving use of isotopes which exhibit isomerism and internal

[2] The asterisk appearing as superscript is used to denote labeled element.

conversion may not be applicable for reasons discussed previously (Chapter I).

3. Abnormalities in metabolism must not be brought about through the action of the isotopic sample on the organism. In the use of radioactive samples, the investigator must proceed with caution, maintaining adequate controls on radiation effects both to himself and to the system being studied (see Chapter IV). It is advisable to keep the concentration of labeled molecules low relative to that of the same molecules which are already present in the unlabeled state. In the case of the stable isotopes complications due to radiation do not enter. In their stead, isotope discrimination may occur if the tracer employed is highly enriched in the rare isotope. This is important in a practical way only for the hydrogen isotopes. No pathological effects can be expected from the use of the nonradioactive isotopes except, of course, for the single exception of deuterium which, in high concentrations, inhibits or poisons respiration and fermentation. The general practice in studies on intermediary metabolism is to avoid deuterium concentrations higher than about 1 part H^2 to 5 parts of H^1.

4. The half-life of the isotope used must be sufficiently long so that decay does not remove tracer faster than it can be extracted, characterized, and assayed. For instance, C^{11} loses intensity by a factor of nearly 1000 in 3½ hr. (10 half-lives). On the other hand, C^{14} suffers no appreciable diminution in activity throughout the lifetime of the observer. This limitation is operative in the cases of oxygen and nitrogen for which no reasonably long-lived radioactive isotopes exist. Fortunately, there are quite adequate tracer isotopes of the rare stable variety for these important elements, namely, N^{15} and O^{18}.

B. Basic Limitations

It has been remarked that the tracer method is based on the validity of the assumptions (1) that elements exhibit constancy of isotopic composition and (2) that chemical identity of isotopes is maintained. In connection with the first assumption it may be noted that minor fluctuations have been reported for some elements such as carbon,[3] oxygen,[4, 5] potassium,[6] and, in particular, hydrogen.[7] These fluctuations determine the ultimate precision of work with samples enriched in rare stable isotopes. Waters of biological origin fluctuate in density as much as 3 parts per million owing to variations in isotopic ratios of hydrogen and oxygen amounting to

[3] Nier, A. O., and Gulbransen, E. A., *J. Am. Chem. Soc.* **61,** 697 (1939).

[4] Gilfillan, E. S., Jr., *J. Am. Chem. Soc.* **56,** 406 (1934).

[5] Birge, R. T., *Repts. Progr. in Phys.* **8,** 90 (1941).

[6] Brewer, A. K., *Ind. Eng. Chem.* **30,** 893 (1938).

[7] For a general discussion see Kamen, M. D., *Bull. Am. Museum Nat. Hist.* **87,** 105 (1946).

several per cent.[8] In the case of carbon, the maximum uncertainty appears to be a variation of ± 0.005 in the normal C^{13}/C^{12} ratio, usually taken as $\frac{1}{100}$. There is no evidence that any marked deviations occur in different specimens from the same organism.[9, 10] However, few refined assays of isotopic fluctuations in hydrogen and oxygen have been reported. It may be concluded that such fluctuations in assay of stable tracers are outside the range of sensitivity possessed by assay instrumentation adequate for most tracer research. Natural radioactivity is negligible in elements lighter than lead, with the exception of potassium, rubidium, and some rare earths.

The validity of the assumption that chemical identity of isotopes is maintained in biochemical systems is questionable only in the case of hydrogen, for which extreme isotope mass ratios exist. Some estimates of the magnitude of isotopic differentiation in purely physicochemical systems can be made from formulas available in the literature.[11] Furthermore there have been tabulated[12] the equilibrium constants of various exchange reactions for isotopic compounds of elements in the first row of the periodic table as well as halogens. Both the ultimate equilibria as well as reaction rates may be affected by variation in mass.

As an example of what effects may be involved there will be considered a study of the decomposition of oxalic acid by concentrated sulfuric acid at 100°C.[13] In this purely physicochemical system, the reaction is practically unidirectional. The products of the decomposition are CO and CO_2, which can be isolated and analyzed for variations in C^{13}/C^{12} ratio. The isotopic reactions involved may be written as follows:

$$(C^{12}OOH)_2 \xrightarrow{k_1} C^{12}O_2 + C^{12}O + H_2O \tag{1}$$

$$\begin{matrix} C^{13}OOH \\ | \\ C^{12}OOH \end{matrix} \xrightarrow{k_2} C^{13}O_2 + C^{12}O + H_2O \tag{2a}$$

$$\begin{matrix} C^{12}OOH \\ | \\ C^{13}OOH \end{matrix} \xrightarrow{k_3} C^{12}O_2 + C^{13}O + H_2O \tag{2b}$$

At the concentrations employed one may assume these reactions to be

[8] Emeléus, H. J., James, F. W., King, A., Pearson, T. G., Purcell, R. H., Briscoe, H. V. A., *J. Chem. Soc.* **136B**, 1207 (1934).

[9] Swendseid, M. E., Barnes, R. H., Hemingway, A., and Nier, A. O., *J. Biol. Chem.* **142**, 47 (1942).

[10] Krampitz, L. O., Wood, H. G., and Werkman, C. H., *J. Biol. Chem.* **147**, 243 (1943).

[11] Bigeleisen, J., and Mayer, M. G., *J. Chem. Phys.* **15**, 261 (1947).

[12] Urey, H. C., *J. Chem. Soc.* 562 (1947).

[13] Lindsay, J. G., McElcheran, D. E., Thode, H. G., *J. Chem. Phys.* **17**, 589 (1949).

first order in the concentration of oxalic acid. The ratio $(C^{12}O_2)/(C^{13}O_2)$ may be shown as equal to $(C^{13}O)/(C^{12}O)$ and also as equal to an expression involving the isotopic concentration of oxalic acid at any time t and the rate constants k_1, k_2, and k_3. The expression when $t = \infty$ (complete reaction) is

$$\frac{(C^{12}O_2)}{(C^{13}O_2)} = \frac{(C^{13}O)}{(C^{12}O)} = \frac{(\text{oxal}^{12})(k_2 + k_3) + k_3(\text{oxal}^{13})}{k_2(\text{oxal}^{13})} \tag{3}$$

where (oxal^{12}) and (oxal^{13}) are the number of moles of the two isotopic species of oxalic acid at the start of the reaction. In one run the per cent C^{13} in CO_2 was 1.087, and in CO 1.050, starting with 1.069% C^{13} in oxalic acid. An average ratio of 1.035 for k_2/k_3 could be calculated from these data. If no isotope effect occurred, k_2/k_3 would be exactly unity.

If the reaction is run a short time so that only a small fraction of the oxalic acid is decomposed, one can assume that the concentration of C^{13} in the oxalic acid is constant. The C^{12} appearing as CO_2 derives from reactions 1 and 3, the C^{13} as CO_2 from reaction 2, so

$$\frac{(C^{12}O_2)}{(C^{13}O_2)} = \frac{k_1(\text{oxal}^{12}) + k_3(\text{oxal}^{13})}{k_2(\text{oxal}^{13})} \tag{4}$$

Thus from isotopic analysis of the first CO_2 evolved, the ratio $k_1/(k_2 + k_3)$ can be determined. The deviation from unity of this ratio indicates the strength of the C^{12}—C^{12} bond relative to the C^{12}—C^{13} bond. The measurements show that the C^{12}—C^{12} bond is more easily split, the isotope effect amounting to 3.5%. The k_2/k_3 value of 1.035 also indicates that an effect of the same magnitude characterizes the relative ease of splitting of C^{12}—O^{16} and C^{13}—O^{16}, the former being the more easily split.

One may expect that with C^{14} somewhat larger isotopic effects might be observed; 5% to 8% per unit process appears to be a maximal estimate. Some results on the thermal cracking of propane,[14] and dissociation of propane by electronic impact[15] would seem to indicate somewhat greater values.

Differences in reaction rates of isotopes can be expected to diminish as the atomic number increases.[16] Thus, for elements heavier than carbon, nitrogen, and oxygen, isotopic differentiation should be much less. In a few cases, an isotope effect has been demonstrated, e.g., in sulfur and in chlorine. As an example, Bartholomew *et al.*[17] have shown that when ter-

[14] Stevenson, D. P., Wagner, C. D., Beeck, O., and Otvos, J. W., *J. Chem. Phys.* **16,** 993 (1948).

[15] Beeck, O., Otvos, J. W., Stevenson, D. P., and Wagner, C. D., *J. Chem. Phys.* **16,** 255 (1948).

[16] Biegeleisen, J., *Science* **110,** 14 (1949).

[17] Bartholomew, R. M., Brown, F., and Lounsbury, M., *Nature* **174,** 133 (1954).

tiary butyl chloride reacts with silver nitrate in 98% alcohol at room temperature, the lighter isotope (Cl^{35}) reacts faster than the heavier isotope (Cl^{37}). The relative rates indicate that the reaction constant for Cl^{37} is only about 0.7% to 0.8% as great as that for Cl^{35}. It can be concluded that no theoretical basis exists for large isotope effects in purely chemical reactions.[18] Experimental results such as those cited on oxalic acid decomposition are in accord with theoretical expectations.

Studies on isotopic differentiation in enzyme-catalyzed reactions are few in number. In one such study, the hydrolysis of C^{14}-labeled urea to CO_2 and NH_3 in the presence of urease was examined.[19] It was found that a slight disproportionation in isotope distribution occurred during the course of the reaction, the first 5% to 10% of the CO_2 evolved being a few per cent richer in C^{14} than the last 10% collected. In another study,[20] malic acid labeled in the β carboxyl with C^{14} was incubated with a heart muscle extract containing succinic dehydrogenase and fumarase. The resulting malic acid was found to be uniformly labeled in both carboxyls within experimental error (1%).

From this discussion, it is conceivable that in some reactions, particularly those which are largely unidirectional and involve a number of products, there exists a possibility that appreciable isotopic effects may occur. From the standpoint of tracer experimentation, this may complicate analysis in two ways. First, there may be differentiation in the organism. It is unlikely that effects of the magnitude observed in the extreme cases cited above will occur under conditions of biochemical function. In fact, no large biochemical effects have been observed, except in the atypical case of the hydrogen isotopes (see Chapter IX). Second, there may be differentiation during chemical analysis of labeled material, particularly in degradation reactions designed to determine placement of isotope in molecules supposedly involved in intermediary metabolism. If a molecule such as oxalic acid were to be examined by the method described above, the experimenter would have to correct for the 3% deviation from randomness in the isotope distribution. Effects such as these would limit the precision obtainable in quantitative analysis of labeling experiments. Of course, no trouble of this nature would arise in degradations which involve complete combustion of the organic material.

The dilution range available in tracer studies is another important factor and is determined by the concentration of isotope available or permissible, the constancy of the isotopic composition in the element

[18] Biegeleisen, J., *Ann. Rev. Nuclear Sci.* **2,** 221 (1953).

[19] Daniels, F., and Myerson, A. L., private communication, University of Wisconsin (1950).

[20] Racusen, D. W., and Aronoff, S., U. S. Atomic Energy Comm. AECU-1490 (1951).

studied, and the precision of assay. Thus the upper limit of dilution for the rare stable isotope of carbon (C^{13}) cannot exceed that resulting in an isotope ratio less than the error involved in determining the normal isotope ratio, which is given in one instance[3] as 0.0110 ± 0.0002. If fluctuations introduced by isotopic differentiation are neglected, it is seen that a sample of 10% C^{13} cannot be diluted more than fiftyfold when a precision of ±5% is desired.

Radioactive isotopes in general offer much greater dilution factors because radioactive assay methods are so sensitive. To compare directly with the example cited for stable isotope C^{13}, a sample 25% concentrated in C^{14} corresponds to an activity of 1.2 mc./mg. The usual assay apparatus employed can detect approximately 1×10^{-8} mc. with a precision of ±5%, so that a dilution of fifty millionfold is possible. However, it must be noted that 25% C^{14} is available only in milligram lots, so that for many researches considerable dilution may be required before the tracer carbon is in the chemical form needed.

For stable isotopes the ultimate concentration of tracer employed is limited by the obvious fact that stable isotope concentrations cannot exceed 100%. For radioactive isotopes the concentration cannot be more than that above which radiation damage occurs in the organism studied. In general, the concentration of radioactive tracers required for most biological studies can be lowered to a value at which normal physiological processes appear to remain undisturbed. The occurrence of both stable and radioactive tracer isotopes for the same element—C^{13} and C^{14}, H^2 and H^3, and so on—affords the possibility of checking the system investigated for possible radiation effects. Thus in researches with C^{14}, occasional repetitions with the stable isotope C^{13} can be made. Deviations in experimental results obtained with the two isotopes may be interpreted as resulting from abnormalities in metabolic activity induced by radiation.

C. THE DYNAMIC STATE OF CELL CONSTITUENTS AND THE CONCEPT OF THE "METABOLIC POOL"

An important concept in modern biochemistry which is based primarily on the researches made possible by the tracer method is that of the "metabolic pool," i.e., the existence of a circulating body of chemical substances in partial or total equilibrium with similar substances derived by continued release and uptake from cellular tissues. More precisely, the metabolic pool is that mixture of compounds derived either from diet or tissue breakdown which the organism uses for synthesis of tissue constituents.[21] The existence of the metabolic pool is a consequence of the general concept of the dynamic state of cell constituents. According to this concept, all cellular materials are in a state of constant flux.

[21] Sprinson, D. B., and Rittenberg, D., *J. Biol. Chem.* **180**, 715 (1949).

Neither the classical comparison of a living being to a combustion engine nor the theory of independent endogenous and exogenous reactions popular with biochemists in the past can be reconciled with the finding that a rapid degeneration and resynthesis of all molecules occurs in biochemical processes involving constant interchange of specific atomic groups. This is true whether attention is focused on the relatively stable structural elements (fats, proteins) or on the relatively unstable energy-yielding substrates. Schoenheimer put it aptly in 1941:

"A simple analogy which may be taken as an incomplete illustration of this concept of living matter can be drawn from a military regiment. A body of this type resembles a living adult organism in more than one respect. Its size fluctuates only within various limits, and it has a well-defined highly organized structure. On the other hand the individuals of which it is composed are continually changing. Men join up, are transferred from post to post, are promoted or broken and ultimately leave after varying lengths of service. The incoming and outgoing streams of men are numerically equal, but they differ in composition. The recruits may be likened to the diet; the retirement and death correspond to excretion. This analogy is necessarily imperfect as it relates to only certain aspects of the dynamic state of biological structure. While it depicts the continual replacement of structural units it takes no account of their chemical interaction."[22]

In elaborating the concept of a metabolic pool it is essential to note that the various components participate at different and characteristic rates. Nitrogen compounds alone participate in the lumped nitrogen pool at markedly different rates, varying from protein to nonprotein nitrogen, and even from protein to protein. Data on protein synthesis suggest that in the rat or human only a fraction of the muscle protein is involved in the dynamic state.[21]

The data which have led to this concept of the dynamic state have been derived for the most part from simple experiments in which a compound suitably labeled has been introduced into the biological system from which, at some later time, various biochemical fractions have been prepared and the location and nature of labeled compounds determined. In principle it is possible by such experiments or extensions of them to obtain data on a number of questions in intermediary metabolism. There are two fundamental general questions:

1. What is the nature of molecules which are used in the synthesis of the structural elements of living cells?

2. What molecular mechanisms are involved in the breakdown of substrates and mobilization of energy sources for synthetic reactions?

[22] Schoenheimer, R., "The Dynamic State of Body Constituents," p. 65. Harvard U. P., Cambridge, Mass., 1946.

In practice three difficulties confound the researcher who attempts to answer such questions. First, it is necessary to isolate the intermediates formed in a pure state so that specific isotopic contents can be determined. Second, it must be assumed that the labeled material administered is in equilibrium with the same material already present in unlabeled form in the organism. Third, it is necessary to prepare the labeled compounds and later to isolate the labeled products in pure form. Further elaboration of these difficulties is given in the sections which follow.

D. Precursor-Product Researches

The demonstration that substance *B* is derived from substance *A*, that is, that a one-to-one correlation exists between appearance of tracer in *B* from *A*, has been recorded repeatedly. There will be recalled the reversible interconversion of the fatty acids, i.e., palmitic acid into stearic acid[23] and the reverse,[24] and the metabolic relation between various amino acids, i.e., the conversion of ornithine into arginine[25] and of phenylalanine into tyrosine.[26] It should be remarked that, in any study of metabolism in which the organism is growing, it is possible to demonstrate possible relations between metabolites by nontracer feeding experiments. In the steady state when growth has ceased, the only method available is a tracer method.

Biological conversion of specific metabolites into normal structural entities as well as waste products has been studied for a large variety of molecules of biochemical interest. A classical instance[22] is the demonstration of the biological synthesis of creatine from methyl (derived from methionine or choline), glycine (from protein degradation), and amidine (derived from arginine).

Two examples from the literature are the demonstration of pregnandiol formation by degradation of cholesterol[27] and the synthesis of the carbon chain of cystine from serine.[28, 28a] A demonstration of a negative nature is the proof using doubly labeled methionine ($H^3C—S^*—C^*H_2C^*H_2$-$CHNH_2COOH$) that methionine contributes sulfur but not its carbon chain in the biosynthesis of cystine.[29]

[23] Stetten, D., Jr., and Schoenheimer, R., *J. Biol. Chem.* **133,** 329 (1941).

[24] Schoenheimer, R., and Rittenberg, D., *J. Biol. Chem.* **120,** 155 (1937).

[25] Clutton, R. F., Schoenheimer, R., and Rittenberg, D., *J. Biol. Chem.* **132,** 227 (1940).

[26] Moss, A. R., and Schoenheimer, R., *J. Biol. Chem.* **135,** 415 (1940).

[27] Bloch, K., *J. Biol. Chem.* **157,** 661 (1945).

[28] Binkley, F., and du Vigneaud, V., *J. Biol. Chem.* **144,** 507 (1942).

[28a] Stetten, D., Jr., *J. Biol. Chem.* **144,** 501 (1942).

[29] du Vigneaud, V., Kilmer, G. W., Rachele, J. R., and Cohn, M., *J. Biol. Chem.* **155,** 645 (1944).

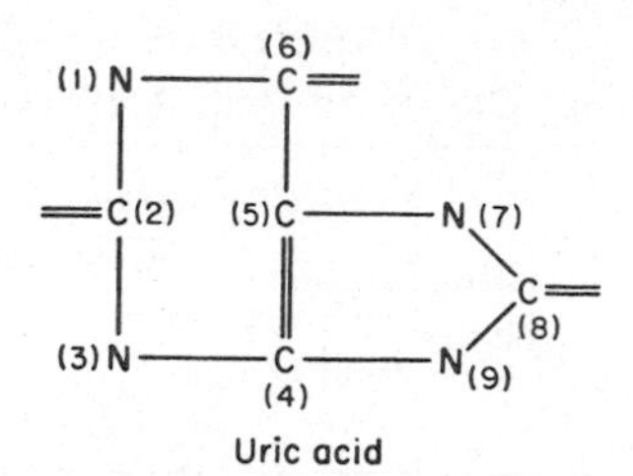

FIG. 39. Uric acid skeleton. The numbers in parentheses distinguish the various carbon and nitrogen atoms.

There are few examples in the literature of what may be called "quantitative isotopy." Much of the literature tends to describe demonstrations of the qualitative relation between any two or more metabolites, but there are few clear-cut analyses of the extent to which the given relationship may be taken to account for the formation of a given metabolite from a given precursor. A fundamental difficulty has already been cited—the extent to which equilibration of ingested and circulating or storage material takes place. Another difficulty, of course, is that the actual number of molecules intervening between precursor and product cannot be determined merely from feeding experiments. Conditions for the precursor-product relationship in the simple case, wherein a steady state obtains and in which there is no discrimination between storage material and similar ingested materials, have been analyzed and resulting formulations applied in a number of cases.

A detailed example of research on biological precursors is an investigation of carbon sources in the formation of uric acid (Fig. 39). In these experiments with the rare stable isotope C^{13}, labeled carbonate, carboxyl-labeled acetate, carboxyl-labeled lactate, α-β-labeled lactate, carboxyl-labeled glycine, and labeled formate were studied as possible precursors for uric acid in pigeon excreta.[30] The isolation procedures and degradation reactions employed are described elsewhere (see pp. 336–337). Results are summarized in Table 5. In this table the isotopic concentrations are expressed in atom per cent excess which, as explained previously (p. 10), is the excess of percentage abundance of isotopes in the labeled element over that in the normal element. Low values (<0.02) probably arise as artifacts during degradation and may be ignored for the present.

It is immediately evident that these compounds are utilized in different ways. CO_2 appears to be a source for carbon 6. Decarboxylation of acetate and lactate yields CO_2 which also appears in carbon 6 and is equilibrated

[30] Sonne, J. C., Buchanan, J. M., and Delluva, A. M., *J. Biol. Chem.* **173,** 69, 81 (1948).

TABLE 5

PRECURSORS OF CARBON CHAIN OF URIC ACID (AFTER SONNE *et al.*)[30]

Precursor	Labeled Carbons	C^{13} Concentration (Atom Per Cent Excess) Uric Acid Carbon No.						
		Resp. CO_2	6	4 + 5	5	4	2	8
C^*O_2	8.13	0.28	0.25	0.04			0.02	0.02
HC^*OOH	3.34	0.01	0.01	0.08			2.41	2.41
CH_3C^*OOH	5.82	0.26	0.22	0.04	0.00	0.00		
$NH_2CH_2C^*OOH$	5.20	0.12	0.11		0.14	1.13	0.00	0.00
DL-CH_3CHOHC^*OOH	8.80	0.25	0.26		0.00	0.31	0.01	0.01
DL-$C^*H_3C^*HOHCOOH$	5.40	0.11	0.09		0.14	0.04	0.10	0.10

with the respiratory carbon dioxide. Formate carbon appears in carbons 2 and 8, a fate not shared markedly by any other carbon. No appreciable decarboxylation of formate analogous to that of acetate occurs under the conditions noted. Glycine carboxyl appears to be a major contributor to carbon 4, as does the carboxyl from lactate. Deductions from previous work on conversion of serine to glycine,[31] make it possible to conclude that lactate may be converted to glycine or a derivative of glycine, thus explaining the contribution of lactate carboxyl to carbon 4. If this is true, it is reasonable to suppose that carbon 5 comes from the α carbon of glycine. Similar results have been obtained with N^{15}-labeled glycine in man.[32]

In work on carbon dioxide utilization in the uric acid decomposition by the anaerobe *Cl. cylindrosporum*,[33] it has been noted that the acetic acid and glycine formed in the presence of labeled CO_2 exhibit different isotopic distribution; the glycine is labeled only in carboxyl carbon, but the acetic acid is labeled in both carbons. The two molecules, then, would appear to arise by different metabolic pathways in the degradation of uric acid. It has also been noted that acetyl carbon is not involved in urea formation in the rat.[34]

It should be remembered that there was relatively little dilution of the isotope in these experiments because of the large quantity of isotopic material fed and the consequent small dilution with nonisotopic material in the animal. Nonetheless, it still appears reasonable to conclude that compounds incorporated with such high resultant isotopic levels in product, i.e., formate carbon into carbons 2 and 8, must be involved as precursors quite directly.

[31] Shemin, D., *J. Biol. Chem*, **162,** 297 (1946).

[32] Shemin, D., and Rittenberg, D., *J. Biol. Chem.* **167,** 875 (1947).

[33] Barker, H. A., and Elsden, S. R., *J. Biol. Chem.* **167,** 619 (1947).

[34] Sonne, J. C., Buchanan, J. M., and Delluva, A. M., *J. Biol. Chem.* **166,** 395 (1946).

TABLE 6
INCORPORATION OF LABELED N INTO RAT HEME (AFTER SHEMIN AND RITTENBERG[37])

Compound Fed			
Compound	N^{15} content (atom % excess)	Heme N^{15} (Atom % Excess)	Heme N^{15} Assuming Compound Fed Contained 100% N^{15} (Atom % Excess)
Glycine	11.6	0.108	0.93
Glycine	19.0	0.169	0.89
Ammonium citrate	13.0	0.012	0.09
DL-Glutamic acid	18.6	0.032	0.17
DL-Proline	11.6	0.031	0.18*
DL-Proline	11.6	0.028	0.15*
DL-Leucine	32.7	0.051	0.07*

* Corrected for ammonia liberation owing to degradation of D-isomers.

Having indicated that the use of α-labeled glycine would result in uric acid labeled primarily in carbon 5, it is of interest to note the results of such an experiment.[35] As expected, most of the label appeared in carbon 5. There was a small amount in carbon 4. However, an unexpectedly large fraction of the label also appeared in carbons 2 and 8, which would seem to indicate that the α carbon of glycine somehow was in a steady-state relationship to formate. Later experiments[36] indicated that it was possible for α-labeled glycine to be degraded to formate with subsequent recondensation of formate and glycine to reform α-β-labeled serine.

To illustrate the possibilities of the more quantitative type of study involving time relations between precursor isotopic content and product isotopic content, researches on the role of glycine in synthesis of blood heme in man and deductions therefrom concerning the life span of the human erythrocyte will be discussed.[37] In these experiments, N^{15}-labeled glycine, glutamic acid, proline, leucine, and ammonia as ammonium citrate were fed to rats on a protein-free diet. Two weeks later these compounds plus the heme were reisolated and tested for isotopic content. In Table 6 the various isotopic concentrations are compared. From this table it can be seen that, regardless of whether N^{15}-labeled ammonia, glutamic acid, proline, or leucine was fed, only a relatively small N^{15} excess appeared in the porphyrin, compared to that found after feeding the labeled glycine. From these experiments and others relating to the isotopic content of circulating glycine compared to heme, it could be concluded that the nitrogen of glycine was directly utilized in the synthesis of the proto-

[35] Karlsson, J. L., and Barker, H. A., *J. Biol. Chem.* **177**, 597 (1949).
[36] Sakami, W., *J. Biol. Chem.* **176**, 995 (1949).
[37] Shemin, D., and Rittenberg, D., *J. Biol. Chem.* **165**, 627 (1946).

porphyrin of heme, the nitrogen of the other compounds being used only indirectly.

With these results established in rats, N^{15}-labeled glycine was used to build up labeled heme in a human subject. After cessation of feeding, the variation of the isotopic concentration of heme as well as blood protein was followed. The isotopic concentration of heme continued to rise after the period of feeding (3 days) to a maximum reached after 25 days, remained relatively constant for nearly 75 days, and then fell slowly along an S-shaped curve (Fig. 40). Usually if labeled material is incorporated during the feeding period into material in a state of continual synthesis and degradation (as would be the case for utilization of most amino acids in tissue protein), there occurs a more or less immediate exponential drop of isotope concentration in the labeled product after administration of isotopic material ceases. In the case of glycine, however, there was incorporation into heme which was fixed in the erythrocyte and not released until the whole cell was broken down. As a consequence, the N^{15} concentration in the heme rose to a maximum value, remained constant for a period approximating the average life of the cells, and then declined.

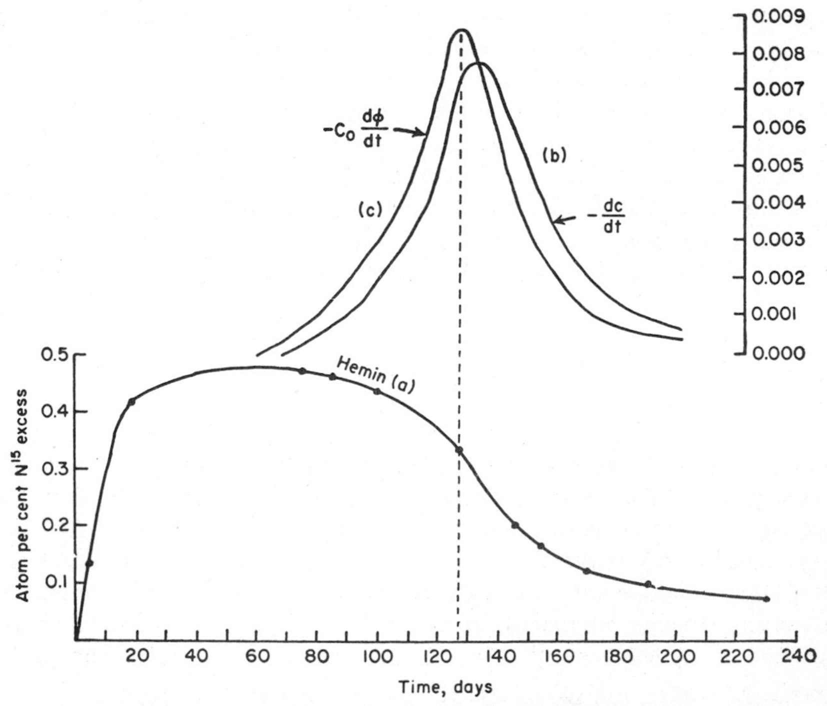

FIG. 40. N^{15} concentration in hemin in human subject fed N^{15}-labeled glycine for 3 days. (After Shemin and Rittenberg.[37])

This behavior was precisely that which could be expected provided that none of the components liberated was reutilized. As a matter of fact, this reutilization does not take place with the glycine nitrogen contributed to heme, although it does for the iron liberated by heme destruction, as demonstrated with labeled iron.[38] The slow and prolonged rise after the end of feeding was due to the stability of freshly synthesized cells in circulation and the preferential destruction of the older unlabeled cells.

In general, any biochemical system involves a number of intermediates in the main reaction sequence. The main sequence is interspersed with side reactions. When the two kinds of reactions are analyzed, integral equations solvable only by approximation methods result.[39] In the researches on the incorporation of glycine into heme cited above, a fortuitous set of circumstances obtained so that a relatively simple analysis resulted. The glycine label was not metabolized away by general transamination reactions but acted as a precursor molecule directly producing a stable product which was neither metabolized during the life of the system (erythrocyte) nor reutilized appreciably in breakdown. It is informative to note the mathematical expressions obtaining in this instance.

It could be shown[37] that the isotope (N^{15}) concentration in the product heme, $C(t)$, was related to the time, t, as in the following integral expression:

$$C(t) = \frac{100}{\overline{T}} \int_0^t f(\theta)\phi(t - \theta)\, d\theta \tag{5}$$

where $f(\theta)$ was the N^{15} concentration in the heme and its direct precursor synthesized at a time $t = \theta$, $\phi(t - \theta)$ was the probability that a given red cell had a life span greater than $(t - \theta)$, and $\overline{T}$ was in the first approximation the average life of the cells. By substitution involving a new variable, x, defined as $t - \theta$, Eq. 6 was obtained:

$$C(t) = \frac{100}{\overline{T}} \int_0^t f(t - x)\phi(x)\, dx \tag{6}$$

The experimental results for the initial period ($t \lessapprox 30$ days) could be fitted by an exponential expression, viz.,

$$C(t) = 0.48(1 - e^{-0.11t})$$

or, in general,

$$C(t) = C_0(1 - e^{-\lambda t}) \tag{7}$$

Substitution in Eq. 6 and differentiation resulted finally in the expression

$$C_0 \frac{d\phi}{dt} = \frac{1}{\lambda} \frac{d^2C}{dt^2} + \frac{dC}{dt} \tag{8}$$

[38] Cruz, W. O., Hahn, P. F., and Bale, W. F., *Am. J. Physiol.* **135,** 595 (1946).
[39] Branson, H., *Science* **106,** 404 (1947).

This expression means that the death rate of the cells ($d\phi/dt$) is proportional not only to the change in isotope concentration with time in the declining portion of the death curve (see Fig. 40), but also to a factor involving the generation time of the cells during the labeling period. Had the labeled cells been introduced into the circulation at one instant, then the second-order term in Eq. 8 would not have been involved.

With the value of $C(t)$ at any time t known, it was possible by graphical integration to evaluate all the terms in Eq. 8 for the time range in which most cells died (released isotope). It could be shown that no cells died for the first 70 days. The maximum value of the death rate occurred at 127 days. The death curve ($d\phi/dt$ vs. t) was found to be symmetrical about the ordinate $t = 127$ days, so that 127 days could be taken for the average survival time of this particular human subject's red cells. Furthermore, one could show that half the cells died in a period of 28 days, taking $t = 127$ days as the mean. Thus, half the cells survived in the period 113 to 141 days, and the rest died before and after this period.

Criteria for the establishment of a precursor-product relation have been given by Zilversmit *et al.*[39a] in the case where a steady-state condition exists; i.e., the amount of substance (product) is constant with time, the rate of appearance (synthesis from precursor) equaling the rate of disappearance of product. It is necessary to define here a concept which is implicit in the establishment of the steady state, namely "turnover." Turnover is the continued renewal of a given substance without over-all change in net concentration. From a labeling standpoint a substance can be "turned over" in any of the following ways: (1) the labeling atoms may be incorporated by synthesis or exchange; (2) the labeled substance may replace unlabeled substance by transport into the tissue site under consideration; (3) a combination of these two processes may be involved. The true turnover rate is the amount of substance turned over in unit time when a steady state is established. A corrollary quantity, the "turnover time," is the time required in the steady state to renew completely the amount of substance initially present in the tissue. If the rate of appearance (or disappearance) is x, and the amount of substance initially present is a, then the turnover time is a/x.

Unfortunately, two different definitions of turnover rate[40] are in use, with consequent confusion in the literature. In one, turnover rate is thought of as the concentration of substance in the tissue divided by the turnover time. In the other, it is conceived as the rate at which all the substance present is replaced in the tissue, which is given by the reciprocal of the turnover

[39a] Zilversmit, D. B., Entenman, C., and Fishler, M. C., *J. Gen. Physiol.* **26,** 323 (1943).

[40] Kleiber, M., *Nature* **175,** 342 (1955).

time. Two tissues could have identical turnover times and very different turnover rates according to the first definition from those given by the second definition. Mawson[41] cites an example from work on turnover of Zn^{65} in the dorsolateral and ventral prostates of rats.[42] For the ventral prostate the mean zinc content in micrograms per gram is 13.5. The time required for elimination of half of the labeled zinc is 7.6 days. The turnover time, which is equivalent to the average life, is 1.44 times this half-life, or 10.9 days. The corresponding quantities for the dorsolateral prostate are 168 μg./g., 9.9, and 14.3 days. By the first definition, the turnover rate is the concentration of zinc in the ventral prostate, 13.5 μg./g., divided by the turnover time, 10.9 days, or 1.24 μg./g. per day. In the dorsolateral prostate, the corresponding value is 168 divided by 14.3, or 11.8. Thus, by the first definition the turnover rate of zinc is some ten times as great in the dorsolateral prostate as it is in the ventral. By the second definition, taking the reciprocals of the two turnover times, there result the values of 0.092 and 0.064 for the fraction of the zinc pool turned over per day in the ventral and dorsolateral prostates, respectively. The second definition is seen to give a result opposed to the first in that it assigns a greater turnover rate to the ventral prostate.

To eliminate confusion, Zilversmit[43] has suggested that the term "turnover rate" should continue to mean the concentration of material being turned over divided by the turnover time. In most of the literature the second definition has been used, however. Mawson suggests, therefore, that the second definition continue to be applied to the turnover rate, and the quantity obtained by the first definition be termed "flux rate." A thorough discussion of the implications of flux rate in tracer experiments is outside the scope of this book. The reader is referred to an excellent discussion by Reiner.[43a]

If one assumes that all molecules of substance, whether newly formed or old, are equally available for synthesis or breakdown, one may arrive at a simple criterion for the precursor-product relation as follows.

Let p = constant rate of conversion of precursor A to product B.

r = amount of B present in tissue (constant).

x = amount of labeled B in tissue.

$f(t)$ = specific activity (s.a.) of A which is a function of time t.

[41] Mawson, C. A., *Nature* **176,** 317 (1955).

[42] Mawson, C. A., Fischer, M. I., and Riedel, B. E., unpublished work, cited in reference 41.

[43] Zilversmit, D. B., *Nature* **175,** 863 (1955).

[43a] Reiner, J. M., *Arch. Biochem. and Biophys.* **46,** 53, 80 (1953).

The amount of labeled material converted into B per unit time is $p \cdot f(t)$. The amount of labeled material lost per unit time is $p \cdot x/r$. The time variation (change) in the specific activity (or labeled content) of B is then

$$\frac{dx}{dt} = pf(t) - p\frac{x}{r} \tag{9}$$

Recalling that r is constant and assembling terms, one arrives at

$$r\left(d\frac{x}{r}\Big/ dt\right) = p\left[f(t) - \frac{x}{r}\right] \tag{10}$$

and

$$\left(d\frac{x}{r}\Big/ dt\right)\Big/\left[f(t) - \frac{x}{r}\right] = \frac{p}{r} = \text{constant} \tag{11}$$

The numerator in Eq. 11 measures the slope of the specific activity-time curve of product B. The denominator is the difference at any given time between specific activity of precursor $f(t)$ and specific activity of product B, x/r. It can be seen (Fig. 41) that, when the specific activity of B is increasing (slope s.a. B positive), the specific activity of A must at all times be greater than B before the specific activity of B reaches a maximum. After B reaches its maximal specific activity, the specific activity relations are reversed, B having a higher specific activity than A. At the maximum specific activity for B, the specific activities of A and B are equal.

It is easy to show that the turnover time, which is r/p, can be obtained merely by determining the area between the specific activity curves for any two times, t_1 and t_2, and dividing by m, the increase in specific activity of B between times t_1 and t_2. Turnover rate is most easily determined from rate of disappearance of label after steady state has been established. Under these conditions, the nature of the immediate precursor and its labeling content does not have to be known.

All these considerations apply to the usual system observed in physiological function in which a dynamic equilibrium involving constant synthesis and degradation obtains. This was not the case in the heme synthesis cited above. Had the blood cells died irrespective of age and had there been a constant interchange of intracellular and extracellular material, then as remarked previously the labeling curve (a) of Fig. 40 would have stopped rising after cessation of feeding (third day) and begun to fall along an exponential death curve. Thus, the two calculations presented in this section deal with the two disparate situations one may encounter in precursor-product researches.

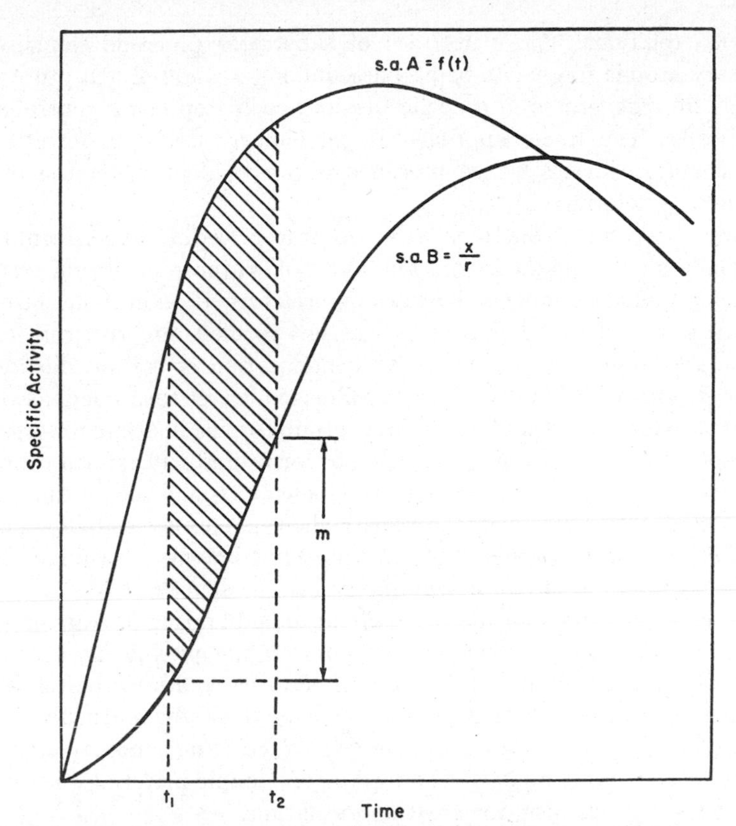

FIG. 41. Specific activity-time relations of precursor A and product B. (After Zilversmit *et al.*[39a])

Examples of calculations involving various situations other than those considered abound in the literature.[44]

It may be concluded that this aspect of tracer research will remain important despite complexities introduced by dynamic flux of cellular constituents and the necessity for obtaining equilibration of ingested material with tissue substance, added to difficulties occasioned by synthesis of labeled material, administration, and then isolation and purification.

E. Metabolic Cycles and Detection of Intermediates

The subject matter of the previous section may be considered a special aspect of the general study of mechanisms involved in anabolic and

[44] See Comar, C. L., "Radioisotopes in Biology and Agriculture," pp. 23–38. McGraw-Hill, New York, 1955.

catabolic relations. The utilization of substrates proceeds stepwise, the necessary atomic fragments being passed along a chain of acceptors which usually are regenerated in a cyclic fashion, each step being controlled enzymatically. The tracer approach is particularly useful in ferreting out possible intermediates whose presence is not evident when the over-all chemistry is determined.

Many successful deductions from nontracer feeding experiments have been made concerning the origins and mode of synthesis of certain excretory products. A good example is the work on urea formation in tissue slices,[45–47] wherein it was deduced that urea was not formed directly from carbon dioxide and ammonia but rather by condensation of carbon dioxide and ammonia with ornithine to form citrulline which in turn condensed with ammonia to form arginine. The amidine group formed in arginine was shown to be split off by an enzyme, arginase, to form urea and reform ornithine. This mechanism was investigated with labeled CO_2 and shown to be essentially correct.[48–50] Later, nontracer methods were employed to demonstrate the additional participation of an enzyme cycle between citrulline and aspartic acid with formation of arginine and malic acid.[51a, b]

The tracer method, of course, can go far in elaborating or supplementing feeding or nutritional balance experiments for uncovering mechanisms of substrate utilization. Thus, the origin of creatine could not be established by feeding various different amino acids and proteins. In the case of creatine, no significant change in the level of creatine concentration could be induced in balance studies. The feeding of isotopic material was required to establish a mechanism for creatine formation.

In connection with creatine formation an important biochemical process, *transmethylation*, should be mentioned. The transfer of methyl groups could be inferred in liver tissue slices by nontracer means because methionine was found to accelerate greatly the rate at which creatine could be formed from guanidoacetic acid.[52] The direct proof of the methyl shift from methionine (to choline, in this case) was supplied by tracer experiments with methionine in which the active methyl group attached to sulfur was labeled with heavy

[45] Krebs, H. A., and Henseleit, K., *Z. physiol. Chem.* **210,** 33 (1932).
[46] Cohen, P. P., and Hayano, M., *J. Biol. Chem.* **172,** 405 (1948).
[47] Grisolia, S., and Cohen, P. P., *J. Biol. Chem.* **176,** 929 (1948).
[48] Evans, E. A., Jr., and Slotin, L., *J. Biol. Chem.* **136,** 805 (1940).
[49] Rittenberg, D., and Waelsch, H., *J. Biol. Chem.* **136,** 799 (1940).
[50] MacKenzie, C. G., and du Vigneaud, V., *J. Biol. Chem.* **172,** 353 (1948).
[51a] Ratner, S., *J. Biol. Chem.* **170,** 761 (1947).
[51b] Ratner, S., and Pappas, A., *J. Biol. Chem.* **179,** 1183, 1199 (1949).
[52] Borsook, H., and Dubnoff, J. W., *J. Biol. Chem.* **132,** 559 (1940).

hydrogen.[53] It appears that a balanced diet requires substrate material capable of supplying transferable methyl groups and that only a few substances such as methionine and choline are available biologically for this purpose in animal metabolism.[54]

Undoubtedly the most extensive contribution to intermediary metabolism resulting from tracer studies has been the firm establishment of carbon dioxide as a metabolite important in an astonishing diversity of processes. That CO_2 may be utilized not only by autotrophic systems but also by heterotrophs in general is an idea with origins deep in the beginnings of microbiology. Definite evidence for CO_2 utilization by systems essentially heterotrophic began accumulating from a variety of nontracer studies in the middle of the 1930's.[55–57] It remained for the tracer method to establish the general role of CO_2 in cellular metabolism. The reader will find numerous reviews of the status of this field.[58–60] For this discussion it will suffice to mention a few experiments which show certain unique features of tracer research.

The majority of heterotrophic organisms which respire or ferment organic substrates produce CO_2 as an end product. The conclusion that CO_2 may also be utilized can be deduced in nontracer studies only when utilization reactions outweigh CO_2 excretion resulting from degradation of cellular material or substrate, as in the glycerol fermentation by propionic acid bacteria.[56] It had been possible to deduce from indirect evidence that CO_2 absorption occurred in some bacterial systems.[61] Direct proof of CO_2 utilization involved maintaining the respiring or fermenting systems in the presence of labeled carbonate. The appearance of labeled carbon in cellular material and in waste products other than carbon dioxide could then be demonstrated with comparative ease.

Thus, consider the fermentation of purines by *Clostridium acidi-urici*. It had been established that uric acid, xanthine, and hypoxanthine were fer-

[53] du Vigneaud, V., Cohn, M., Chandler, J. P., Schenck, J. R., and Simmonds, S., *J. Biol. Chem.* **140,** 625 (1941).

[54] du Vigneaud, V., *Proc. Am. Phil. Soc.* **92,** 127 (1948).

[55] Barker, H. A., *Arch. Mikrobiol.* **7,** 404 (1936).

[56] Wood, H. G., and Werkman, C. H., *Biochem. J.* **30,** 48 (1936).

[57] Woods, D. D., *Biochem. J.* **30,** 515 (1936).

[58] Van Niel, C. B., Ruben, S., Carson, S. F., Kamen, M. D., and Foster, J. W., *Proc. Natl. Acad. Sci. U. S.* **28,** 8 (1942).

[59] Buchanan, J. M., and Hastings, A. B., *Physiol. Revs.* **26,** 120 (1946).

[60] Wood, H. G., *Physiol. Revs.* **26,** 198 (1946).

[61] Barker, H. A., Ruben S., and Beck, J. V., *Proc. Natl. Acad. Sci. U. S.* **26,** 477 (1940).

mented as follows:[62]

```
HN—CO
|    |
OC   C—NH + 5.5H2O → 0.75CH3COOH + 4NH3 + 3.5CO2
|    ||   \
|    ||    CO
|    ||   /
HN—C—NH
(Uric acid)
```

```
HN—CO
|    |
OC   C—NH + 6H2O → CH3COOH + 4NH3 + 3CO2
|    ||   \
|    ||    CH
|    ||   //
HN—C—N
(Xanthine)
```

```
HN—CO
|    |
HC   C—NH + 6.5H2O → 1.25CH3COOH + 4NH3 + 2.5CO2
||   ||   \
||   ||    CH
||   ||   //
N—C—N
(Hypoxanthine)
```

The decrease in CO_2 production from hypoxanthine compared to the other purines together with the apparently abnormal quantity of acetic acid found (one could expect at most only one mole from simple fission of the C-3 chain) indicated that CO_2 absorption might be involved, at least in the case of hypoxanthine.

The organisms were allowed to ferment the three purines in the presence of labeled carbonate. Isolation of cell material and of the acetic acid revealed appreciable utilization of CO_2 to form acetic acid as well as some cellular material. Both carbons of the acetic acid were found to be labeled.

In this type of experiment only qualitative results could be obtained; that is, the precise extent to which CO_2 entered into the synthesis reactions could not be inferred. As an example of a more quantitative study, work on another fatty acid-producing anaerobe may be cited.[63] *Cl. thermoaceticum* ferments glucose almost quantitatively to acetic acid,[64] viz.,

$$C_6H_{12}O_6 \rightarrow 3CH_3COOH \quad (12)$$

[62] Barker, H. A., and Beck, J. V., *J. Biol. Chem.* **141,** 3 (1941).

[63] Barker, H. A., and Kamen, M. D., *Proc. Natl. Acad. Sci. U. S.* **31,** 219 (1945).

[64] Fontaine, F. E., Peterson, W. H., McCoy, E., Johnson, M. J., and Ritter, G. J., *J. Bacteriol.* **43,** 701 (1942).

This reaction, representing a simple dismutation, is unusual because practically no C-1 compound is produced, whereas in the classical fermentation reactions equimolar quantities of C-2 and C-1 compounds are usually formed. From a variety of researches relating to other organisms requiring CO_2 for growth, it might be supposed that the reason for absence of C-1 production is related to some C-1 requirement which results in utilization of C-1 compound as rapidly as it formed. If the precursor of this C-1 fragment is identified with CO_2, there may be written the following reaction scheme:

$$C_6H_{12}O_6 + 2H_2O \rightarrow 2CH_3COOH + 2CO_2 + 8H \quad (13)$$

$$8H + 2CO_2 \rightarrow CH_3COOH + 2H_2O \quad (13a)$$

It will be noted that addition of reactions 13 and 13a yields the over-all reaction 12. Reaction 13 represents the fermentation as a partial oxidation of glucose to acetic acid and carbon dioxide, presumably via the usual glycolytic mechanism. This is coupled (through 8H) with reaction 13a representing reductive condensation of CO_2 or C-1 units derived therefrom, to acetic acid. Another scheme can also be written,

$$C_6H_{12}O_6 + 6H_2O \rightarrow 6CO_2 + 24H \quad (14)$$

$$6CO_2 + 24H \rightarrow 3CH_3COOH + 6H_2O \quad (14a)$$

Reactions 14 and 14a also yield reaction 12 on addition. Although a scheme based on reactions 14 and 14a is less likely than one based on reactions 13 and 13a, reactions 14 and 14a must be considered as possible.

Evidence on these points was obtained with labeled carbon. The organisms were allowed to grow in a normal medium containing carbonate labeled with C^{14}. After a suitable interval (3 to 6 days) the cultures were analyzed with results as exemplified in Table 7.

TABLE 7

FERMENTATION OF GLUCOSE BY *Cl. thermoaceticum* IN PRESENCE OF C^*O_2 (AFTER BARKER AND KAMEN[63])

	Mg./10 ml.	Ct./min./mg.	Total ct./min.
Glucose fermented	54.0		
Initial CO_2 as $BaCO_3$	22.3	117	2610 ± 50
Final CO_2 as $BaCO_3$	22.3	5.7	128 ± 20
Acetic acid formed (as $BaCO_3$)	101.6	19.9	2020 ± 40
Cell material	2.5	12.8	32 ± 4
Trichloroacetic acid extract	4.5	1.5	7 ± 4
Nonvolatile cell-free material			96 ± 30
Total C^{14} in products			2155
Recovery of C^{14} (%)		(2155/2482) × 100 = 87	

It will be noted that, although no change occurred in the amount of CO_2, more than 90% of the C^{14} was lost from the CO_2. Nearly all this C^{14} was recovered in the acetic acid. The acetic acid was carefully purified and characterized by a Duclaux distillation in which it was found that the specific activity remained constant in the Duclaux fractions, proving that the C^{14} was associated with acetic acid. The cellular material remaining after extraction with trichloroacetic acid appeared to have a specific activity similar to that of the acetate, so that it appeared reasonable to conclude that a considerable fraction of the synthesized cell material had incorporated carbon originating from CO_2.

When the acetic acid was examined for C^{14} distribution it was found that the label was equally distributed between the methyl and carboxyl groups. From this it could be concluded that a considerable fraction of the acetic acid was synthesized by condensation of C-1 units derived from CO_2 as shown in reaction 13a. The appearance of C^{14} in acetic acid as well as in both carbons of acetic acid and the disappearance of C^{14} from CO_2 were in excellent accord with the notion that the dismutation of glucose in this fermentation was not simple but involved partial oxidation to C-1 and C-2 units followed by reductive condensation of C-1 units.

Data leading to such a result could be obtained only by the labeling technique because it was required that carbon from the C-1 unit (CO_2) be followed as it entered into the metabolic process. Because no change in total CO_2 occurred in the process, no conventional method based on changes in CO_2 concentration could be applicable. Before regarding the conversion of CO_2 to acetic acid as established, it was necessary to rule out direct reversible exchange reactions of C^{14} between $C^{14}O_2$ and acetic acid. This was done by allowing fermentation of glucose to proceed in the presence of acetate labeled in both the methyl and carboxyl positions, but with no label in the CO_2. It was found that no label appeared in the CO_2, so that no reversible exchange was involved.

It was important to determine the actual turnover of CO_2 in the process. This was possible from the data given in Table 7. It could be supposed that the only reactions of importance in the dilution of CO_2 were

$$(a) \qquad \text{Glucose } (C^{12}) \rightarrow CO_2 \ (C^{12})$$

$$(b) \qquad CO_2 \ (C^{12} + C^{14}) \rightarrow \text{Acetic acid } (C^{12} + C^{14})$$

The exogenous CO_2 was equilibrated in step *b*, with the endogenous CO_2 being produced in step *a*. Hence, the CO_2 in the medium was always a mixture of $C^{14}O_2$ and $C^{12}O_2$. In the calculations which follow, C^{12} refers to normal carbon. The quantity (specific activity) of $C^{14}O_2$ per unit amount of CO_2 at any time during the fermentation is denoted by x. At time $t = O$, x will have the value x_0, and at the end of the fermentation the value x_f. V

denotes the amount of ($C^{12}O_2 + C^{14}O_2$) converted to acetic acid at any time. Because there is no net change in CO_2 amount, V must also equal the $C^{12}O_2$ formed at any time from glucose. V_a represents the constant amount of carbon dioxide present in the medium, and V_f is the total CO_2 utilized, as in reactions 13a and 14a.

When a small quantity (ΔV) of $C^{14}O_2 + C^{12}O_2$ is converted to acetic acid with simultaneous formation of an equal amount of $C^{12}O_2$ from glucose, the decrease in $C^{14}O_2$ ($-\Delta x$) is given by

$$-\Delta x = \frac{\Delta V}{V_a + \Delta V} x$$

This relation states that the fraction of $C^{14}O_2$ lost at any time is equal to the amount present multiplied by the dilution factor for the total carbon dioxide present. By separating variables and proceeding to the limit, there is derived the differential relation

$$-dx/x = \mathrm{d}V/V_a$$

This is integrated between the limits x_0 and x_f for x, and 0 and V_f for V. Thus

$$V_f = 2.3V_a \log (x_0/x_f)$$

V_f is divided by the quantity of glucose fermented to obtain the CO_2 production per unit of glucose. V_a, x_0, and x_f are known experimentally, so that V_f can be calculated. The results obtained in this way gave values for V_f per unit of glucose close to 2. Hence the reaction scheme of 13 and 13a was borne out, rather than that associated with 14 and 14a. This case is a good example of the use of dilution calculations. Similar studies have been carried out in which the total CO_2 changes during the experiment.[65]

According to the suggestion based on the calculations of CO_2 dilution during fermentation by *Cl. thermoaceticum*, the acetate formed should be a mixture of labeled and normal acetate, of which one-third would be formed from labeled CO_2 and would be doubly labeled ($C^{14}H_3C^{14}OOH$) and two-thirds would be from unlabeled glucose and would be unlabeled ($C^{12}H_3C^{12}OOH$). It is plain that it would not be possible to state whether this is actually the case or whether the acetate is a mixture containing one-third of each of the singly labeled acids ($C^{14}H_3 \cdot C^{12}OOH$ and $C^{12}H_3 \cdot C^{14}OOH$) and one-third of the unlabeled acid, because in the latter case the same average concentration of tracer would be found in the methyl and carboxyl carbons as in the former case.

An approach based on methods in which the actual masses of the mole-

[65] Barker, H. A., Kamen, M. D., and Haas, V., *Proc. Natl. Acad. Sci. U. S.* **31**, 355 (1945).

cules are determined is essential to proceed further with this problem. For such measurements large quantities of isotope are required. Therefore, the stable tracer C^{13} must be used. To use the mass spectrometer for C^{13} assay, the acetate must be converted to some molecule which exists in the gaseous form. In addition, this molecule must be free of oxygen in order to avoid the complications introduced by the presence of the various isotopes of oxygen. Finally, the molecule must be a two-carbon compound because it is necessary to test whether labeled carbons are adjacent or not. A convenient molecule for this purpose is ethylene, and the masses dealt with are

$$C^{13}H_3 \cdot C^{13}OOH \rightarrow H_2C^{13}{=}C^{13}H_2 \qquad \text{Mass} = 30$$

$$\left.\begin{matrix} C^{12}H_3 \cdot C^{13}OOH \\ C^{13}H_3 \cdot C^{12}OOH \end{matrix}\right\} \rightarrow H_2C^{12}{=}C^{13}H_2 \qquad \text{Mass} = 29$$

$$C^{12}H_3 \cdot C^{12}OOH \rightarrow H_2C^{12}{=}C^{12}H_2 \qquad \text{Mass} = 28$$

Wood[66] performed experiments in which *Cl. thermoaceticum* fermented glucose in the presence of CO_2 containing 24.57% C^{13}. The conditions for the fermentation were different from those of the earlier studies by Barker and Kamen in that a larger gas space was present. At the end of the fermentation, the CO_2 and acetic acid were collected. The isotopic content of each was determined, as well as the total amounts of each recovered. The acetate was converted to ethylene by reaction with benzoyl chloride to form the acetyl chloride. This was distilled in a stream of nitrogen into lithium aluminum hydride, butyl carbitol was added, and the mixture refluxed. The resultant ethyl alcohol was volatilized and carried with nitrogen gas through boiling 57% HI to form ethyl iodide. The iodide was treated with trimethylamine to form a salt which could be crystallized and dissolved in alkali. An equivalent of silver nitrate was added to remove iodide, leaving a quaternary base from which the ethylene was liberated by drying under vacuum at room temperature, then heating at 190°C. The ethylene was purified by fractional distillation. This complex procedure was required to prepare ethylene free of high molecular weight compounds and suitable for mass spectrometer assay.

The acetate formed during the fermentation was considered as derived from three sources: (1) "unlabeled," in which neither carbon came from CO_2 ; "singly labeled," in which one or the other carbon came from CO_2 ; (3) "doubly labeled," in which both carbons came from CO_2 . For simplicity, assume that the "unlabeled" carbon contained 1.00% C^{13} and that the labeled CO_2 contained 25% C^{13}. Completely random selection of C^{13} and C^{12} from either the labeled or the unlabeled source is also assumed.

The relative contributions of the different types of labeled acetate to the various molecular species could be calculated as follows. In case 1, the

[66] Wood, H. G., *J. Biol. Chem.* **194,** 905 (1952).

percentage of C^{13} in both methyl and carboxyl was 1.00. Doubly labeled acetate (represented by $H_2C^{13}{=}C^{13}H_2$, mass 30) would occur as the product of the abundance of C^{13} in each carbon, or $0.01 \times 0.01 = 0.0001$. Singly labeled acetate, made up equally from $C^{13}H_3 \cdot C^{12}OOH$ and $C^{12}H_3 \cdot C^{13}OOH$ (represented by $H_2C^{13}{=}C^{12}H_2$, mass 29) would occur twice as the product of $0.01 \times 0.99 = 0.0099$. Unlabeled acetate, (represented by $H_2C^{12}{=}C^{12}H_2$, mass 28) would occur as the product $0.99 \times 0.99 = 0.9801$. The percentage of mass 30 relative to mass 28 would be 0.01; the percentage of mass 29 relative to mass 28 would be 2.02. Likewise in case 2 the corresponding percentages would be 0.34 and 34.0; in case 3 they would be 11.1 and 66.7. These would be the percentage abundances for labeled ethylene, if the labeled acetates were derived entirely from the type of synthesis assumed in each of the three cases.

If, for instance, the labeled ethylene were derived from a mixture of A% doubly labeled, B% singly labeled, and C% unlabeled acetates, then each type of acetate would contribute to all these masses and the total amount of each mass would be the sum of the contributions from each type of acetate.

From the experimental observations of the mass ratios found for the various acetylenes derived from the labeled acetate formed, Wood was able to calculate the percentage of each type of acetate formed. The acetate formed during the fermentation of glucose in the presence of $C^{13}O_2$ was found to have the following per cent composition: 26.5 to 31.8, $C^{13}H_3 \cdot C^{13}OOH$; 2.3 to 5.0, $C^{13}H_3 \cdot C^{12}OOH$; 26.7 to 33.1, $C^{12}H_3 \cdot C^{13}OOH$; and 30.1 to 42.0, $C^{12}H_3 \cdot C^{12}OOH$. These data confirmed the earlier conclusions of Barker and Kamen in showing that a synthesis of acetate entirely from CO_2 occurred. They also indicated an appreciable formation of singly labeled acetate by virtue of an exchange between the acetate carboxyl and labeled CO_2, a process not noted under the conditions of the experiments performed by Barker and Kamen.

This series of experiments provides a good illustration of how conclusions inferred from relatively simple experiments can be further tested and established by more refined procedures as the latter become available.

In establishing a precursor-product relation between any two compounds, a complete set of data on the variations with time of the specific activity of the compounds is needed. In addition, if a metabolic chain is postulated, observations must be extended to include such data on all compounds assumed to participate in such a chain. Data on the time variations in specific activity of any one compound provide only collateral evidence that it is involved in the chain. Further, it is desirable to perform properly controlled degradations on the isotopic material to establish the labeling site in all participating molecules, thus testing the validity of postulated mechanisms.

TABLE 8
INCORPORATION OF LABELED CARBON IN GLUCOSE FROM RAT LIVER GLYCOGEN (AFTER WOOD ET AL.[67])

Labeled Compound Fed	% C^{13}	% C^{13} in Carbons: 3,4	2,5	1,6
$NaHC^*O_3$	5.10	0.16	0.00	−0.01
CH_3C^*OOH	2.63	0.14	0.01	0.01
C^*H_3COOH	2.02	0.08	0.18	0.16
$CH_3CH_2C^*OOH$	1.54	0.15	−0.07	−0.01
$CH_3C^*H_2COOH$	1.56	0.07	0.26	0.27
$C^*H_3CH_2COOH$	0.74	0.04	0.17	0.15
$CH_3CH_2CH_2C^*OOH$	0.98	0.13	0.01	0.01
$CH_3CH_2C^*H_2COOH$	0.74	0.05	0.16	0.14
$CH_3C^*H_2CH_2COOH$	1.09	0.16	0.02	0.02

Such extensive and exhaustive work has been performed in few laboratories. (An example will be given on pp. 213–224.)

Data on time variation are not usually so readily available as data on relative incorporation of label at any given time after administration of a labeled substrate.[67, 68] Typical results are shown in Table 8 for the distribution of C^{13} in glucose of rat liver glycogen. As remarked previously in connection with data on uric acid precursors (p. 143) no attention need be given very low values (~ 0.02).

These results indicate distribution of label falling into two categories. The first in which practically all label appears in the 3,4 positions of glucose is characteristic of labeled carbonate, carboxyl-labeled acids, and β-labeled butyric acid. The second in which label appears predominantly in the 2,5 and 1,6 positions arises from α-labeled acetate, α- and β-labeled propionate, and α-labeled butyrate.

In interpreting these results, the qualitative equivalence of carbonate and carboxyl should be noted first. Undoubtedly, this equivalence can be explained most readily by supposing decarboxylation of the acids to occur, thus equilibrating carbonate and carboxyl. On a quantitative basis, however, in relation to respiratory labeled CO_2, much more C^{13} appears in glycogen when labeled carboxyl is fed than when labeled carbonate is fed. It is likely, therefore, that still another mechanism is involved in incorporation of carboxyl carbon.

The results of feeding butyrate and acetate are in good accord with what would be expected with the knowledge that butyrate after oxidation to acetoacetate is cleaved to two molecules of acetate. Carboxyl-labeled bu-

[67] Wood, H. G., Lifson, N., and Lorber, V., *J. Biol. Chem.* **159,** 475 (1945).
[68] Lorber, V., Lifson, N., and Wood, H. G., *J. Biol. Chem.* **161,** 411 (1945).

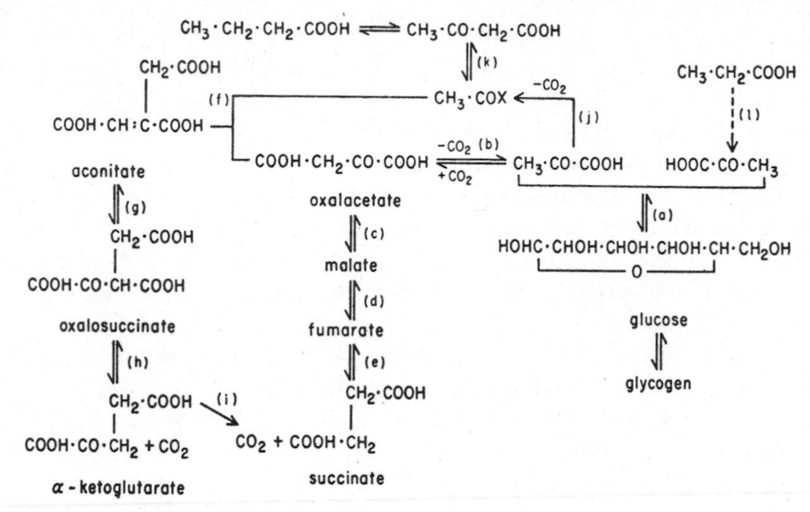

FIG. 42. Abridged tricarboxylic acid cycle and its relation to glycogen synthesis. (After Wood.[68a]) The system citrate-cisaconitate-isocitrate is represented only by aconitate in this scheme.

tyrate, which would be in equilibrium with carboxyl-labeled acetate, gives 3,4 labeling in glucose as does carboxyl-labeled acetate. Likewise β-labeled butyrate gives 3,4-labeling in glucose, because in cleavage it would form carboxyl-labeled acetate. Alpha-labeled butyrate would give methyl-labeled acetate, however, and would result in the appearance of C^{13} in positions other than 3,4.

In relation to these results, it is instructive to follow, in the manner of Wood,[68a] a cyclic series of reactions. In Fig. 42 there is given a simplified scheme for the tricarboxylic acid cycle in relation to glycogen synthesis. Suppose labeled carbonate is fed. One may assume fixation by reversal of reaction *b* so that label enters the carboxyl group β to the keto carbon. Through reversible equilibria with other dicarboxylic acids which are symmetric such as fumarate and succinate, the label appears finally in the carboxyl group α to the keto group of oxalacetate. Decarboxylation results in carboxyl-labeled pyruvate which, undergoing condensation and following reactions characteristic of the reverse of glycolysis, forms glucose with carbons 3 and 4 labeled. It is necessary to suppose that no labeling of pyruvate occurs by any other means.

Conversion of acetate to glycogen in glucose can take place by mechanisms operative in the tricarboxylic acid cycle. If one considers first carboxyl-labeled acetate, then following reactions *f*, *g*, *h*, and *i*, there is ob-

[68a] See Wood, H. G., *in* "Symposium on Use of Isotopes in Biology and Medicine," (H. T. Clarke, ed.), p. 228. Univ. Wisconsin Press, Madison, 1949.

tained carboxyl-labeled succinate; 3,4 labeling of glucose follows as in the above scheme via pyruvate. Methyl-labeled acetate passing through the same reactions would result in α-β-labeled succinate. The glucose resulting from the pathway via pyruvate would give labeling in carbons 1, 2, 5, and 6. Recycling of the central-labeled oxalacetate formed during this conversion through the same reaction chain would result in appearance of carboxyl-labeled succinate, so that eventually 3,4 labeling in glucose would be apparent. The initial concentration of label in 3,4 carbons under these conditions would be less than in the other positions, which is in accord with the experimental findings.

If a mechanism such as direct oxidation of butyrate at the methyl group with direct conversion subsequently to glycogen were invoked, then the same labeling would occur in glucose regardless of whether α- or β-labeled butyrate was fed. This is not in accord with the experimental findings under the particular laboratory conditions employed.

It may be remarked that conversion of butyrate via acetate results in no increase of glycogen, because for each traversal of the cycle by acetate two molecules of CO_2 are lost. Because these are not identical with the carbons of the original acetate, there results an incorporation of label when labeled acetate is fed. The isotopic experiments demonstrate that carbon from fatty acids can be transferred to glycogen, but nothing is deducible merely from these experiments regarding net production of glycogen. The reader should ascertain for himself the results to be expected on the basis of Fig. 42 for a mechanism involving direct conversion of acetate to glycogen via reversal of reaction j, the conversion of various labeled propionates to glycogen, etc.

F. The Method of Isotopic Competition

Roberts and Roberts,[69] at the Carnegie Institution of Washington, observed that cells growing on phosphorylated carbohydrates in the presence of P^{32}-labeled inorganic phosphate incorporated little radioactivity as compared to cells growing in the same medium devoid of phosphorylated carbohydrates. Later, Cowie *et al.*[70] also working in the same laboratories, noted that various sulfur compounds could suppress the incorporation of sulfur from S^{35}-labeled inorganic sulfate. Thus, unlabeled methionine or homocystine prevented the uptake of S^{35} from labeled sulfate into methionine without affecting the appearance of S^{35} in cysteine. On the other hand, cystine prevented the incorporation of S^{35} into both cysteine and methionine.

These observations and many others were elaborated by the group of investigators at the Carnegie Institution to develop a general method for

[69] Roberts, R. B., and Roberts, I. Z., *J. Cellular Comp. Physiol.* **36,** 15 (1950).
[70] Cowie, D. B., Bolton, E. T., and Sands, M., *J. Bacteriol.* **60,** 233 (1950).

establishing synthetic mechanisms in living cells which they called the "method of isotopic competition." The basic postulate is that any compound which can function as a metabolic intermediate in a synthetic pathway can be used directly by the living cell rather than synthesized *de novo* by the cell from endogenous materials. If a metabolic sequence is represented as a series of compounds, $A \rightarrow B \rightarrow C \rightarrow D$, etc., then if A is labeled and metabolized to D through B and C, addition of unlabeled B or C to the medium will suppress labeling of D because the cell will use unlabeled B or C, thus diluting out label coming from A.

In principle, the design of experiments exploiting this method is simple. Two cultures are prepared in which cells growing in the log phase metabolize a labeled compound. To one of these a compound thought to be an intermediate is added. After a period of growth the cells are chemically fractionated; the distribution of label in each culture is determined and compared. Cells growing in the log phase are used in order to study synthetic pathways during actual growth under physiological conditions. Test organisms which grow in relatively simple, well-defined media are preferred.

As an example, when the uptake of labeled sulfur from S^{35}-labeled sulfate by growing cultures of *E. coli* is studied, it is shown that addition of any one of a number of sulfur compounds which permit growth drastically reduces appearance of label in cellular material. In Table 9 there are reproduced the results of such an experiment.

Similar experiments with homocystine, homocysteine, and methionine revealed only partial suppression of sulfate sulfur uptake. Some compounds, including L-cystathionine, L-allocystothionine, djenkolic acid, cysteic acid, taurine, methionine sulfoxide, methionine sulfone, and thiourea, fail completely to affect labeled sulfur uptake from sulfate. The failure of cystathionine to compete when *E. coli* in the test system is particularly interesting

TABLE 9

ELIMINATION OF $S^{35}O_4$ UPTAKE BY NONRADIOACTIVE COMPETITORS (AFTER COWIE ET AL.[70])

Nonradioactive Supplement*	Uptake of S^{35} (% of Control)
None†	100
Na_2SO_3	1.0
Na_2S	1.0
L-Cystine	0.3
DL-Lanthionine	0.3
DL-S-methylcysteine	0.3

* S concentration of unlabeled compound 0.028 mg./ml. medium.
† $S^{35}O_4^{--}$ as S, concentration 0.026 mg./ml. medium.

because cystathionine is known to be a nutritional adjunct in sulfur metabolism in rats,[71] in *Neurospora crassa*,[72] and in *Torulopsis utilis*. Indeed, when isotopic competition studies are made with these systems, cystathionine is found to be an effective competitor.

There are some obvious limitations on interpretation of results obtained by the method of isotopic competition. When the competitor compound is highly unstable, for instance, it is not always possible to provide the needed combination of labeled and unlabeled material. There is also the possibility that the compound as added does not compete in itself, but rather affects the metabolism of the organism indirectly, producing the appearance, but not the fact, of competition. The actual competitor may be a derivative of the compound added, or it may be a totally different compound produced by the cell in response to the addition of the presumptive competitor because a new synthetic pathway is activated. Competition may not be observed because the added compound may fail actually to enter the cell or to mix with the like compound already present in the endogenous pool.

Despite these limitations, the method works very well in studying a surprisingly large number of biosynthetic mechanisms and can be considered a useful procedure, particularly in the study of mutants, nutritional requirements, isolation of labeled intermediates from purified enzyme systems, and so on.

G. Ogston's Hypothesis

The tracer method has provided clear evidence that symmetrical compounds can be intermediates in the enzyme-catalyzed formation of substances which are labeled asymmetrically. In early experiments with labeled carbonate, administered together with unlabeled pyruvate, it was found that label appeared in ketoglutarate only in the carboxyl group α to the keto carbon.[73] It was supposed that a symmetrical molecule like citrate could not be in the pathway from pyruvate, because label would be expected to appear with equal probability in both carboxyls of ketoglutarate. In 1948, however, Ogston,[74] in a penetrating note, provided arguments against the validity of such considerations. His reasoning can be represented briefly by taking up the case of aminomalonic acid as an intermediate in interconversion of glycine and serine. The steric considerations involved are shown in Fig. 43, according to Ogston's notation.

It has been found that doubly labeled serine (N^{15}-labeled amino, C^{13}-

[71] du Vigneaud, V., Brown, G. B., and Chandler, J. P., *J. Biol. Chem.* **143,** 59 (1942).

[72] Horowitz, N. H., *J. Biol. Chem.* **171,** 255 (1947).

[73] Wood, H. G., Werkman, C. H., Hemingway, A., and Nier, A. O., *J. Biol. Chem.* **139,** 483 (1941).

[74] Ogston, A. G., *Nature* **162,** 963 (1948).

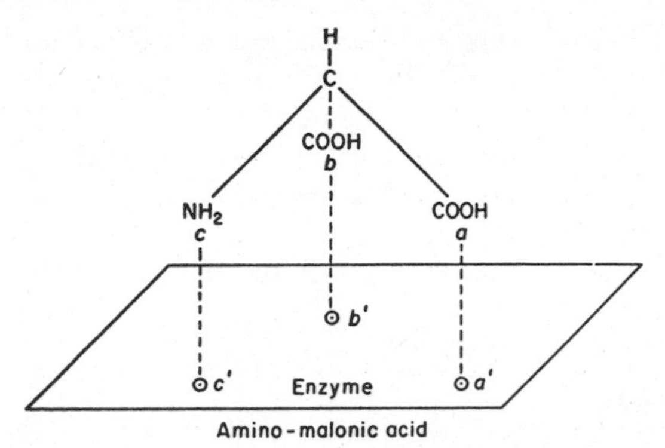

FIG. 43. Placement of aminomalonic acid on enzyme, as postulated by Ogston.[74]

labeled carboxyl) gives rise to glycine with a ratio of N^{15} to C^{13} identical with that of the original serine.[75] Here one could argue also that lack of ability to distinguish carboxyl arising by oxidation of the carbinol carbon from the original carboxyl would result in loss of label from carboxyl if a symmetrical intermediate like aminomalonic acid were formed, so that the N^{15} to C^{13} ratio would change. Suppose, however, that the enzyme can distinguish two identical groups of a symmetrical product arising from one optical form of the metabolite, L-serine. Thus, as in Fig. 43, the aminomalonic acid may be able to come into combination with the enzyme only when the stereochemical relations allow *a*, *b*, and *c* to contact points *a*′, *b*′, and *c*′ on the enzyme surface. Suppose further that decarboxylation can occur only at *a*′ or *b*′, but not at both sites. Restriction of decarboxylation to *b*′ would always result in retention of the original labeled carboxyl. It need only be assumed, then, that the sites *a*′ and *b*′ differ in catalytic properties and that a three-point combination occurs between substrate and enzyme. Both assumptions are quite likely to be valid, so that asymmetry in a product is not conclusive evidence against a symmetrical precursor. Thus, citric acid could be an intermediate in the tricarboxylic acid cycle, and aminomalonic acid could be an intermediate in the conversion of serine to glycine.

Ogston's hypothesis, then, is that an *asymmetric enzyme can distinguish between the identical groups of a symmetric compound*. Proof that this hypothesis is valid depends on tracer experiments. The experiments to test it were performed by Potter and Heidelberger[76] and by Lorber *et al.*[77] It was found[76] that, when pyruvate was oxidized by rat liver homogenate in the

[75] Shemin, D., *J. Biol. Chem.* **162,** 297 (1946).

[76] Potter, V. R., and Heidelberger, C., *Nature* **164,** 180 (1949).

[77] Lorber, V., Utter, M. F., Rudney, H., and Cook, M., *J. Biol. Chem.* **185,** 689 (1950).

presence of unlabeled oxalacetate, malonate, and C^{14}-labeled CO_2, labeled citrate was formed. This labeled citrate was added to a second homogenate which metabolized the citrate to ketoglutarate which was isolated as the 2,4-dinitrophenylhydrazine derivative. This was oxidized with permanganate to CO_2 and succinate. All the radioactivity was found in the CO_2. In this oxidation the CO_2 was derived from the α-carboxyl carbon of the ketoglutarate. The reactions involved in this demonstration were:

$$C^*O_2 + \text{pyruvate} + \text{oxalacetate} \xrightarrow[+ \text{ malonate}]{\text{homogenate}}$$

```
(a) COOH
     |
    CH2
     |
     |    COOH                      (a) COOH                  (a) COOH
     |   /                               |                         |
     C            enzyme                CH2        KMnO4          CH2
     |   \       -------->               |       -------->         |
     |    OH                            CH2                       CH2
     |                                   |                         |
    CH2                                 C=O                       COOH
     |                                   |                         +
(b) C*OOH                           (b) C*OOH                 (b) C*O2
```

In another series of experiments, Wilcox *et al.*[78] chemically synthesized citric acid with label in carbon of carboxyl *a* and found that when this compound was incubated with the same enzyme preparations to form ketoglutarate, the resultant CO_2 on oxidation with permanganate was unlabeled, all the radioactivity being found in the succinate.

Lorber *et al.*[77] began with oxalacetate labeled in the β-carboxyl carbon. Radioactive citrate isolated from one enzyme experiment and added to a second preparation also gave ketoglutarate labeled exclusively in the α-carboxyl carbon.

The mechanism for synthesis of asymmetrically labeled citrate from pyruvate was elaborated shortly thereafter when it was shown that the active agent in condensation with oxalacetate was acetyl-coenzyme A. The purified enzyme catalyzing this citrate synthesis was isolated from *E. coli*.[79]

Ogston's hypothesis clarified some results obtained when racemic dideutero-citrate labeled in the methylene carbons was incubated with pigeon breast muscle preparations in the presence of arsenite to accumulate α-keto-

[78] Wilcox, P. E., Heidelberger, C., and Potter, V. R., *J. Am. Chem. Soc.* **72,** 5019 (1950).

[79] Stern, J. R., Shapiro, B., Stadtman, E. R., and Ochoa, S., *J. Biol. Chem.* **193,** 703 (1951).

glutarate. It had been observed that the keto acid isolated had lost half the deuterium present originally in the labeled citrate. This result was difficult to understand prior to Ogston's suggestion, but with the new insight afforded, Martius and Schorre[80] performed experiments using resolved (+) and (−) forms of the labeled citrate and showed that when the (−) form was degraded to the keto acid all the deuterium was retained, whereas when the (+) form was so treated the product keto acid lost all its label. As expected, the racemic mixture lost half its deuterium. Again, it followed that the enzyme attached itself by three centers to the substrate; otherwise it could not have distinguished between the (+) and (−) acids.

Later demonstrations of the working of the Ogston hypothesis are found in the studies of Schambye *et al.*[81] and of Swick and Nakao.[82] These workers showed that glycerol biologically synthesized from either carboxyl-labeled acetate or glucose-3,4-C^{14} gave rise to glycogen in rat liver containing C^{14} mostly in carbons 3 and 4 of the glucose moiety.

H. Reversibility of Biochemical Equilibria

An important phase of biochemical research concerns the investigation of isolated reaction systems *in vitro*. Most enzymatic studies on single reactions studied *in vitro* are carried out with systems in which the reaction proceeds predominantly in one direction. Thus, in peptide syntheses using component reagents (amino acids), the equilibrium lies far in the direction of dissociation. Some drastic (usually unphysiologic) means of removing product peptides is required to displace equilibria sufficiently to bring about observable utilization of the reagents.

However, it is simple in principle to demonstrate reversibility of degradation reactions by employing labeled reagents (in this case, labeled amino acids) because, even if there is a net decrease of peptide or protein during the course of the reaction, labeled peptide will be formed if the reaction is at all reversible. The appearance of labeled peptide constitutes positive evidence for reversibility, provided, of course, that the proper control chemistry is done to obviate the possibility that the isotopic content of the products as isolated is not due to contamination by mere absorption or reactions other than direct peptide formation. Such an approach has been used in demonstrating protein synthesis *in vitro* with methionine labeled with S^{35}, the test system being rat liver homogenates.[83] An example of spurious results obtained when labeled cystine was employed is also available in the same researches (see p. 359).

[80] Martius, C., and Schorre, G., *Ann.* **570,** 143 (1950).
[81] Schambye, P., Wood, H. G., and Popják, G., *J. Biol. Chem.* **206,** 875 (1954).
[82] Swick, R. W., and Nakao, A., *J. Biol. Chem.* **206,** 883 (1954).
[83] Melchior, J. B., and Tarver, H., *Arch. Biochem.* **12,** 301 (1947).

The direct demonstration of *in vitro* formation of protein peptide bonds, based on incorporation of labeled amino acids into protein or peptide residues, is complicated by the existence of numerous equilibria other than peptide bond formation in the amino acid chain. Carboxyl-labeled amino acids may be decarboxylated and the resultant labeled carbonate incorporated into free carboxyl residues, peptides may be formed on side-chain residues, and labeled compounds may be adsorbed strongly enough to escape removal by precipitation and washing procedures. Thus exhaustive isolation and degradation procedures are required in many instances to establish actual incorporation of any given amino acid into a protein moiety.

A number of reports have appeared which seem to establish the reversibility of protein breakdown.[84, 85] Criteria used to judge true incorporation of carboxyl-labeled amino acid into protein are: (1) unhydrolyzed protein does not yield labeled CO_2 on heating with ninhydrin solution but does yield practically all labeled carbon as CO_2 on treatment of the protein hydrolyzate with ninhydrin; (2) partial hydrolysis with proteolytic enzymes results in liberation of little labeled CO_2 from hydrolyzates with ninhydrin; (3) a major portion of labeled carbon is recovered still in original labeled acids after hydrolysis of protein incubated with these acids.

The study of peptide bond synthesis is more easily and definitely approached by using as a test system a simple natural peptide such as glutathione, a tripeptide of glutamic acid, cysteine, and glycine. A typical experiment by Bloch with this system will be described.[86]

A pigeon liver homogenate was incubated for 1 hr. at 37°C. with a mixture of the following composition: phosphate buffer of *p*H 7.4, 0.05 *M*; KCl, 0.03 *M*; $MgSO_4$, 0.0024 *M*; glutamic acid, 0.01 *M*; cysteine, 0.003 *M*; 1-C^{14}-glycine, 0.016 *M*; and 25 mg. of glutathione. After incubation, the reaction mixture was deproteinized with trichloroacetic acid and the glutathione was separated first as the cadmium salt and then as cuprous mercaptide. Several reprecipitations were made to minimize the possibility that the glutathione C^{14} content was spurious. It was shown that nearly all the C^{14} in the glutathione could be recovered after acid hydrolysis in the glycine residue with some appearing in the cysteine or glutamic acid. In Table 10 are given the results of a similar experiment in which C^{14}-glycine and N^{15}-glutamic acid were used in equimolar amounts.

One feature of this table should be most emphatically emphasized. Results are given as *percentage of incorporation* of original labeled substrate. Thus, the C^{14} data are presented as specific activity of glycine moieties

[84] Greenberg, D. M., Friedberg, F., Schulman, M. P., and Winnick, T., *Cold Spring Harbor Symposia Quant. Biol.* **13,** 113 (1948).

[85] Frantz, I. D., Jr., Loftfield, R. B., and Miller, W. W., *Science* **106,** 544 (1947).

[86] Bloch, K., *J. Biol. Chem.* **179,** 1245 (1949).

TABLE 10

FORMATION OF GLUTATHIONE IN PIGEON LIVER HOMOGENATES (AFTER BLOCH[86])

Isotopic Additions	Relative Isotope Concentrations in Glutathione	
	C^{14} (relative specific activity)	N^{15} (relative atom % excess)
C^{14}-Glycine and N^{15}-glutamic acid	1.40	0.60
	1.34	0.43
C^{14}-Glycine and $N^{15}NH_4Cl$	1.76	0.48

calculated for a specific activity of 100 in added glycine. Likewise the atom per cent excess N^{15} is calculated for the glutamic acid moiety on the basis of 100 atom per cent excess N^{15} in added glutamic acid. In reporting tracer experiments of this type, the most important datum is the relative uptake. If this is very low, the results may be regarded with suspicion. Nevertheless many experimenters fail to report data from which relative uptake may be easily estimated. The mere presentation of counts per minute in a sample of peptide without specifying either the weight of the aliquot determined, total weight of sample, and initial specific activity, or the atom per cent excess of labeled substrate, is totally inadequate.

The data of Table 10 indicate that uptake of glycine carbon is more than twice that of glutamic nitrogen. This arises from the use of DL-glutamic acid of which only half (the L form) is available for synthesis. Also the action of glutamic acid dehydrogenase with resultant loss of labeled nitrogen would act to lower the measured incorporation of N^{15}–glutamic acid. Reversible deamination is shown in the experiment with N^{15}–ammonium chloride. The N^{15} content of glutamic acid from hydrolyzed glutathione after incubation with N^{15}–NH_4Cl is found to be roughly equal to the N^{15} content of glutathione when N^{15}-glutamic acid is used. An alternative explanation is that reversible amination of the glutamyl residue with labeled NH_3 occurred in the glutathione molecule. It may be remarked that incorporation of amino acid into the peptide requires aerobic oxidation or the presence of an equivalent energy source such as adenosine triphosphate. This has been found to be true in a general way for practically all *in vitro* syntheses.

A number of major contributions to our knowledge of mechanisms of enzymatic synthesis have come from studies on the reversibility of decarboxylation reactions. In the fixation of CO_2 by both liver and bacterial extracts, it has been established that, during decarboxylation of oxalacetate to pyruvate and CO_2 in the presence of adenosine triphosphate and labeled bicarbonate, the oxalacetate becomes labeled in the carboxyl β to the ketone carbon.[87, 88] In a similar way it has been shown that the "phosphoroclastic" split of pyruvate to acetyl phosphate and formate is reversible; labeled formate was used, and appearance of label in the carboxyl of pyruvate was observed, viz.,[89]

$$CH_3COC^*OOH + H_3PO_4 \rightleftharpoons CH_3CO(OPO_3H_2) + HC^*OOH \quad (15)$$

However, a reaction of this type cannot be interpreted as a simple equilibrium even though it is so written. It must be remembered that the enzymatic catalysis of an over-all reaction such as reaction 15 may involve unsuspected pathways which differ in the forward and backward processes. With only one labeled component it may appear that a reaction is simply reversible.

It has been found, in both instances cited, that simple reversibility is not involved. Thus, in the case of the oxalacetate decarboxylation, it has been observed[90] that when incorporation of C^{13}-carbonyl-labeled pyruvate into oxalacetate is compared with simultaneous incorporation of C^{14}-labeled carbonate there is little correlation in the two rates of fixation. Some enzyme preparations fix CO_2 rapidly but are inactive with pyruvate. Furthermore, adenosine triphosphate accelerates CO_2 fixation but not pyruvate fixation into oxalacetate. In the phosphoroclastic reaction it has been shown similarly that labeled acetyl phosphate is not a component in the synthesis of pyruvate from formate.[91] These researches show the importance of multiple labeling to avoid drawing premature conclusions from a study of a single component.

I. Mechanism of Enzyme Action

A variety of researches on mechanisms involved in the action of enzymes, particularly oxidases, phosphatases, phosphorylases, and transferases, have been made possible by the application of tracer methods. There may be cited studies on phosphatatic cleavage of P–O bonds in adenosine triphosphate,[92] fission of glycosidic linkages by invertase,[93] and action of notatin[94] and uricase.[95] An example of the action of a transferase is afforded in work on the enzyme sucrose phosphorylase isolated from the bacterium

[87] Krampitz, L. O., Wood, H. G., and Werkman, C. H., *J. Biol. Chem.* **147**, 243 (1943)

[88] Utter, M. F., and Wood, H. G., *J. Biol. Chem.* **160**, 375 (1945).

[89] Utter, M. F., Lipmann, F., and Werkman, C. H., *J. Biol. Chem.* **158**, 521 (1945).

[90] Utter, M. F., and Chenoweth, M. T., *Federation Proc.* **8**, 261 (1949).

[91] Strecker, H., and Wood, H. G., *Federation Proc.* **8**, 257 (1949).

[92] Cohn, M., *J. Biol. Chem.* **180**, 771 (1949).

[93] Koshland, D. E., Jr., and Stein, S. S., *J. Biol. Chem.* **208**, 139 (1954).

[94] Bentley, R., and Neuberger, A., *Biochem. J.* **45**, 584 (1949).

[95] Bentley, R., and Neuberger, A., *Biochem. J.* **52**, 694 (1952).

TABLE 11
EXCHANGE OF P^{32} BETWEEN INORGANIC PHOSPHATE AND GLUCOSE-1-PHOSPHATE (AFTER DOUDOROFF ET AL.[96])

Experiment	Reaction Mixture	Radioactivity (ct./min./μM.)	
		Inorganic P	Glucose-1-P
1	0.1 *M* glucose-1-P + 0.033 *M* labeled inorganic P	98 (±40)	0 (±1)
2	Same as (1) but with enzyme	859 (±40)	119 (±3)
3	Same as (2) but with 0.06 *M* fructose	886 (±40)	99 (±3)
4	Same as (2) but with 0.12 *M* glucose	1096 (±40)	7 (±1)

Pseudomonas saccharophila.[96] This enzyme catalyzes the reversible reaction between glucose-1-phosphate and certain ketoses which results in synthesis of sucrose and various disaccharides. When a preparation of the enzyme is incubated for 1 hr. at 30°C. with glucose-1-phosphate and P^{32}-labeled inorganic phosphate, P^{32} is found in the glucose phosphate, whereas no exchange occurs in the absence of enzyme. Parallel experiments in the presence of fructose or glucose show an inhibition of the exchange between the phosphate ester and free phosphate particularly in the case of glucose. Table 11 is a summary of data from a typical set of exchange experiments.

It appears that the enzyme can liberate inorganic phosphate from the ester without producing an equivalent quantity of glucose. One may postulate the following reaction:

$$\text{Glucose-1-phosphate} + \text{enzyme} \rightleftharpoons \text{Glucose-enzyme} + \text{phosphate} \quad (16)$$

Thus, a glycosidic linkage is substituted for a phosphate ester bond. The enzyme is essentially a transglucosidase rather than a phosphorylase. This has been borne out by observations that phosphate-free enzyme preparations can catalyze the synthesis of the sucrose analog glucosidosorboside from sucrose and sorbose, viz.,

$$\text{Glucose-1-fructoside} + \text{sorbose} \underset{\text{enzyme}}{\rightleftharpoons} \text{Glucose-1-sorboside} + \text{fructose} \quad (17)$$

Likewise, sucrose can be synthesized from its synthetic analog, glucosidoketoxyloside, and fructose, viz.,

$$\text{Glucose-1-ketoxyloside} + \text{fructose} \underset{\text{enzyme}}{\rightleftharpoons} \text{Glucose-1-fructoside} + \text{ketoxyloside} \quad (18)$$

The inhibition of the phosphate exchange by free fructose or glucose mirrors competition of the sugars with phosphate as acceptors of the glucose

[96] Doudoroff, M., Barker, H. A., and Hassid, W. Z., *J. Biol. Chem.* **168,** 725 (1947).

moiety from the glucose-enzyme complex. It does not follow that the more well-known muscle phosphorylase is capable of accomplishing polysaccharide formation by a similar transglucosidase action. It has been shown that muscle-phosphorylase preparations are incapable of effecting exchange between glucose phosphate and free phosphate.[97]

As examples of researches on phosphotransferases there can be cited studies on the interconversion of α-glycerophosphoric acid and β-glycerophosphoric acid, and of glucose-1-phosphate and glucose-6-phosphate.

In acid medium it had been shown[98, 99] that β-glycerophosphoric acid (I) was converted to α-glycerophosphoric acid (II), i.e.,

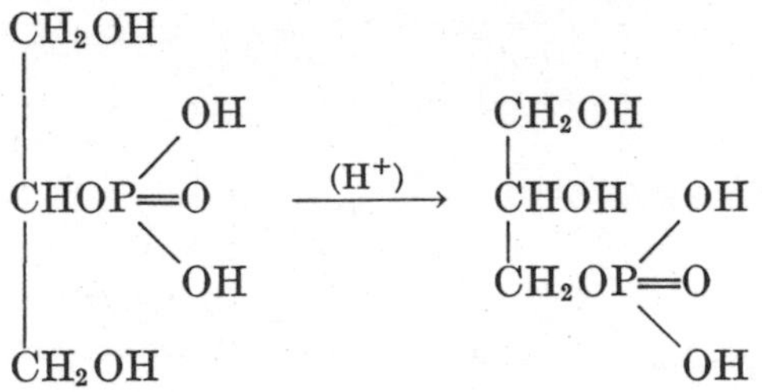

$$\underset{\text{(I)}}{CH_2OH-CH(OP(=O)(OH)_2)-CH_2OH} \xrightarrow{(H^+)} \underset{\text{(II)}}{CH_2OH-CHOH-CH_2OP(=O)(OH)_2}$$

Folch[100] found that this isomerization took place under conditions of hydrolysis of phospholipids. Courtois[101] showed somewhat earlier that this phosphate group migration could be accomplished enzymatically. It could be inquired whether, during the isomerization, the phosphate group is labilized so that it leaves the molecule and comes into equilibrium with inorganic phosphate (if such is present), or whether it is held in the molecule, migrating through the intermediary formation of a cyclic diester (III) as suggested by Verkade *et al.*[102] i.e.,

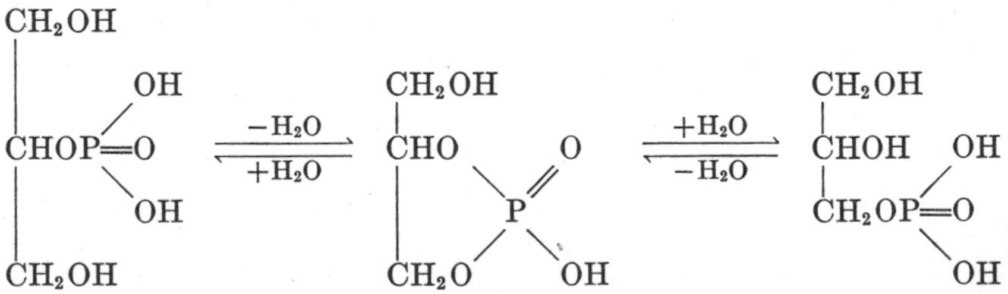

$$\underset{\text{(I)}}{CH_2OH-CH(OP(=O)(OH)_2)-CH_2OH} \underset{+H_2O}{\overset{-H_2O}{\rightleftharpoons}} \underset{\text{(III)}}{CH_2OH-\overset{|}{CHO}-CH_2O-P(=O)(OH)\ \text{(cyclic)}} \underset{-H_2O}{\overset{+H_2O}{\rightleftharpoons}} \underset{\text{(II)}}{CH_2OH-CHOH-CH_2OP(=O)(OH)_2}$$

[97] Cohn, M., and Cori, G. T., *J. Biol. Chem.* **175,** 89 (1948).
[98] Bailly, M. C., *Compt. rend.* **206,** 1902 (1938).
[99] Bailly, M. C., *Compt. rend.* **208,** 443 (1939).
[100] Folch, J., *J. Biol. Chem.* **146,** 31 (1942).
[101] Courtois, J., *Bull. soc. chim. biol.* **20,** 1393 (1938).
[102] Verkade, P. E., Stoppelenburg, J. C., and Cohen, W. D., *Rec. trav. chim.* **59,** 886 (1940).

Chargaff[103] tested this scheme by carrying out the isomerization reaction, both in acid medium and enzymatically, in the presence of labeled inorganic phosphate. For the study on acid-influenced migration, 5 g. of crystalline sodium β-glycerophosphate (5½ moles of water of crystallization) was dissolved in 50 ml. of water containing 0.8 mg. of sodium phosphate with a P^{32} activity of 4 mc. To this solution was added 5 ml. of sulfuric acid. After refluxing for ½ hr., the mixture was treated according to the methods of Verkade *et al.*[102] The crude salt obtained was dissolved in 150 ml. of water, 20 ml. of 10% phosphoric acid carrier was added, and the mixture was made alkaline with barium hydroxide. The precipitate was filtered off through Celite, and residual radioactive phosphate was removed from the filtrate by two further dilutions with carrier phosphate and precipitations. The filtrate was freed of excess barium hydroxide with CO_2. The filtrate was then concentrated *in vacuo* to 70 ml., and barium glycerophosphate was precipitated by addition of 270 ml. of absolute alcohol. The salt showed practically no P^{32} present. The α-glycerophosphate was isolated by the method of Fischer and Pfahler[104] and found to be inactive. It could be concluded that the intramolecular mechanism of Verkade *et al.*[102] was operative, because the alternative mode of migration involving labilization of phosphate would have resulted in equilibration with labeled phosphate of the medium, and introduction of activity into the glycerophosphate formed.

In a similar experiment, carried out with pig kidney phosphatase, it was found that no P^{32}-labeled glycerophosphate was formed during the incomplete hydrolysis of the ester phosphate in the presence of glycerol and labeled inorganic phosphate. It could be concluded that, in these phosphate transfers, no equilibration with inorganic phosphate, resulting from labilization, occurred.

A similar mechanism has been proposed for the interconversion of glucose-1-phosphate to glucose-6-phosphate in the presence of phosphoglucomutase on the basis of experiments employing C^{14}-labeled glucose.[105] However, it has been shown[106] that catalytic amounts of glucose-1,6-diphosphate are involved in the mechanism of action of purified phosphoglucomutase. In experiments[107] with C^{14}- and P^{32}-labeled glucose-1-phosphate incubated with unlabeled glucose-1,6-diphosphate and enzyme, even distribution of label among the three molecules involved has been observed, confirming the participation of glucose-1-6-diphosphate in the mutase reaction.

The mechanisms involved in enzyme-catalyzed transfer of phosphate from various acyl phosphates to hydroxyl compounds have been studied

[103] Chargaff, E., *J. Biol. Chem.* **144,** 455 (1942).
[104] Fischer, E., and Pfahler, E., *Ber.* **53,** 1606 (1920).
[105] Schlamowitz, M., and Greenberg, D. M., *J. Biol. Chem.* **171,** 293 (1947).
[106] Sutherland, E., Posternak, T., and Cori, C. F., *Federation Proc.* **8,** 258 (1949).
[107] Sutherland, E., Posternak, T., and Cori, C. F., *J. Biol. Chem.* **179,** 501 (1949).

with labeled phosphate. The observation that enzyme in addition to inorganic phosphate fails to transfer phosphate is indicative that phosphate transfer involves direct interaction between donor and acceptor rather than transfer via dissociation to inorganic phosphate.[108] When P^{32}-labeled nitrophenylphosphate is used in the presence of purified citrus phosphatase (a phosphoferase enzyme) with methyl alcohol as acceptor, it is found[109] that the specific P^{32} activity of the residual nitrophenyl-phosphate is identical with that of the methylphosphate. Thus no dilution with inorganic phosphate occurs. In another experiment with labeled inorganic phosphate and unlabeled nitrophenylphosphate, little incorporation of labeled phosphate into methylphosphate can be demonstrated. These results confirm the hypothesis of direct transfer of phosphate from donor to acceptor.

An interesting application of tracer technique to the problem of hydrogen transfer has been reported by Kaplan *et al.*,[110] who have studied the interconversion of triphosphopyridine nucleotide (TPN) and diphosphopyridine nucleotide (DPN) brought about by enzyme preparations from *Pseudomonas fluorescens*. The reaction may be written

$$\text{TPNH} + \text{DPN} \rightleftharpoons \text{TPN} + \text{DPNH}$$

by substituting deamino-DPN for DPN, according to the reaction

$$\text{TPNH} + \text{deamino-DPN} \rightleftharpoons \text{TPN} + \text{deamino-DPNH}$$

it was shown[111] that a phosphate transfer could not be involved because, if it were, the products of the reaction would have been DPNH and deamino-TPN. Another possibility was that there was exchange of the two nicotinamide mononucleotide moieties. To probe this possibility, C^{14}-nicotinamide labeled in the $CONH_2$ group was made by the exchange reaction between DPN and labeled nicotinamide catalyzed by beef spleen DNAase. When reduced TPN and labeled DPN were incubated with the *Pseudomonas* preparation, labeled-reduced DPN was formed according to the equation[110]

$$\text{N}_{\text{red}}\text{RPP}\underset{\text{P}}{\text{R}}\text{A} + \text{N}^*_{\text{oxid}}\text{-RPPRA} \rightleftharpoons \text{N}_{\text{oxid}}\text{RPP}\underset{\text{P}}{\text{R}}\text{A} + \text{N}^*_{\text{red}}\text{RPPRA}$$

The symbols indicate the structural components of the nucleotides, e.g., N ≡ nicotinamide moiety, R ≡ ribose moiety, P ≡ phosphate, A ≡ adenyl moiety.

[108] Axelrod, B., *J. Biol. Chem.* **172,** 1 (1948).

[109] Axelrod, B., *J. Biol. Chem.* **176,** 295 (1948).

[110] Kaplan, N. O., Colowick, S. P., Zatman, L. J., and Ciotti, M. M., *J. Biol. Chem.* **205,** 31 (1953).

[111] Kaplan, N. O., Colowick, S. P., and Neufeld, E. F., *J. Biol. Chem.* **195,** 107 (1952).

The method of analysis exploited the fact that the *Neurospora* DNAase attacked only the oxidized forms of DPN and TPN.[112] By incubating the reaction mixtures with this DNAase from *Neurospora*, only the portion of the labeled nicotinamide associated with the oxidized nucleotides was liberated.

It was also possible to show that there was exchange of label between oxidized and reduced forms of DPN. Thus, the mechanism of the enzyme action was shown to be that of a transhydrogenase.

As a final example of research on enzyme action, there will be considered briefly the important studies by Westheimer and his colleagues[113-115] on the stereospecificity of hydrogen transfer by alcohol dehydrogenase and lactic acid dehydrogenase. Negelein and Wulff[116] had assumed that a compound between enzyme and substrate was formed. This conclusion was strengthened when spectrophotometric measurements made in studies by Theorell and Bonnichsen[117] and by Theorell and Chance[118] indicated that alcohol dehydrogenase formed a compound with DPN which exhibited an absorption spectrum different from that of unbound DPN. Westheimer *et al.*[113] equilibrated DPN with CH_3CD_2OH in the presence of yeast alcohol dehydrogenase and found that the reduced DPN formed contained one nonexchangeable deuterium atom per molecule. The acetaldehyde formed also had one nonexchangeable deuterium atom per molecule. Repetition of the experiment with unlabeled alcohol and DPN in the presence of heavy water resulted in the formation of DPN with no nonexchangeable deuterium. From these observations and others, it could be concluded that in the enzymic reduction hydrogen was transferred directly to the DPN from the alcohol, and that the reduction was stereospecific with respect to the reduced position of the dihydropyridine ring.

Similar results were obtained when lactate was isolated from a reaction mixture containing monodeutero-reduced DPN (prepared by enzymic reduction with 1,1-dideuteroethanol), pyruvate, and lactic dehydrogenase.[115] The lactate contained one atom of deuterium. It was shown that chemically reduced DPN, after enzymic reoxidation with acetaldehyde, contained 0.44 atom of deuterium per molecule.[113, 114] This fact indicated that the chem-

[112] Kaplan, N. O., Colowick, S. P., and Nason, A., *J. Biol. Chem.* **193,** 497 (1951).

[113] Westheimer, F. H., Fisher, H. F., Conn, E. E., and Vennesland, B., *J. Am. Chem. Soc.* **73,** 2403 (1951).

[114] Fisher, H. F., Conn, E. E., Vennesland, B., and Westheimer, F. H., *J. Biol. Chem.* **202,** 687 (1953).

[115] Loewus, F. A., Ofner, P., Fisher, H. F., Westheimer, F. H., and Vennesland, B., *J. Biol. Chem.* **202,** 699 (1953).

[116] Negelein, E., and Wulff, H. J., *Biochem. Z.* **284,** 289 (1936).

[117] Theorell, H., and Bonnichsen, R., *Acta Chem. Scand.* **5,** 1105 (1951).

[118] Theorell, H., and Chance, B., *Acta Chem. Scand.* **5,** 1127 (1951).

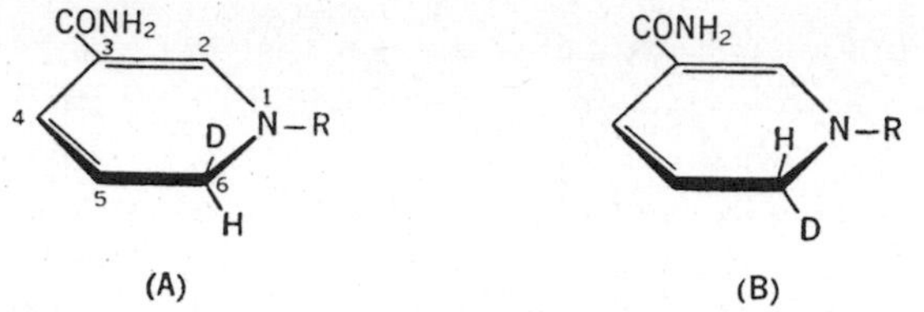

FIG. 44. Diamers of reduced DPN. (After Loewus *et al.*[115])

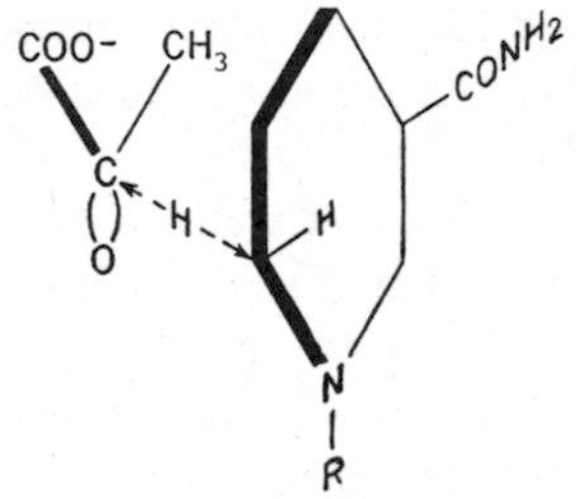

FIG. 45. Proposed mechanism for stereospecific reduction of DPN by pyruvate in presence of lactic dehydrogenase. (After Loewus *et al.*[115])

ically reduced monodeutero-DPN was a mixture of two diamers in the ratio 56:44, differing only in the position of the deuterium, as shown in Fig. 44. When this chemically reduced DPN was used in the experiment with lactic dehydrogenase to reduce pyruvate, the same percentage of deuterium was transferred. This indicated that both the alcohol dehydrogenase and the lactic dehydrogenase had the same stereospecificity toward DPN. It was clear that the enzymes could catalyze addition or removal of hydrogen only from one side of the pyridine ring, as indicated in Fig. 45. These results provided a dramatic confirmation of the suggestion from the early nonisotopic experiments that an enzyme-substrate complex was formed during enzymatic catalysis.

J. ANALYSIS BY ISOTOPE DILUTION

An aspect of quantitative tracer methodology which exhibits increasing importance is concerned with application to analytical problems in biochemistry. The principles involved may be understood best by reference to a typical problem—the analysis of an amino acid mixture obtained by hydrolysis of a protein. Any such mixture may contain as many as twenty of these acids in varying proportions, and each must be detected, separated quantitatively, and assayed. This requires specific reagents which are quantitative for each acid. Such reagents do not exist. It is necessary to isolate each component acid quantitatively in a pure state. Yet, in most analyses, purity and quantitatively complete isolation are mutually contradictory requirements.

The tracer method supplies an answer to this dilemma in the following manner. The amino acid to be assayed is synthesized using an appropriate labeling isotope. A given quantity of the labeled amino acid is then added to the unknown mixture. Any of this particular amino acid component present in the mixture will act as unlabeled diluent for the labeled amino acid. The dilution of labeled amino acid depends on the amount of unlabeled acid present. To detect the amount of dilution requires only that a portion of the added amino acid be recovered in a pure state. Hence, losses in the separation procedure are permissible.

A further extension to the determination of racemization of amino acids is also feasible because, by addition of a DL mixture of labeled carrier, it is possible to assay the amino acid content of both L and D forms. In amino acids derived from natural protein the L form is found almost exclusively, so that a correction factor of 2 is required in general if the carrier used is a racemic mixture of the two isomeric forms.

The error inherent in the method can be reduced below 1%, which is excellent precision for such determinations. The method differs from that described for detection of intermediates in metabolism in that in analysis by isotope dilution it is the carrier (added material) which is labeled. The method is applicable to all types of mixtures and has been developed as a useful analytical tool in particular by Rittenberg, Foster, and their associates, using the stable isotopes of hydrogen, carbon, and nitrogen.[119]

The relation between added and recovered carrier is particularly simple when assay involves a compound labeled with an element possessing two isotopes present in varying amounts in two batches of compound and where, as is usually the case, one isotope is present in normal abundance. If X_2 represents the grams of unknown (unlabeled) compound, X_1 the grams of labeled compound added, C_2 the isotopic content (atom per cent excess or specific radioactivity) of the final product, C_1 the isotopic content of added component, M_1 the molecular weight of added component, and M_2 the normal molecular weight, then

$$X_2 = \left(\frac{C_1}{C_2} - 1\right) X_1 \left(\frac{M_2}{M_1}\right) \tag{19}$$

This relation is identical with that commonly found in the literature[120] except for the term M_2/M_1 which corrects for the change in molec-

[119] Rittenberg, D., and Foster, G. L., *J. Biol. Chem.* **133**, 737 (1940); Graff, S., Rittenberg, D., and Foster, G. L., *ibid.* **133**, 745 (1940); Foster, G. L., *ibid.* **159**, 431 (1945); Shemin, D., *ibid.* **159**, 439 (1945).

[120] See Gest, H., Kamen, M. D., and Reiner, J. M., *Arch. Biochem.* **12**, 273 (1947), for a treatment of the general case wherein there is any number of batches containing any number of isotopes.

ular weight of compound in the two batches as the isotopic composition changes. This term is of practical importance only when the molecular weights of the two isotopes are greatly different and when the atom per cent excess is high in the added carrier, as when compounds of low molecular weight and highly enriched with deuterium are used.

In using radioactive isotopes the atom per cent is so low for specific radioactivities which are easily detectable that there is no necessity to include corrections for variations in molecular weight. Suppose x_1 grams of labeled material with an activity of X counts per minute is mixed with x_2 grams of diluent. The specific activity becomes $X/(x_1 + x_2)$. The initial specific activity is X/x_1. Thus,

$$\frac{\text{Specific activity after mixing}}{\text{Special activity before mixing}} = \frac{x_1}{x_2 - x_1}$$

The specific activity is identified with the atom per cent A so that this relation may be written $A_i/A_f = (x_2 + x_1)x_1$, or $x_2 = [(A_i/A_f) - 1]\cdot x_1$, where A_i and A_f are initial and final atom per cent of added labeled material. As an example, suppose 0.5 g. of sodium phosphate containing 6000 ct./min. is mixed with a sample of blood plasma. One-tenth gram of phosphate is recovered with an activity of 300 ct./min. The initial specific activity is $6000/0.5 = 12{,}000$ ct./min./g. The final specific activity is $300/0.1 = 3000$ ct./min./g. A fourfold dilution has taken place, showing that $3 \times 0.5 = 1.5$ g. of radioactive phosphate was present in the 1000-ml. sample of plasma. From the formula given,

$$x_2 = \left(\frac{12000}{3000} - 1\right) 0.5 \qquad \text{and} \qquad x_2 = 1.5$$

As stated previously, analysis by isotope dilution is the reverse of the method used to isolate biological intermediates in that for detection of intermediates unlabeled material is added, whereas in the isotope dilution procedure it is the added material which is labeled. In this connection it should be noted that reverse isotopic dilution procedures can be extended to precursor-product researches (see pp. 142–151) in the following manner, as suggested by Bloch and Anker.[121] In the usual procedure isolation of metabolite by means of unlabeled carrier yields only a qualitative answer to the question whether a given metabolite is in the reaction chain because neither the amount nor the isotopic concentration of metabolite from a labeled precursor is known. If one divides the metabolite solution into two aliquots and carries out two separate experiments with different quantities of unlabeled carrier in each aliquot, it is possible to obtain data from which one can set up simultaneous equations and solve for the concentration of metabolite present.

[121] Bloch, K., and Anker, H. S., *Science* **107**, 228 (1948).

Keston, Udenfriend, and their collaborators[122-124] have shown that the sensitivity of isotopic dilution procedures can be greatly increased by the use of "derivative" or "indirect," as opposed to "direct," types previously discussed. In the derivative method, and its variants, a mixture of unknowns is converted to some well-characterized derivative using a labeled reagent under conditions which ensure complete conversion to derivative. A large excess of the unlabeled derivative is then added, the mixture separated, a pure sample of the derivative obtained, and the dilution measured.

This method, theoretically applicable to any class of compounds for which stable, well-defined derivatives exist, has been most thoroughly elaborated for analysis of amino acids, using the reagent *p*-iodophenylsulfonyl chloride (more commonly called "pipsyl"), the iodine being labeled with the 8-day isotope I^{131}. In this procedure the sensitivity is greatly increased over the simpler nonderivative type of isotopic dilution analysis because it is operable with trace amounts of unknown. The radioactivity level is determined by the reagent rather than the unknown, and a more versatile choice of radioactive elements is possible. By using racemic carriers, errors arising from partial racemization can be avoided.

The separation of the pipsyl derivatives is best accomplished by paper chromatography. The derivatives are separated as bands on a paper strip and then eluted for radioassay. However, there are a number of disadvantages in this procedure. Thus, (1) bands sometimes overlap; (2) elution of bands must be complete; (3) transfers of extremely small volumes of solution to paper must be accomplished without loss. Also, there exist uncertainties regarding the constancy of R_F values for different amino acids from paper to paper.

These difficulties are obviated by multiple labeling. S^{35}-labeled derivative is added in accurately measured amounts to the mixture after derivatization with I^{131}-pipsyl reagent. From this point the procedure is as before, but losses are now immaterial, because the S^{35}-labeled derivative serves as an "indicator," or internal monitor. The pure portion of any band can be determined by constancy of the S^{35}/I^{131} ratio. With the S^{35} activity and this ratio known, it requires only a simple calculation to estimate the total original I^{131}-pipsyl derivative. The choice of S^{35} and I^{131} is dictated by the convenient differences in radiation characteristics of the two isotopes. An aluminum foil a few mils thick serves to absorb practically all S^{35} radiation without greatly affecting the I^{131} radiation. Standards are employed to calibrate the absorbing foil.

[122] Keston, A. S., Udenfriend, S., and Cannan, R. K., *J. Am. Chem. Soc.* **68,** 1390 (1946); **71,** 249 (1949).

[123] Keston, A. S., Udenfriend, S., and Levy, M., *J. Am. Chem. Soc.* **69,** 315 (1947).

[124] Keston, A. S., Udenfriend, S., and Levy, M., *J. Am. Chem. Soc.* **72,** 748 (1950).

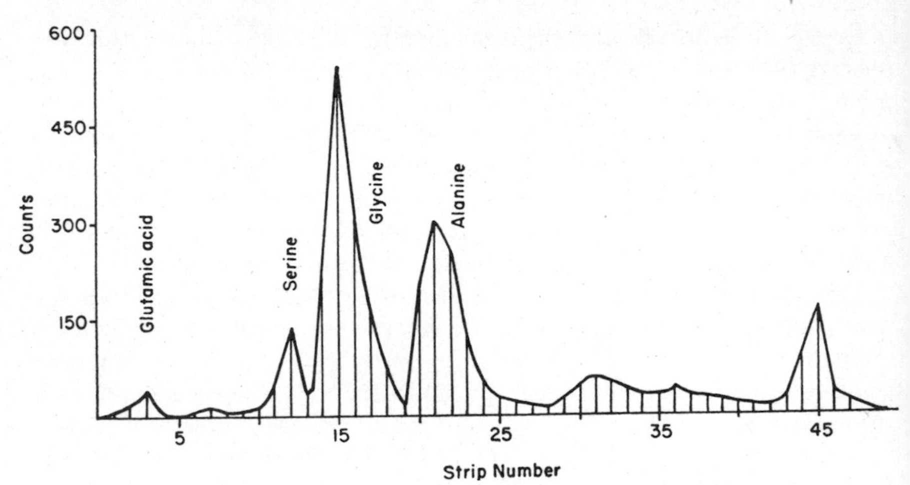

FIG. 46. Activity as a function of strip number in paper chromatogram prepared from silk hydrolyzate. (After Keston *et al.*[124])

A typical procedure in brief is:

1. Convert unknown acids by reaction with excess I^{131}-pipsyl chloride.
2. Add an accurately measured quantity of indicator S^{35}-pipsyl derivative for each amino acid to be assayed.
3. Remove hydrolyzed pipsyl reagent by successive extraction, as in a countercurrent procedure.
4. Concentrate to small volume the solution of ammonium salts of pipsyl derivatives resulting from step 3 by extraction with dilute ammonia.
5. Prepare chromatogram. (A typical chromatogram pattern is shown in Fig. 46.)
6. Autograph chromatogram to determine placement of bands.
7. Cut bands into thin strips, elute, and determine S^{35}/I^{131} ratios.

Calculations involved are as follows. Let y = radioactivity (ct./min.) of indicator (S^{35}), x = radioactivity of I^{131}-pipsyl derivative, m = moles of compound determined. If S is the total radioactivity of added indicator and C the molar activity (ct./min./mole) of the I^{131}-pipsyl reagent, then

$$m = xS/yC$$

The determination of various amino acids in protein hydrolyzates by this procedure is facilitated by preliminary separation of the various amino acids into groups by countercurrent extraction between chloroform and dilute HCl.[125] The method has been applied to the determination of eleven amino acids—aspartic acid, glutamic acid, serine, threonine, hydroxyproline, proline, methionine, phenylalanine, valine, alanine, and glycine—

[125] Velick, S. F., and Udenfriend, S., *J. Biol. Chem.* **190,** 721 (1951).

as well as to end-group analysis of proteins such as horse hemoglobin, rabbit muscle aldolase, and insulin.[126] Data are available showing analyses for amounts of 10 γ or less, with over-all errors of no more than 2%. It is felt that this error can be made less than 1% by further refinements.

Another feature of the method is that upper limits can be set on the quantity of amino acids present in extremely small amounts. This can be done by placing the absorbing aluminum foil over half of the mixed pipsyl band and noting whether all radiation is cut out as compared to the adjacent unfiltered area. Conversely, the presence of extremely small amounts of a given amino acid can be established. Thus, Velick and Udenfriend[125] confirmed the presence of small amounts of glycine in the protein salmine, reported as present on the basis of microbiological assay. They also proved the presence in trace amounts of phenylalanine and threonine.

These isotopic dilution procedures for amino acid analysis are highly specific, relatively rapid, and not prohibitive as regards the manipulative experience or apparatus required to make them routine. As seen, they also enable accurate analyses to be performed on very small amounts of material and are ideally adapted to the elegant countercurrent and paper chromatographic procedures now in such widespread use. Furthermore, they provide a means for checking some of the more conventional procedures.

The derivative method also provides in some cases for an otherwise impossible extension of chromatographic methods to mixtures which are ordinarily inseparable and hence indeterminate by conventional nonisotopic or direct isotopic chromatographic procedures. Thus, leucine and isoleucine yield but one band in chromatograms, as do valine and norvaline. The chromatograms of the mixed pipsyl derivates of leucine and isoleucine give a single band no wider than either of the two acids singly. Marked variations in the isotopic ratio across the band are seen, however. In this way it can be seen that the derivative method is capable of revealing inhomogeneities in what otherwise appear to be homogeneous chromatograms.

The derivative method as described requires complete derivitization of the amino acid with the labeled reagent. As Keston *et al.* point out, however, this requirement can be circumvented by adding a known amount of the compound to be assayed for in the labeled form before derivitization.[127] Thus, if C^{14}-labeled glycine were added to a protein hydrolyzate and the resultant mixture reacted with I^{131}-pipsyl reagent, the fractional recovery of C^{14} in the pipsyl-glycine would give a precise index of the fractional recovery of all the glycine originally present. This suggestion has been followed up recently by Keston and Lospalluto,[127] who have demonstrated the validity of this procedure for glycine analysis, obtaining checks on known amounts

[126] Udenfriend, S., and Velick, S. F., *J. Biol. Chem.* **190,** 733 (1951).

[127] Keston, A. S., and Lospalluto, J., *Federation Proc.* **10,** 207 (1951).

of glycine up to 45 γ within 2%. In this method the index of band purity is constancy of the C^{14}/I^{131} ratio. It is probable that this variant represents the ultimate in simplification of the derivative method. In this form, there is no need to know what is in the mixture or how much has reacted with the reagent. All that is required is to find a part of the chromatogram showing constant isotopic ratio. Its inventors customarily refer to this procedure as "idiot's delight."

The extension of these derivative methods to other classes of compounds appears to be in prospect. A few suggestions for potential application may be offered. Thio esters of fatty acids should provide proper labeled derivatives for fatty acid analysis. A large number of thio derivatives are available for alcohols. Khym and Cohn[128] have shown that monosaccharides such as glucose, fructose and galactose form borate complexes which are readily separable on ion-exchange columns. Such complexes can be applied to the separation of C^{14}-labeled sugars. However, boron is one element which is unique in having no practical isotopic label. Research on homologous complexes such as molybdates or tungstates is indicated. If these act in a similar fashion, appropriate derivatives for derivative analysis will be at hand.

The determination of end groups in polysaccharides, making use of the reactivity of the reducing terminal aldehyde group with radioactive cyanide to form a radioactive cyanohydrin, has been proposed by Isbell.[129] On alkaline hydrolysis the cyanohydrin will yield a carbohydrate with radioactive carboxyl which can be isolated and purified by means of ion-exchange resins. From the combining proportion of the radioactive cyanide, as determined by radioassay, the average molecular weight of the polysaccharide may be inferred. Alternatively, one is enabled to determine the number of reactive carbonyl groups in a substance of known molecular weight. This procedure appears particularly suitable for study of polysaccharides of relatively low molecular weight and for partially hydrolyzed products. This application of radioactive derivative methods to biochemical analysis presents still another kind of potentiality contrasted with those already discussed. It illustrates how quantitative methods involving labeled derivatives may proliferate through the whole field of biochemical analysis.

This discussion may be concluded by considering the application of the derivative method to the positive identification of a rare amino acid found in alcoholic extracts of mouse brain by Roberts and Frankel[130] and tentatively identified on the basis of paper chromatography as γ-aminobutyric acid. This amino acid appears only in very small amounts. Its separation

[128] Khym, J. S., and Cohn, W. E., *J. Am. Chem. Soc.* **75,** 1153 (1953).
[129] Isbell, H. S., *Science* **113,** 532 (1951).
[130] Roberts, E., and Frankel, S., *J. Biol. Chem.* **10,** 207 (1951).

by paper chromatography involves only microgram amounts and would require a prohibitive number of chromatograms to obtain a sufficient quantity for identification by rigorous procedures. The S^{35}-labeled pipsyl derivative was prepared[131] by treating approximately 25 mg. of an authentic sample of the amino acid with the S^{35}-pipsyl reagent. Approximately 0.6 γ of the unknown amino acid, representing 10% of the total available, was treated with 10 mg. of I^{131}-pipsyl reagent. The mixture was then worked up as usual, and aliquots were chromatographed on filter paper strips. It was found that the mixed chromatograms agreed with the behavior to be expected if the unknown was, in fact, γ-aminobutyric acid.

K. Concluding Remarks

In this chapter the nature of tracer methodology as applied to biochemistry has been indicated. To enable the reader to probe more deeply into the powers and limitations of the method, it seems best to continue the discussion by examining how tracer research actually has contributed in a detailed way to the elucidation of specific biochemical problems. In Chapter VII a number of such researches are described. In addition, there is presented a critical evaluation of the methodology as it is applied to the study of intermediary metabolism. General references for Chapters VI and VII are supplied at the end of Chapter VII.

[131] Udenfriend, S., *J. Biol. Chem.* **187**, 65 (1950).

CHAPTER VII

SURVEY OF TRACER METHODS: BIOCHEMICAL ASPECTS, PART II

1. SPECIAL TOPICS IN INTERMEDIARY METABOLISM

Three examples have been chosen for discussion in this section. Each has a long history, as tracer research goes, and each may be considered as a classic illustration of the character of tracer research.

A. CO_2 FIXATION IN PHOTOSYNTHESIS

The fate of carbon during photosynthesis of CO_2 into organic material is an age-old problem. Despite the endeavors of an illustrious list of investigators, nothing positive was known up to 1938. At that time, Ruben, in collaboration with Hassid and Kamen,[1] began experiments using the short-lived carbon isotope C^{11} (half-life, 20.5 min.). The plan of research was simple. Green plants or algae, actively photosynthesizing CO_2, were exposed to tracer quantities of the gas labeled with C^{11}. At various times, determined by the nature of the experiment, the plants were killed by boiling in ethanol or water. The resulting suspension was fractionated, and the fractions were assayed for radioactivity. Efforts were made in separate experiments to identify the radioactive molecules contained in the various fractions. (An example of the procedures possible in the form of a detailed protocol of an experiment of CO_2 fixation in protozoa will be given in Chapter X.) Despite the short half-life and the fact that no methods were then available for detection of micro amounts of organic compounds except those involving the use of "carriers," a considerable amount of information was obtained which supplied a basis for later investigators using long-lived carbon and chromatographic procedures when both of these became widely available.

In their first series of papers, Ruben *et al.*[1, 2] reported that none of the commonly proposed compounds, such as formaldehyde, formic acid, and organic acids, were labeled in the first few minutes of photosynthesis. They could not show appreciable radioactivity in any single organic compound despite several hundred attempts. Certain classes of compounds could not

[1] Ruben, S., Hassid, W. Z., and Kamen, M. D., *J. Am. Chem. Soc.* **61,** 661 (1939).
[2] Ruben, S., Kamen, M. D., and Hassid, W. Z., *J. Am. Chem. Soc.* **62,** 3493 (1940).

be tested because proper analytic procedures had not been worked out and because of the short time available to complete any given experiment. As it turned out later, this inability to test a number of compounds was particularly significant with respect to the stable phosphate esters of hydroxycarboxylic acids.

Ruben and Kamen then turned their attention to the general characterization of the radioactive molecules. They fixed on exposure times of 1 min. as short enough to preclude fixation in more than one or two compounds. (Later researches with C^{14} showed that even 1 min. was not a short enough time!) They measured the appearance of label in functional groups such as carboxyl, carbonyl, and other types of reduced organic carbon and showed that fixation occurred first in the carboxyl carbon and then spread to more reduced forms. By determining the diffusion and sedimentation constants of the 1-min. material formed in *Chlorella* suspensions,[3, 4] they concluded that the first fixation products might have molecular weights as high as 1000. This conclusion was at best tentative, however, in view of the semiquantitative nature of the measurements dictated by the short half-life of C^{11}. It was concluded that the product of fixation in the light was a polyhydroxycarboxylic compound of intermediate molecular weight. Ruben and Kamen, with a number of collaborators,[5-7] also studied CO_2 fixation as a general phenomenon and demonstrated its occurrence in a variety of nonphotosynthetic systems, thus underscoring the definitive researches of Wood and Werkman (see p. 153).

At this stage of the research little more could be accomplished, as already noted, because of the absence of the necessary techniques for isolation and identification of organic material in micro quantities, and the inadequacy of the tracer isotopes available (C^{11} had too short a half-life, and the stable tracer C^{13} could not be diluted to the extent encountered in short-term experiments and still be detected).

In 1940, a long-lived isotope, C^{14}, was discovered.[8, 9] This solved the problem of an adequate tracer material. Shortly after the end of the war, two new groups took up work with C^{14}. One of these, formed around Calvin and Benson, was in California; the other, led by Gaffron and Fager, was in Chicago.

[3] Ruben, S., Kamen, M. D., and Perry, L. H., *J. Am. Chem. Soc.* **62,** 3450 (1940).

[4] Ruben, S., and Kamen, M. D., *J. Am. Chem. Soc.* **62,** 3451 (1940).

[5] Ruben, S., and Kamen, M. D., *Proc. Natl. Acad. Sci. U. S.* **26,** 418 (1940).

[6] Barker, H. A., Ruben, S., and Kamen, M. D., *Proc. Natl. Acad. Sci. U. S.* **26,** 426 (1940).

[7] Van Niel, C. B., Ruben, S., Carson, S. F., Kamen, M. D., and Foster, J. W., *Proc. Natl. Acad. Sci. U. S.* **28,** 8 (1942).

[8] Ruben, S., and Kamen, M. D., *Phys. Rev.* **57,** 549 (1940).

[9] Kamen, M. D., and Ruben, S., *Phys. Rev.* **58,** 194 (1940).

The California group finally identified phosphoglycerate as the compound in which the major fraction of CO_2 carbon entered in photofixation.[10] They demonstrated that appreciable quantities of this compound were present in the algae and in green plants. Their procedure consisted in fractionating the soluble labeled material by use of ion-exchange resins and showing that most of the radioactivity was contained in material which was anionic. In a typical experiment a portion of this material (0.2 ml. containing 9250 ct./min.) was recrystallized repeatedly from water, after addition of 5.4 mg. of barium phosphoglycerate (Ba-PGA) as carrier. The specific activity of the material recovered at each stage remained essentially constant. As discussed later (see p. 228ff.), this kind of assay did not prove that the labeled material was identical with Ba-PGA. The next step was to hydrolyze the labeled material with carrier, recover the free glyceric acid, and prepare the *p*-bromophenylglycerate. The distribution of the radioactive material between phases in a mixture of toluene, acetic acid, and water was measured and found to be identical with that for the carrier derivative.

At first the Chicago group,[11, 12] using precipitation methods and carrier Ba-PGA, was unable to confirm the identity of the labeled material with PGA. Attempts to find activity in the β carbon by periodic acid oxidation and isolation of the formaldehyde were also negative.

Full agreement was reached largely because the method of paper chromatography became available. This analytical procedure, which obviated the need for carrier, established the original contention of the California group that PGA was the first identifiable product of CO_2 fixation. The Chicago group, in the meantime, isolated labeled glyceric acid from the hot-water extract of large quantities of algae.[13] The California group analyzed algal extracts by two-dimensional paper chromatography and showed that in the first few seconds of photofixation practically all the radioactivity was contained in an area of the chromatogram with R_F values identical with those for PGA in the solvents used.[14] Small amounts of radioactivity were also found in a few other regions associated with other phosphate esters. It was then realized that inability to examine such compounds directly had been one of the most serious drawbacks to the early researches with C^{11}.

[10] Benson, A. A., Calvin, M., Haas, V. A., Aronoff, S., Hall, A. G., Bassham, J. A., and Weigl, J. W., *in* "Photosynthesis in Plants" (J. Franck and W. B. Loomis, eds.), Chapter 19. Iowa State College Press, Ames, Iowa, 1949.

[11] Brown, A. H., Fager, E. W., and Gaffron, H., see Chapter 20 in reference 10.

[12] Fager, E. W., see Chapter 21 in reference 10.

[13] Fager, E. W., Rosenberg, J. L., and Gaffron, H., *Federation Proc.* **9,** 535 (1950).

[14] Benson, A. A., Bassham, J. A., Calvin, M., Goodale, T. C., Haas, V. A., and Stepka, W., *J. Am. Chem. Soc.* **72,** 1710 (1950).

Kinetic studies were also made in both laboratories. It was shown that label appeared very rapidly in the carboxyl carbon of PGA. This functional group was found to be saturated in a matter of seconds. Thus, it was shown[13] that in steady-state photosynthesis all the C^{14} was in the carboxyl (extrapolated value, 80% to 90% in 1 sec.) and that in 10 sec. this had dropped to about 60% in the carboxyl and α carbon. In a few minutes most of the total activity was still in PGA but was almost equally distributed between the various carbons. Labeling in the α and β carbons appeared to be symmetric. On this basis, the California group proposed a set of cyclic reactions in which a C-2 compound condensed with CO_2 to form a carboxyl-labeled C-3 compound which was thought to be PGA or a compound in equilibrium with it. For a time, the postulated immediate product was pyruvate, as some activity was found to be associated with it as well as with PGA during the early stages of photosynthesis. It was assumed that reduction of a portion of the pyruvate at the time of a second carboxylation formed a four-carbon compound, presumably malic acid. Further reduction to succinate followed by reductive cleavage of the succinate to the original C-2 compound completed the cycle. A portion of the pyruvate was assumed to be phosphorylated to PGA which could be condensed reductively to form hexose. (It had been demonstrated that after some time labeled hexose was formed and that the label was found first in the 3,4 position and then spread out to the other carbons. The mechanism postulated above, involving condensation of PGA, would be in accord with this sequence of labeling.)

The Chicago group contended that there was only one carboxylation per C-2 unit, mainly on the basis that their kinetic studies indicated PGA as the immediate product of carboxylation, rather than pyruvate. Two of these PGA molecules could be transformed to triose and condensed to hexose. They then suggested a three-way split to three C-2 compounds to reform the original C-2 condensing unit.

It had been noted[14] that PGA was not the sole repository of label, even in very short periods of photosynthesis. Small but appreciable amounts of radioactivity were found in spots associated with other phosphate compounds. These were first ascribed to hexosephosphate, but later Benson *et al.*[15, 16] observed that hydrolysis of these compounds after elution resulted in the appearance of some new free sugars which turned out to have R_F values identical with those of authentic samples of sedoheptulose and ribulose. The discovery that a seven-carbon and a five-carbon sugar were involved in the photofixation along with PGA was critical in elucidating the nature of the cycle actually concerned in the generation of the "C-2"

[15] Benson, A. A., *J. Am. Chem. Soc.* **73**, 2971 (1951).

[16] Benson, A. A., Bassham, J. A., Calvin, M., Hall, A. G., Hirsch, H. E., Kawaguchi, S., Lynch, V., and Tolbert, N. E., *J. Biol. Chem.* **196**, 703 (1952).

compound involved in CO_2 fixation. It took on added significance from the studies of Horecker, Racker, and others, who, at about the same time, were showing that in the oxidation of carbohydrate there were interconversions between pentoses and hexoses involving the formation of sedoheptulose as an intermediate.[17] The key reactions were demonstrated to involve two kinds of enzyme systems. One was a "transketolase" catalyzing, for example, the reaction between two molecules of ribulose-5-phosphate to form glyceraldehyde-3-phosphate and sedoheptulose-7-phosphate, with thiamine pyrophosphate as coenzyme, as in Eq. 1. The other was a "trans-

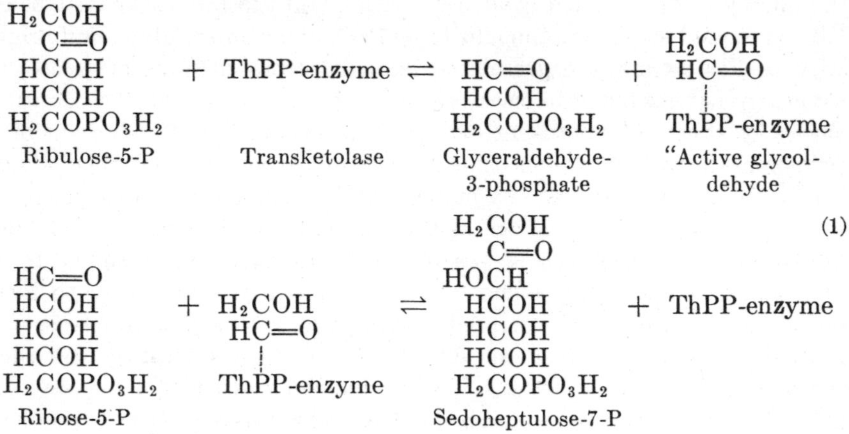

$$
\begin{array}{ccccccc}
\begin{array}{c} H_2COH \\ C{=}O \\ HCOH \\ HCOH \\ H_2COPO_3H_2 \end{array} & + & \text{ThPP-enzyme} & \rightleftharpoons & \begin{array}{c} HC{=}O \\ HCOH \\ H_2COPO_3H_2 \end{array} & + & \begin{array}{c} H_2COH \\ HC{=}O \\ | \\ \text{ThPP-enzyme} \end{array} \\
\text{Ribulose-5-P} & & \text{Transketolase} & & \text{Glyceraldehyde-3-phosphate} & & \text{"Active glycoldehyde} \\
 & & & & & & \\
\begin{array}{c} HC{=}O \\ HCOH \\ HCOH \\ HCOH \\ H_2COPO_3H_2 \end{array} & + & \begin{array}{c} H_2COH \\ HC{=}O \\ | \\ \text{ThPP-enzyme} \end{array} & \rightleftharpoons & \begin{array}{c} H_2COH \\ C{=}O \\ HOCH \\ HCOH \\ HCOH \\ HCOH \\ H_2COPO_3H_2 \end{array} & + & \text{ThPP-enzyme} \\
\text{Ribose-5-P} & & & & \text{Sedoheptulose-7-P} & &
\end{array}
\quad (1)
$$

aldolase" reaction, an example of which is the dismutation of the products of the transketolase reaction to form a four-carbon sugar phosphate, presumably erythrose-4-phosphate and fructose-6-phosphate, as in Eq. 2.

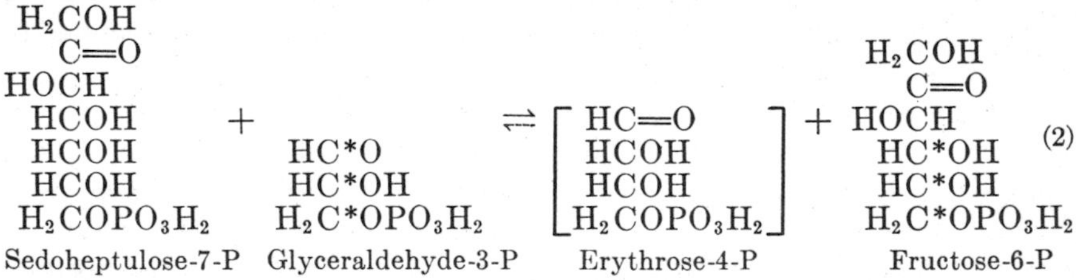

$$
\begin{array}{ccccccc}
\begin{array}{c} H_2COH \\ C{=}O \\ HOCH \\ HCOH \\ HCOH \\ HCOH \\ H_2COPO_3H_2 \end{array} & + & \begin{array}{c} HC^*O \\ HC^*OH \\ H_2C^*OPO_3H_2 \end{array} & \rightleftharpoons & \left[\begin{array}{c} HC{=}O \\ HCOH \\ HCOH \\ H_2COPO_3H_2 \end{array}\right] & + & \begin{array}{c} H_2COH \\ C{=}O \\ HOCH \\ HC^*OH \\ HC^*OH \\ H_2C^*OPO_3H_2 \end{array} \\
\text{Sedoheptulose-7-P} & & \text{Glyceraldehyde-3-P} & & \text{Erythrose-4-P} & & \text{Fructose-6-P}
\end{array}
\quad (2)
$$

The cycle involved in the oxidation of hexose could be represented by the sum of the partial reactions:[17]

$$
\begin{array}{l}
2\ \text{Hexose-P} \rightarrow 2CO_2 + \text{hexose-P} + \text{tetrose-P} \\
\text{Hexose-P} \rightarrow CO_2 + \text{pentose-P} \\
\text{Pentose-P} + \text{tetrose-P} \rightarrow \text{hexose-P} + \text{triose-P} \\
\hline
\text{Sum: Hexose-P} \rightarrow 3CO_2 + \text{triose-P}
\end{array}
$$

[17] See Horecker, B., *Brewer's Dig.* **28,** 214 (1953), for a review of these studies.

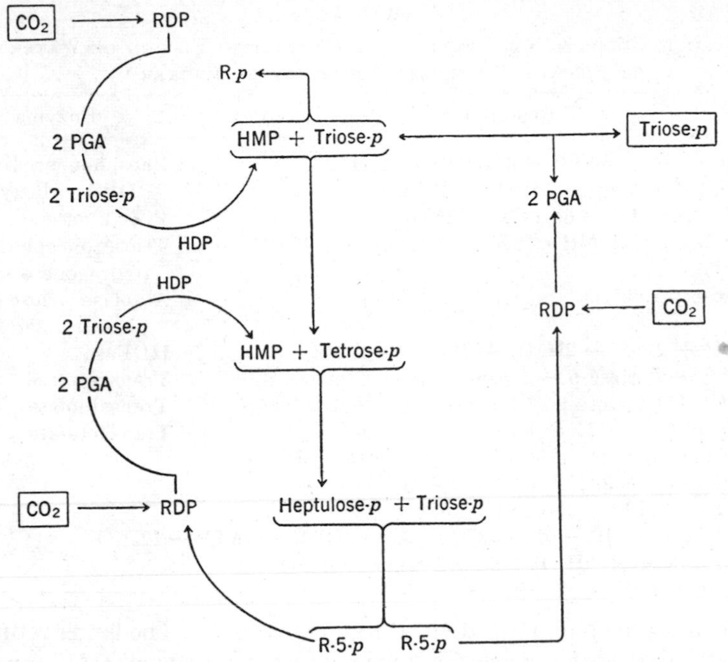

FIG. 47. Representation of pentose phosphate cycle. (After Racker.[18a])

The basis was laid in these studies for postulating a cycle[18]—generally called the "pentose phosphate cycle"—in which the C-2 fragment involved in formation of PGA was not free but rather a portion of a five-carbon sugar phosphate. The first reaction was assumed to be a carboxylation of ribulose-5-phosphate to form a hexose. This was followed by a fission to two C-3 compounds, presumably PGA or compounds from which it could easily be formed. The PGA would then be transformed to triose, after which a series of reactions leading through interconversions involving transketolase and transaldolase and regenerating the ribulose-5-phosphate could be postulated. One form of the cycle is shown in Fig. 47. A reaction scheme is given in Table 12.

The California group[18] studied the labeling of various compounds in this cycle during steady-state photosynthesis in the green alga *Scenedesmus* and in soybean leaves. The algae were suspended in distilled water, placed in a reaction chamber at constant temperature, and allowed to photosynthesize. At the same time, a mixture of 4% unlabeled CO_2 and air was blown through the suspension. The suspension was pumped from the bottom of the vessel to the top by means of a small transparent pump.

[18] Bassham, J. A., Benson, A. A., Kay, L. D., Harris, A. Z., Wilson, A. T., and Calvin, M., *J. Am. Chem. Soc.* **76**, 1760 (1954).

[18a] Racker, E., *Nature* **175**, 249 (1955).

TABLE 12

REACTION SCHEME FOR COUPLING OF PHOTOLYTIC FISSION OF WATER TO PENTOSE PHOSPHATE CYCLE (AFTER RACKER[18a])

Reaction	Enzyme
1. 3 pentose-p + 3ATP → 3"RDP" + 3ADP	Phosphopentokinase
2. 3"RDP" + $3CO_2$ + $3H_2O$ → 6PGA	"RDP" carboxylase
3. 6PGA + 6ATP → 6diPGA + 6ADP	PGA-kinase
4. 6 diPGA + 6DPNH + $6H^+$ → 6 triose-p + 6DPN + 6Pi	Triose phosphate dehydrogenase
5. 4 triose-p → 2HDP	Aldolase, triose phosphate isomerase
6. 2HDP + $2H_2O$ → 2HMP + 2Pi	HDPase
7. 1HMP + 1 triose-p → 1 pentose-p + 1 tetrose-p	Transketolase
8. 1HMP + 1 tetrose-p → 1 heptulose-p + 1 triose-p	Transaldolase
9. 1 heptulose-p + 1 triose-p → 2 pentose-p	Transketolase
Sum (1–9) $3CO_2$ + 9ATP + $5H_2O$ + 6DPNH + $6H^+$ → 1 triose-p + 9ADP + 6DPN + 8Pi	
10. $9H_2O$ + 9DPN → 9DPNH + $9H^+$ + 9O	
11. 3DPNH + 9ADP + 9Pi + $3H^+$ + 3O → 3DPN + 9ATP + $12H_2O$	
Sum (1–11) $3CO_2$ + $2H_2O$ + Pi → 1 triose-p + 6O	

The output of the pump was divided into two streams. The larger returned the greater part of the suspension to the illumination chamber; the smaller gave aliquots for sampling. The apparatus is shown in Fig. 48.

After a period of steady photosynthesis, a solution of labeled CO_2 was injected rapidly by means of the syringe. By picking the point of entry it was possible to control the exposure time of the cells to the label. With the rate of flow and the distance to the point of entry known, the time of exposure could be calculated. As shown in Fig. 48, the period of photosynthesis in labeled CO_2 was terminated when the sampling tube delivered the aliquot into the beaker of boiling methanol. Exposures averaged from 1 to 16 sec.

The methanolic extract was fractionated by paper chromatography, the various purified compounds eluted, and total activity in each determined. Then the labeled ribulose and sedoheptulose were degraded, and the distribution of label was determined. Labeled sedoheptulose, after addition of carrier, was degraded as follows (see Fig. 49). One portion was converted to the osazone with phenylhydrazine. This was oxidized with periodate[19] in bicarbonate buffer to give formaldehyde (carbon 7), formic acid (carbons 4, 5, and 6), and a residue C-3 osazone (carbons 1, 2, and 3). A second portion was converted to the sugar anhydride, oxidized with periodate to yield carbon 4 as formate, and with cerate and perchlorate[20]

[19] Topper, Y. J., and Hastings, A. B., *J. Biol. Chem.* **179**, 1255 (1949).

[20] Smith, G. F., "Cerate Oxidimetry." G. F. Smith Chemical Co., Columbus, Ohio, 1942.

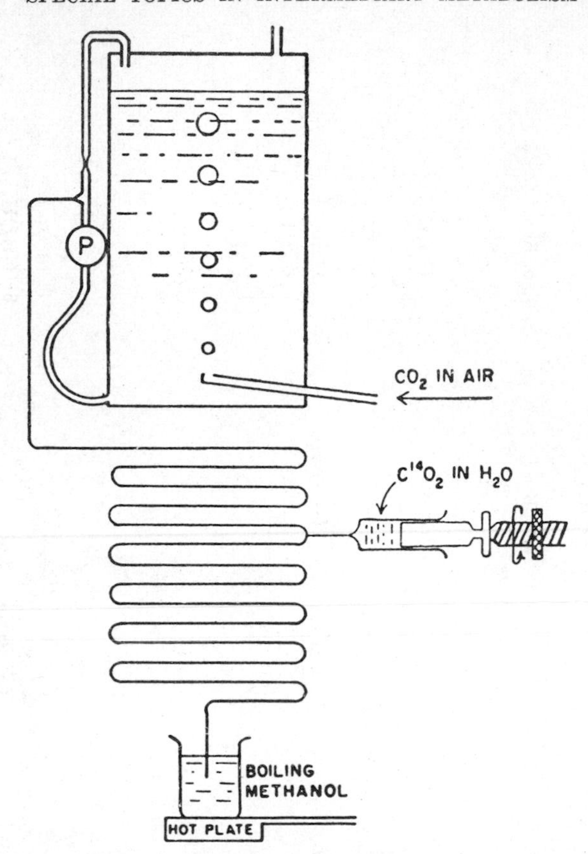

RIG. 48. Apparatus (schematic) for study of short-term $C^{14}O_2$ fixation. (After Bassham *et al.*[18])

to give carbon 2 as CO_2. A third portion was converted to the sugar alcohol by reduction with hydrogen-PtO_2. The alcohol was oxidized by suspensions of *Acetobacter suboxydans*[21] to give guloheptulose, which on oxidation with cerate and perchlorate yielded carbon 6 as CO_2. Other products formed were sedoheptulose and mannoheptulose. Another portion of the products from the bacterial oxidation was oxidized with periodic acid to yield carbons 1 and 7 as formaldehyde and the other carbons as formic acid.

Ribulose, after addition of carrier, was degraded by converting one portion to the osazone, and the other by reduction with hydrogen-PtO_2 to the alcohol. The osazone was oxidized with periodate in bicarbonate buffer to yield carbon 4 as formate, carbon 5 as formaldehyde, and the C-3 osazone. Oxidation of the alcohol with periodic acid gave carbons 1 and 5 as formaldehyde, and carbons 2, 3, and 4 as formic acid. Direct oxidation of

[21] Stewart, L. C., Richtmyer, H. K., and Hudson, C. S., *J. Am. Chem. Soc.* **74**, 2206 (1952).

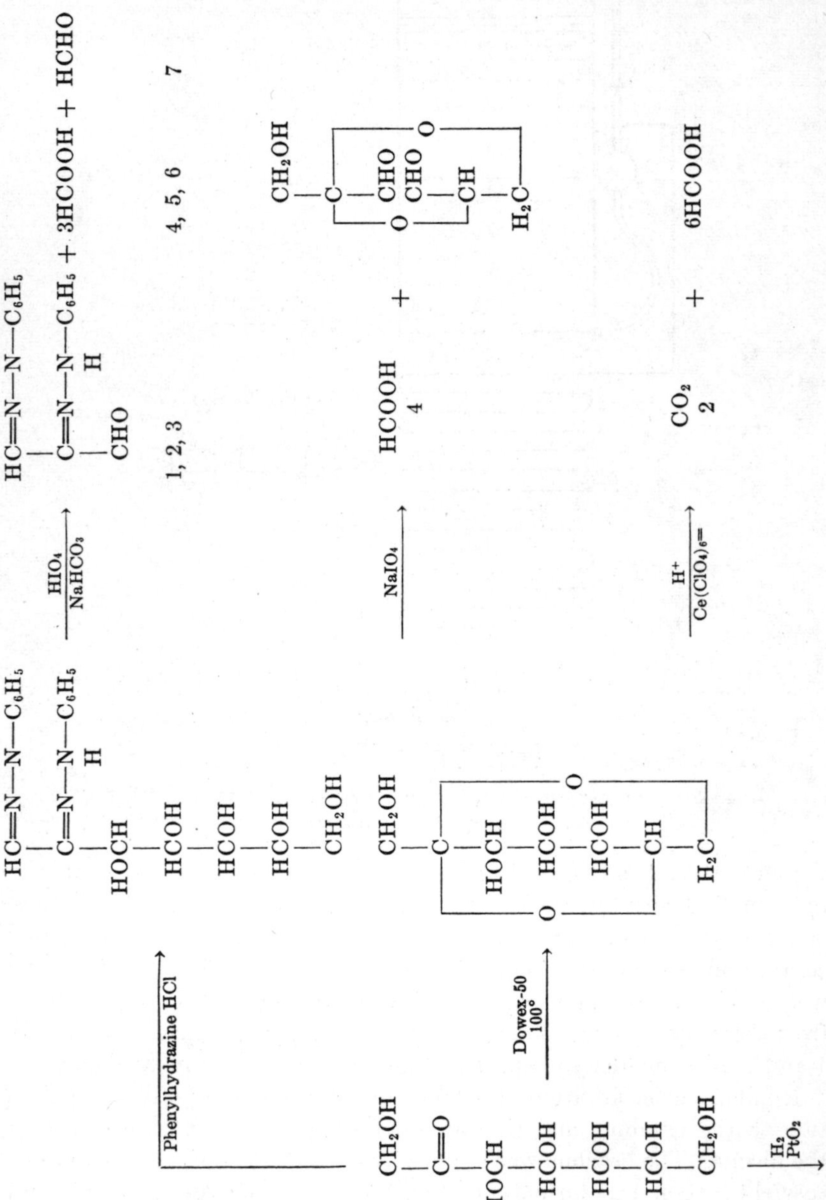

Sedoheptulose (CH_2OH—$C{=}O$—$HOCH$—$HCOH$—$HCOH$—$HCOH$—CH_2OH):

$\xrightarrow{\text{Phenylhydrazine HCl}}$ $HC{=}N{-}NH{-}C_6H_5$ / $C{=}N{-}NH{-}C_6H_5$ / $HOCH$ / $HCOH$ / $HCOH$ / $HCOH$ / CH_2OH

$\xrightarrow[NaHCO_3]{HIO_4}$ $HC{=}N{-}NH{-}C_6H_5$ / $C{=}N{-}NH{-}C_6H_5$ / CHO (1, 2, 3) + $3HCOOH$ (4, 5, 6) + $HCHO$ (7)

$\xrightarrow[100^\circ]{\text{Dowex-50}}$ Anhydride: CH_2OH—C(—O—)—$HOCH$—$HCOH$—$HCOH$—CH(—O—)—H_2C(—O—)

$\xrightarrow{NaIO_4}$ $HCOOH$ (4) + CH_2OH—C(—O—)—CHO CHO—CH(—O—)—H_2C(—O—)

$\xrightarrow[Ce(ClO_4)_6^{=}]{H^+}$ CO_2 (2) + $6HCOOH$

$\xrightarrow{H_2,\ PtO_2}$ ↓

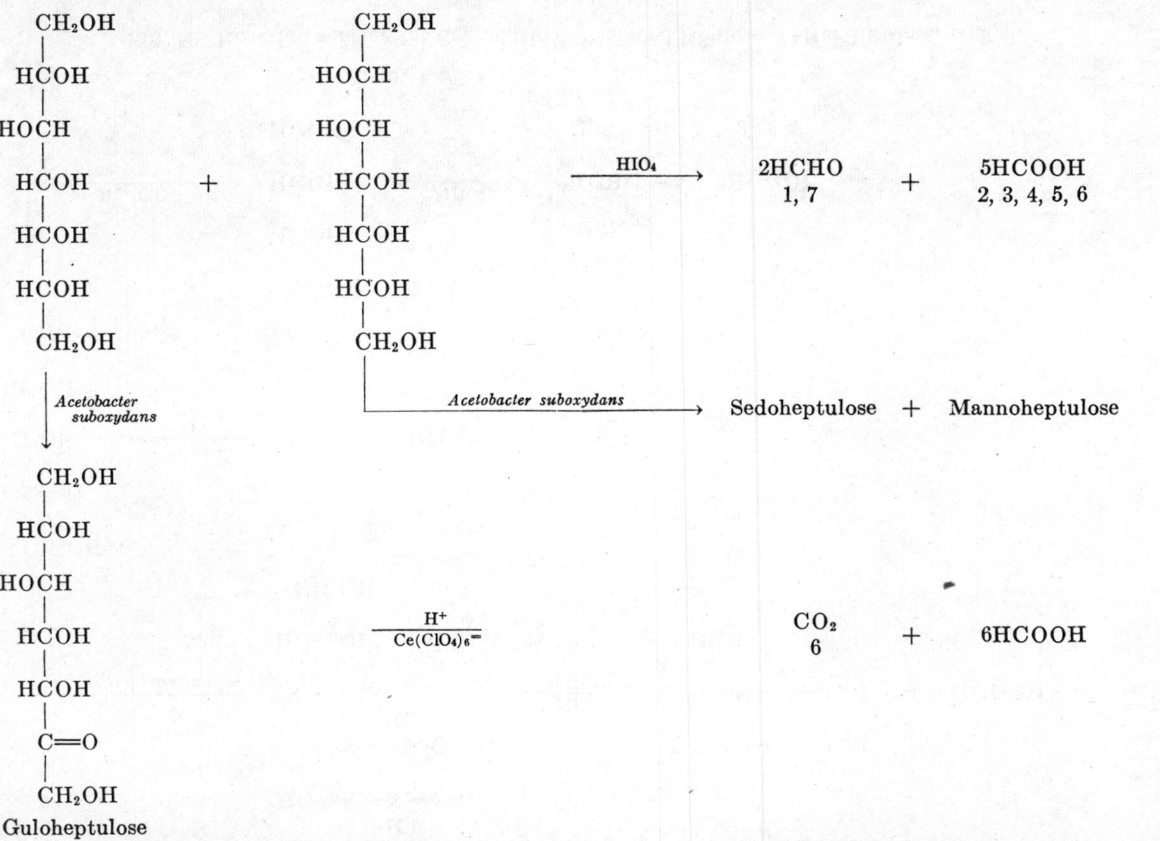

FIG. 49. Degradation procedures for analysis of labeling pattern in seduheptulose. (After Bassham *et al.*[18])

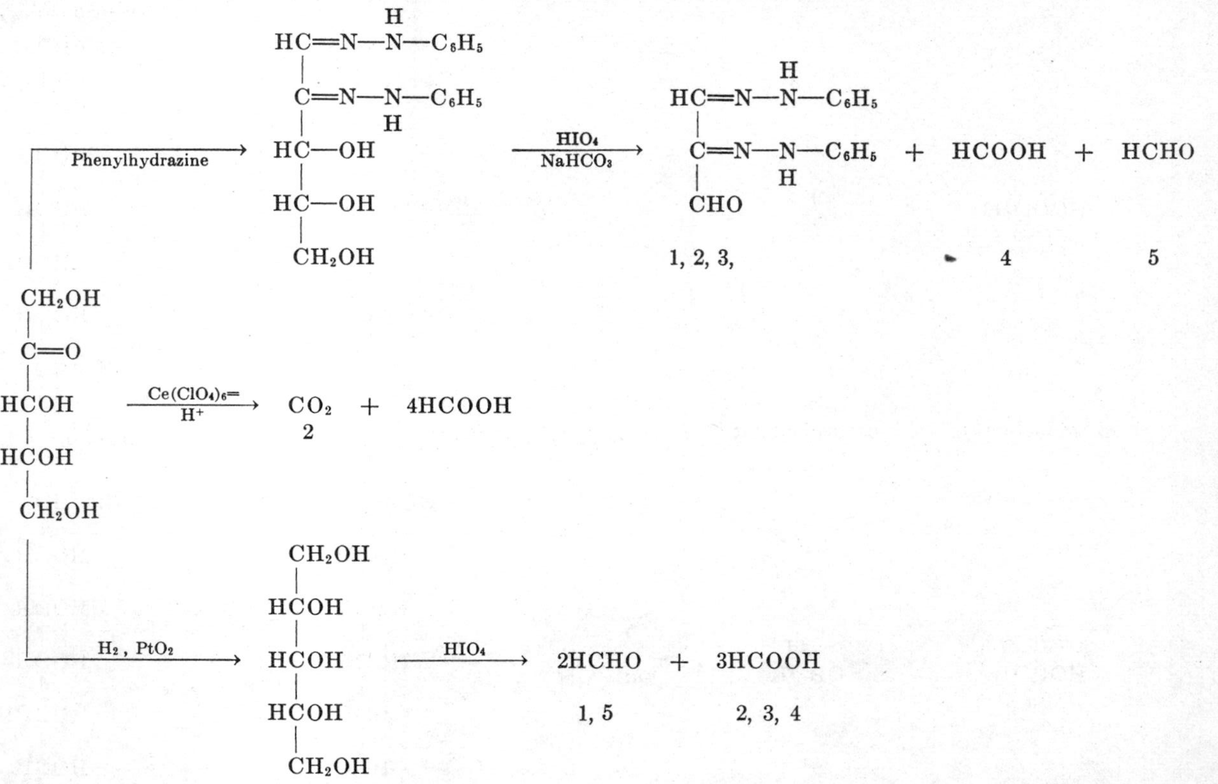

FIG. 50. Reaction scheme for degradation of labeled ribulose. (After Bassham *et al.*[18])

TABLE 13

DISTRIBUTION (IN %) OF C^{14} FROM CO_2 IN LABELED SEDOHEPTULOSE AND HEXOSE FROM SOYBEAN LEAVES. ROW "SEDUM" SHOWS LABELING IN LONG-TERM EXPOSURE OF SEDUM PLANTS (AFTER BASSHAM ET AL.[18])

Time (sec.)	Sedoheptulose							Hexose	
	C-4	C-1,2,3	C-4,5,6	C-7	C-2	C-1,7	C-6	C-1,2,3	C-4,5,6
0.4	8	33	57	0				47	52
0.8	18	43	60	2				48	51
1.5	24								
3.5	26					3			
5.0	29	36	64	2	4	4	4		
8.0	24								
10.0	28								
20.0	21	44	...	5	7				
300	14				12.5				
Sedum	12	37	35	12	12.5	28	15		

the ribulose with cerate and perchlorate gave carbon 2 as CO_2. The reaction scheme is shown in Fig. 50.

During these experiments the amount of label incorporated into the cells was shown to be a linear function of the time. As usual, the PGA showed the fastest incorporation; however, the percentage of label in this compound dropped off with time. The distribution of label in the various parts of a number of compounds from the soybean extracts is shown as a function of time in Table 13.

These results could be accounted for by using the pentose phosphate cycle beginning with carboxyl-labeled PGA. (It will be left as an exercise for the reader to verify this statement, by means of the reaction sequences in Table 12.) From these studies and others by the California group it appeared that the cycle involving initial carboxylation of ribulose diphosphate was in fact the long-sought mechanism involved in the light fixation of CO_2.

It was now possible for biochemists to proceed with some assurance to attempt the isolation of the various partial reactions indicated in this scheme. One observation made by the California group turned out to be of particular significance. When they studied only ribulose diphosphate (RuDP), phosphoglycerate (PGA), and triose phosphate (C_3P), the CO_2 cycle could be represented as follows:

$$\text{RuDP} \xrightarrow[\text{CO}_2]{} \text{PGA} \xrightarrow[h\nu]{} \text{C}_3\text{P} \quad (\text{C}_3\text{P} \rightarrow \text{RuDP})$$

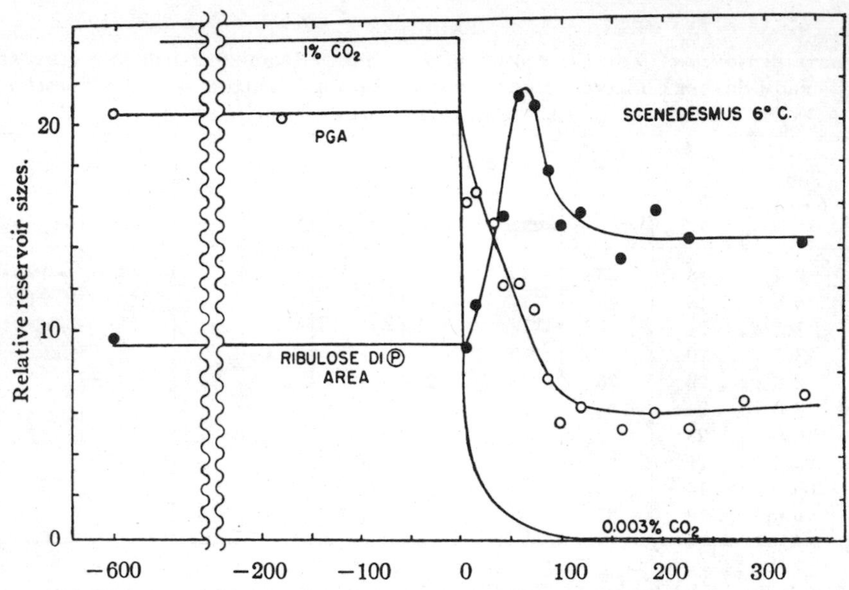

Fig. 51. Kinetic relations between RuDP and PGA. (After Wilson and Calvin.[22])

According to this scheme, RuDP would be the immediate precursor of PGA by way of a carboxylation reaction followed by a split to two PGA molecules. PGA, in turn, would be the source of triose phosphate for synthesis of cell material via reductive processes with light energy. It would be expected that in steady-state photosynthesis with labeled CO_2 sudden reduction of the CO_2 pressure would favor a short-term increase in the effective concentration or "pool size" of RuDP and a corresponding decrease in the pool size of PGA. Conversely, at constant CO_2 pressure a decrease in light intensity would be expected to decrease the RuDP pool and increase the PGA pool. In Fig. 51 are shown, in a simplified way, typical results[22] of experiments to determine relative sizes of the pools for PGA and RuDP consequent on a lowering of CO_2 pressure. In these experiments, algae placed in a flow system arranged to allow sudden changes in CO_2 pressure and sampling at short-time intervals were allowed to reach a steady state of photosynthesis in 1 % labeled CO_2 . Then they were suddenly shifted to a state in which the CO_2 pressure was only 0.003 %. As expected, the PGA dropped and the RuDP rose in amount. The corresponding effect of light in enhancing the RuDP pool at constant CO_2 pressure was also observed in other experiments.[23]

The significance of these observations on the precursor-product relation

[22] Wilson, A. T., and Calvin, M., *J. Am. Chem. Soc.* **77,** 5948 (1955).
[23] Calvin, M., and Massini, P., *Experientia* **8,** 445 (1952).

of RuDP and PGA in the living cell lay in the inference that activation by light was not necessary for the carboxylation reaction. As a result, enzyme chemists were encouraged to look for enzyme systems which would carry out a carboxylation of RuDP to PGA *in vitro*. Earlier, Horecker and Smyrniotis[24] had shown that enzyme preparations could be made from yeast which catalyzed the reversible fixation of CO_2 in phosphogluconic acid according to the reaction

Ribulose-5-phosphate + CO_2 + reduced triphosphopyridine nucleotide

+ $H^+ \rightarrow$ 6-Phosphogluconate + oxidized triphosphopyridine nucleotide

Weissbach *et al.*,[25] began a search for an analogous enzyme in green plants. They soon demonstrated that a pigment-free enzyme could be isolated and purified from spinach extracts which fixed labeled CO_2 in the carboxyl group of PGA with ribose-5-phosphate as substrate. Their preparation contained a pentose phosphate isomerase which brought about a conversion of ribose phosphate to ribulose phosphate. Later they prepared ribulose-1,5-diphosphate by reaction of ribulose-5-phosphate with phosphate in the presence of a specfic phosphokinase from spinach and were able to show that RuDP was the actual substrate in the reaction. Quayle *et al.*[26] in California presented evidence at about the same time suggesting that enzymatic carboxylation of RuDP could be carried out by cell-free preparations from *Chlorella*. As expected, the reaction was not light-sensitive.

Thus, by indicating the kind of reagents and products to be looked for, the studies of the California group made possible the isolation of the enzyme which, as far as present knowledge goes, is apparently responsible for the first step in CO_2 fixation in green plant photosynthesis.

Meanwhile, experiments by Kamen and collaborators[27] on CO_2 fixation during photoassimilation by photosynthetic bacteria had been proceeding along parallel lines. It was shown that fixation in PGA was apparently involved in anaerobic as well as aerobic fixation. In addition, it was demonstrated that phosphate esters were involved only when CO_2 was being fixed by these bacteria. It was found that a different pattern of fixation occurred if the substrate presented was acetate or a fatty acid. Thus, it appeared that the fate of carbon during photoassimilation was not uniquely deter-

[24] Horecker, B. L., and Smyrniotis, P. Z., *J. Biol. Chem.* **196**, 135 (1952).

[25] Weissbach, A., Smyrniotis, P. Z., and Horecker, B. L., *J. Am. Chem. Soc.* **76**, 3611 (1954).

[26] Quayle, J. R., Fuller, R. C., Benson, A. A., and Calvin, M., *J. Am. Chem. Soc.* **76**, 3610 (1954).

[27] Glover, J., Kamen, M. D., and Van Genderen, H., *Arch. Biochem. and Biophys.* **35**, 384 (1952).

mined by the photochemical act. However, these researches strengthened the conclusion that photochemically induced CO_2 fixation in general proceeded through the pentose phosphate cycle. More recently, these results have been confirmed and extended.[28] That this cycle may occur in a variety of nonphotosynthetic systems is indicated by reports[29, 29a] of the existence of the ribulose-PGA fixation system in some chemosynthetic autotrophs (*Hydrogenomonas*, *Thiobacillus*) as well as in a strain of the heterotroph *E. coli* adapted to ribose.[29b]

Racker[18a] has shown that net synthesis of carbohydrate from CO_2 and molecular hydrogen can be accomplished by using a mixture of purified enzymes—specifically phosphopentokinase, "RDP"-carboxylase, PGA-kinase, triose phosphate dehydrogenase, aldolase, triose phosphate isomerase, RDPase, transketolase, transaldolase, and hydrogenase (see Table 12). At this stage, then, tracer methods and enzyme chemistry seem to have combined to indicate a highly probable solution to the problem of CO_2 fixation in photosynthesis and in general autotrophy.

The tracer studies on CO_2 fixation in photosynthesis illustrate how precursor-product problems can be approached and how data derived from kinetic studies with labeled material can be used to infer a reaction sequence in intermediary metabolism.

B. Biosynthesis of Cholesterol

Cholesterol occurs generally in animal tissues. It has been of interest to biochemists and clinicians ever since it was isolated from biliary calculi in 1775. Its structure, which was established definitively about twenty years ago, and its numbering scheme are shown in Fig. 52. The four-ring nuclear structure carries a methyl group at carbons 10 and 13 and an hydroxyl at carbon 3. There is a double bond between carbons 5 and 6. All other carbons are saturated. The researches on the distribution of acetate carbon in cholesterol, which have given biochemists their first inkling of possible steps in the biosynthesis of this sterol and have oriented future research in this important field, afford an excellent example of the interplay of organic chemistry and labeling procedures.

One of the earliest observations in tracer research was the incorporation of tracer deuterium into endogenous cholesterol when adult rats were fed a diet containing heavy water,[30] from which it could be guessed that the

[28] Stoppani, A. O. M., Fuller, R. C., and Calvin, M., *J. Bacteriol.* **69,** 491 (1955)
[29] Vishniac, W., private communication (1955).
[29a] Burris, R., private communication (1955).
[29b] Fuller, R. C., private communication (1955).
[30] Schoenheimer, R., and Rittenberg, D., *J. Biol. Chem.* **114,** 381 (1938).

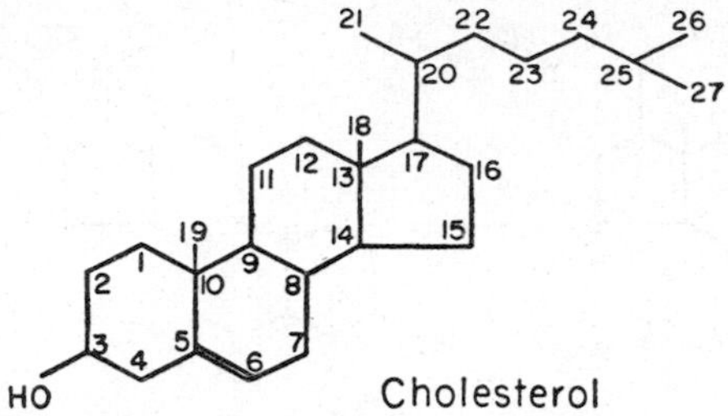

FIG. 52. Numbering scheme for cholesterol.

steady-state synthesis of cholesterol involved turnover of a large number of small molecules.[31]

It is difficult for biochemists of the present generation to imagine the impression this claim made on sterol chemists in the 1930's, when it was first advanced. No reactions which could account for the synthesis of so complex a molecule as cholesterol from small molecules were known. The mystery deepened when Bloch and Rittenberg[32] showed that acetate could serve as an efficient precursor for cholesterol. Since the early work was completed, other research has supplied ample evidence that acetate and substances that can be metabolized to acetate (e.g., acetoacetate, acetone, long-chain fatty acids, ethanol, and similar substances) can supply carbon to the cholesterol molecule.[33] It also appears that some of these compounds, notably acetoacetate[33-35] and the isopropyl moiety of isovalerate,[36] can act as precursors to some extent without preliminary conversion to acetate.

The first major advance after the initial researches of Schoenheimer and his collaborators resulted from the finding by Bloch *et al.*[37] that rat liver slices could incorporate enough acetate carbon to provide material for degradation studies. Little and Bloch[38] reported the first such research in which they examined the distribution of acetate carbon in the cholesterol side chain (carbons 20 to 27). Using doubly labeled acetate ($C^{13}H_3C^{14}OOH$), they found that somewhat more methyl than carboxyl carbon entered the whole molecule. It appeared that of the twenty-seven carbons of cholesterol fifteen originated from the acetate methyl. To degrade the side chain of

[31] See Schoenheimer, R., "The Dynamic State of Body Constituents," p. 21. Harvard U. P., Cambridge, 1946.

[32] Bloch, K., and Rittenberg, D., *J. Biol. Chem.* **143,** 297 (1942).

[33] Gurin, S., and Brady, R. O., *in* "Isotopes in Biochemistry," Ciba Foundation Symposium (G. E. W. Wolstenholme, ed.), p. 17. Blakiston, New York, 1952.

[34] Zabin, I., and Bloch, K., see p. 24 in reference 33.

[35] Curran, G. L., and Rittenberg, D., see p. 26 in reference 33.

[36] Zabin, I., and Bloch, K., *J. Biol. Chem.* **185,** 13 (1950).

[37] Bloch, K., Borek, E., and Rittenberg, D., *J. Biol. Chem.* **162,** 441 (1946).

[38] Little, H. N., and Bloch, K., *J. Biol. Chem.* **183,** 33 (1950).

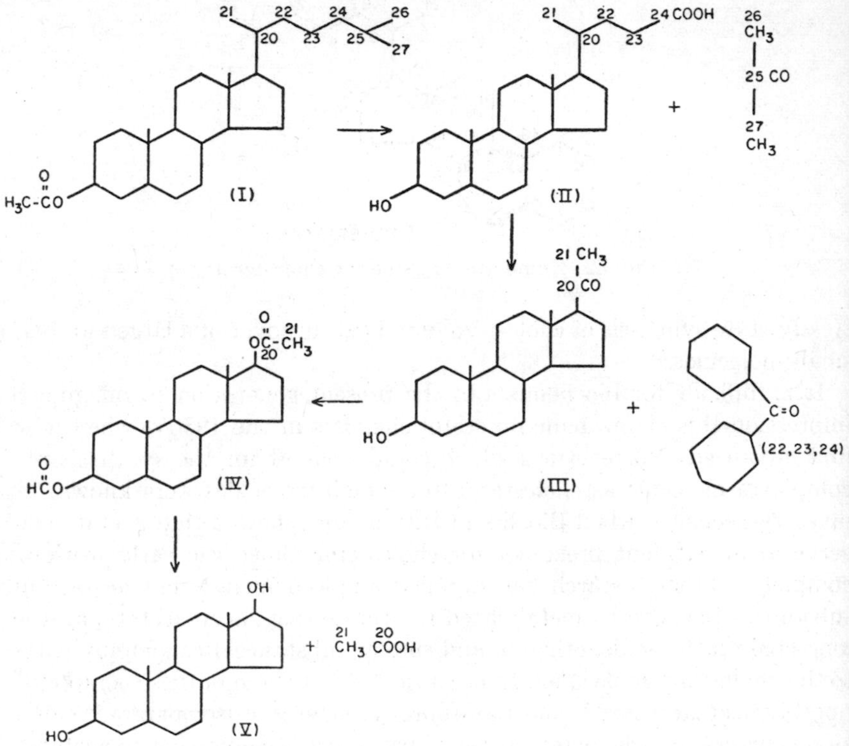

FIG. 53. Degradation scheme to establish origin of side-chain carbons in cholesterol (After Würsch *et al.*[39])

cholesterol so labeled they began by thermal fission of the chloride (chloride on carbon 3) which yielded a mixture of iso-C-8 hydrocarbons from the original side chain and a polycyclic residue from the nucleus. From other results on double labeling of the whole molecule they could predict that the ratios of C^{13} to C^{14} in the side-chain products and in the ring residue would be 5:3 and 10:9, respectively.

Enough earlier work had been done on sterol chemistry to make it possible to devise more definitive degradations. Basing their work on the researches which originally established the structure of the side chain, Würsch *et al.*[39] began with the oxidation of cholestanyl acetate (I) to give acetone-3,β-hydroxyallocholanic acid (II) (See Fig. 53).

The acetone which was derived from the isopropyl end of the chain (carbons 25, 26, and 27) was degraded further by the iodoform reaction (see p. 334). Next, the side chain of the acid was subjected to repeated degradations by the Barbier-Wieland procedure (see p. 336), which removed, in

[39] Würsch, J., Huang, R. L., and Bloch, K., *J. Am. Chem. Soc.* **195**, 439 (1952).

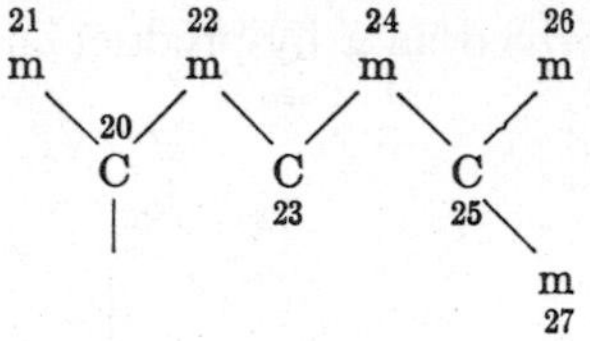

FIG. 54. Labeling sequence in side-chain carbons of cholesterol derived from acetate carbon (m = methyl, C = carbon).

order, carbons 24, 23, and 22, as benzophenone. The remainder of the molecule was isolated as allopregnanolone (III). The formic ester of this compound was oxidized with perbenzoic acid. This yielded the 17-acetyl-3-formyl derivative of androstanediol (IV) which could be hydrolyzed to formic acid. This in turn was oxidized with permanganate to give CO_2, leaving carbons 20 and 21 as acetic acid, and the residue as androstanediol (V).

Labeled cholesterol was synthesized with two batches of acetate. One contained label (C^{14}) in the carboxyl; the other was labeled in the methyl carbon. This procedure was equivalent, of course, to the previous one in which one batch containing doubly labeled carbon (C^{13} methyl, C^{14} carboxyl) had been used. After degradation by the procedures described above, it was found that the pattern of labeling was as shown in Fig. 54. Each carbon label alternated down the chain.

The origin of the two methyl carbons (positions 18 and 19) was indicated when Little and Bloch[38] examined the polycyclic residue they had previously obtained by thermal fission of the cholesteryl chloride. Assuming this residue to have the structure shown (VI), it was expected that oxida-

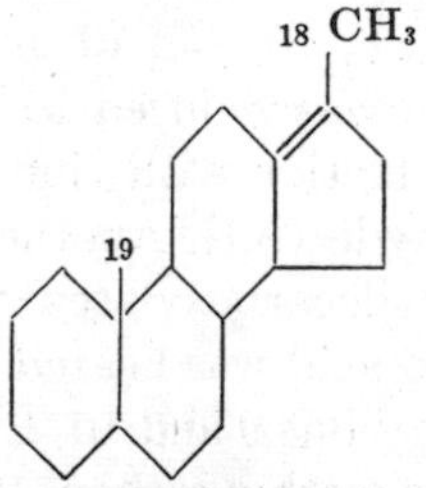

(VI)

tion with chromic-sulfuric acid, under the conditions used for determination of methyl groups attached to carbon, would yield two moles of acetic acid representing carbons 18 and 19 as acetate methyl and carbons 10 and 17 as acetate carboxyl. They found that label from methyl carbon of acetate had entered carbons 18 and 19 and either carbon 10 or carbon 17. Later, Woodward and Bloch[40] treated in a similar fashion a sample of

[40] Woodward, R. B., and Bloch, K., *J. Am. Chem. Soc.* **75**, 2023 (1953).

epiandrosterone (VII) derived as a by-product in the side-chain degrada-

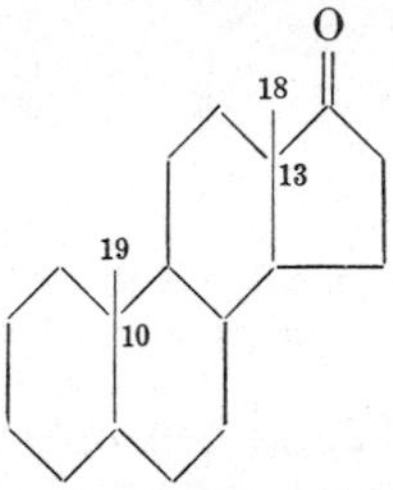

(VII)

tion of cholesterol which had been synthesized by liver slices from methyl-labeled acetate. The acetic acid isolated contained carbons 19 and 18 as methyl and carbons 13 and 10 as carboxyl. Further degradation of the acetate showed that the label in the acetate carboxyl was slightly more than one-half that of the methyl, which led to the conclusion that either carbon 10 or carbon 13 was originally derived from acetate. From other work by Popják and his collaborators,[40a] which will be described below, it was known that carbon 10 was derived from acetate carboxyl, so it could be inferred that carbon 13 was derived from acetate methyl.

Bloch[41] also showed the origin of carbon 7 by oxidative fission of the ring between carbons 7 and 8, leading to formation of a carboxyl group at position 7. The carboxyl group was removed and carbon 7 was isolated as CO_2. Labeled cholesterol treated in this fashion showed carbon 7 to have been derived from acetate methyl.

In parallel researches, Cornforth *et al.*[42] worked out degradation procedures for determining the origin of each carbon in the ring structure. They isolated carbons 1, 2, 3, 4, 5, 6, 10, and 19 in the following way (see Fig. 55). The cholesteryl chloride was reduced to cholest-5-ene (VIII), which was then ozonized (IX). Reduction with zinc in acid gave a product (X) which recyclized to an aldehyde (XI). Another ozonolysis gave an α-keto acid (XII), which could be decarboxylated to yield carbon 6 as CO_2. Another sample of the α-keto acid was heated at elevated temperatures to split the molecule between carbons 9 and 10. Ring carbons 1, 2, 3, 4, 5, and 10 appeared together with the methyl carbon, 19, as 2-methylcyclohexanone (XIII). This ketone was converted to the lactam of ε-aminoheptanoic acid (XIV) with acid azide, and finally the free amino acid (XV) was liberated by acid hydrolysis. Another product of the azide treatment of the ketone was the isomeric ε-aminomethylhexanoic acid lactam (XVI). This com-

[40a] See Popják, G., *Roy. Inst. Chem. (London) Lectures Monographs Repts. No.* **2**, 26 (1955).

[41] Bloch, K., *Helv. Chim. Acta* **36,** 1611 (1953).

[42] Cornforth, J. W., Hunter, G. D., and Popják, G., *Biochem. J.* **54,** 590, 597 (1953).

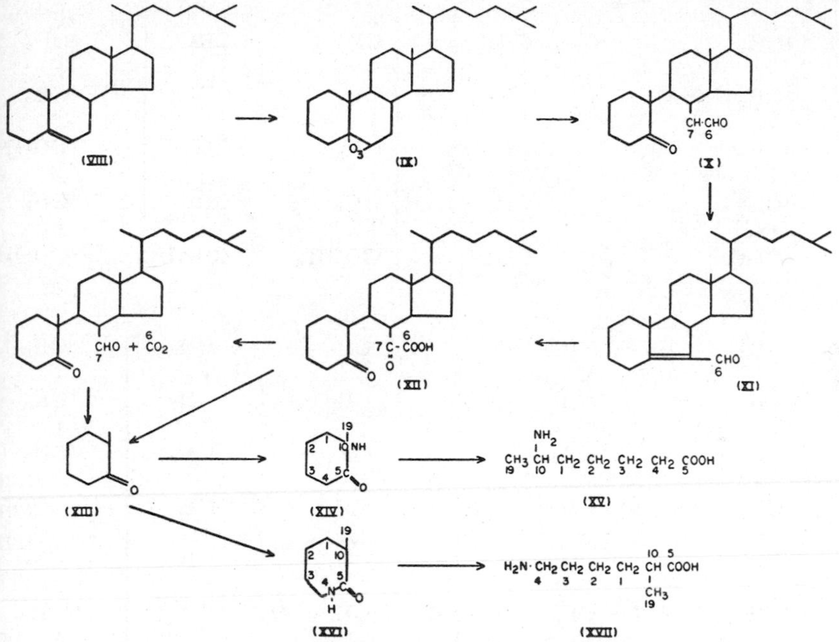

IG. 55. Degradation scheme for determining origin of some ring carbons in cholesterol. (After Cornforth *et al.*[42])

ound was hydrolyzed to the corresponding free acid (XVII). Each of the wo acids was degraded further as shown in the scheme of Fig. 56.

This degradation procedure, applied to a labeled cholesterol sample, howed that carbons 1, 3, 5, and 19 came from acetate methyl, and carons 2, 4, 6, and 10 from acetate carboxyl. The labeling pattern for this ortion of the cholesterol could be represented as in Fig. 57.

In another series of degradations (see Fig. 58), Cornforth and co-workers[42] egan with the benzoate of cholest-14(15)-en-3,β-ol (XVIII), which after zonolysis and reduction yielded a keto aldehyde (XIX). Thermal fission f this compound after solution in cyclohexane gave a cyclic hydroxyketone XX) and a volatile α,β-unsaturated aldehyde (XXI). The aldehyde conained the side chain and the ring carbons 15, 16, and 17. The aldehyde was onverted to the oxime (XXII). This was heated with alkali, whereupon mmonia was evolved. Further heating produced acetic acid and a C-9 cid (XXIII). The acetic acid represented carbons 15 and 16, which could e determined by the usual procedures. The carboxyl carbon of the longhain acid represented carbon 17. This was isolated by decarboxylation. 'he cyclic hydroxyketone was converted to the unsaturated ketone XXIV), which was oxidized to yield a dicarboxylic acid (XXV) and acetic cid from carbons 13 and 18.

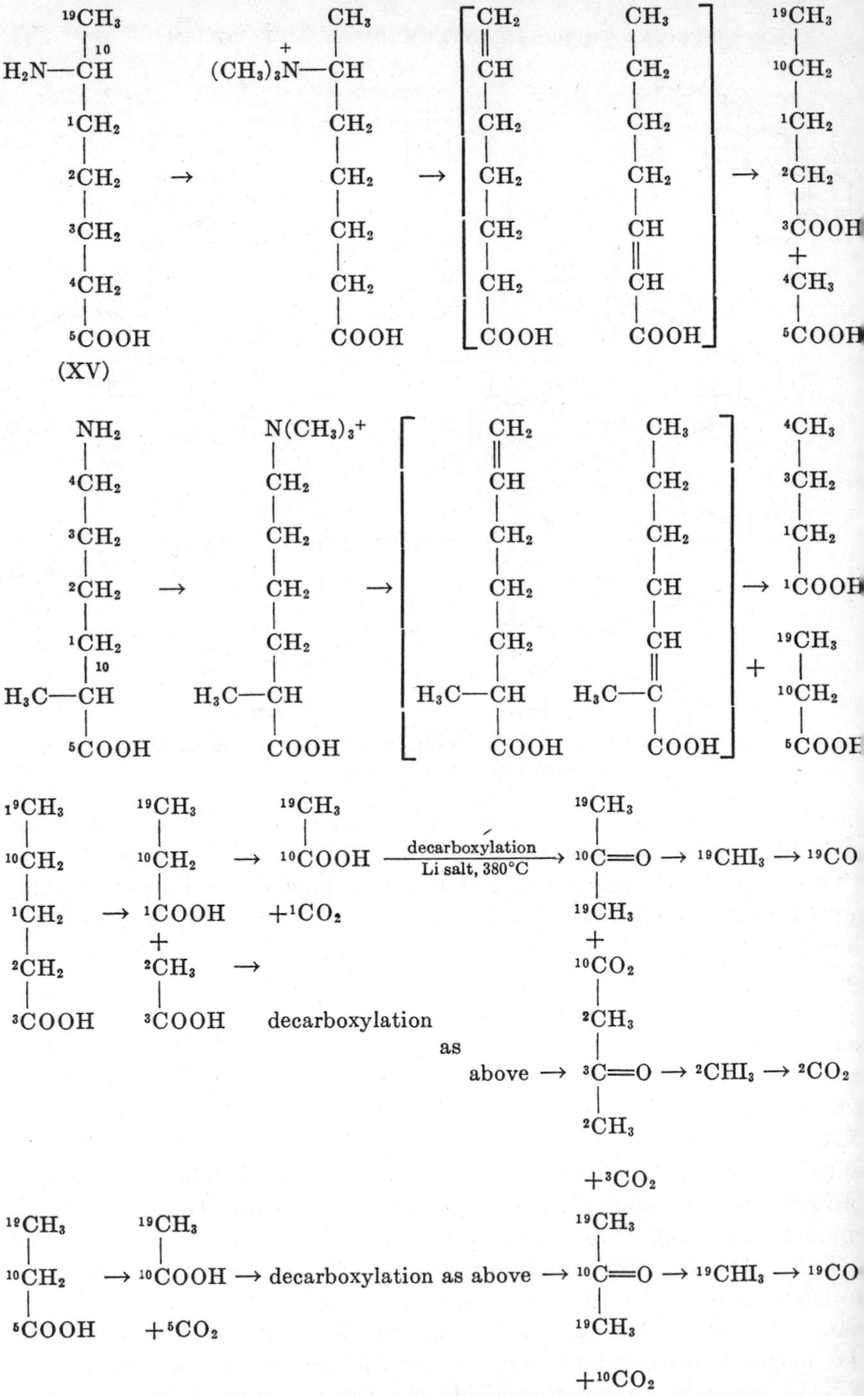

$$H_2N{-}{}^{10}CH({}^{19}CH_3){-}{}^{1}CH_2{-}{}^{2}CH_2{-}{}^{3}CH_2{-}{}^{4}CH_2{-}{}^{5}COOH \text{ (XV)} \rightarrow (CH_3)_3\overset{+}{N}{-}CH(CH_3){-}CH_2{-}CH_2{-}CH_2{-}CH_2{-}COOH \rightarrow$$

$$\left[CH_2{=}CH{-}CH_2{-}CH_2{-}CH_2{-}CH_2{-}COOH \quad CH_3{-}CH_2{-}CH_2{-}CH_2{-}CH{=}CH{-}COOH\right] \rightarrow {}^{19}CH_3{-}{}^{10}CH_2{-}{}^{1}CH_2{-}{}^{2}CH_2{-}{}^{3}COOH + {}^{4}CH_3{-}{}^{5}COOH$$

$$NH_2{-}{}^{4}CH_2{-}{}^{3}CH_2{-}{}^{2}CH_2{-}{}^{1}CH_2{-}{}^{10}CH(H_3C){-}{}^{5}COOH \rightarrow N(CH_3)_3^{+}{-}CH_2{-}CH_2{-}CH_2{-}CH_2{-}CH(H_3C){-}COOH \rightarrow$$

$$\left[CH_2{=}CH{-}CH_2{-}CH_2{-}CH_2{-}CH(H_3C){-}COOH \quad CH_3{-}CH_2{-}CH_2{-}CH{-}CH{=}C(H_3C){-}COOH\right] \rightarrow {}^{4}CH_3{-}{}^{3}CH_2{-}{}^{1}CH_2{-}{}^{1}COOH + {}^{19}CH_3{-}{}^{10}CH_2{-}{}^{5}COOH$$

$${}^{19}CH_3{-}{}^{10}CH_2{-}{}^{1}CH_2{-}{}^{2}CH_2{-}{}^{3}COOH \rightarrow {}^{19}CH_3{-}{}^{10}CH_2{-}{}^{1}COOH + {}^{2}CH_3{-}{}^{3}COOH$$

$${}^{19}CH_3{-}{}^{10}CH_2{-}{}^{1}COOH \rightarrow {}^{19}CH_3{-}{}^{10}COOH + {}^{1}CO_2 \xrightarrow[\text{Li salt, 380°C}]{\text{decarboxylation}} {}^{19}CH_3{-}{}^{10}C({=}O){-}{}^{19}CH_3 + {}^{10}CO_2;\quad {}^{10}C{=}O \rightarrow {}^{19}CHI_3 \rightarrow {}^{19}CO$$

$${}^{2}CH_3{-}{}^{3}COOH \rightarrow \text{decarboxylation as above} \rightarrow {}^{2}CH_3{-}{}^{3}C({=}O){-}{}^{2}CH_3 + {}^{3}CO_2;\quad {}^{3}C{=}O \rightarrow {}^{2}CHI_3 \rightarrow {}^{2}CO_2$$

$${}^{19}CH_3{-}{}^{10}CH_2{-}{}^{5}COOH \rightarrow {}^{19}CH_3{-}{}^{10}COOH + {}^{5}CO_2 \rightarrow \text{decarboxylation as above} \rightarrow {}^{19}CH_3{-}{}^{10}C({=}O){-}{}^{19}CH_3 + {}^{10}CO_2;\quad {}^{10}C{=}O \rightarrow {}^{19}CHI_3 \rightarrow {}^{19}CO$$

FIG. 56. Further degradations of products derived from cholesterol. (After Cornforth *et al.*[42])

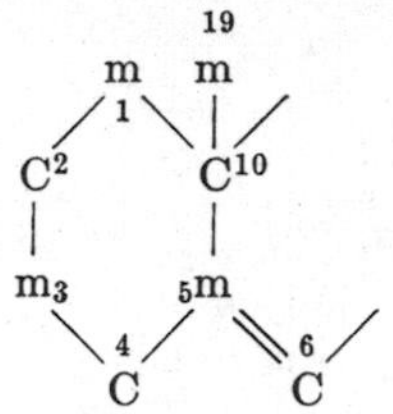

FIG. 57. Origin of some ring carbons from acetate carbon (m = methyl, C = carboxyl).

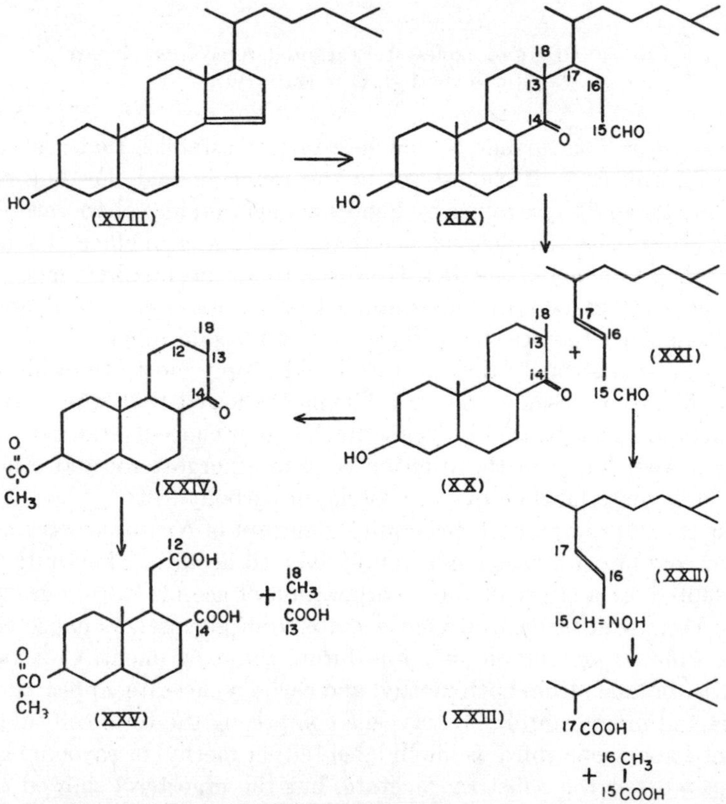

FIG. 58. Continuation of degradation procedures for establishing incorporation of acetate carbons in cholesterol. (After Cornforth *et al.*[42])

When this process was applied to labeled cholesterol, it was found that only carbon 16 came from acetate carboxyl; the other five carbons (15, 16, 17, 18, and 13) were all derived from acetate methyl. The total pattern of labeling indicated from the degradation studies could now be formulated as shown in Fig. 59.

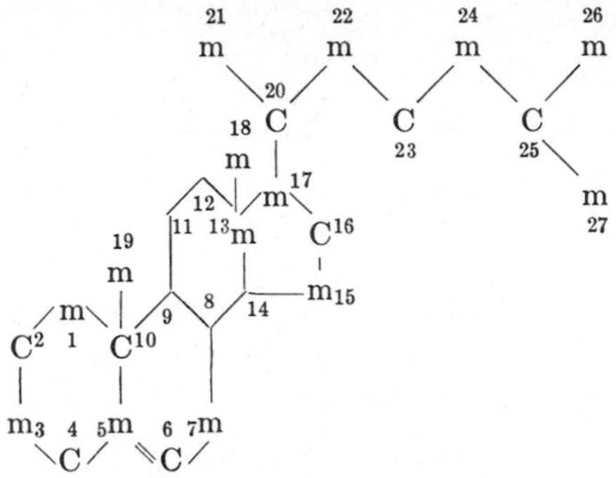

```
                      21         22         24         26
                      m          m          m          m
                       \20    /     \     /     \     /
                     18 C            C            C
                      m |            23           25\
                     /\ |  17                        m
                    /12\|/m\                         27
           19       |11 13m    C16
           m        |     |     |
        /m\ |   /\ 8 |/14  ---m15
      C2  1 C10  9  \/
      |     |        |
      m3 4 5m    6 7m
        \C/    \\C/
```

FIG. 59. Origin of cholesterol carbon derived from acetate (m = methyl, C = carboxyl).

Before proceeding further, it was necessary to establish that acetate was the actual precursor of cholesterol in the living animal. Feeding experiments and the early researches by Schoenheimer and his collaborators cited above had furnished ample evidence that acetate was an efficient source of carbon when supplied in the diet. However, there was no direct evidence to show how many carbons of cholesterol actually came from acetate and how many from other sources. Some data from studies on fungi were available which suggested strongly that acetate could supply most, if not all, of the cholesterol carbon. Sonderhoff and Thomas[43] showed that when deuterioacetic acid was metabolized by yeast much higher concentrations of acetate hydrogen were found in the fraction containing ergosterol and other unsaponifiable material than in fat, protein, or carbohydrate.

Ottke *et al.*[44] grew an acetate-requiring mutant of *Neurospora crassa* on a medium containing glucose and doubly labeled acetate. This mutant was well adapted for a study of the incorporation of acetate during growth on glucose because it could not form acetate from glucose. As expected, the acetate suffered no isotopic dilution during the experiment. Considerable quantities of label, from both methyl and carboxyl acetate, appeared in the mycelia and in ergosterol. The mycelia comprising the total cell substance contained about one-third as much label (either methyl or carboxyl carbon from acetate) as the substrate acetate, but the ergosterol showed an isotope concentration which was almost the same as that of the acetate. This experiment demonstrated that more than twenty-six out of the twenty-eight carbon atoms of ergosterol had been derived from the acetate in the medium.

[43] Sonderhoff, R., and Thomas, H., *Ann.* **530,** 625 (1937).

[44] Ottke, R. C., Tatum, E. L., Zabin, I., and Bloch, K., *J. Biol. Chem.* **189,** 429 (1951).

Although it was not sure that results obtained with one sterol in a fungus could be extrapolated to another sterol in a different system, these data encouraged belief in the hypothesis that acetate could serve as a major precursor of cholesterol in animal tissues. With the results of the labeling experiments showing the origin of the various carbons of cholesterol as either methyl or carboxyl carbon of acetate arranged in an alternating pattern, it was possible to formulate schemes whereby such labeling could be obtained, beginning with reactions involving acetate or compounds in equilibrium with it.

An observation which turned out to be fruitful in this connection was made by Bonner and Arreguin in 1949.[45] They reported that acetic acid and β,β-dimethylacrylic acid stimulated the production of rubber in guayule seedlings. Because the rubber hydrocarbons consist of branched C-5 ("isoprenoid") units, these authors suggested that each of these units could be synthesized by condensation of acetate with itself to form acetoacetate, which could be decarboxylated to acetone. Further reaction between the acetone and more acetate would form the desired branched hydrocarbon. The reaction sequence beginning with doubly labeled acetate (M—C—) can be shown schematically as follows:

2M—C— → M—C—M—C— → M—C—M + CO_2 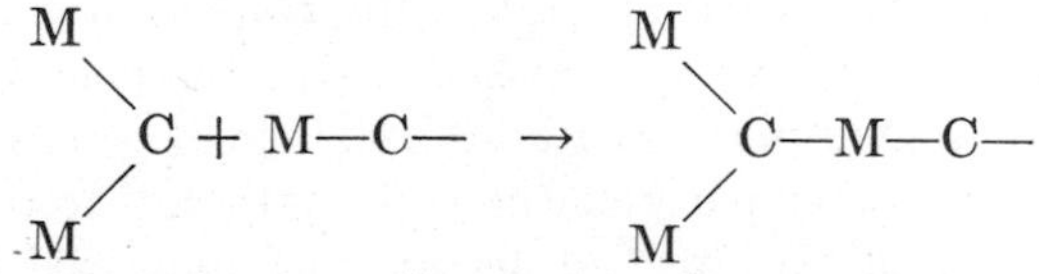
Acetic acid   Acetoacetic acid   Acetone

```
M                          M
 \                          \
  C + M—C—    →              C—M—C—
 /                          /
M                          M
Acetone    Acetic acid    Isoprenoid unit
```

Polymerization to rubber would give a labeling sequence of the type

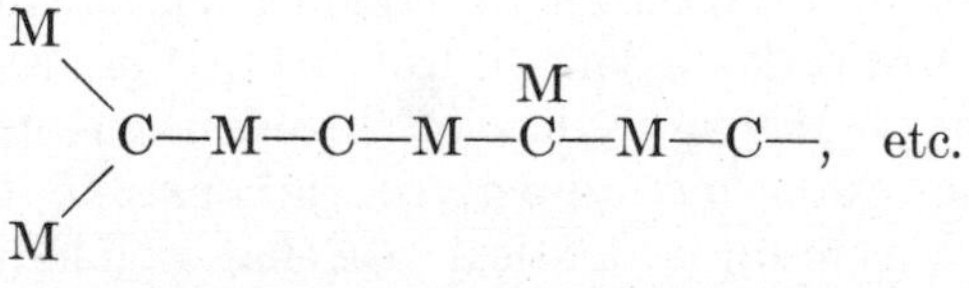

```
M
 \          M
  C—M—C—M—C—M—C—, etc.
 /
M
```

Bloch[46, 47] noted the identity of such a labeling pattern with that seen in cholesterol. He also recalled that Robinson[48] had pointed out in 1932 a remarkable relation between squalene, a triterpenoid hydrocarbon ($C_{30}H_{50}$), and cholesterol—namely, that squalene had a structure based on isoprenoid

[45] Bonner, J., and Arreguin, B., *Arch. Biochem. and Biophys.* **21,** 109 (1949).

[46] Bloch, K., *in* "Isotopes in Biochemistry," Ciba Foundation Symposium (G. E. W. Wolstenholme, ed.), p. 24. Blakiston, New York, 1952.

[47] Referred to by Bloch, K., *Harvey Lectures Ser.* **48,** 68 (1953).

[48] Robinson, R., *J. Soc. Chem. Ind. London* **51,** 464 (1932).

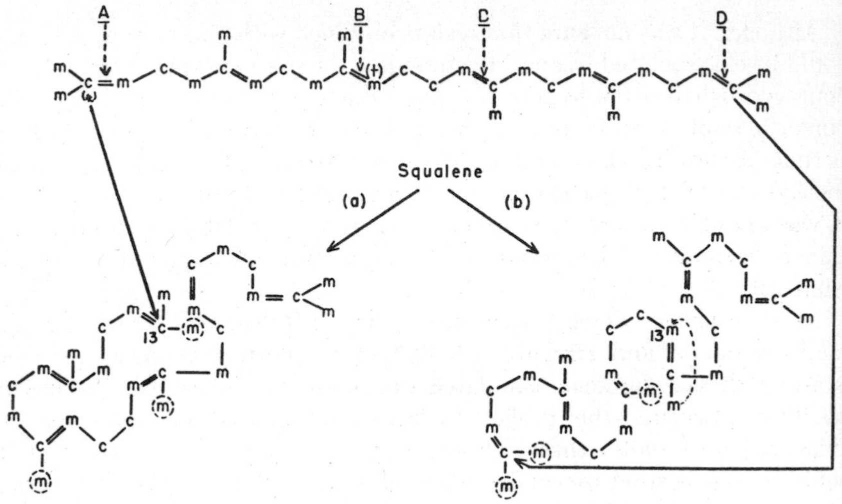

FIG. 60. Scheme to show how cyclization of squalene could lead to formation of cholesterol (see text).

units which could be coiled into a structure identical with that of cholesterol. If this process actually occurred in living tissue, then squalene might be considered an intermediate in the biosynthesis of cholesterol.

To see the structural relations clearly, the reader is referred to Fig. 60. The squalene molecule is shown labeled in the manner expected if it is utilized directly in cholesterol synthesis. The cyclization process, as suggested by Robinson, takes place as shown in path *a*. The terminal carbon, originally derived from carboxyl of acetate and indicated by an asterisk, is folded over to make carbon 13 of cholesterol. The carbons (all derived from acetate methyl) which are indicated in dotted circles are eventually lost. A consequence of this mechanism is that each isoprene unit of squalene must consist of three carbons derived from methyl of acetate and two derived from carboxyl. Moreover, loss of three carbons during cyclization leaves a total of fifteen methyl-derived carbons and twelve carboxyl-derived carbons. This result is identical with that originally found by Little and Bloch[38] (see p. 119). In one respect, however, the scheme fails. Degradation data show that carbon 13 should be derived from methyl carbon (see Fig. 59), whereas the method of folding shown in path *a* leads to placement of a carboxyl-derived carbon in position 13.

Bloch[47] suggested that another kind of folding would result in a structure in harmony with the findings of the degradation experiments. This mechanism is shown in path *b*, wherein carbon 4 comes from the terminal carboxyl-derived carbon at the right end of the chain. This results in the methyl-derived carbon, marked by the dagger, being placed in position 13.

The rest of the labeling conforms to the experimental results, as do those reached by path *a*. It is necessary in path *b* that one of the branched methyl-derived carbons of squalene migrate to carbon 13 during cyclization.

Cornforth and Popják[49] tested the squalene hypothesis more directly by determining the labeling pattern of squalene isolated from rat liver slices incubated with doubly labeled acetate. (The fact that such labeled squalene was synthesized from labeled acetate was first demonstrated by Langdon and Bloch,[50] who fed rats squalene and labeled acetate. A small amount of the squalene was found in the liver, isolated, and found to contain significant amounts of radioactivity. These workers also showed that in mice this labeled squalene was converted to cholesterol more efficiently than was acetate. Later, biosynthesis of squalene from acetate was observed in rat liver slices[51] and in perfused pig liver.[52] Cornforth and Popják degraded the labeled squalene from rat liver slices incubated with labeled acetate in the following manner. The squalene was subjected to oxidative ozonolysis. Referring to Fig. 60, the reader will see that there would result one molecule of acetone, from the terminal isopropyl group broken off at point *A*. Between *A* and *B*, ten more carbons would be split off to give two molecules of levulinic acid ($CH_3COCH_2CH_2COOH$). Between *B* and *C*, four more carbons would yield one molecule of succinic acid. Similarly, fission between *C* and *D* would give two molecules of levulinic acid and one molecule of acetone. The determination of label distribution in the levulinic acid fragments would have sufficed to establish the validity of the hypothesis, but the acetone and succinate were also degraded to check the results obtained from study of the origin of the levulinic acid carbon.

The levulinic acid was converted to the phenylhydrazone, which was reduced with Al-Hg amalgam to 4-aminopentanoic acid. Methylation to the betaine, followed by fusion with KOH at 350°C., gave an almost quantitative yield of acetic acid representing the carboxyl and α carbons of the levulinic acid, and propionic acid corresponding to the remainder. The two acids were degraded further, carbon by carbon, in the usual manner. Succinic acid was converted by means of the Curtius rearrangement according to the following reactions:[49]

$$(a)\quad HOO\overset{1}{C}—\overset{2}{C}H_2—\overset{2}{C}H_2—\overset{1}{C}OOH \xrightarrow{CH_2N_2} CH_3OO\overset{1}{C}—\overset{2}{C}H_2—\overset{2}{C}H_2—\overset{1}{C}OOCH_3$$

[49] Cornforth, J. W., and Popják, G., *Biochem. J.* **58,** 403 (1954).

[50] Langdon, R. G., and Bloch, K., *J. Biol. Chem.* **200,** 129, 135 (1953).

[51] Popják, G., *Arch. Biochem. and Biophys.* **48,** 102 (1954).

[52] Schwenk, E., Todd, D., and Fish, C. A., *Arch. Biochem. and Biophys.* **49,** 187 (1954).

$$(b)\quad CH_3OO\overset{1}{C}—\overset{2}{C}H_2—\overset{2}{C}H_2—\overset{1}{C}OOCH_3 \xrightarrow{H_2N—NH_2} H_2N—NH—O\overset{1}{C}—\overset{2}{C}H_2—\overset{2}{C}H_2—\overset{1}{C}O—NH—NH_2$$

$$(c)\quad H_2N—NH—OC—CH_2—CH_2—CO—NH—NH_2 \xrightarrow[HCl]{NaNO_2} N_3O—\overset{1}{C}—\overset{2}{C}H_2—\overset{2}{C}H_2—\overset{1}{C}ON_3$$

$$(d)\quad N_3O—\overset{1}{C}—\overset{2}{C}H_2—\overset{2}{C}H_2—\overset{1}{C}ON_3 \xrightarrow{C_2H_5OH} C_2H_5OO\overset{1}{C}—NH—\overset{2}{C}H_2—\overset{2}{C}H_2—NH—\overset{1}{C}OOC_2H_5$$

$$(e)\quad C_2H_5OO\overset{1}{C}—NH—\overset{2}{C}H_2—\overset{2}{C}H_2—NH—\overset{1}{C}OOC_2H_5 \xrightarrow[\Delta]{HBr} HBr—H_2N—\overset{2}{C}H_2—\overset{2}{C}H_2—NH_2—HBr + 2\overset{1}{C}O_2 + 2C_2H_5OH$$

Acetone was treated with alkaline hypoiodite in the usual way to yield iodoform and acetate. The latter in turn was degraded by pyrolysis to acetone and carbonate, the acetone being again converted to iodoform, and so forth.

It was found from these studies that the carbon was distributed in the squalene molecule exactly in the manner expected (see Fig. 60). These data provided substantial support for the hypothesis of a close relation between squalene and cholesterol, but they did not prove that squalene was a direct precursor.

The difficulty arising in this connection is one which occurs frequently in the interpretation of tracer researches on precursor-product relations and is occasioned by *uncertainty about degree of equilibration between different endogenous pools which may exist for any given metabolite.* In the squalene-cholesterol case, it is true that feeding of labeled squalene in experiments with rat liver slices results in much more highly labeled cholesterol than when the source of the label is acetate.[50] However, this could happen if both squalene and cholesterol were derived from a common precursor according to the following scheme:

$$\begin{array}{c} \text{Acetate} \rightarrow \text{C-5 unit} \rightarrow \text{“x”} \rightarrow \text{Cholesterol} \\ \downarrow\uparrow \\ \text{Squalene} \end{array}$$

The experiments of Popják[51] are illuminating in this regard. He observed that in certain membranes of ovarian tissue from the laying hen there is a marked activity in synthesizing cholesterol from acetate. These membranes (granulosa and theca interna) were incubated in a physiological medium (Krebs-Henseleit solution, pH 7.4) in the presence of 0.3 % glucose and the

TABLE 14
BIOSYNTHESIS OF SQUALENE AND CHOLESTEROL *IN VITRO* FROM 1-C^{14}-ACETATE (AFTER POPJÁK[51])

Experiment	Tissue	Fresh weight (g.)	Acetate added: μM.	Acetate added: μc. C^{14}	Carrier squalene (μM.)	Cholesterol recovered (μM.)	Specific activity (1 × 10^{-3} μc./μM.): Squalene	Specific activity (1 × 10^{-3} μc./μM.): Cholesterol	Acetate incorporated (1 × 10^{-3} μM./g. fresh tissue): Squalene	Acetate incorporated (1 × 10^{-3} μM./g. fresh tissue): Cholesterol
1	Ovarian membranes	1.52	26	21.5	122	9.6	1.25	40.6	121.00	310
2	Ovarian membranes	1.50	26	21.5	122	11.8	0.035	39.8	3.33	379
3	Ovarian membranes	1.60	19	54	122	12.4	0.099	70.0	2.69	191
4	Ovarian membranes	1.62	19	54	122	16.8	0.134	54.3	3.58	198
5	Ovarian slices	7.30	38	108	244	53.6	0.205	10.1	2.41	26
6	Rat liver slices	5.20	18	51	122	32.7	0.165	94	1.35	208
7	Rat liver slices	5.20	18	51	122	30.0	119.0	2.50	2.50	242

amounts of acetate shown in Table 14. In experiments 1, 3, 5, and 6, carrier squalene was added at the start of the incubation. In experiments 2, 4, and 7, it was added at the end. The incubation period was 4 hr. The experiment was ended by addition of ethanol to give a final concentration of 50%. The squalene and cholesterol fractions were isolated and purified.

It is evident that both tissues produced appreciable quantities of labeled squalene and cholesterol. With the exception of the first experiment, the amount of acetate carbon appearing in the squalene was only about one-hundredth that in the cholesterol. This could be explained by assuming that there was a very rapid turnover of squalene (see below). On the assumption that the amount of endogenous squalene in the liver slices was about 0.3 μM. (based on data obtained from estimates by Langdon and Bloch[50]), the original specific activities of the endogenous squalene would have been greater than those found in the two experiments with liver slices (experiments 6 and 7) in the ratio of the amounts of exogenous and endogenous squalene, i.e., 122/0.3 times the values given in Table 14, or 0.066 and 0.123 μc./μM. for experiments 6 and 7, respectively. From the data in the table, it would follow that in the sixth experiment, where there was 0.208 μM. of acetate incorporated into cholesterol per gram wet weight of tissue, there would have been (0.208 × 51 × 5.20) 18, or 3.06 μc. of C^{14} in cholesterol. Similarly, in experiment 7, there would have been 3.56 μc. To provide this amount of label through squalene with the specific activities estimated above would have required 3.06/0.066 = 46, and 3.56/0.123 = 29 μM. of squalene, respectively. These amounts are greater than those actually present by 100- to 150-fold. (This result merely reflects the original finding that the specific activities of the squalene were low by amounts of the same magnitude compared to cholesterol activities.) To obtain these quantities of endogenous squalene, it would be necessary to postulate several hundred turnovers of squalene. Then it would follow that the squalene would have become saturated with C^{14} from acetate and would have exhibited the maximum specific activity, which could be calculated, assuming formation of isoprenoid units from acetate in the manner described previously (see p. 207), to be 34 μc./μM.

If it is assumed that only a very small fraction of the endogenous pool of squalene was available, then the result obtained could fit the postulate that squalene is a direct precursor of cholesterol. This notion can be rephrased from the standpoint of enzyme chemistry as a postulate that the newly synthesized squalene exists only as substrate bound firmly to enzyme and is not in equilibrium with a "free" squalene pool in the tissue. In any case, the reader will see that a negative result in an experiment of the kind illustrated in Table 14 cannot be interpreted unambiguously. Other examples of this type of uncertainty abound in the literature.

The status of the cholesterol problem is not so well advanced as that of the topic discussed in the previous section—CO_2 fixation in photosynthesis—where a definite compound, PGA, has been established as part of a cycle in which ribulose diphosphate is a precursor for the fixation of CO_2 into PGA. On the other hand, there are indications that certain small-molecule intermediates could be involved in the general biosynthesis of terpenes on the one hand and sterols on the other. Bloch[47] has suggested that β,β-dimethylacrylic and β-hydroxy; β-methylglutaric acids may be such intermediates. It has been reported that these substances are used efficiently in cholesterol synthesis when included in the diet of rats.[53] Furthermore these acids can be synthesized in rat liver homogenates from acetate and show a label distribution which fits into the general scheme for C-5 unit synthesis and incorporation into terpenes and sterols.[53a]

[53] Bloch, K., Clarke, L. C., and Harary, J., *J. Am. Chem. Soc.* **76**, 3859 (1954).
[53a] Rudney, H., *J. Am. Chem. Soc.* **76**, 2595 (1954).

More recently, it has been found, with rat liver slices, that acetate can be incorporated into the C-30 sterol, lanosterol, which in turn can be converted to cholesterol.[53b] Intact microsomes from rat liver homogenates also appear able to cyclize squalene to lanosterol and then to demethylate it to cholesterol. If such preparations are further split into a particulate and a soluble fraction, the particulate fraction resulting can cyclize squalene to lanosterol but cannot finish the synthesis of cholesterol by demethylation. These results indicate that C-30 sterols are intermediates in the conversion of squalene to cholesterol.[53c]

C. Biosynthesis of Porphyrin

Porphyrins are derivatives of the cyclic *porphin*, which is a ring system containing four pyrrole units linked by methene carbons. The whole structure comprises a sixteen-membered ring with twelve carbons and four nitrogens (see Fig. 61). The porphyrins are made from porphin by addition of various side groups. A number of porphyrins are of unique importance in biology; *protoporphyrin* (XXVIII), which is characterized by the structure shown in Fig. 61, is especially important. Other porphyrins found in nature are *coproporphyrin* (XXIX) and *uroporphyrin* (XXVII). The importance of protoporphyrin lies in the fact that its iron salt is *heme*, the prosthetic group for a series of proteins and enzymes (catalases, peroxidases, hemoglobins, and certain cytochromes) required in transport and utilization of molecular oxygen. The numbering system used is shown in Fig. 61. Uroporphyrin is taken as the prototype because it has the largest number of carbons in the side chains. A large number of isomers is possible, resulting from various placements of the side chains. Those indicated in the figure are the natural isomers *uroporphyrin* III, *protoporphyrin* IX, and *coproporphyrin* III.

As an example of a skillful interweaving of biochemical and organic chemical research with tracer methodology, the studies on the biosynthesis of porphyrin have not been surpassed.

It had already been established (see p. 145) that glycine nitrogen was an efficient precursor of pyrrole nitrogen in porphyrin in the intact animal. When these experiments were repeated with carboxyl-labeled glycine fed or injected,[54, 55] it was shown that glycine carboxyl was not utilized in porphyrin synthesis even though it was incorporated in the globin associated with the porphyrin. Methyl-labeled glycine, on the other hand, proved to be an efficient precursor of porphyrin carbon.[56]

[53b] Clayton, R. B., and Bloch, K., *Federation Proc.* **14,** 194 (1955).

[53c] Tchen, T. T., and Bloch, K., *J. Am. Chem. Soc.* **77,** 6083 (1955).

[54] Grinstein, M., Kamen, M. D., and Moore, C. V., *J. Biol. Chem.* **174,** 101 (1948).

[55] Radin, N. S., Rittenberg, D., and Shemin, D., *J. Biol. Chem.* **184,** 755 (1950).

[56] Altman, K. J., Casarett, G. W., Masters, R. E., Noonan, T. R., and Salomon, K., *J. Biol. Chem.* **176,** 319 (1948).

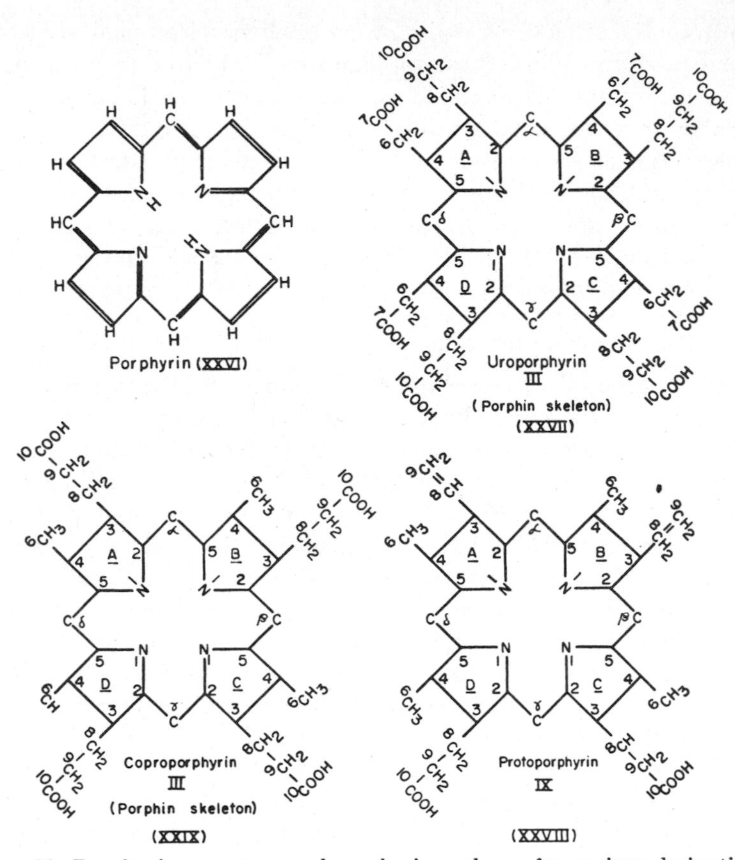

FIG. 61. Porphyrin structure and numbering scheme for various derivatives.

With the demonstration that an *in vitro* system—a suspension of nucleated red cells from duck blood—could incorporate glycine carbon and nitrogen just as in the intact animal,[57] it was possible to begin quantitative work on the actual nature of the precursor material. The major advantage of the *in vitro* system lay in the absence of many side reactions which could occur in the intact animal, notably conversion of glycine to some C-2 compound such as acetate which might also contribute carbon to porphyrin. In one experiment, Radin *et al.*[58] incubated duck blood with doubly labeled glycine, $N^{15}H_2C^{14}H_2 \cdot COOH$. The fraction of heme synthesized could be determined from the ratio of the N^{15} content in heme compared to glycine fed. When this value was known, it was relatively simple to calculate the number of carbon atoms originating from the α carbon of glycine.

[57] Shemin, D., London, I. M., and Rittenberg, D., *J. Biol. Chem.* **173,** 799 (1948).
[58] Radin, N. S., Rittenberg, D., and Shemin, D., *J. Biol. Chem.* **184,** 745 (1950).

There are thirty-four carbon atoms in heme. The fraction of newly synthesized heme was equal to the ratio of C^{14} in heme to that in the α carbon of glycine corrected for the factor A/N where N was the number of glycine carbons utilized and A was the total number of heme carbons (thirty-four). N was determined to be 8. Since it was known that all four nitrogens could be derived from glycine nitrogen, this result showed that for every glycine nitrogen used, two α carbons could be incorporated. Wittenberg and Shemin[59] and Muir and Neuberger[60] in England conducted some parallel researches which showed that glycine nitrogen was used for all the pyrrole nitrogens symmetrically and that half of the glycine carbon was used for the bridge methene carbons.

An early observation had been made by Bloch and Rittenberg[61] showing that deuteroacetate fed to rats resulted in the production of deuteroheme. With this result in mind, both the American[55] and British workers[62] were able to show that both carbons of acetate could contribute to the synthesis of heme. From the results already at hand, however, it was not expected that any of this acetate carbon would find its way into positions other than on the side chains or in the heme carbons 3, 4, and 5. This expectation was confirmed by means of some partial degradations of heme labeled by incubation of duck blood with labeled acetate.[55, 57, 62] From these studies the origin of fourteen carbon atoms in the heme was established. The four bridge-methene carbons came from the α carbon of glycine, the carbons in position 10 of rings C and D from acetate carboxyl, and the eight carbon atoms of positions 4 and 6 in the four rings from acetate methyl.

Shemin and Wittenberg[63, 64] proceeded to elaborate a procedure for the complete degradation of porphyrin, carbon by carbon. The reactions are shown in Fig. 62. Heme, split from hemoglobin, was treated with powdered iron in boiling formic acid (method of Fischer and Pützer[65]) to form protoporphyrin (XXVIII) which was then reduced with a colloidal palladium catalyst in hydrogen (method of Granick[66]) to yield mesoporphyrin (XXX). This compound was split oxidatively, according to the procedure of Küster,[67] to α-ethyl-α-methyl maleimide (XXXI) arising from rings A and B, and to hematinic acid (XXXII) arising from rings C and D. The hematinic acid was decarboxylated to CO_2 representing carbons in position 10 from

[59] Wittenberg, J., and Shemin, D., *J. Biol. Chem.* **178,** 47 (1949).
[60] Muir, H. M., and Neuberger, A., *Biochem. J.* **45,** 163 (1949).
[61] Bloch, K., and Rittenberg, D., *J. Biol. Chem.* **159,** 45 (1945).
[62] Muir, H. M., and Neuberger, A., *Biochem. J.* **47,** 97 (1950).
[63] Wittenberg, J., and Shemin, D., *J. Biol. Chem.* **185,** 103 (1950).
[64] Shemin, D., and Wittenberg, J., *J. Biol. Chem.* **192,** 315 (1951).
[65] Fischer, H., and Pützer, B., *Z. physiol. Chem.* **154,** 39 (1926).
[66] Granick, S., *J. Biol. Chem.* **172,** 717 (1948).
[67] Küster, H., *Ber.* **45,** 1935 (1912).

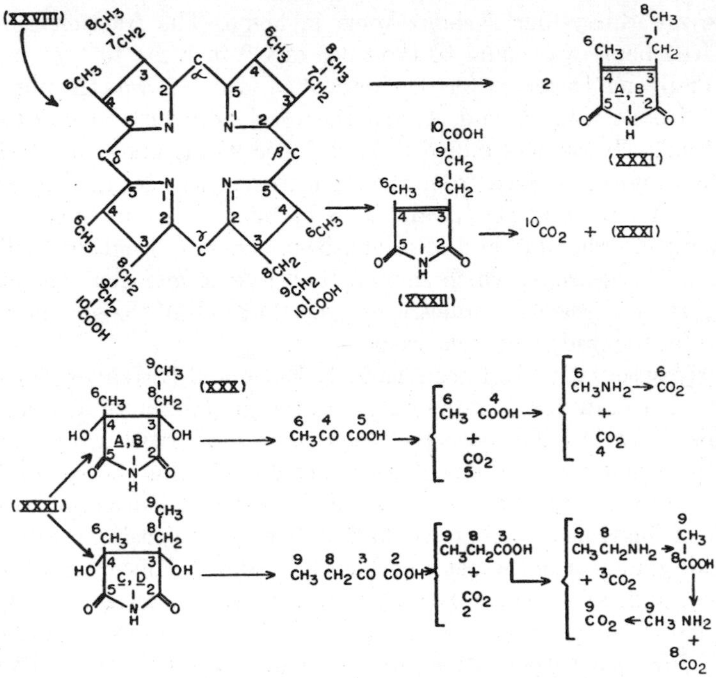

FIG. 62. Degradation scheme to establish origin of heme carbons. (After Shemin and Wittenberg.[63, 64])

rings *C* and *D*. The residue ethyl methyl maleimide was oxidized to α-ethyl-α-methyl tartarimide.[68] The ethyl methyl maleimide derived originally from rings *A* and *B* was oxidized separately in the same way to give the same tartarimide derivative.

The tartarimides were treated with periodic acid[69] to give pyruvic and α-ketobutyric acids, which were resolved chromatographically. The pyruvic acid was derived from carbons 4, 5, and 6 of rings *A* and *B* and from the corresponding carbons of rings *C* and *D*. The pyruvic methyl came from carbon 6, the carbonyl from carbon 4, and the carboxyl from carbon 5. The α-ketobutyric acid was derived from carbons 2, 3, 8, and 9 of the corresponding rings from the two halves of the porphyrin, with the carbonyl coming from carbon 3 and carboxyl from carbon 2. Further degradation of the short-chain acids is shown in Fig. 62.

In this fashion, each carbon from the two halves of the porphyrin except the methene bridge carbons could be collected as CO_2. Label in the methene bridge carbons could be calculated readily, however, from previous degra-

[68] Milas, N. A., and Terry, E. M., *J. Am. Chem. Soc.* **47**, 1412 (1925).
[69] Fleury, P., and Lange, J., *Compt. rend.* **195**, 13 (1932).

dation data or from the application of the complete degradation scheme of Wittenberg and Shemin to heme from doubly labeled glycine ($N^{15}H_2C^{14}H_2 \cdot COOH$) or doubly labeled acetate ($C^{14}H_3COOH$ and $CH_3C^{14}OOH$).

When these procedures were employed to study the label distribution of heme formed by biosynthesis from $N^{15}H_2C^{14}H_2COOH$, it was found that the two halves of the porphyrin represented by the maleimide derivative (XXXI) and the hematinic acid (XXXII) had equal amounts of radioactivity and together accounted for half of the total activity in the heme. The four bridge carbons accounted for the remainder. Only the keto acid contained activity in the fragments from these molecules, and all the radioactivity was found in the carboxyl carbon of this acid representing carbons in position 2 of the pyrrole. The label in the keto acid from both halves of the heme was equal in amount. Thus, the four carbons in position 2 came equally from the α carbon of glycine.

The results obtained when labeled samples of heme from studies employing doubly labeled acetate ($C^{14}H_3COOH$ and $CH_3C^{14}OOH$) were degraded can be summarized as follows.

1. All the C^{14} was in carbons other than those in positions 2.
2. The total C^{14} content of rings *A* and *B* equaled that in rings *C* and *D*, when heme labeled from methyl-labeled acetate was used.
3. The total C^{14} content in the two pairs of rings was the same, when carboxyl-labeled acetate was used, provided that the activity in the carboxyl groups of the propionyl side chains in rings *C* and *D* was subtracted.

All these observations supported the suggestion[70] that the two halves of the porphyrin ring system arose from a common precursor. The most important idea arising from these studies, however, was that of Shemin and Wittenberg, who postulated that the citric acid cycle might be supplying the reagent which with glycine formed the pyrrole precursor. To see how these workers arrived at this hypothesis it is most convenient to reproduce their original argument (see Fig. 63).[70a]

It has been seen that the same activities were found in rings *A* and *B* and in *C* and *D*, regardless of whether methyl- or carboxyl-labeled acetate was the starting material. On further degradation of the keto acids (Fig. 62), it was found that the C^{14} content of comparable carbon atoms was similar. In the porphyrin arising from methyl-labeled acetate not only did the methyl group carbon atoms (carbons in position 6 of rings *A*, *B*, *C*, and

[70] Turner, W. J., *J. Lab. Clin. Med.* **26**, 323 (1940–41).

[70a] Shemin, D., and Wittenberg, J., *in* "Isotopes in Biochemistry," Ciba Foundation Symposium (G. E. W. Wolstenholme, ed.), p. 53, Fig. 4. Blakiston, New York, 1952.

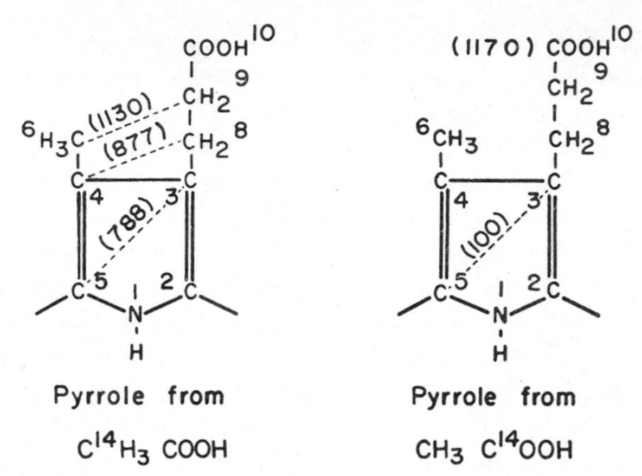

FIG. 63. Labeling pattern of pyrroles from experiments with labeled acetate. (After Shemin and Wittenberg,[70b] see text.)

D) of each pair of pyrroles have similar C^{14} content, but this content was also the same as that of the terminal carbons of the vinyl side chains (carbons 9 of rings *A* and *B*) and also the corresponding carbons of the propionyl side chains (carbons 9 of rings *C* and *D*). The carbons in position 4, bearing the methyl groups, had the same label content as the carbons in position 8 attached proximally to carbons 3. The carbons in position 5 had the same C^{14} content as all the ring carbons in position 3. These relationships are indicated by the dotted lines in Fig. 63. Finally, the carboxyl carbon of the keto acid fragment corresponding to ring carbons in position 2 contained no label, in agreement with the previous finding that this position was derived from methyl carbon of glycine.

Similar findings are indicated in Fig. 63 for heme derived from carboxyl-labeled acetate. The additional datum was that the carboxyl group of the propionyl side chain found only in rings *C* and *D* came from acetate carboxyl. The conclusion seemed inescapable that in each pyrrole ring the same compound was utilized for the methyl side of the system and for the vinyl and propionyl sides. In agreement with this suggestion, it was found that the pyruvic and keto acid fragments of the pyrrole units had the same C^{14} contents in each of the experiments using methyl- and carboxyl-labeled acetate. Data illustrating these findings are given in Table 15.

The radioactivities of carbons in position 10 arising from acetate carboxyl were equal to those of carbons 9 coming from acetate methyl. Since these carbons were adjacent, it followed that the acetate most probably entered as a unit. Entry of C^{14} by way of CO_2 from decarboxylation and secondary fixation could be ruled out on the basis that radioactive carbonate was known not to be fixed in heme.[55, 71] The equality in activities of carbons

[70b] Shemin, D., and Wittenberg, J., see p. 52 in reference 70a.

[71] Bufton, A. W. J., Bentley, R., and Rimington, C., *Biochem. J.* **43,** xlix (1948).

TABLE 15

LABELING OF FRAGMENTS OF PYRROLE RINGS (AFTER SHEMIN AND WITTENBERG[70b])

Pyrrole Ring Fragments	Total Activity: From methyl-labeled acetate (ct./min.)	Total Activity: From carboxyl-labeled acetate (ct./min.)
I. Rings *A* and *B*	11,620	456
(*a*) Pyruvic acid	5,440	206
(*b*) α-Ketobutyric acid	5,560	208
II. (*a*) + (*b*)	11,000	414
III. Rings *C* and *D*	11,840	450
(*a*) Pyruvic acid	5,500	204
(*b*) α-Ketobutyric acid	5,480	190
IV. (*a*) + (*b*)	10,980	394

10 and 9 also indicated strongly that a simple hypothesis involving incorporation of acetate through a four-carbon compound would be a valid basis for further experiment.

It was natural to think of the Krebs cycle as a mechanism for incorporation of acetate as a unit into some four-carbon fragment because of the oxidative character of the red-cell system and the fact that four-carbon dicarboxylic acids are formed from acetate during operation of the cycle (see Fig. 42). Furthermore, it could be seen that the pattern of labeling in the vinyl and propionyl side chains was that to be expected if acetate were being taken up into a Krebs cycle four-carbon compound which could then condense directly with glycine in the manner shown in Fig. 64. For example, the reader can see by referring to Fig. 42 that, beginning with methyl-

```
                    COOH
                     |
        COOH         C
         |           |
         C           |
         |           C
         |           |
         C-----------COX
         |           :
        COX        CH2COOH
           .       /
            H2N
```

FIG. 64. Hypothetical condensation scheme for incorporation of glycine into pyrrole (see text).

TABLE 16

RELATIVE DISTRIBUTION OF C^{14} IN CARBONS OF α-KETOGLUTARIC ACID RESULTING FROM UTILIZATION OF C^{14}-LABELED ACETATE IN THE KREBS CYCLE (AFTER SHEMIN AND WITTENBERG[71a])

	From $C^{14}H_3COOH$ (10 units in methyl) (ct./min.)				From $CH_3C^{14}OOH$ (10 units in carboxyl) (ct./min.)			
	Number of cycles							
	1st	2nd	3rd	∞	1st	2nd	3rd	∞
COOH	0	0	0	0	10	10	10	10
| CH_2	10	10	10	10	0	0	0	0
| CH_2	0	5	7.5	10	0	0	0	0
| C=O	0	5	7.5	10	0	0	0	0
| COOH	0	0	2.5	5	0	5	5	5

labeled acetate, α-ketoglutarate would arise after one turn of the cycle with only the γ carbon labeled. Recycling of the oxalacetate derived from this γ carbon-labeled keto acid would result in eventual redistribution of the C^{14} in the α-keto acid, as shown in Table 16 (if it is assumed that there are ten arbitrary units of label to begin with in the labeled acetate carbon). Similarly, the fate of C^{14} starting with carboxyl-labeled acetate is shown.

If the α-keto acid, or some asymmetrically labeled compound derived from it, were used in heme synthesis, the carbon atoms corresponding to the γ carbon of the ketoglutarate would have the highest radioactivity and the two adjacent atoms somewhat lower but equal radioactivities. This is precisely the pattern of labeling shown in Fig. 63 for heme labeled by use of methyl-labeled acetate. Comparable carbons 6 and 9 have the highest radioactivity, whereas carbons 4 and 5 on the methyl side and 8 and 3 on the opposite side have somewhat lower ratio activities. The somewhat higher C^{14} content associated with carbons 4 and 8, as compared to 5 and 3, comes about because carbons 3 and 5 are derived in part from the acetate carboxyl (Fig. 63). In fact, adding the 100 ct./min. which are contributed on the average to carbons 5 and 3 by acetate carboxyl to the 788 ct./min. actually found in the methyl-labeled acetate experiment, one finds a corrected activity for these carbons of 888 ct./min., which is very close to the 877 ct./min. found for carbons 4 and 8.

[71a] Shemin, D., and Wittenberg, J., see p. 57 in reference 70a.

It appeared from arguments of this nature that the intervention of the Krebs cycle could be assumed. It now remained to establish the nature of the reagent involved in the condensation with glycine. It could be argued that on the propionic side of the pyrrole a four-carbon compound should be involved, because the carbons 2 had been shown to arise from α carbon of glycine. The labeling data appeared to exclude the intervention of a three-carbon compound, such as pyruvate, on the following basis. If pyruvate had been used, then from the distribution of label expected (see Table 16) the methyl-group carbon 6 and carbon 4 to which it was attached should have had equal radioactivities, whereas carbon 5 would have been much less radioactive than carbon 4. The same findings would have been expected for carbons 3, 8, and 9. Also, in the experiment with carboxyl-labeled acetate, carbons 5 and 3 would have been labeled to the same extent as carbons 10.

The simplest hypothesis fitting the actual label distribution was the utilization of two molecules of some unsymmetrical compound arising from the Krebs cycle into which acetate entered by the reactions shown schematically in Fig. 42. Shemin and Wittenberg[70b] suggested that the four-carbon compound might be a coenzyme complex of succinate formed from the decarboxylation of ketoglutarate in a manner analogous to the formation of the acetyl coenzyme A made from the oxidative decarboxylation of pyruvate. Such a succinyl derivative would be labeled asymmetrically, as the reader will recall from the discussion on the Ogston hypothesis (see p. 164). It developed a little later, during researches on the mechanism of the action of α-ketoglutaric acid oxidase, that during the oxidative decarboxylation of this keto acid, succinyl-coenzyme A was formed.[72, 73]

Shemin and Kumin[74] tested the suggestion that an "active" form of succinate was involved in heme synthesis by incubating duck blood with either carboxyl- or methylene-labeled succinate, in the presence and in the absence of malonate. They found that labeled heme was formed from either type of succinate but that the incorporation of label into heme from the methylene-labeled succinate was markedly depressed by malonate. This could be understood on the following basis. No labeling of pyrrole could occur when carboxyl-labeled succinate was used because in the cycle the labeled carboxyls would be lost by oxidative decarboxylation (see Fig. 42). On the other hand, a direct utilization of the succinate before it got into the cycle would lead to labeled heme. The same arguments applied to the methylene-labeled succinate indicated that such a substrate would give

[72] Sanadi, D. W., and Littlefield, J. W., *J. Biol. Chem.* **201,** 103 (1953).

[73] Kaufmann, S., Gilvary, C., Cori, O., and Ochoa, S., *J. Biol. Chem.* **203,** 869 (1953).

[74] Shemin, D., and Kumin, S., *J. Biol. Chem.* **198,** 827 (1952).

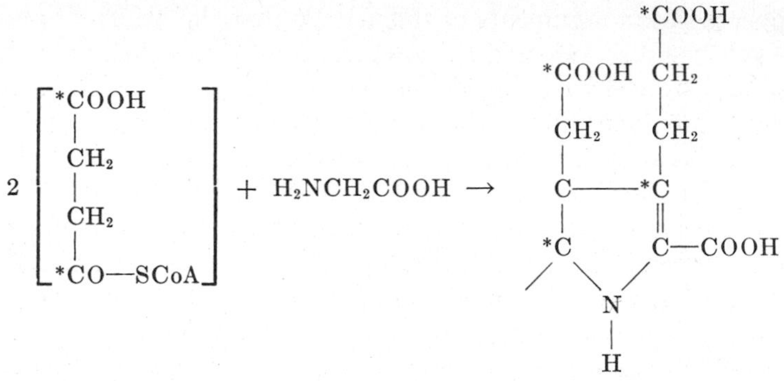

$$2\left[\begin{array}{l} {}^{*}COOH \\ | \\ CH_2 \\ | \\ CH_2 \\ | \\ {}^{*}CO{-}SCoA \end{array}\right] + H_2NCH_2COOH \rightarrow \text{(XXXIII)}$$

(XXXIII): pyrrole ring with $^{*}COOH{-}CH_2{-}CH_2{-}$ on C, $^{*}COOH{-}CH_2{-}CH_2{-}$ on $^{*}C$, $^{*}C$ adjacent to N, $C{-}COOH$ adjacent to N, N–H

$$4\ \text{(XXXIII)} \xrightarrow[-2C_{10}]{-4C_7} \text{Protoporphyrin (label in } C_3, C_5, C_{10})$$

FIG. 65. Hypothetical condensation scheme for incorporation of glycine into pyrrole by way of succinyl-CoA (see text).

rise to labeled heme through the cycle as well as by direct utilization. On addition of malonate, no effect would be seen with carboxyl-labeled succinate because malonate affects mainly the oxidation of succinate. However, malonate would suppress the entry of label from methylene-labeled succinate through the cycle although it would not affect its direct utilization.

These experiments seemed to leave no alternative but to search for the asymmetric form of succinate and for the enzyme system catalyzing its condensations with glycine to yield the pyrrole precursor of porphyrin. The scheme for these reactions could be written as in Fig. 65, with the label distribution to be expected from carboxyl-labeled succinate shown. (The pyrrole precursor indicated was similar to that postulated some time previously by Radin *et al.*[58] as well as by Neuberger *et al.*[75] The origin of the bridge carbons from the α carbons of glycine still remained unexplained.

At this point developments from a completely independent set of researches shed light on another possible precursor which ultimately resolved this question and provided a firm basis for further enzyme research. The compound, *porphobilinogen*, was finally isolated[76] in pure form in 1952 after having been known for some twenty years as a product of derangements in porphyrin metabolism.[77] Its structure was shown[78] to be as represented by

[75] Neuberger, A., Muir, H. M., and Gray, C. H., *Nature* **165**, 948 (1950).
[76] Westall, R. G., *Nature* **170**, 614 (1952).
[77] Sachs, S., *Klin. Wochschr.* **10**, 1123 (1931).
[78] Cookson, G. H., Rimington, C., and Kennard, O., *Nature* **171**, 1185 (1953).

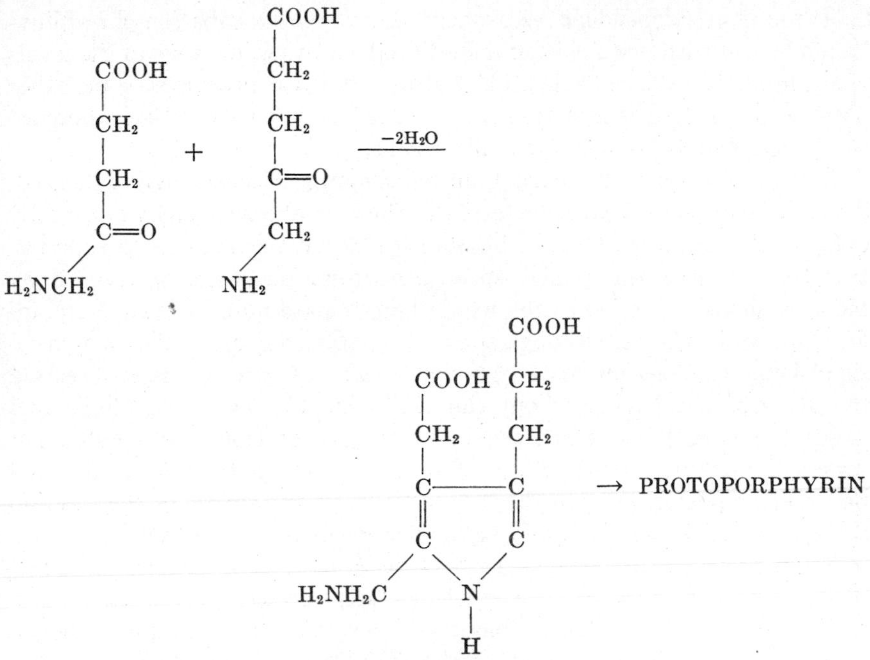

$$H_2NCH_2-C(=O)-CH_2-CH_2-COOH \;+\; NH_2-CH_2-C(=O)-CH_2-CH_2-COOH \xrightarrow{-2H_2O}$$

COOH
COOH CH_2
CH_2 CH_2
C——C → PROTOPORPHYRIN
C C
H_2NH_2C N
H

(XXXIV)

FIG. 66. Condensation of δ-aminolevulinic acid to form porphobilinogen in porphyrin synthesis (see text).

XXXIV in Fig. 66. As the reader will note, porphobilinogen turned out to resemble the hypothetical precursor (XXXIII) very closely.

Very shortly after this series of researches, Shemin and Russell[79] showed that δ-aminolevulinic acid could replace "active" succinate and glycine in the heme synthesis by duck blood. They found that addition of this compound to a reaction mixture containing the erythrocytes, glycine and succinate—with either the glycine or succinate labeled—depressed the labeling of heme nearly tenfold. Addition of [δ-C^{14}]aminolevulinic acid resulted in the appearance of some forty times as much label in heme as from an equimolar amount of glycine. Shemin and Russell proposed a reaction scheme (Fig. 66) in which two molecules of levulinic acid condensed with glycine to give porphobilinogen. They also suggested that the primary product of the condensation between glycine and succinate was aminoadipic acid which was decarboxylated to δ-aminolevulinic acid, thus solving the troublesome problem of how to remove the carboxyl carbon (see Fig. 64) and derive the bridge carbon from α carbon of glycine. Neuberger and

[79] Shemin, D., and Russell, C. S., *J. Am. Chem. Soc.* **75**, 4873 (1953).

Scott found in independent researches[80] that *in vitro* incubation of δ-aminolevulinic acid with lysed chicken blood produced a net increase in the levels of all three natural porphyrins. They also excluded as precursors some other possible products of succinate-glycine condensation such as succinamidoacetic acid and δ-succinamidolevulinic acid.

With the nature of the reagents in heme synthesis so strongly indicated, it might be expected that elucidation of the partial reactions by enzymatic analysis should be accelerated. Shemin *et al.*[81] have shown that a supernatant fraction obtained by high-speed centrifugation of homogenized duck blood is almost as active as the whole homogenate in converting δ-aminolevulinic acid to heme, although such a preparation cannot effect a synthesis of heme from succinate and glycine. Degradation of heme synthesized by the supernatant fluid from the [δ-C^{14}]-labeled δ-aminolevulinic acid yields the same labeling pattern in heme as that from the α carbon of glycine in duck blood suspensions. This is as expected, because in the reaction scheme postulated (see Fig. 66) α carbon is the source of the δ-amino carbon of the δ-aminolevulinic acid. The original notion that there is a common precursor in porphyrin synthesis (which may now be identified with porphobilinogen) has been experimentally verified by Falk *et al.*,[82] who have found that lysed chicken blood suspensions convert porphobilinogen mainly to uroporphyrin III and coproporphyrin III under anaerobic conditions and to porphyrin under aerobic conditions. The scheme which appears to be involved in the partial reactions during the synthesis of uroporphyrin can be summarized as follows.

Succinyl-CoA + glycine → [α-amino-β-ketoadipic acid] →
δ-Aminolevulinic acid → Porphobilinogen → Uroporphyrin III

The conversion of uroporphyrin may proceed directly through coproporphyrin III to protoporphyrin, or protoporphyrin may be formed more directly from porphobilinogen.

It can be expected that there will be a great acceleration in the clarification of all aspects of porphyrin synthesis arising primarily from the basic discoveries made possible by the employment of tracer techniques.[82a]

D. Concluding Remarks

The topics considered in this section are only a minute part of the material available for discussion. Many important subjects have had to be

[80] Neuberger, A., and Scott, J. J., *Nature* **172,** 1093 (1953).

[81] Shemin, D., Abramsky, T., and Russell, C. S., *J. Am. Chem. Soc.* **76,** 1204 (1954).

[82] Falk, J. E., Dresel, E. I. B., and Rimington, C., *Nature* **172,** 292 (1953).

[82a] For recent references to the "succinate-glycine" cycle, see Shemin, D., Russell, C. S., and Abramsky, T., *J. Biol. Chem.* **215,** 613 (1955). Medical aspects are discussed by Schwarz, S., *Federation Proc.* **14,** 717 (1955).

arbitrarily left out, even though they represent equally valid examples of the manifold applications of tracer methodology in biochemistry. It is hoped that the reader will augment his knowledge by consulting the general references given at the end of this chapter.

2. SOME CRITICAL REMARKS ON TRACER METHODOLOGY IN STUDIES OF INTERMEDIARY METABOLISM

The application of the tracer method to the study of intermediary metabolism involves essentially the following procedure. A pure labeled compound is administered to an active metabolizing organism. At some appropriate time after ingestion, the organism is sacrificed and chemical extracts prepared from various tissues. These extracts, containing whatever labeled compounds have resulted from the participation of the administered labeled material in the metabolism of the organism, are fractionated until a series of pure compounds has been isolated. The specific labeled content of these fractions is determined at each stage of fractionation and, finally, on the pure compound. The actual identity of the labeled material is confirmed by further purification. If no change in specific labeled content (atom per cent excess of labeling atom, specific activity) occurs as a result of repeated purification, the compound isolated is considered pure from the tracer standpoint. The criterion of purity characteristic of the tracer method is *constancy of specific labeled atom content under all possible conditions of chemical purification.*

Under optimal conditions, high sensitivities are characteristic of the tracer method. This is most particularly so when radioactive tracers are used. High sensitivities imply that extremely small quantities of labeled material can be detected easily. They also imply that even the most exacting purification procedures may be inadequate to remove tracer impurities unless special precautions are observed. The best chemical purifications of organic compounds rarely remove impurities present in an amount of one part per million. In general, impurities bulk considerably more than one part per thousand. Most biochemical extracts contain carbohydrates, fats, and proteins. The ultimate separation of such constituents is one of the most difficult the chemist is required to accomplish.

The situation is very similar to the isolation and purification of biologicals such as vitamins and hormones. In such isolations the criterion of constant biological activity per unit weight does not prove finally that the molecule isolated is actually involved in biological function. There is always the possibility that a minute impurity is responsible. The proof in this case consists in the synthesis of the compound and the demonstration that the synthetic compound exhibits the expected biological activity. In an analogous fashion, the tracer chemist can use the criterion of constant specific activity or atom per cent excess to trace and purify a particular fraction of an extract. Unfortunately, the synthetic proof available for biologicals is not available for establishing the participation of a given molecule as an intermediate in biological reactions. However, the tracer criterion may be a useful adjunct in isolation of biologicals.

After exhaustive fractionation fails to increase specific labeled content of the material in the extract, its nature is determined as well as the labeled content of its constituent parts. Comparison with similar values obtained for other compounds isolated in the same fashion provides a basis for deducing mechanisms for participation of such compounds in metabolism. However, unambiguous proof of such deductions depends on the assumption that the criterion of constant specific labeled content is adequate to specify the intermediates involved.

In general, purification procedures using tracers of high sensitivity (labeled substrates with high specific labeled content) will be reliable only when all components in the mixture are known. With a mixture of known composition it can be ascertained whether it is possible to devise a foolproof procedure—and it usually is possible. In elaborating such a procedure, the chemist can make use of a technique often employed in the tracer method, namely "washing out." In such a procedure excess unlabeled material is added. Thus, in separating a small quantity of highly radioactive carbonate from some propionic acid of low radioactivity, a large quantity of non-radioactive carbonate is added. The solution is acidified, whereupon the contaminating radioactive carbonate, diluted with the added unlabeled carbonate is swept out of solution. The procedure may be repeated until the contaminating radioactivity is reduced to negligible proportions.

However, the use of washing-out techniques is predicated on knowledge of the nature of the impurities present. Much tracer research results in extracts of known compositions. But a much greater proportion of tracer research is concerned with systems which yield extracts containing intermediates not suspected to be present and concerning the nature of which nothing is known.

Two courses can be pursued. In the first, the extract is fractionated until all fractions show extreme (maximal or minimal) values for specific labeled content. These fractions are then analyzed. Such a procedure is similar to vitamin and hormone isolation. In the second type of procedure, judicious guesses as to the nature of the intermediates are made, and unlabeled material is added to facilitate separation. This procedure is dangerous because the actual intermediates involved may not be the same molecules as those of material added and yet, owing to certain circumstances to be discussed briefly later, they may accompany such material through all fractionation procedures. Its feasibility, in other words, depends on how accurate the investigator's guesses are. The method has the advantage that less material is required to work up and purify intermediates than when no unlabeled material is employed.

The major difficulty of the tracer method arises when compounds are present in small amounts but possess very high specific labeled content as compared to that of the extracted material. If washing-out procedures cannot be used, it is necessary to employ every available chemical fractionation scheme to show that specific isotopic content is invariant under all conditions of isolation. A "final" proof of purity will not be forthcoming, but it is very rarely that biochemical mixtures will display such gross disproportion in specific labeled content as to make satisfactory isolation and purification impossible within reasonable limits no matter how exhaustive the procedure employed. In most research, compounds of interest will be present in varying amounts but with specific labeled contents not different by several orders of magnitude. It is difficult to conceive of mechanisms beginning with a single labeled substrate which would result in a mixture of compounds having specific labeled contents differing by many orders of magnitude.

Ultimately, the purity of the compound obtained, with its labeled content, will depend on the excellence of the chemical methods employed. This is a difficulty not of tracer methodology but rather of quantitative chemistry to which the tracer method is an adjunct.

A thorough treatment of tracer methods in biochemistry would require a detailed description of how each and every procedure available for the analysis of the constituents of carbohydrate, fat, and protein, separately or in various combinations, is affected by the use of tracers—a manifestly impossible task. What follows, then, is not so much a general critique as a recapitulation of the advantages and the occasional hazards of the tracer method as applied to biochemical problems.

The procedures used to precipitate biochemical mixtures may be carried out either with the original constituents, with derivatives of these constituents, or with specific degradation products. The methods are based on fractional distillation, differential extraction, fractional crystallization, and differential adsorption. If these methods are to be effective the components must not interact; if interaction cannot be avoided, it should be minimal. Thus in isoelectric precipitation of amino acids, which may be taken as an example of purification by fractional crystallization, mutual solubility of the solid phases in each other will result in abolishing separation based on differential solubility of the solid in the liquid solvent employed. Similarly, minimal or maximal boiling point mixtures will complicate fractional distillation, and addition complexes will vitiate adsorption procedures.

Such complications become particularly obnoxious in the face of the sensitivities so often found in tracer research. A few laboratory experiences will serve to illustrate. In a study of phosphate turnover in yeast[83] the appearance of labeled phosphate in the various organophosphate compounds was determined. The labeled phosphate was administered as exogenous inorganic orthophosphate labeled with P^{32}. Among the various fractions isolated was one presumed to contain only deoxyribonucleic acid (DNA). Because information about the role of nucleic acids in protein synthesis was needed, the rate of phosphate entry into DNA with that in other fractions of yeast phosphate had to be measured and compared.

A major difficulty arose immediately because the normal content of DNA in yeast is about 0.01% on a dry weight basis. As a result an extremely small quantity of DNA was recovered in semipure form even though the experiment was carried out with 10 g. of yeast (wet weight). Actually the DNA fraction contained 1.6 mg. as phosphorus, but only 0.3 mg. of this was DNA phosphate. The rest of the fraction (80%) was some other phosphate-containing contaminant. The specific activity of the lumped phosphate was 12 ct./min./γ.

Enough carrier-inactive DNA was added to bring the specific activity to 6.8 ct./min./γ. The solid was dissolved in neutral aqueous medium, the *p*H brought to 2.5 with acid in acetate buffer, and the resulting precipitate (which represented only two-thirds of the carrier material added as phosphorus) was assayed. The specific activity was now 6.27 ct./min./γ. Two more reprecipitations resulting in a final recovery of only about 10% of carrier added gave substantially the same values, 6.7 and 6.28.

It might be expected that the sample was now pure DNA on the basis of constant specific isotopic content. However, analysis revealed that there was little DNA present. In fact, the ratio of impurities to DNA in the precipitate appeared to have been altered but little by the isoelectric precipitation.

It was speculated the impurities either possessed the same solubility as DNA at *p*H 2.5 or formed solid solutions with the solid DNA phase. If the latter case were true, the quantity of impurity included in the DNA precipitate (or vice versa) depended on a distribution of one solid phase between the liquid solvent and the other solid phase and not on the solubility coefficient for each solid phase in the liquid solvent.

It was impossible to unravel this situation because the nature and specific activity of the impurities present were unknown. An apparent constant specific isotopic content at each precipitation could have been explained as resulting from (1) a single impurity which had formed a solid solution with DNA; (2) numerous impurities with different specific activi-

[83] Juni, E., Kamen, M. D., Reiner, J. M., and Spiegelman, S., *Arch. Biochem.* **18**, 387 (1948).

ties which had been precipitated at random, producing the effect of apparent constancy in specific activity; or (3) a complex of many impurities, including DNA which formed a solid solution phase. It is plain that, with so many possibilities, constancy of specific activity told nothing about the purity of the sample being isolated. Fortunately, in this case adequate chemical tests could be made to reveal the existence of impurities in the DNA.

When the same DNA fraction was subjected to repeated precipitation in the presence of DNA carrier at much lower *p*H (1), no constancy was observed in the specific activity, although only a small fraction of the material in solution was recovered at each precipitation. The specific activity kept dropping at each precipitation and by the sixth fractionation reached a value of only 1.5 ct./min./γ. At this point the fractionation had to be discontinued because there was practically no material left. This experiment showed only that in strong acid there was insoluble material with a low specific activity present. Finally, by investigating the behavior of the specific activity curve at various acid strengths and checking all fractions obtained for DNA content, it was shown that it was possible to work out a procedure in which DNA could be isolated in a reasonably pure form so that its true P^{32} content of DNA could be deduced.

This experiment demonstrates one of the cases in which repeated use of a single fractionation method may not be adequate to ensure the validity of the criterion of constant specific activity.

A quantitative study of the factors involved in vitiation of this criterion for the case of precipitation from a single solvent has been reported in a study of procedures for purification of amino acids or their *p*-iodophenylsulfonyl derivatives.[84, 85] Experiments were made in the following manner. Varying amounts of C^{14}-labeled glycine were added to aqueous solutions containing about 10 mg. of unlabeled alanine. The solutions were warmed. Alcohol was added and alanine allowed to crystallize slowly so that approximately half of it precipitated. The radioactivity and amino nitrogen were measured in the supernatant liquid and in the crystals. In Table 17 some typical results are presented. The quantity K is a distribution coefficient calculated on the basis that the impurity (glycine) partitions itself between the solid phase (alanine crystals) and the liquid phase as though it were being distributed between two immiscible solvents. "Carrier" in this instance is total amino acid present. Thus

$$\frac{\left(\dfrac{\text{Amount of impurity}}{\text{Amount of carrier}}\right)_{\text{supernate}}}{\left(\dfrac{\text{Amount of impurity}}{\text{Amount of carrier}}\right)_{\text{crystals}}} = K.$$

[84] Keston, A. S., Udenfriend, S., and Cannan, R. K., *J. Am. Chem. Soc.* **71,** 249 (1949).

[85] Keston, A. S., Udenfriend, S., and Levy, M., *J. Am. Chem. Soc.* **72,** 748 (1950).

TABLE 17

Coprecipitation of Glycine with Alanine (after Keston et al.[85])

Sample	Alanine (γ)	Glycine (γ)	Impurity (%)	K
1	7350	2.4	0.03	1.9
2	7350	4	0.05	1.8
3	7350	24	0.3	1.8
4	8350	75	0.9	1.6
5	8350	100	1.2	1.8
6	8350	125	1.5	1.6
7	8350	190	2.3	1.7
8	8350	305	3.7	1.6
9	8350	600	7.2	1.6
10	8350	1075	12.9	1.8

The amount of impurity is determined from the radioactivity and the amount of carrier from the amino nitrogen.

It will be noted that, although the amount of impurity varies from several hundredths of a per cent to nearly 13%, the distribution coefficient, K, remains fairly constant (1.6 to 1.9). The *p*-iodophenylsulfonyl derivatives exhibit a similar behavior, with values of K practically the same as for the corresponding amino acids. Thus, this solid solution behavior during crystallization results in a coprecipitation which is independent of the quantity of contaminant present or of whether derivatives instead of the original amino acids are used. Hence repeated crystallizations in this case would afford little assurance of purity, nor would preparation of derivatives help.

It is instructive to note that in the system, glycine contaminating alanine ($K = 1.6$), almost 200 recrystallizations losing 5% of the alanine at each step would be required to bring the glycine impurity to about 1% of its original level. Very little of the alanine would be left in such a procedure, and very probably the associated labeled content would be too low to be significant. Over a large number of recrystallizations, the change in specific activity would be so small as to be thought constant by most workers. Few researches in the past have been characterized by any such exhaustive recrystallization procedures.

It may be appreciated that in these circumstances addition of carrier to wash out impurities is ineffective. A variety of fractionation procedures must be used. It has been shown[84, 85] that the use of countercurrent distribution between solvents, in addition to paper chromatography, bolstered by a multiple-labeling technique, suffices to achieve a satisfactory resolution of labeled amino acid derivatives.[86]

In the examples cited involving the participation of CO_2 in metabolism, most of the troubles detailed above were obviated by the unique chemical character of the metabolite which, being a gas or readily isolated in pure form as a gas, offered no difficulty in isolation and purification. Therefore, a certain disarming simplicity seemed to characterize the use of tracers in metabolism studies using CO_2. The appearance or disappearance of label could be interpreted rather directly as evidence for the intervention of the metabolite in the system considered.

From the foregoing discussion it should be evident that blind adherence to rule-of-thumb purification procedures, inflexibility in devising analytical methods, and failure to make adequate control experiments are invitations to disaster. Tracer techniques badly exploited can result in mistakes of a magnitude hardly possible with conventional chemical procedures. It is important to emphasize that there is no substitute for good chemistry.

[86] It is even possible to push the precision of chromatographic procedures to a degree far beyond that which can be tolerated in tracer research. Thus, K. Piez and H. Eagle [*Science* **122,** 968 (1955)] have observed that, when C^{14}-labeled glycine or alanine of high specific activity is adsorbed on columns of the ion-exchange resin Dowex 50, with a length much greater than those usually employed, the peak obtained on elution of the acid is different depending on whether the radioactivity or the color developed with ninhydrin is used to determine the amino acid. The two elution curves are identical in shape and height but are slightly displaced from each other, the C^{14} component moving slower. There is some indication that the effect depends not only on the total mass difference of the carbon isotopes but also on the placement of the C^{14}. Such effects can be observed only with procedures of extremely high resolution and are not to be feared with the usual chromatographic procedures, such as paper chromatography or methods in which columns of the size usually recommended are employed.

GENERAL REFERENCES (CHAPTERS VI AND VII)

A. Books

1. Comar, C. L., "Radioisotopes in Biology and Agriculture." McGraw-Hill, New York, 1955.
2. Francis, G. E., Mulligan, W., and Wormall, A., "Isotopic Tracers." Athlone Press, Univ. London, 1954.
3. Hevesy, G., "Radioactive Indicators." Interscience, New York, 1948.
4. Schoenheimer, R., "The Dynamic State of Body Constituents." Harvard U. P., Cambridge, 1946.
5. Tubiana, M., "Les Isotopes Radioactifs en Medecine et en Biologie." Masson and Cie, Paris, 1950.
6. Wolstenholme, G. E. W. (ed.), "Isotopes in Biochemistry," Ciba Foundation Symposium. Blakiston, New York, 1952.

B. Monographs and Pamphlets

1. Popják, G., Chemistry, biochemistry and isotopic tracer technique. *Roy. Inst. Chem. (London) Lectures Monographs Repts. No.* **2,** 1–59 (1955).
2. Roberts, R. B., Abelson, P. H., Cowie, D. B., Bolton, E. T., and Britten, R. J., Studies of biosynthesis in *Escherichia coli. Carnegie Inst. Wash. Publ. No.* **607,** 1–521 (1955).
3. S. Gurin, Chairman, Symposium on cholesterol metabolism. *Federation Proc.* **14,** 752 (1955).

CHAPTER VIII

SURVEY OF TRACER METHODOLOGY: PHYSIOLOGICAL AND MEDICAL ASPECTS

1. INTRODUCTION

The labors and interests of the physiologist and medical researcher merge continuously with those of the biochemist. The physiologist, however, who is interested primarily in the organism as a whole, is not concerned so much with isolated enzyme systems as with interactions between enzyme systems and their functions in the cellular economy and the mechanisms which regulate them. The importance of such knowledge in elaborating medical procedures will be indicated briefly.

2. PHYSIOLOGICAL APPLICATIONS

A. PERMEABILITY, ABSORPTION, AND DISTRIBUTION STUDIES

Beginning with the relation between living cells and their environment, the physiologist is confronted immediately with processes grouped under the term "permeability." The rate at which metabolites and, in particular, mineral constituents diffuse into and through the cell interface must be determined as a function of cellular metabolism and environment.

The tracer method affords an almost ideal approach to the problem of cellular permeability because it makes possible the study of diffusion processes under conditions where there is no net transfer of metabolite. For instance, the passage of sodium ions back and forth through such cells as mammalian erythrocytes can be investigated even when no net transfer of sodium ions is demonstrable by conventional chemical procedures. Cohn and Cohn[1] injected into dogs, intravenously, samples of isotonic saline solution containing labeled sodium (Na^{24}) and withdrew blood samples at various intervals. The plasma was separated from the red cells by centrifugation and washing. The concentration of the radioactive sodium in the cells relative to the plasma showed a steady increase over a period of a few minutes, rising to nearly 80% of the value to be expected for complete equilibration. The rate of sodium penetration *in vitro* was also measured by suspending the cells in isotonic saline containing labeled sodium ion and was found to be nearly the same as *in vivo*. The appearance of radioactive sodium in the corpuscles followed the course

[1] Cohn, W. E., and Cohn, E. T., *Proc. Soc. Exptl. Biol. Med.* **41,** 445 (1939).

predicted on the basis of a simple diffusion process. Equilibrium was approached at a rate dependent on the difference in Na^{24} concentration. In this simple way, it was shown that cells were permeable to sodium ion both *in vivo* and *in vitro*.

Biologists familiar with the literature on the subject of erythrocyte permeability to cations will appreciate the acceleration of research made possible in this field where no such direct approach has hitherto been available. Furthermore, such an established phenomenon as the intracellular accumulation of potassium can no longer be posed in terms of selective permeability because sodium ion has been proved unequivocally to penetrate rapidly into erythrocytes. In this connection it is interesting to note that Hevesy and Hahn,[2] using potassium chloride labeled with K^{42}, found that in rabbit erythrocytes about 25% of the cellular potassium was replaced by plasma potassium.

Further work[3] indicates that the penetration of ions into cells is associated with metabolic processes which govern interchange equilibria between extracellular and intracellular ions. This conclusion fits well with the claim that no strictly physical potentials generated at a cell wall can account for selective permeability of the type shown in the case of potassium and sodium.[4]

An example of both the power and limitations of tracer studies in ion transport across the membranes is afforded by the work of Ussing and Levi on the uptake of Na^+ and Cl^- by Ringer's solution through isolated frog skin.[5] In a typical experiment, a piece of abdominal skin was bathed on the outside by a salt solution containing labeled Na^+ and Cl^-, and on the inside by unlabeled Ringer's solution. The appearance of labeled Na^+ and Cl^- in the Ringer's solution was then determined as a function of time, concentration of Na^+ and Cl^- in the outside cup, pH, potential difference, CO_2 tension, adrenaline, cyanide poisoning, and other factors. The rate at which Na^{24} entered the inside solution was measured. From this datum and from the net change of the amount of Na^+ in one of the solutions, the flux in or out—that is, the total amount of Na^+ (Na^{23} and Na^{24}) which crossed a unit area of membrane in either direction—could be calculated.

[2] Hevesy, G., and Hahn, L., *Kgl. Danske Videnskab. Selskab, Biol. Medd.* **16,** 1 (1941).

[3] Brooks, S. C., *J. Cellular Comp. Physiol.* **14,** 383 (1939); Brooks, S. C., *Proc. Soc. Exptl. Biol. Med.* **42,** 557 (1939); Mullins, L. J., and Brooks, S. C., *Science* **90,** 256 (1939).

[4] Spiegelman, S., and Reiner, J. M., *Growth* **6,** 367 (1942).

[5] Ussing, H. H., *Cold Spring Harbor Symposia Quant. Biol.* **13,** 193 (1948); *Acta Physiol. Scand.* **17,** 1 (1949); Levi, H., and Ussing, H. H., *Acta Physiol. Scand.* **16,** 232 (1948); for the background material, see Krogh, A., *Proc. Roy. Soc.* **13,** 133, 140 (1946).

It was desired to ascertain whether active uptake was a specific response to need for salt, or whether it was a continuous process concealed by compensation due to diffusion. In these researches influx was identified with the actual uptake across the membrane because conditions were such that inward passage of ions by simple diffusion could be assumed to be negligible compared to the total flux. The experimental results indicated that: (1) influx of Na^+ was normally much higher than outflux even when the outside NaCl concentration was only 10^{-3} *M*; (2) the active Na^+ uptake depended directly on the *p*H of the outside solution, high *p*H values giving high Na^+ influx; (3) the Cl^- influx was less than the simultaneous Na^+ influx; (4) adrenaline added to the inside compartment increased Na^+ outflux enormously and Na^+ influx considerably, whereas cyanide reduced Na^+ influx somewhat and Na^+ outflux not at all. When Ringer's solution was applied to both sides of the skin, the mean outflux was less than 10% of the mean influx, the influx taking place against a potential difference of 30 to 110 mv. Thus, at least 90% of the influx was due to active uptake.

Confining discussion to the experiments in which diffusion played a negligible role in influx (low outside concentrations), the authors showed that Cl^- transport was most probably determined by the potential difference generated across the skin, the inside of the skin always being positive relative to the outside. The transport of Na^+ ions required the overcoming of the electrical potential gradient as well as a concentration gradient.

Under these circumstances outflux due to free diffusion should have been much greater than influx. But actually the reverse was true. The authors postulated, therefore, that the active transport of Na^+ required, first, formation of a chemical complex (possibly uncharged) in the outer skin layer, then diffusion of this complex to the inner skin layer, and finally liberation of Na^+ because of a change in chemical environment relative to that encountered in the outer skin layer. An interesting conclusion was that the potential difference generated across the skin arose from the active Na^+ transport. It was argued that the inside skin assumed a positive charge relative to the outside because, owing to metabolic and chemical action, Na^+ flowed in faster than Cl^- could follow.

Detailed consideration of other experimental phenomena in these studies suggested that regulation of internal cellular *p*H was correlated with rate of Na^+ transport, which in turn required a forced exchange of Na^+ with H^+ ion. The transport of Na^+ was regulated in turn by the amount of the complex formed on entry. For the present discussion it is relevant only to note that such conclusions derive from added insight into the validity of assumptions about mechanisms of ion transport brought about by the availability of labeled material and establishment of experimental conditions under which transport of labeled ion can be correlated with true influx. The tracer method provides little information in these matters when influx and outflux as measured are nearly equal in magnitude. Under these conditions exchange diffusion as well as active transport contributes to the flux measured so that difficulties in interpretation arise.[6]

The transport of ionic species in and out of muscle and nerve fibers has been studied in a number of laboratories in efforts to relate movement of ions to physiological activity. A representative few of these researches will be described. Keynes and Lewis[7, 8] reported that in the resting crab leg nerve there was a rapid leakage of potassium ion. They estimated that the flux outward was 22×10^{-12} mole/cm.^{2}/sec., and the flux inward was 19×10^{-12} mole/cm.^{2}/sec. They found that the rate of loss was roughly proportional to the external potassium concentration.

Stimulation of the nerve resulted in increasing the leakage rate of potassium ions. It was estimated that the extra outward movement provoked by activity involved 1.6×10^{-3}% of the internal potassium per impulse. The extra inward movement was about one-tenth as large. The net effect of the stimulation was to cause a leakage of between 2 and 3×10^{-12} mole/cm.^{2}/impulse. These data indicate the nature of the information which can be obtained only through tracer methods.[9]

An interesting experiment on determination of ionic exchange in nerve tissue by "activation analysis" has been reported.[10] In this method, the tissue is irradiated with neutrons; the different amounts of radioactive isotopes produced are then determined.[11, 12] From a knowledge of yields of radioactive material from neutron capture reactions it is possible to estimate the amounts of stable target material present in the tissue. This method is particularly inviting in work with isolated nerve axons because it makes possible the direct measurement of outward as well as inward fluxes.[10]

Single 200-μ cuttlefish axons were dissected, dried, and sealed in quartz tubes. These were irradiated in a neutron-producing pile together with standard samples of pure sodium and potassium carbonate. The content of sodium and potassium in the axons was then determined by assay of the radioactivity of the nerves with a G-M counter after they had been transferred to nickel dishes. After values in resting nerve were established, the experiment was repeated with nerves which had been stimulated one

[6] Ussing, H. H., *Acta Physiol. Scand.* **17,** 1 (1949); **19,** 44 (1949).
[7] Keynes, R. D., and Lewis, P. R., *J. Physiol.* **113,** 73 (1951).
[8] Keynes, R. D., *J. Physiol.* **113,** 99 (1951).
[9] For details of other studies, see Solomon, A. K., *J. Gen. Physiol.* **36,** 57 (1952).
[10] Keynes, R. D., and Lewis, P. R., *Nature* **165,** 809 (1950).
[11] Tobias, C. A., and Dunn, R. W., *Science* **109,** 109 (1949).
[12] Brown, H., and Goldberg, E., *Science* **109,** 347 (1949).

hundred times a second for 20 min. before being stored for analysis. Sodium content had increased from a resting value of 40 mM./kg. to about 130 mM./kg. Corresponding values for potassium were 325 in resting nerve and 245 in stimulated nerve. In a series of sixteen stimulated axons, the average gain of sodium was 3.8×10^{-12} mole/cm.^{2}/impulse, and the average loss of potassium was 3.4×10^{-12} mole/cm.^{2}/impulse. These data agreed with previous results obtained with labeled sodium and potassium. With the values for sodium and potassium concentration established by activation analysis, complete calculations could be made on the movement of ions during stimulation.[13] Thus, it was possible to establish the actual transport of sodium and potassium ions as affected by stimulation of nerve tissue.

An example of the use of tracer techniques which calls attention to its limitations may be taken from work on entry of exogenous phosphate into the intracellular phosphate cycle. In general, two kinds of mechanism had been proposed earlier than this work. One was based on the notion that exogenous phosphate entered as inorganic orthophosphate by diffusion and then mixed with intracellular orthophosphate. This phosphate was assumed to be the source from which various organic phosphates in the cell derived phosphate. The second mechanism was thought to involve esterification at the cellular interface. Intracellular inorganic orthophosphate would then arise primarily from the breakdown of organic phosphate.

In principle it appeared that a choice between these two proposals could be made by means of labeling techniques in which the specific isotopic contents of ortho and organic phosphate fractions were compared after suspension of metabolizing cells in a medium containing labeled inorganic orthophosphate. There arose, however, the question of what to regard as cellular "ortho-P." A popular extraction procedure called for the use of cold trichloroacetic acid (TCA) with subsequent precipitation of the resulting mixture with magnesia in alkaline *p*H, the precipitate being considered to represent cellular ortho-P.

Another uncertainty was the identity of the organic cellular phosphate with which the labeled content of such ortho-P was to be compared. Such comparison required isolation of the organic P in a reasonably pure state. Furthermore one could never be certain that an organic P fraction would remain undetected either because of lability or low concentration. It might nonetheless, have a higher isotopic label content (specific activity) than any of the other cellular fractions, including the ortho-P.

Some investigations were made on the significance of the ortho-P fraction prepared by the usual TCA extraction and magnesia precipitation

[13] See also Hodgkin, A. L., and Huxley, A. F., *Cold Spring Harbor Symposia Quant. Biol.* **17**, 43 (1956).

method.[14] It was noted[15] from the variation with time of the relative specific activity of all cellular fractions referred to exogenous phosphate that no cellular phosphate, including ortho-P, ever fully equilibrated isotopically with exogenous phosphate. For instance, in yeast cells fermenting glucose in the presence of labeled phosphate, a rapid rise in the relative specific activity of cellular ortho-P occurred in the first 30 min., the value reached representing 20% of the equilibrium value, after which no further significant increase in specific activity could be noted. One explanation for this lack of equilibration could be that the method of cold TCA extraction actually created a portion of the orthophosphate fraction by hydrolysis of labile organic phosphate esters. From this point of view, the 20% which did equilibrate rapidly could be either the original orthophosphate or that which had been formed by hydrolysis. In any event, this phenomenon indicated that the phosphate fraction prepared in this manner was not derived from a homogeneous cell fraction. One might also suppose that the ortho-P was made up of inorganic phosphate originating from different sources. Some of these might be actively involved in the phosphate cycle, whereas others, because of intracellular steric effects, could not enter appreciably into the phosphate metabolism.

The penetration of molecules active in metabolism is also amenable to the tracer approach, with the important reservation that entry of labeled molecules is not established with certainty by appearance of labeled atoms in the cell after exposure to solutions containing labeled molecules. For example, adenosine triphosphate labeled with P^{32} in the terminal phosphate groups may be dissociated to free phosphate which can enter as such. In general, dissociation of the labeled portion of the molecule in the complex region defined by the cell interface can take place, so that constituent atomic groupings, rather than the molecule as a whole, may enter the cell.

The simple phenomenon of absorption, i.e., penetration of material without regard to mechanism, is readily amenable to the tracer approach. The procedure for investigation of such problems under conditions impossible for conventional chemical methods—as when absorption is studied under steady-state conditions—is exceedingly simple. This is true particularly if the absorption of metal salt constituents is under investigation. The element, labeled by admixture with a radioactive isotope, is introduced into the medium. At appropriate times thereafter the metabolizing organisms are withdrawn, washed, and examined for radioactivity. If any is found, the element is proved to have been ingested by the organism. Conversely, the excretion of the ions can be shown by suspending organisms

[14] Juni, E., Kamen, M. D., Reiner, J. M., and Spiegelman, S., *Arch. Biochem.* **18**, 387 (1948).

[15] Spiegelman, S., and Kamen, M. D., *Cold Spring Harbor Symposia Quant. Biol.* **12**, 211 (1947).

containing labeled ions in a nonradioactive medium and examining the medium for radioactivity. Localization of the ions can be detected by assay of various fractions of the organism. Experiments of this nature are among the earliest of tracer applications. Thus, Hevesy[16] studied the uptake of lead by plants using as tracer the radioactive lead isotope ThB (Pb^{212}). The medium employed contained solutions of lead nitrate admixed with ThB nitrate. After intervals ranging from 1 hr. to 2 days, the plants were ashed and the amounts of ThB present were determined by radioactive assay.

The extension of such studies to a host of problems in plant nutrition is indicated, and numerous studies of this type have been reported.[17] The general result has been to strengthen the conviction that absorption is not governed simply by diffusion laws but is influenced primarily by cellular metabolism. The living cell does not accumulate minerals like potassium, for example, because of a physicochemical "selective" permeability effect based on membrane potentials, but as a result of specific chemical processes in the cell which are most probably directly mediated by enzyme action.[18]

On the other hand, Cowie *et al.*[19] have made many observations on bacteria, especially on *E. coli*, which support the view that the cell wall is highly permeable to many molecules and that transport across the membrane occurs by simple diffusion into a "water space" inside the cell. They have shown that when cells are immersed in solutions containing labeled substances and then centrifuged, the cell pellet formed invariably contains a predictable minimum quantity of labeled material. The residual quantity of labeled material in the water space of the *E. coli* cells is equal to the label content of 0.75 ml. of labeled solution, after correction for material which has been fixed by metabolic processes and for solution adherent between the cells. This volume of solution is equal to the water lost when the pellet is dried. It can be concluded that the water space has material dissolved or suspended in it at the same concentration as in the outer immersion fluid. Cowie *et al.* measured permeability for the cations Na^+, K^+, Rb^+, Cs^+, and Mn^{++}; the anions SO_4^{--} and PO_4^{---}; and the compounds glucose-1-phosphate, fructose-1,6-diphosphate, glutamic acid, methionine, cystine, and glutathione.

Much of the data available from tracer researches concerns distribution

[16] Hevesy, G., *Biochem. J.* **17,** 439 (1923).

[17] See Hevesy, G., *Ann. Rev. Biochem.* **9,** 641 (1940); Stout, P. R., Overstreet, R., Jacobson, L., and Ulrich, A., *Soil Sci. Soc. Amer. Proc.* **12,** 91 (1947).

[18] For a general discussion see Teorell, T., *Progr. Biophys. and Biophys. Chem.* **3,** 305 (1953).

[19] Cowie, D. B., Roberts, R. B., and Roberts, I. Z., *J. Cellular Comp. Physiol.* **34,** 243 (1949); see also Cowie, D. B., and Roberts, R. B., *in* "Electrolytes in Biological Systems" pp. 1–53. American Physiological Society Monograph, Washington, D. C., 1955.

TABLE 18

DISTRIBUTION OF LABELED MANGANESE IN THE RAT (PER CENT TOTAL DOSE, 75.5 HR. AFTER ADMINISTRATION) (AFTER GREENBERG AND CAMPBELL[21])*

Tissues	Oral administration			Intraperitoneal injection		
	Weight	Contents of whole tissue	Contents per gram fresh weight	Weight	Contents of whole tissue	Contents per gram fresh weight
	g.			g.		
Muscle.....	95	0.7 ± 0.12	0.007 ± 0.0012	97	0.8 ± 0.16	0.008 ± 0.0016
Bone......	8.6	0.7 ± 0.11	0.081 ± 0.013	9.8	2.0 ± 0.17	0.20 ± 0.017
Skin.......	30.66			28.94	3.7 ± 0.22	0.13 ± 0.008
Whole blood	5.14	0.5 ± 0.14	0.097 ± 0.027	5.07		
Heart.......	0.50			0.56		
Liver.......	6.49	0.9 ± 0.15	0.14 ± 0.023	8.54	1.2 ± 0.20	0.14 ± 0 023

* Blank spaces in the table indicate no significant retention. Total administered in each case = 1 mg. Mn.

of labeled material in organisms, especially with regard to study of mineral metabolism.[20] Investigations of mineral metabolism which concern elements for which requirements of the living organism are exceedingly small are considerably expedited and, in some cases, can be carried out only by use of tracer methods.

Ordinarily, studies of distribution of mineral metabolites are difficult because of the extremely low concentrations present. Tracer techniques are particularly valuable in solving such a problem as the manner in which a given dose of a "trace" element is ingested.

Manganese, for instance, is known to be an essential element in organic function, although it occurs only to the extent of a few micrograms per gram of tissue. Greenberg and Campbell[21] have used samples of manganese labeled with the radioactive isotope Mn^{54} ($\tau_{1/2}$ = 310 days) to study absorption, excretion, and localization of manganese in the rat. In these preliminary studies 0.001 mg. of manganese was easily detectable. In Table 18 some typical results obtained on manganese distribution are shown.

These results are cited merely to illustrate the ease with which small amounts of ingested mineral elements can be detected. It is obvious that extension of such studies to animal and plant organisms suffering abnormalities in metabolic function may throw light on mechanisms in disease processes and prove of clinical value in studies on humans.

[20] Greenberg, D. M., *in* "Symposium on the Use of Isotopes in Biology and Medicine" (H. T. Clarke, ed.), p. 261. Univ. Wisconsin Press, Madison, 1949.

[21] Greenberg, D. M., and Campbell, W. W., *Proc. Natl. Acad. Sci. U. S.* **26,** 448 (1940).

The major difficulty in experiments on uptake and retention of elements is to make sure that the organism is adequately separated from the surrounding medium in sampling procedures. Working procedures must be good enough to remove contaminating radioactivity. It is usually effective to wash the cells several times with solutions of nonlabeled element so that on each washing a large fraction of the adherent radioactive solution is removed by isotopic equilibration and dilution. Thus, as an example, red blood cells after exposure to radioactive potassium (K^{42}) are removed, centrifuged, and the supernate decanted off. Suppose that 0.5 ml. adheres to the cells. The cells are resuspended in 5 ml. of *N* KCl solution and, after a short wait to assure equilibration of the absorbed solution with the wash liquid, centrifuged, and the supernate again decanted. This procedure removes 90% of the original contamination by simple dilution. Washings may be repeated until the calculated removal of contamination is sufficient so that any absorbed contamination is negligible. As a check on the procedure, the activity appearing in the supernate at each washing is also determined to ascertain that activity calculated to be removed is actually being removed.

Some researches call for elaborate washing procedures which are well standardized.[22] Such a case may arise if the absorption is very critically dependent on the internal metabolism. The labeled element which represents true absorption and retention can be lost during the washing procedure because metabolism continues during washing. In this case it may prove better to use distilled water or even to include a metabolic poison. The retention may also depend markedly on the concentration of salt used in washing. It is, of course, essential to avoid the use of washing solutions which can damage the organisms studied. It is apparent from these remarks that even a simple procedure, such as washing organisms to remove adherent nutrient, may be complicated by factors which require careful study and control.

It may be noted that extensive data have become available for the first time on distribution of "trace" elements and factors affecting their absorption.[23] Certain toxic elements have also been studied, notably among the newer elements resulting from nuclear fission.[24] Although many interesting facts about the circulatory mechanisms involved in distribution of such elements have been recorded, it is premature to attempt any correlation of the data with metabolic patterns in the various organs. It may be expected that a large number of tracer researches in the future will center

[22] See Gest, H., and Kamen, M. D., *J. Biol. Chem.* **176**, 299 (1948), for an example of variation in cellular composition deriving from washing procedures.

[23] References for various elements are given in Chapter XII.

[24] Hamilton, J. G., *Radiology* **49**, 325 (1947).

around the integration of such distribution studies with studies of metabolism.

The successful application of tracer techniques to absorption phenomena requires careful consideration of the physiology of the system in addition to observation of coexisting metabolic patterns. Much of importance can be learned from properly designed experiments, however simple. There are many examples in the literature, especially in matters relating to agriculture such as the effect of various physical factors on uptake of nutrient minerals.[24a]

B. Determination of Intracellular and Extracellular Space by Isotope Dilution Techniques

A critical quantity which must be evaluated in many physiological studies is the fluid volume (space available for solution of electrolytes and metabolites) inside and outside the cells of an organism. An obvious extension of isotope dilution methods affords a relatively simple means for measuring this space. A known quantity of fluid containing a known amount of labeled material is injected into the organism. In a time which varies with the material injected and which depends on diffusion rate, this material reaches dilution equilibrium with all body fluid available. The time required is determined by removing and examining aliquots until the specific activity ceases to decrease markedly. The constant value so obtained can be used to calculate the diluting volume of the space available for mixing.

As an example, consider an experiment by Pace *et al.*[25] Into the antecubital vein of a human being weighing 70.78 kg. was injected 5.09 ml. of a tritium H^3-labeled water sample containing 30,400 ct./sec./ml. At intervals of 30 min., 1, 2, and 3 hr., blood samples were withdrawn from the same vein. The plasma was separated and stored for measurement of H^3. The four values for specific activity of the plasma corresponding to the four times given were found to be 3.02, 3.49, 3.21, and 3.40 ct./sec./ml. Thus, equilibration had occurred in about 1 hr., the average of the last three measurements being 3.37 ct./sec./ml.

The simple isotope dilution formula $A_1V_1 = A_2V_2$ relates the initial activity, A_1, of the injected sample, the final activity of the aliquot taken, A_2, and the corresponding volumes, V_1 and V_2. It follows that the injected triterated water underwent a dilution of approximately 10^4-fold. The diluting volume (total body water space), V_2, of the subject was 45,900 ml.,

[24a] See, for example, Spinks, J. N. T., and Barber, S. A., *Sci. Agr.* **27,** 145 (1947).

[25] Pace, N., Kline, L., Schachman, H. K., and Harfenist, M., *J. Biol. Chem.* **168,** 459 (1947).

or (30,400/3.37) (5.09). From the specific gravity of the subject (1.077) and the total body weight (70.78 kg.), the per cent total body water was 64.7.

In a similar fashion, the tracer method can be extended to determination of intracellular and extracellular spaces for all ions of biological interest, i.e., to the determination of "sodium space," "chloride space," and "bromide space." When the ion is H^+, essentially all the cellular space is determined because H^+ is equilibrated rapidly with water hydrogen inside and outside the tissues. Ions like Na^+, Cl^-, and Br^-, which are confined mainly to extracellular space, can be used to determine extracellular space. These ions injected in a labeled form attain equilibrium rapidly with the same ions present in the organism. In addition the extracellular volume of individual organs can be assessed. Ions like HPO_4^{--} and K^+, which are readily metabolized or concentrated in tissue, will yield values for apparent extracellular volume which are higher than for ions confined mainly to extracellular space.[26] In the case of an ion like Na^+, a small percentage may enter slowly into certain tissues, such as bone and muscle, so that a correction must be made.[27] Estimates of the sodium retained in a given tissue are made by direct assay of the Na^{24} content of the tissue and calculation of total sodium held up, knowing the total weight of the tissue.

The obvious advantage of the isotope dilution method is that the indicator employed can be a normal component of the extracellular space and one not lost perceptibly by metabolism in various organs.

C. Metabolic Turnover in Relation to the Intact Organism

The dynamic flux of metabolites in and out of the structural elements of the living cell is regulated so that no net change occurs in composition or structure. To understand how this comes about, it is necessary to study the interaction between enzyme systems under physiological conditions. The tracer method may be used effectively to aid in unraveling regulation mechanisms by following transfer of atomic groupings from one cellular fraction to another and by validating inferences drawn from researches conducted in unregulated isolated enzymic systems, as exemplified by cell extracts and homogenates.

Many researches dealing with the sites of synthesis of various metabolites have been reported. An example is the study of the origin of plasma phospholipid. When labeled inorganic phosphate was administered to rabbits it was found that the liver phospholipid reached a higher isotopic content than phospholipid from any other organ in the relatively short time of

[26] For a full discussion, see Hevesy, G., "Radioactive Indicators," Chapter 7. Interscience, New York, 1948.

[27] Davies, R. E., Kornberg, H. L., and Wilson, G. M., *Nature* **170**, 979 (1952).

10 hr.[28, 29] In other experiments, labeled phospholipid produced in one rabbit was isolated and introduced into the plasma of another[28] and found to disappear from the plasma rapidly, reappearing at the greatest rate in the liver.

Although these experiments could be interpreted as evidence for the primary involvement of the liver in phospholipid synthesis (turnover), a more direct approach was tried in later researches with hepatectomized animals.[30] The animals (dogs) with excised livers showed a very low rate of plasma phospholipid turnover compared to control animals, despite the appearance of large amounts of labeled phospholipid in kidney and small intestine of both groups. Apparently escape of phospholipid from these organs was blocked, so that, although there was plenty of lipid synthesized, little got into the circulation. It appeared, therefore, that liver was the major contributor to phospholipid in plasma. The synthesis of lipid in the organs other than liver proceeded at about the same rate whether the animals had livers or not, so that one might conclude that phospholipid synthesis in these organs did not require a liver factor.

Further researches on *in vitro* systems (surviving liver and kidney slices) demonstrated that phospholipid formation in the isolated synthetic system required coupled oxidation reactions.[30] This latter type of research is another example of tracer investigation into reversibility of enzymatic degradation. In these experiments the synthesis of phospholipid was detected in a system in which there was a net degradation of lipid. Much work on phospholipid metabolism has appeared in the literature, which should be consulted for further details.[31, 32]

The investigation of metabolic mechanisms in intact cellular systems has been often undertaken with various agents as more or less specific inhibitors for one or another type of cellular reaction. The mechanisms involved in the operation of such inhibitors are amenable to the tracer approach. An example of such work is the demonstration of inhibition of phospholipid synthesis by oxidation inhibitors such as azide and cyanide.[30]

The interpretation of turnover experiments in which the labeled material must enter the cell from the surrounding substrate is complicated by the ever-present factor of nonequilibration. For instance, the specific activity of labeled orthophosphate in yeast never attains the same value as that

[28] Hevesy, G., and Hahn, L., *Kgl. Danske Videnskab. Selskab, Biol. Medd.* **16,** 1 (1941).

[29] Artom, C., Sarzana, G., Perrier, C., Santangelo, M., and Segrè, E., *Arch. intern. physiol.* **45,** 32 (1937).

[30] Fishler, M. C., Taurog, A., Perlman, I., and Chaikoff, I. L., *J. Biol. Chem.* **141,** 809 (1941).

[31] Chaikoff, I. L., *Physiol. Revs.* **22,** 291 (1942).

[32] Bloch, K., *Cold Spring Harbor Symposia Quant. Biol.* **13,** 29 (1948).

of the outside labeled orthophosphate[14, 33] unless the cell is killed. The physiological heterogeneity of cellular fractions isolated as identical chemical fractions has already been shown for the case of yeast metaphosphate.[34] Such limitations must be thought of as inherent in the use of tracer techniques.

Tracer methods may also be used to study the effect of radiation on metabolism in the intact organism.[35] One may mention in this connection the pioneering investigations on x-ray inhibition of deoxyribonucleic acid synthesis as measured by labeled phosphate incorporation.[36, 37] Here the obvious advantage of the measurement of turnover in tissues metabolizing in the steady state before and after irradiation may possibly be turned to good account.

D. Transport Studies

The word "tracer" implies that the method is applicable to the localization and study of transport of metabolites. The distribution of the various elements in plants and animals and the dynamic conditions involved in transportation of nutrients through the living organism are problems for which the tracer method is well adapted. The procedure involved is similar to that described in the discussion of permeability. After exposure to labeled metabolite, the organism is assayed at various places for content of labeled material. This may be done with the intact organism or by sectioning.

Special methods have been developed for the cases when it is necessary to keep the whole organism intact, as in clinical investigations.

As an example of research on movement of metabolite, there may be cited the early work of Stout and Hoagland[38] on the upward and lateral movement of salt in plants. The plants used were cotton, geranium, and willow. These were grown in culture solutions from the seedling stage. In a typical experiment, the roots of a young plant were placed in a beaker containing dilute nutrient solution which was aerated. A longitudinal slit was made in the bark of one branch, and then the bark was carefully pulled away from the wood and paraffined paper inserted between the bark and the wood. In this way the two components were segregated over a length of some 9 inches. To prevent loss of moisture the whole stripped section was then wrapped with paraffined paper. Radioactive potassium salt was

[33] Mullins, L. J., *Biol. Bull.* **83,** 326 (1942).
[34] Juni, E., Kamen, M. D., Spiegelman, S., and Wiame, J. M., *Nature* **160,** 717 (1947).
[35] Kamen, M. D., *Radiology* **49,** 223 (1947).
[36] Euler, H. V., and Hevesy, G., *Svenska Vet. Akad. Arkiv Kemi* **17A,** No. 10 (1944).
[37] Ahlstrom, L., Euler, H. V., and Hevesy, G., *Svenska Vet. Akad. Arkiv Kemi* **19A,** No. 13 (1945).
[38] Stout, P., and Hoagland, D. R., *Am. J. Botany* **26,** 320 (1939).

added to the culture solution to make the concentration of potassium 5 meq./l. After exposure for 5 hr., the plant was removed and sectioned for analysis. Similar experiments were carried out with sodium, phosphorus, and bromine tracers.

In all cases it was found that movement of tracer was extremely slow where the bark was isolated from the wood. In the central portion of the bark strip, no tracer could be detected although large amounts were present in the wood. Wherever wood and bark touched, there was rapid lateral movement of tracer. The investigators concluded that xylem was the path of rapid upward movement of salt.

Researches on plant nutrition have dealt mainly with the movement of such elements as potassium, sodium, phosphorus, iron, bromine, and rubidium.[39] The movements of labeled growth stimulators and inhibitors are, of course, also amenable to study. Even when this work was in its infancy, a number of rather surprising observations were reported—movement of ions from roots upward many feet in very short time intervals ($\sim$15 min.), rapid lateral transfer of solutes from wood to bark, and dynamic flux of nutrient in and out of roots even during intervals of accumulation.

Turning for a single example to animal physiology, it may be noted that deutero-labeled fatty acids have been used to establish the flow of these compounds across rat placenta. It has been shown that, when body fluids of pregnant rats are enriched with heavy water, there results rapid incorporation of deuterium into the glycogen, fatty acids, and cholesterol of the fetus, the conclusion being that synthesis of these compounds occurs in the fetus.[40]

3. APPLICATIONS TO CLINICAL RESEARCH

It has been rather obvious for a number of years that the development of tracer research must inevitably exert a profound effect on biology, because it makes possible elaboration of fundamental phenomena at the biochemical and physiological levels. Practical applications, particularly in the medical arts, may be expected to continue to increase in coming years. In this section an attempt will be made to survey briefly contributions to human physiology in health and disease.

A. Determination of Circulation Time: Capillary Transport

Determinations of the speed of diffusion or peripheral circulation of various ions in the animal or human organism are among the earliest data gathered by the use of tracers.[41, 42] The availability of a simple technique

[39] See also Gustafson, F. G., and Darken, M., *Am. J. Botany* **24,** 615 (1937); *Science* **85,** 482 (1937).

[40] Goldwater, W. H., and Stetten, D. W., Jr., *J. Biol. Chem.* **169,** 723 (1947).

[41] Blumgart, H. L., and Weiss, S., *J. Clin. Invest.* **4,** 15 (1927).

[42] Hamilton, J. G., and Stone, R. S., *Radiology* **28,** 178 (1937).

for measuring circulating time has resulted in a number of applications of tracer methods to the study and diagnosis of a variety of conditions obtaining in congestive heart disturbances and peripheral vascular disease.[43-46]

A typical procedure involves injection of a Na^{24}-labeled saline solution ($\sim$ 100-μc. equivalents) into the vein of the antecubital fossa and measurement of the γ radiation as it appears in the foot, using a shielded G-M tube.[47] The build-up in labeled sodium content of the extremity is a function of the rapidity of mixing or interchange of plasma sodium and extracellular fluids of the foot. The curve of uptake rises to a saturation value representing a steady-state condition in which the amount of labeled sodium leaving the blood vessels is equal to that entering from extravascular fluid. It appears that human subjects suffering from circulatory disturbances produce a curve with lower equilibrium value and slower approach to equilibrium than do normal subjects.

The effect of treatment can be followed by determining build-up curves at various times during therapy. Thus the effect of vasodilators such as histamine and papaverine can be assessed.[44] It was shown that patients with scleroderma, thromboangitis obliterans, and other peripheral circulatory disturbances showed a "subnormal" curve. A number of drugs used for alleviation of circulatory diseases were tested, but only histamine proved effective in causing an immediate rise in the rate of diffusion through capillaries.

It has been noted that extremely large fluctuations in circulation time after intravenous injection occur in normal individuals.[45, 46] The determination of circulation time appears to be useful in determining the advisability of continuing conservative therapy rather than drastic surgery and in deciding on the site of amputation if surgery must be performed.[47] The efficacy of various procedures in restoring circulation has also been tested. Thus it has been shown that in dead but heparinized animals mechanical inflation and deflation of the lungs is sufficient to re-establish a partial circulation of blood, but that no movement is possible once post-mortem clotting occurs.[48]

A large number of studies in capillary permeability have been made.[49]

---

[43] Quimby, E. H., *Am. J. Roentgenol. Radium Therapy* **58,** 741 (1947).

[44] Mufson, I., Quimby, E. H., and Smith, B. C., *Am. J. Med.* **4,** 73 (1948).

[45] Elkin, D. C., Cooper, F. W., Rohrer, R. H., Miller, W. B., Jr., Shea, P. C., and Dennis, E. W., *Surg. Gynecol. Obstet.* **87,** 1 (1948).

[46] Burch, G. E., Threefoot, S. A., Cronvich, J. A., and Reaser, P., *Cold Spring Harbor Symposia Quant. Biol.* **13,** 63 (1948).

[47] Smith, B. C., and Quimby, E. H., *Radiology* **45,** 335 (1945).

[48] Thompson, S. A., Quimby, E. H., and Smith, B. C., *Surg. Gynecol. Obstet.* **83,** 387 (1946).

[49] See Flexner, L. B., Cowie, D. B., and Vosburgh, G. J., *Cold Spring Harbor Symposia Quant. Biol.* **13,** 88 (1948).

The particular subject of transfer across placenta has received much study.[50] The major findings are that the transfer is more rapid, the smaller the number of tissue layers intervening between maternal and fetal circulation, and that there is a correlation between uptake per unit weight of fetus and fetal growth rate.

The circulation in the lower extremities in women in labor and during puerperium has been investigated by determining the rate of venous blood flow with labeled sodium.[51] The procedure in brief was as follows: the limb of the patient was immersed in a water bath at 40°C. for 10 min. to ensure vasodilation. With the patient recumbent, 1 ml. of labeled sterile saline solution was injected rapidly into one of the veins in the dorsum of the foot. The appearance of radioactivity in the femoral vein of the groin was then recorded. It was found that women during labor showed a great retardation of venous flow in the leg as compared to nonpregnant women; in the puerperium, the rate of the venous flow returned rapidly to normal.

Elaboration of external assay procedures has led to fabrication of a rapid ink-writing G-M tube recorder which can be applied in cardiographic studies.[52, 53] In this type of "radiocardiography" the G-M tube is placed over the precordium and 0.1 to 0.2 mc. of Na^{24} solution is injected into one of the antecubital veins. The time pattern of the flow of active blood in and out of the heart is determined. This technique has been used in diagnosis of the role of the heart in a variety of circulatory disturbances.

B. Uptake, Retention, and Excretion, Particularly in Relation to Extension of Radiation Therapy and Diagnosis

The uptake and distribution of radioactive tracers has been studied in much detail. The results have provided information of value in devising radiotherapeutic procedures. The therapist can take advantage of localization of tracer, depending on the particular metabolic characteristics of the elements involved. There is an extensive literature on uptake and retention of labeled phosphate administered as Na_2HPO_4 in normal and tumor tissue. In general, tissues which exhibit a high metabolic rate of turnover, such as liver, lymph, and muscle, will show a rapid uptake followed by a rapid loss of labeled phosphate, whereas slowly metabolizing tissue, such as bone and brain, will retain labeled phosphate for a much longer time, although the uptake is also much less rapid than in the more active tissues.

[50] Flexner, L. B., and Gellhorn, A., *Am. J. Physiol.* **136,** 750 (1942).

[51] Wright, H. P., Osborn, S. B., and Edmonds, D. G., *J. Obstet. Gynaecol. Brit. Empire* **56,** 35 (1949).

[52] Prinzmetal, M., Corday, E., Bergman, H. C., Schwartz, L., and Spritzler, R. J., *Science* **108,** 340 (1948).

[53] Prinzmetal, M., Corday, E., Bergman, H. C., Spritzler, R. J., and Flieg, W., *J. Am. Med. Assoc.* **139,** 617 (1949).

As in the case of phosphate, the available literature on iodine is so large as to render futile any attempt to elaborate its content in a limited space. Among the numerous contributions to knowledge of the physiology of iodine,[54, 55] one may note (1) the demonstration of the remarkable efficiency of the thyroid gland in utilization of very small doses of iodine; (2) that iodine excretion and distribution outside the special thyroid fraction parallels that of other halides; (3) that under physiological conditions iodine appears to be incorporated into thyroxine via diiodotyrosine; (4) that large doses of iodine are less well assimilated than small doses—i.e., thyroid can fix 50% or more of a physiological dose of iodine (< 1 mg. in a normal man) but less than 10% of doses in the 10-mg. range; and (5) that large doses of iodine remain in inorganic form in the thyroid, being synthesized slowly into thyroid metabolites while storage occurs mainly in the colloid follicle.

One further example of the peculiar advantages of the tracer approach which may be cited in connection with iodine metabolism is the demonstration that appreciable synthesis of thyroxine and diiodotyrosine may occur in organs other than thyroid, such as muscle and intestine.[56]

The manner in which considerations such as these may be exploited can be illustrated by a brief survey of the physiology and therapeutic applications of the two most widely employed elements, phosphorus and iodine. The rationale for various forms of radiation therapy employing labeled phosphate is based on the observation that malignant tissue often exhibits a higher metabolic turnover than normal tissue with which it is associated. It may be expected that rapidly growing tissue will absorb more labeled metabolite than normal tissue which is growing slowly or is in a stationary state. Many examples may be cited—the concentration of P^{32} in tumor nuclei[57] and leukemic cells,[58] and the increased uptake of I^{130} and I^{131} by hyperplastic thyroid tissue.[59, 60]

Because of its selective uptake in bone and rapidly growing tissue, P^{32} has been exploited extensively in treatment of various hematologic dyscrasias and malignant neoplastic diseases. In an early report by Reinhard

[54] LeBlond, C. P., *Rev. can. biol.* **1,** 1402 (1942).

[55] Chaikoff, I. L., and Taurog, A., *in* "Symposium on the Use of Isotopes in Biology and Medicine" (H. T. Clarke, ed.), p. 292. Univ. Wisconsin Press, Madison, 1949.

[56] Morton, M. E., Chaikoff, I. L., Reinhardt, W. O., and Anderson, E., *J. Biol. Chem.* **147,** 757 (1943).

[57] Marshak, A., *J. Gen. Physiol.* **25,** 275 (1941).

[58] Tuttle, L. W., Scott, K. G., and Lawrence, J. H., *Proc. Soc. Exptl. Biol. Med.* **41,** 0 (1939).

[59] Hertz, S., and Roberts, A., *J. Am. Med. Assoc.* **131,** 81 (1946).

[60] Hertz, S., Roberts, A., and Salter, W. T., *J. Clin. Invest.* **21,** 25 (1942).

et al.,[61] diseases treated included polycythemia vera, myelogenous leukemia, lymphatic leukemia, monocytic leukemia, Hodgkin's disease, various lymphosarcomas, and assorted neoplastic conditions. Data on a total of 155 patients were analyzed. The advantages of P^{32} irradiation cited were:

1. It was selectively concentrated in organs with a high phosphorus content, such as bone, and in tissue the cells of which were multiplying rapidly.
2. It was easy to administer. Therapeutic doses (1 to 2 mc. of β equivalents in 5 to 10 ml. of isotonic sodium phosphate) rarely caused radiation sickness.
3. The 14-day half-life permitted steady radiation of tissues for several weeks, yet was short enough so that destruction of tissues could be controlled.

P^{32} was retained in growing tissues by incorporation into nucleoprotein. Rapidly growing cells encountered in organs principally involved in polycythemia vera, the leukemias, and the lymphomas attained relatively high concentrations of P^{32}. The differential between these tissues and normal tissues in the same organs and elsewhere in the human was not sufficiently great to avoid some damage to normal tissue. Hence dosages had to be controlled carefully.

General experience indicated that intravenous injection was the most efficient form of administration although oral administration was also employed. The conclusions reached were:

1. P^{32} was probably the best therapeutic agent available for polycythemia vera. Nearly all patients examined showed complete remission of symptoms with no recurrence for periods up to years after a single treatment. No conclusions could be drawn on the life expectancy as a result of P^{32} therapy.
2. No marked effect on the clinical course of acute or subacute myelogenous leukemia was noted. The P^{32} treatment was at least as good as the x-ray procedures, with the added advantages cited above. Similar statements applied to acute lymphatic leukemia and monocytic leukemia.
3. Hodgkin's disease, lymphosarcoma, reticulum cell sarcoma, and multiple myeloma did not respond as well to P^{32} treatment as to x-ray treatment. In a variety of other malignant neoplastic diseases only a few patients were studied, but there were no indications that P^{32} treatment would prove beneficial.

[61] Reinhard, E. H., Moore, C. V., Bierbaum, O. S., and Moore, S., *J. Lab. Clin. Med.* **31,** 107 (1940); see also Low-Beer, B. V. A., Lawrence, J. H., and Stone, R. S., *Radiology* **39,** 573 (1942).

It was pointed out that the localization of P^{32} in bone marrow could have a profound effect on the blood picture, leading, in some instances, to therapeutic complications, such as severe leucopenia, thrombocytopenia, and anemia. Unfortunately, no general statement on the dosage required to produce such effects was possible, because there was a wide variation in response of patients to P^{32}. Treatment had to be individualized, and the dosage requirement could be determined only by repeated study of blood and bone marrow and by clinical observation.[62] In chronic leukemia, it was advisable in some cases to use x-radiations because there might be a necessity for rapid reduction of spleen and lymph nodes, especially if there was excessive pressure on vital organs. No appreciable changes in the conclusions drawn from these early studies have been necessary as the result of later work.

Therapy with radioiodine is based on the fact that active thyroid tissue absorbs the major fraction of labeled iodine administered in low concentrations. Thus its use in treatment of hyperthyroidism is suggested and in actuality has become somewhat routine in many hospitals.

In researches[59, 60] with I^{130} (12.6-hr. half-life) labeled iodine was administered either in single or multiple doses, the total amount of carrier iodide being kept below 2 mg. In some cases patients were kept on an iodine-free diet for some time before administration of the labeled iodine. The dosage in roentgens could be calculated if the following data were known: (1) the fractional uptake of labeled iodide by the thyroid, (2) the energy of the I^{130} radiations, (3) the weight of thyroid, and (4) the pattern for uptake and retention of labeled iodine by hyperplastic thyroid. Experience showed that the total labeled iodine administered could be accounted for almost entirely as the sum of the I^{130} retained in thyroid and I^{130} excreted in the urine. Urinary excretion could be used, therefore, to determine I^{130} uptake and retention in thyroid. The weight of thyroid could be estimated from clinical observations. A rough indication of daily variations in uptake was available by means of an external G-M tube counter which had been calibrated with glands excised from patients previously scheduled for surgery. The total dosage (R) for I^{130} could be shown to be equal to 10,000 times the I^{130} retention in millicuries, divided by the weight of the thyroid in grams. For I^{131}, the constant involved was 117,000 (see, however, p. 377). Dosages ranged from 5 to 25 mc. per patient, depending on the size of the goiter. Calculated dosages agreed well with those found to be useful in x-ray therapy (1000 to 1200 r.).

[62] See also Doan, C. A., Wiseman, B. K., Wright, C. S., Geyer, J. H., Myers, W., and Myers, J. W., *J. Lab. Clin. Med.* **32,** 943 (1947); Hall, B. E., *in* "Symposium on the Use of Isotopes in Biology and Medicine" (H. T. Clarke, ed.), p. 353. Univ. Wisconsin Press, Madison, 1949.

In these researches, the effect of I^{131} was quite negligible compared to I^{130} because it was washed out by the later dosage of patients with unlabeled iodine. Patients were maintained on a normal iodine diet for 2 to 4 months after the I^{130} treatment. The iodine treatment was then discontinued, and the metabolic rate was checked. If no rise occurred, the condition was considered alleviated, but no cure was claimed until a prolonged follow-up (6 months to a year) showed no relapse. The results of these investigations, as well as those of Chapman and Evans,[63] definitely have established radiotherapy with radioactive iodine as a recognized accessory treatment for hyperthyroidism.

The extension of such therapy to certain limited types of adenocarcinoma has been reported.[64] In general, therapy with radioactive iodine is ineffective for carcinoma or metastases of thyroid because such tissue most probably has not the ability to accumulate iodine. Procedures which stimulate thyroid function wherever latent in metastases, such as thyroidectomy and use of hormones, can be employed as a basis for improving responses to radioactive iodine.[65, 65a]

A summary of possible applications for radiation therapy with radioactive isotopes has been prepared by Hahn and Sheppard[66] and may be reproduced here in abridged form:

1. General body radiation, corresponding to spray x-radiation. Use of an element such as sodium which is distributed generally in body water.
2. Exploitation of selective absorption as in use of iodine for hyperthyroidism and thyroid tumors.
3. Exploitation of element which may be substituted for one which is taken up selectively and which has more suitable radiation, i.e., Sr^{85} as a substitute for calcium.
4. Utilization of semispecific uptake where some degree of generalized radiation is not a disadvantage, as in use of P^{32} for treatment of myelogenous leukemia and polycythemia vera.

[63] Chapman, E. M., and Evans, R. D., *J. Am. Med. Assoc.* **131,** 86 (1946).

[64] Seidlin, S. M., *Med. Clin. N. Amer.* **36,** No. 3 (1952).

[65] Review articles by the following should be consulted: Marinelli, L. D., Trunnel, J. B., Hill, R. F., and Foote, F. W., *Radiology* **51,** 563 (1948); Frantz, V. K., Quimby, E. H., and Evans, T. C., *ibid.* **51,** 532 (1948); Werner, S. C., Quimby, E. H., and Schmidt, C., *ibid.* **51,** 564 (1948); Rawson, R. W., and Skanse, B. N., *ibid.* **51,** 529 (1948); Hertz, S., *in* "Symposium on the Use of Isotopes in Biology and Medicine" (H. T. Clarke, ed.), p. 377. Univ. Wisconsin Press, Madison, 1949; Rawson, R. W., Rall, J. E., and Peacock, W. C., *J. Clin. Endocrinol.* **11,** 1128 (1951).

[65a] See Owen, C. A., Jr., McConahey, W. M., Keating, F. R., Jr., and Orvis, A. L., *Federation Proc.* **14,** 723 (1955), for a review of investigation of diseases of the thyroid by means of I^{131}.

[66] Hahn, P. F., and Sheppard, C. W., *Ann. Internal Med.* **28,** 598 (1948). This article includes numerous references for the items listed.

5. Utilization of specific functions of certain tissues as they respond to administered material.

a. Phagocytosis of colloidal particles with retention in reticuloendothelial system, as in use of colloidal gold, manganese dioxide, and chromic phosphate for irradiation of liver and spleen, or as in therapy in leukemias, Hodgkin's disease, or lymphoma.

b. Intraperitoneal injection for removal by abdominal lymph nodes in diseases such as lymphoma and abdominal Hodgkin's.

c. Intrathoracic injection for removal by mediastinal and hilar nodes in pulmonary tumors, etc.

d. Localization of dyes labeled appropriately (see below).

6. Topical application of isotopic material in therapy of superficial lesions as in use of P^{32} Au^{198}, and $Ag^{111}NO_3$.

7. Treatment of tumors of hollow viscera by proximal radiation.

a. Bladder tumors by instillation of nonabsorbable material such as suspension of colloidal Au^{198}.

b. Instillation of material which would readily bind to mucosal surface, i.e., $Ag^{111}NO_3$.

c. Instillation of radioactive jelly in uterine cervical tumors.

d. Retention enema therapy of colon and rectum.

e. Localized radiation of various areas in carcinoma of stomach and of segments of intestinal tract by instillation of short-lived insoluble material and plugging pylorus followed by removal of plug and rinsing.

8. Selective therapy of organs or tissues exploiting specific action of drugs to control placement of isotope.

9. Therapy of pulmonary carcinoma by direct inhalation of radioactive material as aerosol or as radioactive inert gas, i.e., krypton.

10. Direct irradiation of tumor masses by infiltration of tumor with inert, insoluble material.

11. Physical manipulation in localization of radioactive material.

This list indicates at least in part the potentialities of generalized radiotherapy which are feasible because of availability of artificially radioactive isotopes in conjunction with conventional x-ray procedures and mechanical methods.

C. Applications in Hematology

1. Blood Physiology. Extensive contributions to blood physiology have resulted from tracer studies, particularly those utilizing the labeling isotopes for nitrogen, carbon, phosphorus, and iron. It has already been noted (Chapter VI) that glycine is utilized specifically for the biosynthesis of hemoglobin protoporphyrin, a fact which has been exploited in studying erythrocyte dynamics. It will be remembered that the average life span of the circulating red blood cell in the normal human being is somewhat greater than 100 days, the cells dying as a function of age, rather than indiscriminately. The availability of a simple labeling technique based on the absence of turnover of protoporphyrin nitrogen or carbon once it is incorporated into the erythrocyte has afforded a valuable method for studying blood dyscrasias.

The peculiar advantage of this approach is that it is possible to determine the rate of formation of hemoglobin, and thus of red cells, and the pattern of destruction of the red cells in the same individual in whom the cells are being made, without altering the physiological state of the organism. The method has been applied to study of two normal adults, male and female, and to subjects with pernicious anemia, sickle-cell anemia, and polycythemia vera.[67] In the normal subjects the curve of incorporation of N^{15}-labeled glycine into protoporphyrin rises rapidly after ingestion of the glycine for approximately 20 days, remains constant for a long period (50 to 70 days), and then falls along an S-shaped curve (Fig. 40).

Analysis of these data led to the result that the average erythrocyte life span of the human male investigated was 120 days, that of the female 109 days. Although survival times of the erythrocytes showed a rather wide range, the time span in which half the cell population died was relatively short (106 to 141 days in the male and 91 to 123 days in the female).

With these data as a basis for comparison one may turn to results in polycythemia vera. Here it was found that, despite the greatly elevated hematocrit characteristic of this disease, the survival curve obtained was normal. The average life span was 131 days, and the shape of the death curve was similar to those found for normal subjects. Hence the elevated content of circulating erythrocytes was correlated with an elevated rate of hemoglobin synthesis and cell formation; the rate of cell production in the particular subject studied was found to be about 2.5 times that of the normal subject. Thus polycythemia vera, at least in this case, was characterized by an abnormally high rate of hematopoiesis and a normal red cell life span. This implied a functional hyperactivity in the blood-forming apparatus, a finding consistent with the pathological demonstration of hyperplasia of bone marrow elements as well as hematologic evidence for increased bone marrow activity (polychromatophilia, basophilic stippling of erythrocytes, and leucocytosis with an increase of immature cells of the myeloid series). The increased hemoglobin synthesis required that hemoglobin degradation be increased correspondingly to maintain the steady state characteristic of polycythemia vera. However, no evidence of an increased rate of degradation of hemoglobin as indicated by the excretion

[67] London, I. M., Shemin, D., West, R., and Rittenberg, D., *J. Biol. Chem.*, **179**, 463 (1949).

of bile pigment (fecal urobilinogen) was noted in this disease. Later studies indicated that the major portion of bile pigment in normal man was derived from the hemoglobin of mature erythrocytes.[68]

In sickle-cell anemia, a much different picture was presented. The N^{15} content of the erythrocytes began to drop precipitately along an exponential curve very shortly after feeding of N^{15}-labeled glycine. In other words, cell destruction was indiscriminate and not a function of age, as in the normal case. The mean survival time in the subject studied was 42 days. The rate of red cell formation and destruction was found to be 2.8 times the normal rate. The curve of N^{15} incorporation and loss in this type of anemia could have resulted from (1) random destruction of the cells with loss of heme from circulating blood, (2) random degradation and synthesis of heme in circulating erythrocytes maintained intact morphologically, or (3) random synthesis and degradation of heme in cells also undergoing random destruction.

An interesting finding was that when whole blood from sickle-cell anemia patients was incubated with N^{15}-labeled glycine there resulted incorporation of N^{15} into heme, indicating that heme was synthesized from glycine *in vitro*.[69] The rate of *in vitro* formation observed corresponded to a survival time of labeled circulating hemoglobin of 500 to 1000 days, whereas the actual *in vivo* survival time was 40 days. Thus, random synthesis of heme in the peripheral blood of these patients was not considered to play an important role in hemoglobin turnover in sickle-cell anemia. Rather it appeared necessary to assume that random destruction of cells was the major cause for disappearance of heme.

Again, the pathological and hematological evidence was in good accord with the finding that increased synthesis and destruction of heme were involved. It appeared, therefore, that the characteristic deficiency in sickle-cell anemia was some defect in capacity for survival.

In pernicious anemia there was disclosed an abnormal pattern of red cell destruction and a somewhat low survival time. The rate of formation of circulating heme was found to be about 80% of normal, the rate of formation of circulating cells about 50% of normal. On treatment with liver extract, the pattern of red cell destruction and the life span were restored to normal.

It may be remarked in passing that neither iron nor phosphorus is a suitable indicator for studying life span of mammalian erythrocytes. Iron released during hemoglobin catabolism is used preferentially for resynthesis.[70] In the case of phosphorus, mammalian erythrocytes do not contain

[68] London, I. M., West, R., Shemin, D., and Rittenberg, D., *J. Biol. Chem.* **194,** 351 (1950).

[69] London, I. M., Shemin, D., and Rittenberg, D., *J. Biol. Chem.* **173,** 797 (1948).

[70] Hahn, P. F., Bale, W. F., and Balfour, W. M., *Am. J. Physiol.* **135,** 600 (1942).

sufficient amounts of metabolically stable phosphorus compounds into which P^{32} can be incorporated. This is in contrast to the erythrocytes in the hen which have a sufficient quantity of deoxyribonucleic acid. Thus, the life spans of both red and white corpuscles in the hen have been determined by labeling of erythrocytes through incorporation of P^{32}-labeled phosphate into the deoxyribonucleic fraction of the cells.[71]

It should be possible to extend labeling methods for determination of life time of blood components by exploiting enzyme reactions for labeling. An indication of the potential value of such an approach is found in researches based on the fact that the compound diisopropylfluorophosphonate (DPFP) combines irreversibly with esterases present on the red cell membrane.[72] By labeling DPFP with P^{32}, it has been shown that a satisfactory procedure can be developed for studying the survival of red cells. More significantly, it is expected that such a procedure can be extended to the determination of the life of platelets which have been shown to contain esterases also. No satisfactory assay for platelet life has hitherto been available. DPFP can be used because the only metabolic product, diisopropylphosphate, is not further absorbed but is quickly excreted in the urine. Some optimism with regard to the future of this method appears warranted from preliminary results obtained on four normal persons, one patient with chronic myeloid leukemia, and one patient with polycythemia vera.[73]

In other directions, one may note a few of the important contributions resulting from the use of labeled iron. For example, the factors involved in blood storage have been assessed.[74–77] It has been found that the labeling technique can be used to determine survival of preserved human erythrocytes under a variety of conditions and to investigate the merits of various preservative solutions proposed in the past. Optimal conditions for storage of whole blood have been established: ratio of whole blood to diluent not less than 4:1; concentration of citrate in diluted plasma between 0.4 and 0.6 g. per 100 ml.; dextrose concentration about 0.5 g. per 100 per ml.; final pH not higher than 7.0 in plasma and 6.8 in cells. Other

[71] Ottesen, J., *Nature* **162,** 730 (1948).

[72] Cohen, J. A., and Warringa, M. G. P. J., *J. Clin. Invest.* **33,** 459 (1954).

[73] Leeksma, C. H. W., and Cohen, J. A., *Nature* **175,** 552 (1955).

[74] Gibson, J. G., II, Aub, J. C. Evans, R. D. Peacock, W. C. Irvine, J. W., Jr., and Sack, T., *J. Clin. Invest.* **26,** 704 (1946).

[75] Gibson, J. G., II, Evans, R. D., Aub., J. C., Sack, T., and Peacock, W. C., *J. Clin. Invest.* **26,** 715, 739 (1946).

[76] Gibson, J. G., II, Peacock, W. C., Seligman, A. M., and Sack, T., *J. Clin. Invest.* **25,** 838 (1946).

[77] Gibson, J. G., II, Weiss, S., Evans, R. D., Peacock, W. C., Irvine, J. W., Jr., Good, W. M., and Kip, A. F., *J. Clin. Invest.* **25,** 616 (1946).

conclusions relate to the rapid removal of nonviable stored erythrocytes after transfusion, necessity for refrigeration, and the reliability of hematocrit determinations. An accurate measurement of the total circulating blood volume by means of the two available radioactive iron isotopes has been described.[76]

Tracer experimental methods in iron physiology have been extended to the study of iron absorption in various clinical conditions.[78] It has been found that pregnancy induces increased iron uptake. Diseased states in which iron stores are abundant, such as pernicious anemia and hemachromatosis, show much less than normal absorption. Chronic infections, in spite of associated anemia, result in no utilization of radioiron. It appears that reserve stores, not anemia, control iron absorption.

Other researches have dealt with the mechanism of anemia of infection which appears ascribable to impaired hemoglobin production.[79] An interesting finding with which to conclude this list is that the efficiency of iron absorption depends on the valence form in which it is administered in man but not in dogs. The former can use ferrous iron much more efficiently than ferric.[80] The literature relating to studies with tracer iron is too extensive for complete review.

2. Determination of Blood Volume. Isotope dilution techniques afford a comparatively simple approach to the problem of determination of circulating red cell volume free from the shortcomings inherent in conventional procedures involving use of dyes and the hematocrit figure.[76] The method depends on determination of dilution of administered isotope with the unlabeled element in the patient, corrections being made for excretion during the interval between administration and sampling.[81–85] Thus Moore[85] has found that in 1 hr. after administration of deuterium water equilibration with body fluids is complete, so that samples for isotopic assay can be taken 1 hr. after administration of the isotopic water. By direct comparison of body water of rabbits determined by the usual desiccation methods, it has been shown that the isotopic assay method is reliable. Similar researches

[78] Balfour, W. H., Hahn, P. F., Bale, W. F., Pommerenke, W. T., and Whipple, G. H., *J. Exptl. Med.* **76,** 15 (1942).

[79] Wintrobe, M. M., Greenberg, G. R., Humphreys, S. R., Aschenbrucker, H., Worth, W., and Kraemer, R., *J. Clin. Invest.* **26,** 103 (1947).

[80] Moore, C. V., Dubach, R., Minnich, V., and Roberts, H. K., *J. Clin. Invest.* **23,** 755 (1944).

[81] Nylin, G., *Arkiv Kemi Mineral. Geol.* **A20,** No. 17 (1945).

[82] Chapin, M. A., and Ross, J. F., *Am. J. Physiol.* **137,** 447 (1942); Anderson, R. S., *ibid.* **137,** 539 (1942).

[83] Nylin, G., and Hedlund, S., *Am. Heart J.* **33,** 770 (1947).

[84] Menerly, G. R., Wells, E. B., and Hahn, P. F., *Am. J. Physiol.* **148,** 531 (1947).

[85] Moore, F. D., *Science* **104,** 157 (1946).

have been carried out with radioactive potassium.[86] Attainment of equilibrium in this case required 36 hr., the total body potassium being determined as approximately 70 meq./kg. in human beings.

A simple procedure for labeling erythrocytes *in vitro* is possible with P^{32}-labeled phosphate. Red cells suspended in isotonic solution containing the labeled phosphate incorporate as much as 30% of the radioactivity in incubation times of 60 to 90 min.[87] A typical procedure involves the following manipulations.[88] Into a clean, dry 25-ml. Pyrex tube is placed 1.2 ml. of isotonic saline solution containing 50 μc. of radioactive phosphate. The mouth of the tube is plugged with cotton. All syringes and needles are washed thoroughly with distilled water. A 10-ml. pipet and a rubber stopper are similarly prepared. Since aseptic technique must be maintained throughout, all equipment is sterilized in an autoclave for 30 min. at 15 lb. of steam pressure.

From the subject, a sample of approximately 15 ml. of heparinized blood is withdrawn without stasis through the antecubital vein (with a 19-gauge needle) and transferred to the test tube containing the radioactive solution. The tube is sealed with a rubber stopper and, after the contents are mixed thoroughly by inversion, it is placed in an incubator maintained at 37°C. and agitated for 2 hours. A motor-driven stirrer with an eccentric shaft to which the samples are attached provides adequate mixing and prevents settling of the red cells.

At the end of the 2-hr. period of incubation and agitation, exactly 10 ml. of blood is removed from the tube with a pipet, transferred to a syringe, and injected into the subject. Blood is drawn back and forth into the syringe several times in order to ensure the complete injection of all the active material. Specimens are then removed at various intervals from the opposite arm after a 5-min. period has elapsed to allow for adequate mixing.

In this method, both labeled corpuscles and labeled plasma are injected. Samples withdrawn from the patient are centrifuged, and the activity of the corpuscles relative to those injected initially is determined. Possible sources of error are (1) adhering of plasma to centrifugal corpuscles, (2) presence of P^{32} in plasma, (3) entry of some P^{32} into the cells during the experiment, (4) release of some P^{32} from the cells during the experiment. The magnitude of these errors is only a few per cent and to a large extent mutually compensatory.[89] Hence removal of active plasma before injection is unnecessary.

If it is necessary to ascertain the fate of labeled erythrocytes for many

[86] Hevesy, G., and Nylin, G., *Acta Physiol. Scand.* **24,** 285 (1951).

[87] Hahn, L., and Hevesy, G., *Acta Physiol. Scand.* **3,** 193 (1942).

[88] Kelly, F. J., Simonsen, D. H., and Elman, R., *J. Clin. Invest.* **27,** 795 (1948).

[89] Hevesy, G., and Zerahn, K., *Acta Physiol. Scand.* **4,** 376 (1942).

hours, neither P^{32}- nor K^{42}-labeled material can be used, because of loss of label due to turnover. The disadvantage of using $Fe^{59,55}$-labeled cells is that the preparation of the labeling sample is tedious, requiring administration of labeled iron for some weeks to volunteer donors and subsequent transfusion into experimental subjects. A procedure which avoids these difficulties is based on the use of Cr^{51}-labeled anionic hexavalent chromium in the form of Na_2CrO_4. It has been shown[90, 91] that addition of as much as 80 γ of chromium in this form, equivalent in radioactivity to about 20 μc., is taken up to the extent of 80 to 90% in a 5-ml. suspension of erythrocytes. This chromium is bound to the protein portion of the hemoglobin, and significant amounts are not released into the circulation for as long as 24 hr. Difficulties in assay arising from the soft radiation emitted by Cr^{51} can be minimized by use of scintillation detectors. An alternative labeling procedure employing a radioactive nuclide with harder radiations (ThB, $\tau_{1/2}$ = 10.6 hr.) has been suggested, based on the observation that erythrocytes in blood plasma absorb and retain radioactivity when exposed to a stream of air containing thoron (thorium emanation).[92, 93]

A typical procedure for determining red cell survival by radiochromium is as follows:[94] Three milliters of sterile standard acid citrate-dextrose solution (ACD) is added to a 50-ml. pyrogen-free, rubber-topped, screw-capped bottle. The bottle is immediately autoclaved and is then stored at room temperature. Immediately prior to use, 76 to 100 μc. of $Na_2Cr^{51}O_4$ is added to the bottle. Approximately 15 ml. of blood is obtained from the patient and injected through the rubber top into the bottle containing the radiochromium and ACD solution. The blood is mixed well by swirling. It is then incubated in a water bath at 37°C. for 30 min., or at room temperature for 45 min. with occasional swirling. The blood is again mixed well, and 10 to 15 ml. is removed through the rubber-topped cap and injected into the patient.

Approximately 20 and 40 min. after injection of the chromated blood, 5-ml. samples of heparinized or oxalated blood are obtained from another vein as baseline samples for the survival curve. An aliquot of these samples is used for cell volume determinations and 3-ml. samples are pipetted into a special plastic tube, hemolyzed with saponin, and kept for scintillation counting. Further samples are obtained at daily or 2-day intervals, depending on the expected survival time, and treated in the same fashion.

[90] Gray, S. J., and Sterling, K., *J. Clin. Invest.* **29,** 1604 (1950).

[91] Sterling, K., and Gray, S. J., *J. Clin. Invest.* **29,** 1614 (1950).

[92] Hevesy, G., *Arkiv Kemi* **3,** 425 (1951).

[93] Alexander, E., *Arkiv Kemi* **4,** 363 (1951).

[94] This procedure is employed routinely in the Department of Internal Medicine, Washington University, St. Louis, Missouri. Dr. E. Brown, private communication (1956).

Nylin and Hedlund,[83] using P^{32}-labeled erythrocytes, have made extended investigations into effects on circulating blood volume of injection of adrenalin, of latent shock, of postural hypertension, and of severe muscular exercise. These studies have shown that in normal persons no appreciable change occurs in circulating blood volume during muscular exercise. An interesting experiment made possible by the propensity of the Scandinavians for long-distance running has been performed. A noted runner was induced to run up and down stairs at such a rate that he effectively raised and lowered his own weight 100 meters in 5 min.[83] No significant effect on blood volume was found to occur despite repeated measurements on venous blood. Administration of adrenalin also made no change in blood volume,[95] an observation also noted by other workers.[96]

There is little doubt that many ingenious applications can be made in directions indicated by the work already done, so that the task of the medical practitioner may be lightened.

D. Immunological Studies

Among the early researches on metabolic turnover there may be noted inquiries into the formation and circulation of "active" and "passive" immune bodies.[97-99] Investigations[100, 101] have been reported which signalize the beginning of attempts to elaborate immunological mechanisms using isotopic tracers. These studies deal with analyses of specific precipitates formed by union of antibody with labeled antigen. The proteins have been phosphorylated, the label being P^{32}. It has been remarked that determinations of P^{32} in conjunction with total N determinations could be useful in analyses of antigen-antibody complexes. Thus, in experiments with a mixture of two closely related antigens, i.e., phosphorylated egg albumin and phosphorylated serum globulin, one being labeled and one unlabeled, it should be a simple matter to determine relative amounts of each antigen in a precipitin reaction by assay of the specific P^{32} content of the precipitate. In a similar way, relative amounts of antigen and substances which inhibit the reaction might be determined.

In the first experiments reported,[100] it was shown that labeled antigens

[95] Nylin, G., *Acta Cardiol.* **1,** 225 (1946).

[96] Parson, W., Mayerson, H. S., Lyons, C., Porter, B., and Trautman, W. V., Jr., *Am. J. Physiol.* **155,** 239 (1948).

[97] Schoenheimer, R., Heidelberger, M., Rittenberg, D., and Ratner, S., *J. Biol. Chem.* **140,** cxii (1941).

[98] Schoenheimer, R., Ratner, S., Rittenberg, D., and Heidelberger, M., *J. Biol. Chem.* **144,** 545 (1942).

[99] Heidelberger, M., Treffers, H. P., Schoenheimer, R., Ratner, S., and Rittenberg, D., *J. Biol. Chem.* **144,** 555 (1942).

[100] Boursnell, J. C., Dewey, H. M., Francis, G. E., and Wormall, A., *Nature* **160,** 339 (1947).

[101] Francis, G. E., and Wormall, A., *Biochem. J.* **42,** 469 (1948).

could be prepared suitable for study either by phosphorylation of plasma proteins, with labeled phosphorylating reagents, or by sulfonation with S^{35}-labeled mustard gas sulfone.

Extensive studies have been made on the preparation of P^{32}-labeled hen ovovitellin.[101] Hens were injected intramuscularly with labeled phosphate solutions, and eggs were collected for a month after injection. The yolks were separated and the vitellin protein isolated after as much lipid as possible was removed by cold ether extraction. Efficient extraction of lipid by hot ethanol or by cold ethanol-ether mixtures rendered the vitellin preparation insoluble. Labeled lecithin was recovered from the ether extracts and used in some experiments. In the immunization procedure, rabbits were injected with unlabeled vitellin. Both "soluble" and "insoluble" vitellin were used. When precipitin reactions indicated good antisera had been produced, the sera were withdrawn and treated with labeled antigen.

When antigen was used in low concentrations, the fraction of total antigen precipitated was constant (about 37%). At high concentrations, the fraction precipitated decreased, and with a large excess of antigen no labeled antigen precipitated, confirming visual observations. The vitellin preparations, particularly the soluble variety, contained considerable amounts of phospholipid, so experiments were performed on possible participation of phospholipids in the precipitin reaction. The results indicated a preferential precipitation of phospholipids in the vitellin-antivitellin complex which was not owing to mechanical carrying down of lipid. The combination of phosphatide with the antigen-antibody complex appeared to be specific because a heavy precipitate, formed between human serum proteins and their antibodies in the presence of labeled lipovitellin, failed to occlude any significant amount of labeled phosphorus. It was difficult to account for the ability of antibodies to remove phosphatide from the lipoprotein complex when ether extraction did not suffice.

The P^{32}-labeled immunologically active proteins have been used in a variety of researches. In a study of the fate of labeled vitellin injected into rabbits, for example, it was shown that only one-eighth of the protein could be detected in the blood 5 min. after injection. Much of the protein could be recovered in the lungs and liver.[102] That this result was not merely the result of the insolubility of the protein was shown when similar results were obtained with the more soluble lipovitellin.[103] Other studies were made on the nature of antibody precipitates produced by labeled lipovitellin.[104]

[102] Banks, T. E., Boursnell, J. C., Dewey, H. M., Francis, G. E., Tupper, R., and Wormall, A., *Biochem. J.* **43,** 518 (1948).

[103] Banks, T. E., Francis, G. E., Franklin, K. J., and Wormall, A., *Biochem. J.* **47,** 374 (1950).

[104] Francis, G. E., and Wormall, A., *Biochem. J.* **47,** 380 (1950).

When carrier-free isotopes of various nuclides such as S^{35}, P^{32}, and I^{131} became available, methods for efficiently labeling serum globulins and other proteins without significant alteration could be developed and researches extended to a great variety of problems in immunology.[105-108] Among these were investigations of the quantitative aspects of the antibody-antigen reaction,[109] precipitin inhibition,[110] and antibody valence.[110] Other studies have dealt with the zone of localization of antisera to mouse kidney and rat kidney.[111] It is anticipated that this aspect of tracer research will provide a basis for improved methods in the clinical applications of immunological research.

4. SPECIAL TOPICS AND CONCLUDING REMARKS

Continuing this survey of physiological and medical aspects of tracer methodology, one may note briefly new insights made possible by the tracer approach in categories such as metabolism and mode of action of metabolic accelerators and inhibitors (drugs, vesicants, etc.), mechanism of interaction of virus and host cell, and enzymatic mechanisms in chemotherapy.

A. Interaction of Vesicants with Protein

In an inquiry into the reaction of mustard-type vesicants with tissue components, du Vigneaud and his collaborators[112, 113] have studied the interaction of vesicants such as benzyl-β-chloroethyl sulfide ("benzyl H") and *n*-butyl-β-chloroethyl sulfide ("butyl H") with amino acids and with some well-characterized and highly purified proteins. The ultimate objective in these researches has been to elaborate the mechanism of vesicant action. The scheme for synthesis of the vesicants from S^{35}-labeled benzyl or butyl mercaptan is shown.

$$\mathrm{RMgBr} \xrightarrow{\mathrm{S}^*} \mathrm{RS^*MgBr} \xrightarrow[\mathrm{(HOH)}]{\mathrm{H}^+} \mathrm{RS^*H} \xrightarrow[\mathrm{ClCH_2CH_2OH}]{\mathrm{NaOH}}$$

$$\mathrm{RS^*CH_2CH_2OH} \xrightarrow{\mathrm{HCl}} \mathrm{RS^*CH_2CH_2Cl}$$

[105] Butement, F. D. S., *Nature* **162,** 731 (1948).

[106] Francis, G. E., Mulligan, W., and Wormall, A., *Nature* **167,** 748 (1951).

[107] Pressman, D., and Sherman, B., *J. Immunol.* **67,** 15 (1951).

[108] Melcher, L. R., and Masouredis, S. P., *J. Immunol.* **67,** 393 (1951).

[109] Banks, T. E., Francis, G. E., Mulligan, W., and Wormall, A., *Biochem. J.* **48,** 180 (1951).

[110] Banks, T. E., Francis, G. E., Mulligan, W., and Wormall, A., *Biochem. J.* **48,** 371 (1951).

[111] See, for instance, Pressman, D., and Sherman, B., *J. Immunol.* **67,** 216 (1951).

[112] du Vigneaud, V., Stevens, C. M., McDuffie, H. F. Jr., Wood, J. L., and McKennis, H., Jr., *J. Am. Chem. Soc.* **70,** 1620 (1948).

[113] Wood, J. L., Rachele, J. R., Stevens, C. M., Carpenter, F. H., and du Vigneaud, V., *J. Am. Chem. Soc.* **70,** 2547 (1948).

Details of the various steps are given in the original paper. The radioactivity in the final products was large enough so that with the assay methods used[114] as little as 5 γ of vesicant residue per milligram of protein could be detected.

Insulin (crystalline) was treated with amounts of vesicant ranging from 0.25 to 4.0 mg. per 100 mg. of protein. It was found that the amount of vesicant which combined with protein depended on the amount of vesicant used. Over a sixteenfold variation in the amount of vesicant applied, the per cent of vesicant combined was essentially constant (about 50%). After recrystallization, the product contained an average of 1.1 vesicant residue per molecule of protein. The product still retained considerable hypoglycemic activity when tested in rabbits.

Similar experiments with crystalline pepsin showed that pepsin combined with approximately 35% of the vesicant. Crystalline tobacco mosaic virus attached from 25 to 40% of the vesicant. These preparations contained hundreds of vesicant residues per protein molecule even at the lowest level of vesicant applied. Two of the preparations were tested for biological activity against half-leaves of *N. glutinosa*. One preparation with 1500 molecules of vesicant per molecule of protein possessed 93% of the activity of a control sample; another with 3200 molecules of vesicant per molecule of protein showed 52%. In other words, addition of 1500 vesicant residues caused little inactivation, whereas addition of another 1700 vesicant residues caused inactivation of half the protein. The addition of 1500 residues also was found ineffective in producing mutations in *N. glutinosa*. It appeared that a relatively enormous number of molecules of vesicant could be attached to reactive centers in the protein without affecting appreciably its biological activity.

B. Mode of Action and Biosynthesis of Penicillin

It was noted early that penicillin-sensitive cells did not appear to take up appreciable amounts of the antibiotic.[115] Later researches showed that very small amounts below the detection limit in early work were in fact incorporated.[116, 117] Estimates ranged from 10 to 1000 molecules of penicillin per bacterial cell in the case of *S. aureus*. Maass and Johnson used S^{35}-labeled penicillin to study the extremely small amounts of binding.[118] The labeled antibiotic was made by a fermentation procedure, growing

[114] Henriques, F. C., Jr., Kistiakowsky, G. B., Margnetti, C., and Schneider, W. G., *Ind. Eng. Chem. Anal. Ed.* **18,** 349 (1946).

[115] Hobby, G. L., Meyer, K., and Chaffee, E., *Proc. Soc. Exptl. Biol. Med.* **50,** 281 (1942).

[116] Pasynskii, A., and Kastorskaya, T., *Biokhimiya* **12,** 465 (1947).

[117] Rowley, D., Miller, J., Rowlands, S., and Lester-Smith, E., *Nature* **161,** 109 (1948).

[118] Maass, E. A., and Johnson, M. J., *J. Bacteriol.* **57,** 415 (1948).

the penicillin-producing mold *P. chrysogenum* in a lactose-glucose medium of the usual composition except for the addition of highly radioactive carrier-free $S^{35}O_4^{--}$ and lowering the sulfur content to 0.2 mg./ml. to avoid excessive dilution of the label. A penicillin precursor, sodium phenyl acetate, was added at intervals during the fermentation (see below). The fermentation mixture was extracted with ether at *p*H 2 to recover the labeled penicillin. This extract was treated with 2% K_2HPO_4 to remove the penicillin. The phosphate extract contained a penicillin which was shown by paper chromatography to be 90 to 95% penicillin G with specific activity relative to that of the sulfur in the medium such that it could be calculated that the maximum radioactive impurity was no more than 8% of the total.

When suspensions of *S. aureus* were equilibrated with equal volumes of penicillin G solution, two types of uptake were noted. First, there was a specific uptake corresponding to 750 molecules of the antibiotic which was independent of the external penicillin concentration. A second kind of uptake resulted from a simple diffusion of penicillin into the cell. Both types were independent of time of equilibration. The penicillin taken up specifically by the cell was firmly bound so that it could not be removed by washing or by equilibration with solutions of unlabeled penicillin. Control experiments with yeast, a nonsensitive cell, revealed no uptake or penetration of penicillin. The irreversible binding of penicillin by *S. aureus* was also observed later by Cooper and Rowley.[119]

Further experiments[120] revealed that the penicillin bound by resting *S. aureus* cells remained bound during subsequent multiplication of cells in penicillin-free media. In a medium capable of supporting growth, cells in the presence of penicillin continued to bind penicillin. This indicated that there was a more rapid synthesis of some compound which could bind penicillin than of cell substance.

The demonstration that there was an extremely firm binding of a very small but constant amount of penicillin by sensitive cells pointed to the existence of some cell component present in small amounts. This component presumably was essential for division because of other evidence showing that the antibiotic inhibited some reaction involved in cell division rather than in respiration or synthesis of cell components. The experiments suggested that resynthesis of this substance present in catalytic amounts was necessary for resumption of growth. On this basis it could be postulated that the concentration of penicillin which was the lowest for bacteriostatic action was that at which blocking of the penicillin-binding compound was slightly more rapid than its resynthesis by the cell.

The addition of sodium phenyl acetate to the fermentation medium in

[119] Cooper, P. D., and Rowley, D., *Nature* **163**, 480 (1949).

[120] Maass, E. A., and Johnson, M. J., *J. Bacteriol.* **58**, 361 (1949).

the experiments of Maass and Johnson requires further explanation. It had been shown by Behrens and his co-workers that penicillin production by fermenting molds was stimulated by addition of certain compounds containing the phenyl acetyl group.[121] To determine whether this effect was the consequence of direct utilization of the compounds as a whole or only of the phenyl acetyl moiety, experiments were made with deuterophenyl acetyl-N^{15}-valine as a "precursor."[122] This particular molecule was chosen because considerable specificity was shown by both acyl and amide portions of the molecule and because the amide portion contained a carbon skeleton similar to that in penicillamine. The labeled precursor was made from deuterophenyl acetic acid and N^{15}-valine by procedures which can be summarized in the following reaction scheme.

$$N^{15}H_3 + \alpha\text{-ketoisovaleric acid} \xrightarrow[\text{(ethanol)}]{(H_2 + Pd)} N^{15}\text{-DL-valine}$$

$$N^{15}\text{-DL-valine} + \text{deuterophenyl acetyl chloride} \xrightarrow[\text{(benzene)}]{\text{(NaOH)}} \text{Deuterophenyl acetyl-}N^{15}\text{-valine}$$

The labeled precursor was added to an actively synthesizing broth of *P. notatum*, and the penicillin subsequently was isolated and assayed for label. The H^2 analyses demonstrated that 92.5% of the benzyl penicillin isolated was derived from the phenyl acetyl portion. In contrast, only 2.7% of the penicillin came from the amide portion containing the N^{15}.

C. Effect of a Metabolic Accelerator

Labeled material can be used to investigate effects of drugs on the incorporation and utilization by Beeckmans *et al.*[123] of the rate of incorporation of C^{14} from carboxyl-labeled acetate in mice, with and without administration of dinitrocyclopentylphenol (DPP). This compound, like other dinitro compounds, causes an increase in oxygen consumption. Five groups of adult mice were injected with a solution of the sodium salt of DPP, equivalent to a dose of 20 mg./kg. Five other groups were injected with a similar volume of physiological saline solution. Ten minutes later, all groups were injected intraperitoneally with a solution containing the labeled acetate. At various times the animals were killed, and various fractions assayed. Administration of the drug was found to increase the

[121] Behrens, O. K., Corse, J., Jones, R. G., Mann, M. J., Soper, Q. F., Van Abeele F. R., and Chiang, M. C., *J. Biol. Chem.* **175,** 751 (1948).

[122] Behrens, O. K., Corse, J., Jones, R. G., Kleiderer, E. C., Soper, Q. F., Van Abeele, F. R., Larson, L. M., Sylvester, J. C., Haines, W. J., and Carter, H. E., *J. Biol. Chem.* **175,** 765 (1948).

[123] Beeckmans, M. L., Casier, H., and Hevesy, G., *Arch. intern. pharmacodynamie* **86,** 33 (1951).

incorporation of C^{14} into liver fat and total liver tissue in the first 10 min. and to decrease it at later times, as expected for a metabolic accelerator. There was very little effect observed in brain fat or total tissue, however. There was a suggestion of a possible specific effect of the drug on heart metabolism, because the appearance of C^{14} in heart muscle showed a different variation with time as compared with skeletal muscle.

D. Virus Metabolism and Biosynthesis

The tracer approach has been particularly fruitful in its application to problems in the structure and interaction of virus components with host cells. Many insights into mechanisms of attachment of bacteriophages,[124-129] origins of viral material,[130, 131] development of virus in the host cell.[132] etc., have been gained largely by ingenious coupling of tracer techniques with the elegant methods for viral assay developed in work with the bacteriophages. Two types of investigation stemming from the use of labeled phages will be described briefly.

First, there are experiments of Hershey and his collaborators[132-138] on the independent functions of viral protein and nucleic acid during growth of bacteriophage which have ramified into many other researches on nucleic acid economy in host bacteria during infection by phage. Radioactive phage T2 was made by growing bacteria in a glycerol-lactate medium containing either S^{35}-labeled sulfate or P^{32}-labeled phosphate.[133] The radioactive bacteria were then infected with the phage which became radioactive growing in the host cells. The phage particles were isolated from the lysate by centrifugation. Suitable immunological tests were employed to establish the radiochemical purity of the phage particles.

Phage prepared in this manner could be disrupted by osmotic shock into two fractions—one consisting of "ghosts" comprising the outer membrane,

[124] Puck, T. T., Garen, A., and Cline, J., *J. Exptl. Med.* **93,** 65 (1951).
[125] Garen, A., and Puck, T. T., *J. Exptl. Med.* **94,** 177 (1951).
[126] Tolmach, L. J., and Puck, T. T., *J. Am. Chem. Soc.* **74,** 5551 (1952).
[127] Puck, T. T., *Cold Spring Harbor Symposia Quant. Biol.* **18,** 149 (1953).
[128] Puck, T. T., and Lee, H. H., *J. Exptl. Med.* **99,** 481 (1954).
[129] Puck, T. T., and Lee, H. H., *J. Epxt. Med.* **101,** 151 (1955).
[130] Putnam, F., and Kozloff, L. M., *Science* **108,** 386 (1948).
[131] Cohen, S. S., *Cold Spring Harbor Symposia Quant. Biol.* **18,** 221 (1953).
[132] Hershey, A. D., *Ann. inst. Pasteur* **84,** 99 (1953).
[133] Hershey, A. D., and Chase, M., *J. Gen. Physiol.* **36,** 39 (1952).
[134] Hershey, A. D., *Cold Spring Harbor Symposia Quant. Biol.* **18,** 135 (1953).
[135] Hershey, A. D., *J. Gen. Physiol.* **37,** 1 (1953).
[136] Hershey, A. D., *J. Gen. Physiol.* **38,** 145 (1954).
[137] Hershey, A. D., Garen, A., Fraser, D. K., and Hudis, J. D., *Carnegie Inst. Wash. Yearbook No.* **53,** 210 (1953–1954).
[138] Hershey, A. D., *Virology* **1,** 108 (1955).

and the other of extruded material from inside the phage. The two fractions were found to show a remarkable distribution of radioactivity. All the labeled sulfur was in the phage coat and all the radioactive phosphorus was in the soluble juice containing practically all the nucleic acid of the phage. The ghosts contained the principle antigens as detected by specific antisera. The deoxyribonucleic acid (DNA) was released as the free acid, or at least, if combined with other material, was linked only to sulfur-free nonantigenic compounds.

Thus, the ghosts represented protein coats surrounding the DNA of the intact particles. This finding made it possible to establish in other experiments that when phage attached to bacterial cells there ensued a series of events in which the phage acted very much like a syringe. It attached itself by its tail portion and then the inner DNA was squirted into the cell, leaving the empty tail and body outside.

The other type of experiment deals with radiation-induced mortality of phage as an index of phage development. Hershey *et al.*[139] showed that bacteriophages of types T2 and T4 were unstable if grown so as to have a high content of P^{32}. It was shown that the logarithm of the number of surviving phages was linearly related to an exponential function of the time and to the specific P^{32} content as was to be expected on the following basis.

If the disintegration of a single P^{32} atom was assumed to be sufficient to inactivate a phage particle, the rate of change in the fraction, S, of surviving phages as a function of time (t) in days could be expressed as

$$-dS/dt = \alpha \mathrm{N}^* \lambda$$

where α was the efficiency of killing per disintegration event, N^* the number of radioactive P atoms per phage, and λ the fractional decay of P^{32} per day. Integration of this expression led to the relation

$$\log S = -1.48 \times 10^{-6} \, \alpha A_0 N(1 - e^{-\lambda t})$$

where A_0 was the specific radioactivity in millicuries per milligram of phosphorus of the medium in which the phages were grown, and N the total number of phosphorus atoms per phage particle.

It was found that the killing of the phage by the P^{32} disintegrations occurred with an efficiency of about 0.086 per atom. In other words, one out of twelve events led to inactivation of one phage particle. In experiments with unlabeled phage in solutions containing P^{32} this efficiency was shown to be much higher than could be attributed to the effects of the β particles emitted by the P^{32}. Accordingly, inactivation was attributed to some con-

[139] Hershey, A. D., Kamen, M. D., Kennedy, J. W., and Gest, H., *J. Gen. Physiol.* **34,** 305 (1951). Analogous experiments using S^{35} incorporation in *Neurospora crassa* have been reported; see Hungate, F., and Mannell, T. J., *Genetics* **37,** 709 (1952).

sequence of the recoil reaction concomitant with the formation of S^{32} in the phage DNA during disintegration of the P^{32}. The experiments were originally designed to test the hypothesis that duplication of the phage could result from a template mechanism in which no material from the phage parent passed over to the progeny. The results of these experiments decisively ruled out this mechanism and indicated that duplication involved dispersion of most of the original parental material among progeny.

This technique of inducing "suicide" in phage by incorporation of highly-labeled material appears capable of extension to many problems in phage development and has been exploited most ingeniously by Stent. As an example, in one set of experiments Stent[140] investigated the temperature dependence of the killing efficiency and showed that at −196°C. the value for α was decreased almost a factor of 2 relative to its value at 4°C. Thus at 4°C. one disintegration in eleven was effective, whereas at −196°C one disintegration in twenty led to inactivation. It was also found that when the bacteria were infected with phage and then stored at −196°C. the phage gradually developed resistance to inactivation by P^{32} decay. In explanation it was suggested that phage reproduction took place in steps each of which could be imagined to depend on the integrity of some independent structure which was destroyed by irradiation. Blockage of any one step led to inactivation. As the process of reproduction continued, more and more steps were completed so that a larger and larger fraction of targets could be destroyed without affecting the ultimate appearance of completed phage.

The loading of atoms in strategic sites in biologically important compounds with radioactive isotopes of the atoms concerned may prove to be a helpful technique in evaluating the validity of proposed structures in a great variety of researches relating biological function to structure.[140a]

E. Enzymatic Bases for Use of Chemotherapeutic Agents

It had been known for a number of years that diphosphopyridine nucleotide (DPN) could be cleaved at the nicotinamide-ribose link by enzymes from animal tissues. Representing the structure of DPN by the symbol ARPPRN, where A stands for adenine, R for ribose, P for phosphate, and N for nicotinamide, the reaction could be written

$$\text{ARPPRN} + H_2O \rightarrow \text{ARPPR} + \text{N}$$

An important observation was made by Mann and Quastel[141] and by

[140] Stent, G. S., *Cold Spring Harbor Symposia Quant. Biol.* **18,** 255 (1953).

[140a] Preliminary experiments on chromosome aberrations in onion root tips with C^{14}-labeled thymidine incorporated specifically in deoxyribonucleic acid have been reported by McQuade, H. A., Friedkin, M., and Atchison, A. A., *Nature* **175,** 1038 (1955).

[141] Mann, P. J. G., and Quastel, J. H., *Biochem. J.* **35.** 502 (1941).

Handler and Klein[142] that nicotinamide inhibited fission of DNA by the enzyme in animal tissues. Zatman *et al.*[143] found, however, that the enzyme derived from the mold *Neurospora crassa*[144] was relatively insensitive to the presence of nicotinamide. They then investigated the mechanism of splitting of DNA by enzymes as derived from the two different sources.

The DPNase of beef spleen was purified and obtained as a particle preparation. The enzyme from *Neurospora* was obtained as a soluble preparation. In studying the inhibition by nicotinamide at varying concentrations of DPN it was found that the inhibition was competitive with the *Neurospora* enzyme and noncompetitive with the beef spleen enzyme. This result led to the suggestion that the beef spleen enzyme functioned by forming a complex with the ARPPR moiety concomitant with splitting away of the nicotinamide. The next step was assumed to be a hydrolysis of the complex to yield ARPPR, free enzyme, and H^+ ion. The scheme could be written

$$\text{ARPPR}\overset{+}{\text{N}} + \text{enzyme} \rightleftharpoons \text{ARPP}\overset{+}{\text{R}}\text{-enzyme} + \text{N}$$
$$\downarrow H_2O$$
$$\text{ARPPR} + H^+ + \text{enzyme}$$

According to this mechanism, the inhibition of the enzyme would depend on a competition between the nicotinamide and water for the enzyme complex. Furthermore, if sufficient nicotinamide were present to inhibit appreciably the disappearance of DPN, then incubation of enzyme with DPN and this amount of nicotinamide labeled with C^{14} would result in incorporation of the label in the nicotinamide moiety of the DPN.

In a typical experiment the incubation mixture consisted of 660 μM. of DPN (final concentration, 1.32×10^{-2} *M*, *p*H 7), 3 mM. of C^{14}-nicotinamide (6×10^{-2} *M*), 2240 units of spleen DNAase preparation, and water to a final volume of 50 ml. After 165 minutes of incubation at 37°C., the DNA had been split to the extent of 30%, as compared to a control without the nicotinamide in which it had disappeared to the extent of 90%. The main reaction mixture was heated to 70°C. for 10 min., terminating the experiment by abolishing the enzyme activity without materially affecting the remaining DNA. The DNA was isolated and purified by column chromatography and found to contain C^{14} in an amount that indicated close to 100% exchange with the labeled nicotinamide. The labeled DNA was then split by using the *Neurospora* DNAase, and the nicotinamide was recovered, and assayed. It was found that all the activity initially in the DPN could be recovered in the nicotinamide moiety.

[142] Handler, P., and Klein, J. R., *J. Biol Chem.* **143**, 49, (1946).
[143] Zatman, L. J., Kaplan, N. O., and Colowick, S., *J. Biol. Chem.* **200**, 197 (1953).
[144] Kaplan, N. O., Colowick, S., and Nason, A., *J. Biol. Chem.* **191**, 473 (1951).

Similar experiments with the *Neurospora* enzyme showed no appreciable exchange of the nicotinamide with the nicotinamide moiety of DPN.

These experiments strongly supported the mechanism suggested because a noncompetitive type of inhibition based on the model shown in the reaction scheme should have resulted in exchange with nicotinamide (as it did in the beef spleen experiments), and a competitive type of inhibition should have given no exchange (as with the *Neurospora* enzyme).

It was established that the exchange reaction between the nicotinamide moiety of DPN and free nicotinamide could be extended in general to compounds related to nicotinamide, such as isonicotinic hydrazide,[145] "marsalid" (the isopropyl derivative of isonicotinic acid hydrazide[145]), 3-acetylpyridine,[146] and ethyl nicotinate.[147] Kaplan and his co-workers[148] have suggested that the insertion of nicotinic acid analogs into DPN resulting in coenzyme analogs brought about by these exchange reactions could be used as a working hypothesis for the pharmacological action of nicotinic acid analogs. By extension they point out that exchange reactions involving coenzymes could be a general means of producing coenzyme analogs which could inhibit or alter cellular metabolism.

Zatman and co-workers,[149] as an example, have shown that isonicotinic acid hydrazide, which has been used as a therapeutic agent in the treatment of tuberculosis, is a much more effective inhibitor of DNAase from beef spleen than is nicotinamide.

F. Concluding Remarks

From the material presented in this section the reader can appreciate that there is a strong temptation to continue indefinitely with the many fascinating aspects of tracer methodology which flow from these examples as well as from a host of topics not even touched at all. It should be possible, however, for the reader to use the material of Chapters VI and VII together with the general references as a base for excursions into areas of particular interest. In following chapters special data on individual isotopes will be presented. Separate chapters will be devoted to elements of particular interest to biologists, e.g., hydrogen, carbon, nitrogen, and oxygen. Other elements will be treated briefly in the final chapter.

[145] Zatman, L. J., Kaplan, N. O., Colowick, S. P., and Ciotti, M. M., *J. Biol. Chem.* **209,** 453 (1954).

[146] Kaplan, N. O., and Ciotti, M. M., *J. Am. Chem. Soc.* **78,** 1713 (1954).

[147] Kaplan, N. O., and Ciotti, M. M., quoted in reference 148.

[148] Kaplan, N. O., Goldin, A., Humphreys, S. R., Ciotti, M. M., and Venditti, J. M., *Science* **120,** 437 (1954).

[149] Zatman, L. J., Colowick, S. P., Kaplan, N. O., and Ciotti, M. M., *Bull. Johns Hopkins Hosp.* **91,** 211 (1952).

GENERAL REFERENCES

A. Books

1. Clarke, H. T. (ed.), "Symposium on the Use of Isotopes in Biology and Medicine" Univ. Wisconsin Press, Madison, 1949.
2. Hahn, P. F. (ed.), "Manual of Artificial Radioisotope Therapy", Academic Press, New York 1951.

B. Monographs and Articles

1. Cowie, D. B., and Roberts, R. B., Permeability of microorganisms to inorganic ions, amino acids and peptides. *In* "Electrolytes in Biological Systems," pp. 1–53. American Physiological Society Monograph, Washington, D. C. (1955).
2. Johnson, J. E. (ed.), "Radioisotope Conference, 1954," Vol. I. Academic Press, New York, 1954.
3. Tabern, D. L., Taylor, J. D., and Gleason, G. I., Radioisotopes in pharmaceutical and medical studies. *Nucleonics* **7,** No. 5, 3; No. 6, 40 (1950); **7, 8,** No. 1, 60 (1951).
4. Harris, G. M., Oddie, T. H., and Gresford, G. B. (eds.), "Proceedings of Conference on Applications of Isotopes in Scientific Research," Commonwealth Scientific and Industrial Research Organization, University of Melbourne, Australia, 1951.

Chapter IX

THE ISOTOPES OF HYDROGEN

1. INTRODUCTION

Hydrogen was once thought to have only one isotope, H^1. When Aston determined the atomic weight of hydrogen in 1927 by means of the mass spectrometer, he found a value[1] in good agreement with that previously measured by chemical means. There seemed, therefore, to be no reason to think of hydrogen as a complex element. In 1929, however, Giauque and Johnson[2] showed that ordinary oxygen, the standard for the chemical scale of atomic weights, was not a simple element with one isotope, O^{16}, but also possessed two rare isotopes, O^{17} and O^{18}. If the weights of these isotopes were included in the calculations for the weight of hydrogen, then, as Birge and Menzel showed,[3] there arose a discrepancy between the chemical and physical scale weights for hydrogen. The difference could be accounted for only by assuming either that one or another of the atomic weight determinations was in error, or that a rare stable isotope of hydrogen existed. In 1932, Urey *et al.*[4] found evidence in the spectrum of residues from exhaustive distillation of water which did indeed indicate the existence of an isotope with mass number 2, deuterium, symbolized D. Large quantities of the isotope were soon made available by electrolysis of water which concentrated the heavy hydrogen in the electrolytic residues. This process is still the basis for the production of deuterium.

The possibility that another isotope of hydrogen, one with mass number 3, might also exist was speculated about during the same period. It was uncertain, however, which of the two isobars, H^3 or He^3, would be stable, because neither had been detected nor isolated. Moreover, the binding energies associated with the combination of two neutrons and a proton (H^3), on the one hand, and with the combination of two protons and a neutron (He^3), on the other, were not known. There were some claims for the detection of a stable species of hydrogen with mass 3 based on mass spectrometer studies. (In the light of what is known now, these results would have required radioactivities of the order of millicuries per milliliter

[1] Aston, F. W., *Proc. Roy. Soc.* **A115,** 487 (1927).
[2] Giauque, W. F., and Johnson, H. W., *Nature* **123,** 318 (1929).
[3] Birge, R. T., and Menzel, D. H., *Phys. Rev.* **37,** 1670 (1931).
[4] Urey, H. C., Brickwedde, F. G., and Murphy, G. M., *Phys. Rev.* **39,** 164 (1932).

in ordinary water!) In 1939, however, the uncertainties were resolved when Alvarez and Cornog[5] proved that a radioactive isotope of hydrogen was formed by a (d,p) reaction when heavy water was bombarded with deuterons in the cyclotron. They also detected the presence of stable He^3 in helium, using the cyclotron accelerator as a mass spectrometer to locate a He^3 component in the particles accelerated when helium was used as an ion-source gas.

Hydrogen, then, consists of two stable isotopes—protium (H^1) and deuterium (H^2). The atom per cent of the lighter isotope in ordinary hydrogen is 99.98. Minute amounts of the radioactive isotope, H^3 (tritium, symbolized T), exist,[6] because neutrons in cosmic rays form tritium by nuclear reactions with atmospheric nitrogen,[7] e.g., $N^{14}(n,H^3)C^{12}$. The main source of tritium is the uranium pile reactor in which it is produced by the $Li^6(n,\alpha)H^3$ reaction. In the cyclotron, the reaction employed is $Be^9(d,2\alpha)$ because of the routine use of beryllium as a target in cyclotron operation. During bombardment of the beryllium, some of the tritium formed diffuses into the fore vacuum line as a gaseous triteride and can be recovered by leading the exhaust gas through a combustion tube in which the radioactive hydrogen is burned over hot CuO to water. At present, tritium can be had in bulk from the various atomic establishments at relatively low cost so that there is little need to produce it in cyclotrons.

Tritium is a negative β-ray emitter. The energy of the radiations is the lowest found among the artificial radioactive elements. Considerable work has been done on the upper limit (E_{max}) of the spectrum. Values ranging from 0.01795 to 0.0194 Mev. have been reported.[8] Measurement of the amount of He^3 arising from tritium β^- decay gives a value of 12.46 yr. for the half-life.[9] Measurement of the specific activity results in a value of 12.4 yr.[10]

2. ASSAY OF DEUTERIUM

The presence of deuterium can be detected quantitatively either with the mass spectrometer or by methods based on measurement of density of water samples. The latter procedures are preferred by most workers because they are relatively simpler and cheaper.

The first step in an assay based on determination of density is the combustion of labeled samples. The standard technique is that described by

[5] Alvarez, L. W., and Cornog, R., *Phys. Rev.* **56,** 613 (1939).
[6] Johnston, W. H., Wolfgang, R. L., and Libby, W. F., *Science* **113,** 1 (1951).
[7] Cornog, R., and Libby, W. F., *Phys. Rev.* **59,** 1046 (1941).
[8] Hollander, J. M., Perlman, I., and Seaborg, G. T., *Revs. Mod. Phys.* **25,** 476 (1953).
[9] Jenks, G. H., Sweeton, F. H., and Ghormley, J. A., *Phys. Rev.* **80,** 990 (1950).
[10] Jones, W. M., *Phys. Rev.* **83,** 537 (1951).

Keston *et al.*[11a] It can be set up as a simple combustion train, as in the original procedure, or made part of a permanent vacuum line. Details can be found in the original article or in the monograph by Glascock.[11b]

After the water sample has been prepared it can be assayed by the methods discussed in Chapter III (pp. 110–111).

The mass spectrometric determination of deuterium is advantageous when the sample is too small for convenient measurement of density. There are a number of difficulties peculiar to deuterium which arise when it is assayed with the mass spectrometer. The most annoying of these is the "memory effect," which results from exchange of hydrogen isotopes between the samples to be assayed and hydrogen-containing materials in the spectrometer. Absorbed water may be present in the ion source, and there may be exchangeable hydrogen in OH groups of glass silicates, and so on. When hydrogen ions are produced in the ion source prior to mass analysis, they are activated. In consequence, they are likely to exchange with whatever exchangeable hydrogen is present on the walls or other parts of the ion source. As a result, the apparent isotopic content of a given sample may be distorted by that of one analyzed previously.

Other troubles may occur because of the large mass difference between deuterium and protium when hydrogen gas is used as the assay compound. For instance, there can be isotopic fractionation between $(H^1H^1)^+$ and $(H^1H^2)^+$ because these molecules have different rates of diffusion. Or there can be formation of singly charged triatomic protium ion $(H^1H^1H^1)^+$ which has the same mass and charge as the positive ion $(H^1H^2)^+$. The two ions can be differentiated only by observing the variation in ion beam with pressure of hydrogen (see below). Water is not a good compound to use in the mass spectrometer because so many different kinds of ions are formed. As a result, a given mass peak may be made up of an unknown number of components. Procedures for the analysis of water hydrogen in the mass spectrometer are available, however.[12]

Hydrogen remains the most widely used compound for deuterium assay in the mass spectrometer. The first step in deuterium assay with the mass spectrometer, then, is the reduction of the water obtained from combustion of labeled material to molecular hydrogen. In Table 19 there are listed various procedures. Method 4 is designed specifically for small quantities of water.

In the mass spectrometer, peaks in the ion current will be seen at mass

[11a] Keston, A. S., Rittenberg, D., and Schoenheimer, R., *J. Biol. Chem.* **122,** 227 (1937).

[11b] Glascock, R. F., "Isotopic Gas Analysis for Biochemists," Academic Press, New York, 1954.

[12] Washburn, H. W., Berry, C. E., and Hall, L. G., *Anal. Chem.* **25,** 130 (1953).

TABLE 19
REDUCTION OF HEAVY WATER TO MOLECULAR HYDROGEN FOR ISOTOPIC ASSAY IN MASS SPECTROMETER

Method	Reference
1. Reduction of steam with magnesium turnings at 620°C	13
2. Reduction of steam with magnesium amalgam at 400°C	14
3. Reduction with lithium aluminum hydride in anhydrous carbitol	15
4. Reduction with zinc at about 400°C	16, 17

numbers 1, 2, and 3. The mass 1 peak arises from singly charged H^1 and is of no interest. The mass 2 peak comes from singly charged H^1H^1, and the mass 3 peak from singly charged H^1H^2 and $H^1H^1H^1$. The triatomic protium ion arises from collisions between molecular ions and neutral molecules, and its amount varies as the square of the pressure. Because of this secondary origin of the triatomic contaminating peak, it is assumed that its contribution will be negligible at reduced pressures, so it is the practice to plot the ratio of mass peak 3 to mass peak 2 at various pressures of hydrogen and then extrapolate this linear plot to zero pressure. The intercept at zero pressure is taken as the true H^1H^2/H^1H^1 ratio. It is necessary to make up standard mixtures of deuterium and hydrogen so that the apparatus can be calibrated to obviate other effects arising from isotopic fractionation in the source.[18a]

The use of hydrocarbons as assay gases has been urged because of the unavoidable complications involved in the use of molecular hydrogen. In these compounds, the hydrogen should be stably bound in essentially unexchangeable form, and the relative mass differences in isotopic species would be much smaller than in hydrogen gas. Fractionation due to differential diffusion would be minimized. The major disadvantage in using hydrocarbons is that the mass peak due to a deuterium-containing ion would be coincident with a mass peak due to a corresponding protium-containing ion into which the heavy carbon isotope, C^{13}, is incorporated; i.e., $C^{12}H^2X$ . . would have the same mass as $C^{13}H^1X$ . . . The two gases which have been made for purposes of deuterium assay are methane and ethane. The method used is hydrolysis of the methyl Grignard reagent[18b]

[13] Allen, M. G., and Ruben, S., *J. Am. Chem. Soc.* **64,** 948 (1942).

[14] Henriques, F. C., Jr., and Margnetti, C., *Ind. Eng. Chem. Anal. Ed.* **18,** 417–421 (1946).

[15] Biggs, M. W., Krichevsky, D., and Kirk, M. B., *Anal. Chem.* **24,** 223 (1952).

[16] Graff, J., and Rittenberg, D., *Anal. Chem.* **24,** 878 (1952).

[17] Dubbs, C. A., *Anal. Chem.* **25,** 828 (1953).

[18a] Nier, A. O., *in* "Use of Isotopes in Biology and Medicine" (H. T. Clarke, ed.), pp. 99–103. Univ. Wisconsin Press, Madison, 1949.

[18b] Orchin, M., Wender, I., and Friedel, R. A., *Anal. Chem.* **21,** 1072 (1949).

or zinc diethyl[19] with isotopically labeled water according to the general reaction $2XMgI$ (or X_2Zn) + $H_2O \rightarrow 2XH + MgO \cdot MgI_2$ (or ZnO), where X is the alkyl radical.

The atom per cent of deuterium must be determined from the ratio of intensities at mass peak 3 to mass peak 2. This may be done on the basis of the following considerations.

Suppose there are, in general, A atoms of mass M' and B atoms of mass M ($M' > M$). If the molecular species used for assay contains N atoms of the element, and the ratio of mass peak M' to mass peak M is given as R, then

$$A/B = R/N \tag{1}$$

(It will be left as an exercise for the reader to verify this expression.) The atom per cent of the isotope with mass M' will be equal to $100A/(A + B)$. On rearranging, this expression is equal to $100(A/B)/(1 + A/B)$, or $100(R/N)/(1 + R/N)$. Multiplying numerator and denominator by N results in the expression

$$\text{Atom per cent (mass, } M') = 100R/(N + R) \tag{2}$$

In the case of the hydrogen isotopes, the factor N is equal to 2 when hydrogen gas is used in the assay. Other factors apply when other gases are used.

Certain correction factors must be applied to the observed intensity ratios because of incomplete resolution of peaks. The correction is best determined with standard samples for calibration. In addition, as remarked previously, the mass peaks used may not represent one ion but several. The nature of this complication in the case of the hydrogen isotopes and the method of correction have been discussed above.

3. ASSAY OF TRITIUM

Because of the unusually low energy of its radiations, tritium is difficult to assay. The maximum range of the β particles is hardly more than 1 cm. in helium at atmospheric pressure. To minimize or obviate self-absorption, therefore, it is necessary to assay tritium in the gaseous form.

If the sample to be determined has an activity exceeding 0.005 μc./cm.^{3} as hydrogen gas, a Lauritsen electroscope can be used. The electroscope chamber must be made airtight and equipped with inlet and outlet stopcocks. The tritium sample contained in a bulb is connected to the chamber, and the hydrogen gas is equilibrated either by diffusion or by a flow method. Obviously it is important that complete equilibration is attained so that the amount of hydrogen in ionizing volume can be calculated. Gases and vapors such as H_2O, CH_4, C_2H_6, C_4H_{10}, CH_3OH, CH_3I, CH_3OCH_3,

[19] Friedman, L., and Issa, A. P., *Anal. Chem.* **24,** 876 (1952).

CH_3COCH_3, C_6H_6 and $C_6H_5CH_3$ can also be used. In all cases it is necessary to sweep out the tritium compound with an inactive sample of the same compound so that "memory" effects are avoided and the instrument is decontaminated. Vapors are highly objectionable for this purpose, and it is best in all cases to convert the tritium to hydrogen gas for assay.

Similar procedures can be used with any ionization chamber–electrometer combination if weaker samples are to be assayed. Alternatively, the tritium as water vapor can be condensed in a side arm from whence it is allowed to diffuse into the ionization chamber. The tritium present in the

TABLE 20

ASSAY PROCEDURES FOR TRITIUM

Method	Gas preparation	Reference
Molecular hydrogen in ionization chamber–Lindemann electrometer	Reduction of water to hydrogen	20
Molecular hydrogen in vibrating-reed electrometer	Reduction of water to hydrogen	15
Molecular hydrogen in proportional counter with added methane	Reduction of water to hydrogen	21, 22
Molecular hydrogen in G-M counter with helium-ethanol counting mixture	Reduction of water to hydrogen	23
Tritiomethane in proportional counter	Reaction between water and aluminum carbide to form tritiomethane	24
Tritiomethane in proportional counter	Hydrolysis of methyl magnesium iodide with tritio-water to give tritiomethane	25
Tritiobutane in proportional counter	Hydrolysis of *n*-butyl magnesium iodide with tritio-water	11b
Mixture of tritium-containing hydrogen and methane in vibrating-reed electrometer (or proportional counter)	Direct reduction of tritium-containing organic compound with zinc powder in presence of Ni_2O_3 and water	26

[20] Berstein, I. A., Bennett, W., and Fields, M., *J. Am. Chem. Soc.* **74,** 5763 (1952).

[21] Bernstein, W., and Ballentine, R., *Rev. Sci. Instr.* **21,** 158 (1950).

[22] Verly, W. C., Rachele, J. R., du Vigneaud, V., Eidinoff, M. L., and Knoll, J. F., Jr., *J. Am. Chem. Soc.* **74,** 5941 (1952).

[23] Reid, A. F., *in* "Preparation and Measurement of Isotopic Tracers" (D. W. Wilson, A. O. Nier, and S. P. Reimann, eds.), p. 104. Edwards, Ann Arbor, Michigan, 1946.

[24] White, D. F., Campbell, I. G., and Payne, P. R., *Nature* **166,** 628 (1950).

[25] Robinson, C. F., *Rev. Sci. Instr.* **22,** 353 (1951).

[26] Wilzbach, K. E., Kaplan, L., and Brown, W. G., *Science* **118,** 522 (1953).

chamber is then easily calculated from the known vapor pressure of water and the volume of the chamber.

In general, the low energy of the H^3 β particles leads to a lowered efficiency for assay when electroscopes or gas-filled ionization chambers are used. However, a modified Lauritsen electroscope for tritium assay has been described. It appears to possess a sensitivity comparing favorably with that of G-M tube counters as well as greater stability.[14]

The majority of workers appear to favor the G-M or proportional counters to assay tritium as molecular hydrogen rather than as water. The difficulties encountered with water vapor are formidable. Gas counters cannot tolerate more than a few millimeters of Hg pressure without breakdown. Memory effects are the rule rather than the exception. For these reasons, it is probably most economical of time and samples to use methods like those described in Section 2 for the reduction of deuterated water to make the radioactive triterated hydrogen for counting.

There is one special attribute of tritium which renders its assay somewhat easier than deuterium: it can be determined as hydrocarbon without the complication of interference by C^{13} (see Section 2, p. 275). Reactions for the production of tritiomethane or tritiobutane are relatively simple to run and provide good assay gases. Tritiobutane has the added advantage that it can be easily condensed and thus readily manipulated for quantitative transfer.

Various methods for assay of tritium are summarized in Table 20.

It may be appreciated that at present the assay of tritium is tedious and requires experience in precision vacuum technique as well as in handling of complex assay apparatus. This state of affairs may not persist, however, if present attempts to adapt scintillation counting to tritium assay succeed. By means of fast coincidence circuitry and research on appropriate liquid scintillation media, it has been shown that a low-energy β-emitting isotope like C^{14} can be assayed in the scintillation detector with efficiencies comparable to those obtained with proportional counters.[27-29] The extension of these techniques to the assay of tritium as water or in organic compounds has been reported.[30, 31] As an example, various tritiated sterols have been dissolved in scintillating media made up of mixtures of xylene, ethanol, and 2,5-diphenyloxazole, and counted with a lower limit in sensitivity of 1.7×10^{-3} μc./ml.[30] The apparatus involved is complex, but it does not seem impossible that it can be developed so that it can be made available for use by laboratory personnel in the average biological laboratory.*

[27] Reynolds, G. T., Harrison, F. B., and Salvini, G., *Phys. Rev.* **73,** 488 (1950).
[28] Hayes, F. N., Hiebert, R. D., and Schuch, R. L., *Science* **116,** 140 (1952).
[29] Hiebert, R. D., and Watts, R. J., *Nucleonics* **11,** No. 12, 38 (1953).
[30] Hayes, F. N., and Gould, R. G., *Science* **117,** 480 (1953).
[31] Farmer, E. C., and Berstein, I. A., *Science* **117,** 279 (1953).

* See Appendix 5.

4. DEUTERIUM AND TRITIUM AS TRACERS FOR HYDROGEN

The large mass ratio of deuterium to protium leads to many marked differences in chemical and physical properties of the two isotopes.[32] These differences are accentuated for tritium, which has an even greater mass ratio with respect to protium. As an example, the vapor pressures of various isotopic combinations in molecular hydrogen are given in Table 21.[33]

In most tracer research involving hydrogen, the major interest is in labilization of carbon-hydrogen bonds. The carbon-protium bond is somewhat looser than the carbon-deuterium bond, which in turn is looser than the carbon-tritium bond. This is owing to the fact that, as the effective mass of the isotope increases, the "zero-point" energy is lowered. The zero-point energy is the vibrational energy residual in the bond at the absolute zero of temperature and is the main term in the expression for the total internal energy of bonding at normal temperatures. By the quantum theory, the energy of vibration is equal to Planck's constant, h, multiplied by a characteristic vibration frequency, ν. This frequency, for a simple oscillator, is an inverse function of the effective mass of the oscillating atoms. Consequently, the higher the mass, the lower is the zero-point energy.

Before the isotope can be induced to split away from the carbon and react, it must be given activation energy. The lower the zero-point energy, the more activation energy is required. Hence, tritium is more difficult to activate than protium and so is more sluggish in reactions involving transfer from carbon to carbon. Since hydrogen-transporting enzymes are extremely sensitive to small gradations in activation energy of C-H bonds, there exists a possibility, not at all remote, that isotope separation reactions can occur in enzyme-catalyzed reactions which can invalidate the use of tritium as a tracer for hydrogen. Such effects are known for deuterium, which, in high concentrations, is a respiratory poison. Deuterium can slow such enzymic hydrogen transfer reactions as the oxidation of glucose. The difference in activation energy between protium and deuterium for certain

[32] A compilation of the thermodynamic properties of the hydrogen isotopes and their compounds can be found in a brochure prepared by T. G. Fox and issued by the U. S. Atomic Energy Commission, Oak Ridge, Tennessee, as Doc. MDDC-1496, 1942.

[33] The value given for the vapor pressure of HT is probably in error. It has been shown experimentally that the vapor pressures of isotopic molecules obey the so-called "rule of the geometric mean," i.e., $p_{HD} = (p_{H_2}p_{D_2})^{1/2}$. The theoretical proof for this rule involves the demonstration that the ratio of the vapor pressures of isotopic molecules is a function of the reciprocal masses of the atoms in the molecule and not of the total molecular weight. The rule of the geometric mean predicts an appreciable difference between the vapor pressures of HT and D_2. That this is actually the case seems apparent from recent experiments by J. Bigeleisen and E. C. Kerr, reported in *J. Chem. Phys.* **23,** 2442 (1955).

TABLE 21

VAPOR PRESSURES OF ISOTOPIC HYDROGEN MOLECULES AT 20.4°K. (AFTER LIBBY AND BARTER[34a])

Molecule	Vapor pressure (mm. Hg)	Reference
H_2	760	34b
HD	438	34c
D_2	256	34d
HT	254 ± 16	34e
DT	123 ± 6	34f
T_2	45 ± 10 (extrap.)	34a
		34a
		34a

photochemical reactions involving chlorine ranges from 600 to 1600 cal.[35a, b] The use of tritium as a hydrogen tracer is attended, therefore, with considerable uncertainty. Careful control experiments to rule out isotope effects must be devised whenever it is used as a tracer for protium.

A few examples of the isotopic differentiation which occurs in metabolizing systems may be mentioned. Thorn[36a] prepared succinic acid containing 77 atom per cent D in the methylene groups by reducing acetylene dicarboxylate with sodium amalgam in D_2O. This tetra-D-substituted acid was oxidized by a succinic acid-oxidase preparation at 40% of the rate found for normal succinic acid. The activation energy for the deuterium-loaded acid was shown to be 1450 ± 450 cal. higher than the activation energy of the reaction involving the normal acid.

A number of investigators have reported appreciable isotopic effects in living organisms. Glascock and Duncombe have found[36b] that the ratio T/D in mammary fat of lactating rats was about 0.8 of the initial T/D ratio in water administered orally. Under certain conditions, however,

[34a] Libby, W. F., and Barter, C. A., *J. Chem. Phys.* **10,** 184 (1942).

[34b] Scott, R. B., and Brickwedde, F. G., *Phys. Rev.* **48,** 483 (1935).

[34c] Scott, R. B., Brickwedde, F. G., Urey, H. C., and Wahl, M. H., *J. Phys. Chem.* **2,** 454 (1934).

[34d] Clusius, K., and Bartholomé, E., *Z. physik. Chem.* **30B,** 237 (1935).

[34e] Bartholomé, E., *Z. physik. Chem.* **33B,** 387 (1936).

[34f] Megaw, H. D., and Simon, F., *Nature* **138,** 244 (1936).

[35a] See Rollefson, G. K., *J. Chem. Phys.* **2,** 144 (1934).

[35b] An example of the isotope effect in hydrolysis reactions from the recent literature is found in the work of L. O. Assarsson [*Acta Chem. Scand.* **9,** 1399 (1955)] on relative rates of hydrolysis of Grignard reagents by tritio-water and light water. The ratios of the velocity constants for T and H in producing methane and benzene from the corresponding alkyl and aryl Grignard reagents were 0.67 and 0.61, respectively.

[36a] Thorn, M. B., *Biochem. J.* **49,** 602 (1951).

[36b] Glascock, R. F., and Duncombe, W. G., *Biochem. J.* **51,** xl (1952).

isotope effects can be minimized, apparently, as Thompson and Ballou[37] report that in similar experiments much less fractionation resulted. The discrepancy in results has been ascribed by Thompson and Ballou to the shorter experimental period used by Glascock and Duncombe. Eidinoff *et al.*[38] have reported that deuterium was incorporated 8 and 18% faster than tritium from water into rat liver glycogen and fatty acids. Reitz and Bonhoeffer[39] and Weinberger and Porter[40] have found lowered incorporation of deuterium or tritium compared to protium in growing cultures of green algae.

Verly *et al.*[22] administered a mixture of methanols containing C^{14}, D, and T to rats and isolated the methyl groups from tissue creatine and choline. The isotopic content of these methyl groups was compared with that of the methyl group in the administered methanol. The ratio of D to C^{14} in choline methyl was 22% of that in the methanol, and the ratio of T to C^{14} in choline methyl was 69 to 75% of that in the methanol. It is apparent that in the process of methyl transfer more D was lost from the methyl than was T.

There are other types of experiment in which such uncertainty about isotopic fractionation does not arise. For example, suppose one wishes to investigate the labilization of C-H bonds during oxidation of a molecule, such as fumaric acid. The distribution of tritium in the oxidation products after reaction in a water medium containing tritium can be studied. Allen and Ruben[13] have shown that when fumaric acid

$$\begin{matrix} HOOC—C=C—COOH \\ \quad H \;\; H \end{matrix}$$

is oxidized with permanganate in tritium water, the resultant formic acid, which is known to originate from one of the methene (middle) carbons of the fumaric acid, contains no tritium. If the C-H bond were labilized during oxidation, the H would dissociate into the water, its place being taken by labeled hydrogen from the water dissociation. Since no such result is obtained, the conclusion may be drawn that no such labilization occurs. The literature on deuterium contains many examples of similar researches.

5. DEUTERIUM AND TRITIUM AS AUXILIARY TRACERS FOR CARBON

Since tritium can form stable linkages with carbon like those between deuterium and carbon, it follows that tritium can be used to label carbon atoms in organic molecules in the same way that deuterium has been employed. This is all the more true, since, as pointed out in the previous

[37] Thompson, R. C., and Ballou, J. E., *Arch. Biochem. and Biophys.* **42,** 219 (1953).
[38] Eidinoff, M. L., Perri, G. C., Knoll, J. E., Marano, B. J., and Arnheim, J. J., *J. Am. Chem. Soc.* **75,** 248 (1953).
[39] Reitz, O., and Bonhoeffer, K. F., *Z. physik. Chem.* **A172,** 369 (1935).
[40] Weinberger, D., and Porter, J. W., *Arch. Biochem. and Biophys.* **50,** 160 (1954).

section, the C-T bond is somewhat stronger than the C-D bond because of the heavier mass of tritium. In this connection, it is of interest to recall briefly the results obtained with deuterium as an auxiliary tracer for carbon. Deuterium cannot be stably linked to oxygen or nitrogen atoms as in $—NH_2$, —OH, because, in aqueous media, sufficient dissociation occurs so that rapid equilibration of deuterium results between the labeled linkage and the water. Compounds labeled with deuterium in this way (CH_3COOD, CHDNHDCOOD, etc.), lose deuterium more or less rapidly by exchange with dissociable hydrogen of body fluids or culture media, and are, in general, of no value in tracer experiments.

Under certain special conditions, however, labile deuterium linkages can be employed.[41] Thus deuteroglycine, which contains no completely stable D, has been used to demonstrate the appearance in urine of deuterohippuric acid after administration of deuteroglycine and benzoic acid to mice. Mandelic and atrolactic acids, containing only labile deuterium, have been employed in studies on optical rotation in media containing no exchangeable hydrogen. Studies in such media have little biological significance, however, because in all cases aqueous media are encountered.

"Semilabile" linkages occur when a C-H bond is attached adjacent to a carbonyl group as in CH_2DCOCH_2D. Under these circumstances an enolization equilibrium resulting in formation of a labile O-D linkage is established i.e.,

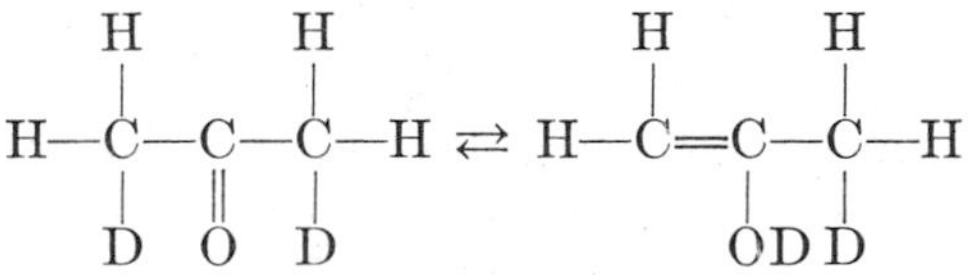

$$\mathrm{H{-}\overset{\displaystyle H}{\underset{\displaystyle D}{C}}{-}\underset{\displaystyle O}{\overset{\|}{C}}{-}\overset{\displaystyle H}{\underset{\displaystyle D}{C}}{-}H \rightleftarrows H{-}\overset{\displaystyle H}{C}{=}\underset{\displaystyle OD}{C}{-}\overset{\displaystyle H}{\underset{\displaystyle D}{C}}{-}H}$$

As a consequence, loss of deuterium occurs by exchange of the dissociable hydrogen in the medium with the deuterium in the alcoholic linkage. It is possible to utilize such a semilabile deuterium as a tracer if, in the metabolic process studied, a biological reduction of the keto group proceeds at a more rapid rate than the exchange due to enolization, as shown by Anchel and Schoenheimer[41] for the conversion of coprostanone to coprosterol. It is usually desirable, however, to avoid the use of labile or semilabile deuterium in tracer experiments. Such deuterium can be removed from any labeled molecule by preliminary treatment with water, leaving only stable deuterium markers.

In most labeling procedures, the molecules should be isolated and the point of entry of the deuterium established by suitable degradation procedures. With respect to labeling procedures themselves, a very large

[41] Anchel, M., and Schoenheimer, R., *J. Biol. Chem.* **125,** 23 (1938).

literature on deuterium is available for guidance in handling tritium. The general methods employed are:

1. Exchange between the organic compound and D_2O, D_2SO_4, or D_2. Some catalyst, such as activated platinum, is generally required for exchange reactions involving D_2O.
2. Synthesis of deuterium into the molecule by a procedure such as hydrogenation of a double bond.
3. Biosynthesis involving isolation of metabolites from organisms grown in D_2O or on deuterium-containing substrates.

Exchange can occur by direct substitution or through enolization. The former process is responsible for the very rapid exchange of deuterium in water with the hydrogen of polar groups such as —OH, —COOH, —NH_2, =NH, *o*-H, and *p*-H atoms in phenols. Direct substitution from D_2SO_4 into paraffin hydrocarbons can also be accomplished.[42] As remarked above, the acidic H of enolic forms is exchangeable, and D-C bonds can be formed in compounds for which existence in both keto and enol tautomers is possible. In fact, the view is held that during formation of deutero amino acids, isolated from protein of mice kept on a D_2O-containing diet, deuterium may not be introduced by direct substitution from body fluid.

TABLE 22

DEUTERIUM CONTENT OF LEUCINE PREPARED BY DIFFERENT METHODS

Method	Deuterium content (atom %)	Reference
1. Exchange with D_2SO_4	0.86	43
2. Synthesis from isovaleraldehyde (aminonitrile reaction)	13.7	44
3. Synthesis from deuteroisocaproic acid (K-phthalimide reaction)	3.87	45
4. Synthesis from deuteroisocaproic acid via amination of the bromo acid	37.02	46
5. Isolation from mouse protein (D_2O in diet)	0.12	47

[42] Ingold, C. K., Raisin, C. G., and Wilson, C. L., *J. Chem. Soc.* **1936,** 1643; see also Kharasch, M. S., Brown, W. G., and McNab, J., *J. Org. Chem.* **2,** 36 (1937); Brown, W. G., Kharasch, M. S., and Sprowls, W. R., *ibid.* **4,** 442 (1939).

[43] Schoenheimer, R., Rittenberg, D., and Keston, A. S., *J. Am. Chem. Soc.* **59,** 1765 (1937).

[44] Kinney, C. R., and Adams, R., *J. Am. Chem. Soc.* **59,** 897 (1937).

[45] Schoenheimer, R., and Ratner, S., *J. Biol. Chem.* **127,** 301 (1939).

[46] Bloch, K., *J. Biol. Chem.* **155,** 255 (1944).

[47] Foster, G. L., Rittenberg, D., and Schoenheimer, R., *J. Biol. Chem.* **125,** 13 (1938).

Rather, it may enter the amino acid through preliminary deamination to a keto acid, which incorporates deuterium by enolization, after which re-amination occurs. There is little evidence for direct enzymatic labilization of C-H bonds. The absence of any such labilization has been demonstrated for lysine.[42]

Many compounds, such as stearic acid, leucine, and succinic acid, can be prepared by all the general procedures described. In Table 22 comparative results for the various methods are listed for leucine.

Deuterium can be introduced into the amino acid valine at the β and γ positions either by exchange with D_2SO_4,[48] or by synthesis from isobutyraldehyde.[44] Synthesis of valine from deuteroisovaleric acid results in nearly uniform distribution of D along the carbon chain.[46] Glutamic acid prepared by exchange in deuterium water contains slowly exchangeable deuterium in the γ position.[49] Using the Knoop reaction, glutamic acid is synthesized with stable deuterium in the α and β positions, but with no deuterium in the γ position.[49]

With fatty acids, it is observed that treatment with D_2SO_4 leads to placement of deuterium, stably bound with respect to acid or alkaline digestion, almost exclusively on the α carbon. Exchange in alkaline D_2O results in the random distribution of stable deuterium along the carbon chain.[50, 51, 51a]

A partial summary of deuterium compounds, prepared according to the three general procedures, is presented in Tables 23, 24, and 25, as compiled by Dr. H. Gest. From these tables some notion of the versatility of deuterium as an auxiliary label for carbon can be obtained.

It is to be expected that all labeling procedures applicable to deuterium are also applicable with little change to tritium. In all instances in which the deuterium and tritium have been studied together, it has been found that the two isotopes parallel each other closely in exchange characteristics.[64a] In prospect, tritium may possess advantages over deuterium, stemming from the greater dilution range available with tritium. With present sources of tritium, samples of H_2 are available with a total radioactivity of 2.63 curies per cubic centimeter of gas (N.T.P.). One cubic centimeter of H_2 from such a sample can be introduced into the counting tube so that, if 10^{12} ct./min. represents the initial activity and 10^2 ct./min. represents the lower limit of precision counting (1 to 5% error in 30 min.

[48] Rittenberg, D., Keston, A. S., Schoenheimer, R., and Foster, G. L., *J. Biol. Chem.* **125**, 1 (1938).

[49] Ratner, S., Rittenberg, D., and Schoenheimer, R., *J. Biol. Chem.* **135**, 357 (1940).

[50] van Heyningen, W. E. Rittenberg, D., and Schoenheimer, R., *J. Biol. Chem.* **125**, 495 (1938).

[51] van Heyningen, W. E., *J. Biol. Chem.* **123**, lv (1938).

[51a] Bloch, K., and Rittenberg, D., *J. Biol. Chem.* **155**, 243 (1944).

TABLE 23

ISOTOPIC COMPOUNDS PREPARED BY HYDROGEN EXCHANGE

With D_2O		With D_2SO_4		With D_2	
Compound	Reference	Compound	Reference	Compound	Reference
1. Acetone	52, 53	1. *n*-Hexane	48	1. Ethylene	62
2. Acetic acid	54	2. Cyclohexane	48	2. Methane	63, 64
3. Acetamide	54	3. Methylcyclohexane	48	3. Ethane	63*
4. Malonic acid	54, 55	4. *n*-Heptane	48		
5. Succinic acid	54, 55	5. Phenylalanine	60a		
6. Urea	55	6. Leucine	43		
7. Benzene	56	7. Valine	51a		
8. Benzamide	54	8. Alanine	43 51a,		
9. Phenol	57	9. Methionine	51a, 65		
10. Resorcinol	58	10. Benzene	61		
11. Pyrogallol	55	11. Palmitic acid	43, 50		
12. Hydroquinone	55				
13. Isatin	54				
14. Isobutyric acid	46				
15. *n*-Valeric acid	51a				
16. Isocaproic acid	45				
17. Caprylic acid	50				
18. Capric acid	50				
19. Lauric acid	50				
20. Myristic acid	46, 49				
21. Palmitic acid	50				
22. Stearic acid	50				
23. Acetanilide	57				
24. Mandelic acid	59				
25. Atrolactic acid	59				
26. Glycine	55				
27. Glutamic acid	49				
28. Cystine	60				
29. Arginine	60				
30. Lysine	60				
31. Histidine	58, 60				
32. Tyrosine	49				
33. Coprostanone	41				
34. Vitamin B_1	55				

* Exchange with free D atoms.

[52] Halford, J. O., Anderson, L. C., and Bates, J. R., *J. Am. Chem. Soc.* **56,** 491 (1934).

[53] Bonhoeffer, K. F., and Klar, R., *Naturwissenschaften* **22,** 45 (1934).

[54] Wynne-Jones, W. F. K., *Chem. Revs.* **17,** 115 (1935).

[55] Hamill, W. H., *J. Am. Chem. Soc.* **59,** 1152 (1937).

TABLE 24

ISOTOPIC COMPOUNDS PREPARED BY SYNTHETIC METHODS

Compound	Method	Reference
1. Acetylene	Reaction of CaC_2 with D_2O	65
2. Ethylene	Reduction of deuteroacetylene with chromous chloride	65
3. Ethane	Reduction of ethylene with D_2	62
4. Ethyl alcohol	Reduction of deuteroacetaldehyde	51a
5. Acetaldehyde	Degradation of deuteroalanine with chloramine-T; oxidation of acetylene in D_2O, H_2SO_4, $HgSO_4$ mixture	51a
6. Acetic acid	(*a*) Reaction of carbon suboxide with D_2O	66, 67
	(*b*) Oxidation of deuteroacetaldehyde (or deutero-alcohol)	51a
7. Propionic acid	Reduction of allylacetate or methyl acrylate with D_2	68
8. Fumaric acid	Dehydrogenation of deuterosuccinic acid with SeO_2	69
9. Succinic acid	(*a*) Reaction of ethane $\alpha,\alpha',\beta,\beta'$-tetracarboxylic acid (tetraethyl ester) with D_2O	69
	(*b*) Reduction of fumaric acid (diethyl ester) with D_2	69, 70
10. Butyric acid	(*a*) Reduction of ethyl crotonate with D_2	68, 71
	(*b*) Reduction of ethyl vinylacetate with D_2	71
11. Caproic acid	Reduction of ethyl sorbate with D_2	68
12. Ethyl undecylate	Reduction of ethyl undecylenate with D_2	51a
13. Stearic acid	(*a*) Reduction of methyl linoleate with D_2	72
	(*b*) Reduction of oleic and linoleic acids mixture with D_2	73
	(*c*) Reduction of linseed oil with D_2	51a
14. Benzene	(*a*) Polymerization of dideuteroacetylene	74
	(*b*) Reaction of benzene with DCl (AlC_3 catalyst-Friedel-Craft reaction)	75
	(*c*) Decarboxylation of calcium mellitate with calcium deuteroxide	76, 77
15. Benzoic acid	(*a*) Friedel-Craft synthesis starting with hexadeuterobenzene	77
	(*b*) Grignard synthesis starting with hexadeuterobenzene	78
16. Alanine	Knoop reaction: reduction of pyruvic acid—NH_3 mixture with D_2 (Pd catalyst)	49
17. Ornithine	Synthesis from α-pyridone via α-piperidone, δ-aminovaleric acid, and α-bromo-δ-aminovaleric acid	79

TABLE 24 (CONTINUED)

Compound	Method	Reference
18. Valine	(*a*) Synthesis from deuteroisovaleric acid via ammonolysis of the α-bromo acid	46
	(*b*) Synthesis from isobutyraldehyde via isobutanol-α,β-D_2, and the amino nitrile reaction	45
19. Glutamic acid	Knoop reaction: catalytic reduction of α-keto glutaric acid-NH_3 mixture with D_2	49
20. Lysine	Synthesis from phenol via deuterocyclohexanone, deuterocyclohexanone oxime, benzoyl-ϵ-amino-caproic acid, the α-bromo derivative, and the phthalimide reaction	80
21. Homocystine	Synthesis from dideuteroacetylene via dideuteroethylene, dideuterobenzylthioethylbromide, and dideuterobenzylhomocysteine	65
22. Leucine	(*a*) Synthesis from deuteroisocaproic acid via ethyl deutero-α-bromoisocaproate and the potassium phthalimide reaction	56
	(*b*) Synthesis from deuteroisocaproic acid via ammonolysis of the α-bromo acid	46
	(*c*) Synthesis from isovaleraldehyde via isopentanal-α,β-D_2, and the aminonitrile reaction	44
23. Proline	Synthesis from methyl coumalate via hydroxynicotinic acid, α-pyridone, α-piperidone, dichloropiperidone, and α,α-dichloro-δ-aminovaleric acid	81
24. Methylmethionine	Methylation of homocysteine with deuteromethyl iodide in liquid NH_3	82
25. Phenylacetic acid	Degradation of deuterophenylalanine via benzyl cyanide	60a
26. Phenyllactic acid	Reaction of deuterophenylalanine with HNO_2	83
27. Choline (halides)	Reaction of deuteromethyliodide and aminoethanol	82
28. Tributyrin	Condensation of deuterobutenyl chloride (prepared from deuterobutyric acid) with glycerol	84
29. Coprostanone (4,5-D_2)	Reduction of cholestenone with D_2	85
30. Δ^4-Cholestenone	Rearrangement of Δ^5-cholestenone in alkaline alcohol-D_2O mixture	41

[56] Bowman, P. I., Benedict, W. S., and Taylor, H. S., *J. Am. Chem. Soc.* **57,** 960 (1935).

[57] Small, P. A., and Wolfenden, J. H., *J. Chem. Soc.* **1936,** 1811.

[58] Geib, K. H., *Z. physik. Chem.* **180A,** 211 (1937).

[59] Erlenmeyer, H., and Schenkel, H., *Helv. Chim. Acta* **19,** 1199 (1936).

TABLE 25
ISOTOPIC COMPOUNDS PREPARED BIOLOGICALLY

Compound	Source	Reference
1. Fumaric acid	Dehydrogenation of deuterosuccinic acid with succinic dehydrogenase	68
2. Succinic acid	Yeast metabolic product—deuteroacetate substrate	66
3. Azelaic acid	Rat tissues—D_2O diet	86
4. Stearic acid	Depot fats of mice—D_2O diet	87
5. Palmitic acid	Depot fats of mice—D_2O diet	87
6. Heptoic acid	Rat tissues—D_2O diet	86
7. Pelargonic acid	Rat tissues—D_2O diet	86
8. β-Hydroxybutyric acid	Rat urine—deuterobutyric acid diet	71
9. Glycine	Mouse protein—D_2O diet	47
10. Aspartic acid	Mouse protein—D_2O diet	47
11. Glutamic acid	(*a*) Mouse protein—D_2O diet	47
	(*b*) Rat protein—deuteroproline diet	81
	(*c*) Rat protein—D_2O diet	86
12. Proline	Mouse protein—D_2O diet	47
13. Ornithine	Rat protein—deuteroproline diet	81
14. Arginine	(*a*) Mouse protein—deuteroörnithine diet	79
	(*b*) Mouse protein—D_2O diet	47
15. Histidine	Mouse protein—D_2O diet	47
16. Cystine	(*a*) Mouse protein—D_2O diet	43
	(*b*) Rat protein—D_2O diet	88
17. Leucine	Mouse protein—D_2O diet	47
18. Tyrosine	(*a*) Mouse protein—D_2O diet	47
	(*b*) Rat protein—deuterophenylalanine diet	60a
	(*c*) Rat protein—D_2O diet	86, 88
	(*d*) Digestion of casein in D_2O with trypsin	60
19. Hippuric acid	Rat urine—deuteroglycine and benzoic acid diet	89
20. Citric acid	Yeast metabolic product—deuteroacetate substrate	66
21. Glycogen	Rat liver and carcass—D_2O diet	86, 90, 91
22. Cholesterol	(*a*) Rat tissues—D_2O diet	86
	(*b*) Rat tissues—deuteroleucine or deutero-isovaleric acid diet	46
	(*c*) Rat tissues—deutero-alcohol, butyric acid, alanine, *n*-valeric acid, or myristic acid diets	49
23. Coprosterol	Human stool—deuterocholestenone diet	41

[60] Stekol, J. A., and Hamill, W. H., *J. Biol. Chem.* **120,** 531 (1937).
[60a] Moss, A. R., and Schoenheimer, R., *J. Biol. Chem.* **135,** 415 (1940).
[61] Ingold, C. K., Raisin, C. G., and Wilson, C. L., *J. Chem. Soc.* **1936,** 915.
[62] Farkas, A., and Farkas, L., *J. Am. Chem. Soc.* **60,** 22 (1938).

of counting), then dilution ratios of 10^9 to 10^{10} are available for precision work.

With such dilutions, many types of syntheses denied with deuterium become available with tritium. Thus histidine, prepared by biosynthesis in the mouse, assays only 0.24 atom per cent deuterium beginning with a deuterium level in the diet of 1.50 to 1.60 atom per cent.[47] Such histidine has a deuterium content only some twelve times that of normal histidine. At best, a dilution range of twentyfold is available for precision work with density assay. Not even this dilution can be tolerated when precision of 1 to 5% is desired, if the spectrometer method of assay is used. Conse-

---

[63] Steacie, E. W. R., and Phillips, N. W. F., *J. Chem. Phys.* **4,** 461 (1936).

[64] Taylor, H. S., Morikawa, K., and Benedict, W. S., *J. Am. Chem. Soc.* **57,** 383, 592 (1935).

[64a] Fontana, B. J., *J. Am. Chem. Soc.* **64,** 2503 (1942).

[65] Patterson, W. I., and du Vigneaud, V., *J. Biol. Chem.* **111,** 393 (1938).

[66] Sonderhoff, R., and Thomas, H., *Naturwissenschaften* **24,** 570 (1936).

[67] Wilson, C. L., *J. Chem. Soc.* **1935,** 492.

[68] Rittenberg, D., Schoenheimer, R., and Evans, E. A., Jr., *J. Biol. Chem.* **120,** 503 (1937).

[69] Erlenmeyer, H., Schoenauer, W., and Sülmann, H., *Helv. Chem. Acta* **19,** 1376 (1936).

[70] Leffler, M. T., and Adams, R., *J. Am. Chem. Soc.* **58,** 1551 (1936).

[71] Morehouse, M. G., *J. Biol. Chem.* **129,** 769 (1939).

[72] Schoenheimer, R., and Rittenberg, D., *J. Biol. Chem.* **111,** 163 (1935).

[73] Schoenheimer, R., and Rittenberg, D., *J. Biol. Chem.* **120,** 155 (1937).

[74] Murray, J. W., Squire, C. F., and Andrews, D. H., *J. Chem. Phys.* **2,** 714 (1934).

[75] Klit, A., and Langseth, A., *Nature* **135,** 956 (1935); also *Z. physik. Chem.* **176A,** 65 (1936).

[76] Erlenmeyer, H., and Lobeck, H., *Helv. Chim. Acta* **18,** 1464 (1935).

[77] Erlenmeyer, H., Lobeck, H., Gartner, H., and Epprecht, A., *Helv. Chim. Acta* **19,** 336 (1936).

[78] Erlenmeyer, H., Lobeck, H., and Epprecht, A., *Helv. Chim. Acta* **19,** 793 (1936).

[79] Clutton, R. F., Schoenheimer, R., and Rittenberg, D., *J. Biol. Chem.* **132,** 227 (1940).

[80] Weissman, N., and Schoenheimer, R., *J. Biol. Chem.* **140,** 779 (1941).

[81] Stetten, M. R., and Schoenheimer, R., *J. Biol. Chem.* **153,** 113 (1944).

[82] du Vigneaud, V., Cohn, M., Chandler, J. P., Schenck, J. R., and Simmonds, S., *J. Biol. Chem.* **140,** 625 (1941).

[83] Moss, A. R., *J. Biol. Chem.* **137,** 739 (1941).

[84] Morehouse, M. G., *J. Biol. Chem.* **155,** 33 (1944).

[85] Schoenheimer, R., Rittenberg, D., and Graff, M., *J. Biol. Chem.* **111,** 183 (1935).

[86] Boxer, G. E., and Stetten, D., *J. Biol. Chem.* **153,** 607 (1944).

[87] Bernhard, K., and Schoenheimer, R., *J. Biol. Chem.* **133,** 713 (1940).

[88] Stekol, J. A., and Hamill, W. H., *Proc. Soc. Exptl. Biol. Med.* **35,** 591 (1937).

[89] Rittenberg, D., Foster, G. L., and Schoenheimer, R., *J. Biol. Chem.* **123,** cii (1938).

[90] Stetten, D., Jr., and Boxer, G. E., *J. Biol. Chem.* **155,** 231 (1944).

[91] Boxer, G. E., and Stetten, D. W., Jr., *J. Biol. Chem.* **155,** 237 (1944).

quently, such a synthetic method for preparing histidine, although more convenient than some others, is not acceptable.

Consider the same situation with tritium. Sufficient tritium with the specific activity given above as readily available (2.63 curies per milliliter) could be obtained to make water with a specific activity of hundreds of curies per gram. Enough of this water could be injected to bring the level of tritium in total body fluids of rats or mice to hundreds of millicuries per milliliter. This fluid would be in equilibrium with tissue hydrogen after a short period so that a specific activity of the same order of magnitude could be expected for tissue hydrogen. About 4 mM. of histidine hydrogen could be recovered from 100 g. of mouse. This hydrogen would have a total activity of at least 10 mc., all of which could be introduced as gas in a G-M or proportional counter. Ten millicuries would correspond to approximately 10^8 to 10^9 ct./min. This material could be diluted 10^7-fold and still allow assays with a precision of 1 to 5%.

Spectacular increases in dilution factors obtainable also attend the use of tritium in chemical syntheses. Thus, Bloch and Rittenberg[92] have obtained cholesterol containing 5.70 atom per cent deuterium, using a platinum-catalyzed exchange reaction between cholesterol and deuterium oxide in acetic acid. The original concentration of deuterium in the water was 88%. In this case, the dilution of the sample for deuterium assay could be many hundredfold. Using tritium with an initial specific activity of 10^{12} ct./min./mole of hydrogen, one should be able to obtain, under the same conditions, $\frac{5}{88} \times 10^{12}$, or 6×10^{10} ct./min./mole of hydrogen in cholesterol. One millimole of such cholesterol burned and converted to hydrogen would yield 6×10^7 ct./min. in 23 mM. of hydrogen. This could be diluted at least 10^6-fold with a precision of 1 to 5%. Samples with a specific activity of thousands of curies per mole of hydrogen are now obtainable. Such samples would give a dilution range of some 10^7-fold with the relatively insensitive but extremely rugged and stable Lauritsen electroscope, so that assay difficulties would be obviated to a large extent.

The importance of such considerations for the general synthesis problem is evident. Carcinogens, such as methylcholanthrene, can be labeled in a manner analogous to that discussed for cholesterol. The labeling of such a compound with tracer carbon rather than tracer hydrogen might require a synthesis which would either be wasteful or impossible and, in any case, might not give a product with the dilution range obtainable when tritium is used.

It can be concluded that the potentialities of tritium as an auxiliary tracer for carbon are such that a considerable extension in carbon tracer research can be expected over and above that already made possible by the many magnificent studies with deuterium.

[92] Bloch, K., and Rittenberg, D., *J. Biol. Chem.* **149,** 505 (1943).

6. USE OF HYDROGEN ISOTOPES IN CLINICAL RESEARCH WITH REMARKS ON INCIDENTAL RADIATION HAZARDS

The importance of hydrogen compounds in biology and of water in particular renders superfluous remarks about the many conceivable researches based on tracer hydrogen isotopes. Many physiological studies have been reported which indicate directions in which further elaboration can be expected. Most of these have to do with the use of H^2-labeled water to determine fluid spaces, investigate capillary permeability and tissue metabolism, and follow transport of metabolites.

All these studies have application to clinical research. As an example, there are the extensive researches of Gallagher, Dobriner, and others[93] on steroid metabolism in man, with isotopically labeled steroid hormones, in which it has been shown possible (1) to distinguish between endogenous glandular production of hormone and administered hormone, (2) to detect transformations of these hormones which could not be detected otherwise, and (3) to set limits on occurrence of any postulated transformation. Some important results have been obtained even when isotope is placed in such a position on the sterol that appreciable loss of label by exchange with water hydrogen occurs.

Extension of studies in synthesis of metabolic products exploiting added sensitivity of tritium in the human in normal and diseased states is obviously indicated. There would appear to be little to concern the investigator in using triterated water, as far as radiation hazards are concerned. The rapid equilibration of any ingested sample of labeled water with body fluid, the removal of labeled material by steady-state turnover, and the very low energy of the tritium radiations would appear to indicate a large tolerance dose.

If one assumes that 100% equilibration occurs and that elimination of labeled water parallels the daily excretion, then the biological half-life is at most 100 days. This is a very conservative estimate. If one desires to maintain the accepted daily dose rate of 0.1 rep. as a maximum, then by calculations described in Chapter IV, using formula 2, the permissible initial concentration, C, in microcuries per gram is given by $C = 0.1/(60E)$, where E, the average energy of the tritium β radiations is 0.0057 Mev. Thus, $C = 0.3\ \mu c/g$. The continued excretion of tritium water with a half-life of 100 days drops this daily dosage rate $\sim 1\%$ a day, so that this is an additional safety factor. The effective half-life is given as 19 days.[94] There is no evidence that selective absorption takes place in appreciable amounts to invalidate this result. Of course, administration of stably

[93] See Gallagher, T. F., *in* "Isotopes in Biochemistry," Ciba Foundation Symposium (G. E. W. Wolstenholme, ed.), pp. 28–40. Blakiston, New York, 1952.

[94] *Nat'l Bur. Standards Handbook* 52 (1953).

labeled compounds which are selectively absorbed is subject to calculated dosage rates revised to take into account the characteristic tissue concentrations and biological half-life involved. The natural decay of tritium is too small to be taken into account.

For a 70-kg. human being, the total tritium content on this basis should not exceed 20 mc. given in one dose or divided into many doses. Some laboratories adopt a more conservative approach. Rather than maintain a daily dose of 0.1 r. as a maximum, they limit the total dose to 1.0 r. On this basis one may calculate that the total tritium content should be roughly 14 mc. for a 70-kg. human subject. These values can be considered to indicate roughly the maximum dosage and are not offered as hard and fast limits, particularly in view of the large fluctuations in radiosensitivity encountered in humans.

GENERAL REFERENCES

1. Clarke, H. T. (ed.), "Symposium on the Use of Isotopes in Biology and Medicine." Univ. Wisconsin Press, Madison, 1949.
2. Glascock, R. F., "Isotopic Gas Analysis for Biochemists." Academic Press, New York, 1954.
3. Kimball, A. H., "Bibliography of Research on Heavy Hydrogen Compounds." McGraw-Hill, New York, 1949.
4. Wiberg, K. B., The deuterium isotope effect. *Chem. Revs.* **55**, 713 (1955).
5. Wilson, D. W., Nier, A. O., and Reimann, S. P. (eds.), "Preparation and Measurement of Isotopic Tracers." Edwards, Ann Arbor, Michigan, 1946.

CHAPTER X

THE ISOTOPES OF CARBON

1. INTRODUCTION

There are six known isotopes of carbon which range in mass number from 10 to 15. Two of these isotopes, C^{10} ($\tau_{1/2}$ = 8.8 sec.)[1] and C^{15} ($\tau_{1/2}$ = 2.25 sec.),[2] are too short-lived for use as tracers. C^{11} ($\tau_{1/2}$ = 20.5 min.)[3] has been used in the past but is now chiefly of historical interest. The two major carbon tracers are (1) the rare stable isotope C^{13} (abundance, 1.10 atom per cent), and (2) C^{14} ($\tau_{1/2}$ = 5568 yr.).[4]

2. SHORT-LIVED RADIOACTIVE CARBON, C^{11}

C^{11} is a positron emitter. The upper energy of the β^+ spectrum is 0.981 ± 0.005 Mev.[5] This corresponds to a maximum range in aluminum of approximately 400 mg./cm.^{2}. Accompanying the positrons is the usual annihilation radiation (E = 0.5 Mev.), with a small percentage of other γ rays. Bombardment of B^{10} with deuterons gives C^{11} by a (d,n) reaction and is the procedure indicated as most efficient for production of the isotope. Details of the preparation can be found in the literature.[6]

The hard quality of the radiations from C^{11} makes it simple to assay. Measurements can be made with equal ease on solid or liquid samples. Any thick-walled cylinder-type G-M tube can be used. Owing to the high intensities available, relatively insensitive electroscope and electrometer ionization chambers can also be employed.

A simple procedure adaptable to most research with C^{11} utilizes the cylinder-type G-M tube discussed in Chapter III. Liquid samples containing nonvolatile C^{11} compounds are pipetted onto thin blotting paper and dried over a hot plate. The solution being assayed is added slowly, and the drops are allowed to dry between applications, so that no liquid is lost by spillage. The blotter is then covered with thin cellophane and cemented

[1] Barkas, W. H., Creutz, E. C., Delsasso, L. A., Fox, J. G., and White, M. G., *Phys. Rev.* **57**, 562 (1940).

[2] Douglas, R. A., Gasten, B., Downey, J., and Mukerji, A., *Bull. Am. Phys. Soc.* [II] **1**, 21 (1956).

[3] Crane, H. R., and Lauritsen, C. C., *Phys. Rev.* **45**, 497 (1934).

[4] Ruben, S., and Kamen, M.D., *Phys. Rev.* **57**, 549 (1940).

[5] Townsend, A. A., *Proc. Roy. Soc.* **177A**, 357 (1941).

[6] See previous editions of this book (1948, 1951).

on one edge with Duco household cement. The smallest practicable amount of cement is used so that it will not spread on the blotter surface when the cellophane is pressed to it. The protected blotter is wrapped around the center of the G-M tube counter and pressed lightly against the tube with a spring clip. Smith and Cowie have given an excellent alternative procedure for C^{11} assay, using liquid samples and an electrometer chamber.[7]

It is obvious that the short half-life of C^{11} is the main limitation on its use as a carbon tracer. Counteracting this to some extent are the enormous intensities available. Samples are obtainable with a specific activity at initial time (end of bombardment) of about 10^{12} ct./min./mg. of carbon. Experiments which can be carried out in 4 hr. (12 half-lives) result in a residual activity of about 10^8 ct./min./mg. of carbon. Such samples can be diluted 10^5-fold for assays, with a resultant precision of a few per cent.

For experiments of short duration requiring high dilutions, C^{11} is an excellent carbon tracer. Unfortunately, use of C^{11} is predicated on the availability of a cyclotron installation and so cannot become very widespread. Neverthless, it has been employed intensively for studies in plant and animal assimilation of carbon dioxide as well as some simple carbon compounds.[8]

Detailed protocols of two experiments with C^{11} to show how such a short-lived isotope is handled will follow. The examples chosen are taken from investigations on assimilation of CO_2 by the protozoön *Tetrahymena gelii*.[9] The purpose of the experiments was to ascertain the site of CO_2 assimilation during fermentation of glucose to lactic, acetic, and succinic acids.

Since it had already been ascertained that suspensions of the protozoa could assimilate measurable quantities of CO_2 in phosphate-bicarbonate buffer, it was decided to test the hypothesis of CO_2 assimilation through the dicarboxylic acid cycle. This involved establishment of reversible equilibria between administered CO_2 and the carboxyl groups of the four acids—oxalacetic, fumaric, malic, and succinic.

The organisms were grown in yeast extract media with 2% glucose under oxidative conditions (aeration). For the experiments, the cells were centrifuged, washed, and resuspended in phosphate buffer (*p*H 7.5) with 1% glucose. The suspension was shaken in the presence of a few milliliters of $C^{11}O_2$ in an oxygen-free atmosphere for 30 min. at 30°C. The $C^{11}O_2$ was prepared for use at 4:35 P.M., at which time the G-M tube background was

[7] Smith, J. H. C., and Cowie, D. B., *J. Appl. Phys.* **12,** 78 (1941).

[8] See Buchanan, J. M., and Hastings, A. B., *Physiol. Revs.* **26,** 120 (1946), for an extensive discussion of C^{11} applications.

[9] Van Niel, C. B., Ruben, S., Carson, S. F., Kamen, M. D., and Foster, J. W., *Proc. Natl. Acad. Sci. U.S.* **28,** 8 (1942).

19 ct./min. For purposes of calculation, 6:35 P.M. was taken as the initial time, because it was not until then that counting began.

After the incubation period, a few milligrams each of inactive lactic, acetic, pyruvic, succinic, and fumaric acids were added as carrier. Repeated boiling with $NaHCO_3$ was used to remove completely all $C^{11}O_2$ from the suspension. The suspensions were centrifuged, resuspended, and centrifuged again. The supernates were collected for further analysis. The residual cell material was suspended in 35 ml. of water, and 1 ml. was pipetted on a blotter. At 7:08 P.M., this sample showed an activity of 183 ct./min. The background count at this time was 30 ct./min. The cell aliquot contained, therefore, 153 ct./min. At 7:30 P.M. (one half-life later) the sample showed an activity of 83 ct./min. corrected for background. Corrected for decay, this was 83/0.50 = 166 ct./min., which agreed well with the measurement at 7.08 P.M.

In the same way, other fractions were isolated and measured at various times. Pyruvic and fumaric acids were precipitated from separate aliquots of the original supernatant solution with 2,4-dinitrophenylhydrazine and mercurous nitrate, respectively. At 8:48 P.M. the hydrazone precipitate of the pyruvic acid formation gave only 5 ct./min. At 9:03 P.M. the fumarate precipitate assayed 27 ct./min.

Volatile acids were determined in another aliquot by vacuum distillation. At 6:48 P.M., 1 ml. out of 54 ml. of distillate gave an activity of 8 ct./min. The nonvolatile residue from the distillation, containing succinic and fumaric acids, retained practically all the activity. One milliliter from 34 ml. at 8:17 P.M. yielded 825 ct./min. At 8:56, the count was 240/min. This, multiplied by the decay correction, 1/0.275, as read from the decay curve for C^{11}, gave an effect of 875 ct./min., referred to the same time as the first assay which had given 825 ct./min. All assays were based on 3-min. counts. A second sample averaged 990 ct./min.

Another fraction of the nonvolatile supernate was oxidized with acid permanganate. The aliquot (1 ml. from 40 ml.) was counted at two different times with results averaging 20 to 30 ct./min.

The calculations of the relative activities in the various fractions follow: $t = 0$ taken at 6:35.

a. Cell material: Counted at 7:08, decay correction = 1/0.335. Average of two assays (153 + 166)/2 = 160 ct./min. in 1 ml. out of 35.

$$\therefore \frac{160 \times 35}{0.335} = 1.68 \times 10^4$$

b. Nonvolatile residue: Counted at 8:17, decay correction = 1/0.035. Average of two assays (850 + 990)/2 = 920 ct./min. in 1 ml. out of 34. This sample was made up from the residue left after distillation of the original supernate containing 55 ml. from which two 1-ml. samples had been removed for assay. Hence the final dilution factor was 34 × 55/53.

$$\therefore \frac{920 \times 34 \times 55/53}{0.035} = 9.3 \times 10^5$$

c. Total suspension: Counted at 7:43, decay correction = 1/0.106. Average of three assays (1400 + 1140 + 1120)/3 = 1220 ct./min. in 2 ml. from 55, this diluted again by 3.

$$\therefore \frac{1220 \times 3 \times 55/2}{0.106} = 9.5 \times 10^5$$

d. Volatile acid (acetic and lactic): Counted at 6:48, decay correction = 1/0.65. One assay yielded 8 ± 4 ct./min. in 1 ml. from 54.

$$\therefore \frac{8 \times 54}{0.65} = 6.65 \times 10^2 \text{ (essentially zero effect)}$$

e. Nonvolatile residue after permanganate oxidation: Counted at 6:59, decay correction = 1/0.45. Two assays averaging 2030 ct./min., from same aliquot as (*b*), diluted further by a factor of 4.

$$\frac{2030 \times 4 \times 34 \times 55/53}{0.45} = 6.4 \times 10^5$$

f. Pyruvate precipitate: Counted at 8:48, decay correction = 1/(0.0375 × 0.335). One assay showed <5 ct./min. in 10.5 ml. from 55.

$$\therefore \frac{5 \times 55/10.5}{0.0375 \times 0.335} \leq 2.1 \times 10^3$$

g. Fumarate precipitate: Counted at 9:03, decay correction = 1/(0.0375 × 0.21). One assay showed 27 ct./min. in 1 ml. from 22 taken from 10.5 ml. originally removed from 55.

$$\therefore \frac{27 \times 55/10.5 \times 22}{0.0375 \times 0.21} = 3.6 \times 10^4$$

The results are summarized in Table 26.

Examination of these results shows that practically all the assimilated $C^{11}O_2$ appeared in the nonvolatile fraction, which contained all the fumaric and succinic acids. It appeared likely that all the activity was in the succinic acid, rather than the fumaric, since the small percentage in the fumaric acid might be due to occlusion of mercurous succinate in the mercurous fumarate precipitate. Boiling the nonvolatile residue with acid permanganate had affected the C^{11} content somewhat too markedly to exclude fumarate, however.

TABLE 26

RELATIVE ACTIVITIES OF CELL FRACTIONS IN *Tetrahymena geleii*, AFTER INCUBATION IN PRESENCE OF $C^{11}O_2$ (AFTER VAN NIEL, *et al.*[9])

Fraction	Ct./min. (corr.)	Per Cent of Total $C^{11}O_2$ Assimilated
Cell material	0.17×10^5	1.8
Nonvolatile residue	9.3×10^5	98
Volatile acid	$<0.006 \times 10^5$	0
Pyruvic acid	$<0.021 \times 10^5$	0
Fumaric acid	0.36×10^5	3.8
Total	9.83×10^5	
Original suspension	9.5×10^5	

Per cent recovery $= \frac{9.83}{9.5} \times 100 = 103\%$

In a second experiment the protozoa were suspended as before in the presence of $C^{11}O_2$ and allowed to ferment glucose for 30 min. at 30°C. The cells were then separated from the supernate. The supernate was freed of labeled carbonate by acidification in the presence of unlabeled bicarbonate. The solution was then neutralized to pH 7.17 and treated with a succinic-dehydrogenase preparation from beef heart, so that labeled succinic, if present, would be converted to labeled fumaric. The supernate from this step, after being freed of protein, was divided into three portions. In the first, the total radioactivity was measured; in the second, the fumaric acid present was precipitated with mercurous nitrate. The third fraction was treated with potassium permanganate in 1.5 N H_2SO_4 at 40°C. The gas evolved was passed through $Ba(OH)_2$ in a stream of nitrogen, the $BaCO_3$ obtained being weighed and assayed for C^{11}. The solution remaining after the oxidation was steam distilled, the distillate made alkaline and assayed.

The calculations follow: $t = 0$ taken at 8:00 P.M.

a. Total C^{11} in supernate: Counted at 8:18, decay correction = 1/0.55. Three assays averaged 678 ct./min. in sample diluted 77-fold.

$$\therefore \quad 678 \times 77/0.55 = 9.5 \times 10^4 \text{ ct./min.}$$

b. Fumarate precipitate: Two samples weighing 0.269 and 0.263 g., respectively. Average assay for both was 1278 ct./min. Decay correction = 1/0.085. Total activity in 0.562 g. calculated as 6.92×10^4 ct./min.

c. $BaCO_3$ precipitate: Two samples totaling 0.632 g. Total activity = 7.8×10^4 ct./min.

d. Formate from oxidation: Counted at 9:08, decay correction = 1/0.105. Assayed <4 ct./min. in 1/5 total.

$$\therefore \quad 4 \times 5/0.105 \leq 2 \times 10^2 \text{ ct./min.}$$

e. Stoichiometry of oxidation: Reaction involved was:

$$\begin{array}{l} \text{COOH} \\ | \\ \text{CH} \\ \| \\ \text{CH} \\ | \\ \text{COOH} \end{array} + 2\,MnO_4^- + 6H^+ \rightarrow 3CO_2 + HCOOH + 2Mn^{++} + 4H_2O$$

Total moles $BaCO_3 = 0.632/197 = 3.21 \times 10^{-3}$.

Total moles mercurous fumarate $= 0.502/515 = 1.09 \times 10^{-3}$.

$\therefore$ $3 \times 1.09 \times 10^{-3} = 3.27 \times 10^{-3}$ mole $BaCO_3$ should have been recovered, as in reaction shown. Recovery was 3.21×10^{-3} mole or 98.2%, which was quite satisfactory.

From these results the following conclusions should be drawn. It had been shown that in the supernate of the first experiment precipitation as mercurous fumarate brought down <4% of the total activity. In the second experiment, $6.9/9.5 \times 100 = 73\%$ of the activity could be precipitated after treatment with succinic dehydrogenase, showing that most of the C^{11} was incorporated into succinic acid. From the labeled fumarate, $7.8/9.5 \times 100 = 82\%$ of the total C^{11} that could be recovered in carbonate, and none in the formate, after permanganate oxidation. This meant that all C^{11} was in the carboxyl groups of succinic acid, since control experiments with carboxyl-labeled succinic acid had shown that carbonate came only from carboxyl groups.[10]

The small difference between the C^{11} percentages in total fumarate (73%) and carboxyl (83%) was attributable to the fact that the fumarate and carbonate precipitates were bulky; therefore, inaccuracies due to self-absorption entered. Calculations to correct for these effects could be made in the following manner. The G-M tube counter wall thickness was 0.055 g./cm.^{2}. The cellophane contributed 0.004 g./cm.^{2}. The total C^{11} count was made in a blotter with thickness 0.026 g./cm.^{2}. The effective thickness of the blotter could be assumed as roughly half this, or 0.013 g./cm.^{2}. The total absorption was equivalent, therefore, to $0.055 + 0.004 + 0.013 = 0.072$ g./cm.^{2}. From absorption data on C^{11} radiation taken with the assay geometry used, this corresponded to an absorption correction of 1/0.78. The corrected total C^{11} activity was therefore

$$9.5 \times 10^4/0.78 = 1.21 \times 10^5 \text{ ct./min.}$$

The fumarate precipitate averaged 0.068 g./cm.^{2}, so that the absorption was equivalent to $0.059 + 0.068/2 = 0.093$ g./cm.^{2}. The fumarate activity

[10] Allen, M. B., and Ruben, S., *J. Am. Chem. Soc.* **64**, 948 (1942).

corrected for absorption was thus $6.92 \times 10^4/0.69 = 1.01 \times 10^5$ ct./min. The carbonate had an average absorption thickness of $0.059 + 0.050/2 = 0.084$ g./cm.^{2}. The corrected carbonate activity could be calculated as

$$7.8 \times 10^4/0.73 = 1.07 \times 10^5 \text{ ct./min.}$$

Hence, the percentage recovery in fumarate was $(1.01 \times 10^5)/(1.21 \times 10^5) \times 100 = 83\%$, and in carbonate $(1.07 \times 10^5)/(1.21 \times 10^5) \times 100 = 89\%$. The agreement is satisfactory. The percentage recovered in formate was negligible.

It will be noted that the total recovery of C^{11} in these experiments varied from 90% to 100%. Precision attainable with C^{11} is usually no better than 5% in individual samples, and recoveries better than 90% are not to be expected. Under some conditions precisions better than 1% to 2% can be achieved in checking individual samples. Accuracy is usually best when the experiment is so devised that ratios of activities are obtained, so that decay corrections and fluctuations in assay due to the necessarily short assay periods are minimized.

3. HEAVY STABLE CARBON, C^{13}

A. Introduction

The existence of C^{13} was first established by observation of characteristic isotope shifts in the band spectra of diatomic carbon produced at high temperature in the carbon arc.[11] The amount of the rare form of carbon, C^{13}, found in natural carbon varies slightly, depending on the source. The fluctuations are large enough to limit dilutions of the isotope to factors of about 1000 for samples available.[12] It can be prepared in small amounts[13] by separation in a thermal diffusion column, by means of gaseous isotope exchange reactions like.

$$C^{13}O + C^{12}O_2 = C^{12}O + C^{13}O_2$$

or in bulk amounts[14] in two-phase fractionation columns by means of two-phase isotope exchange reactions such as

$$HC^{12}N_{(g)} + C^{13}N^{-}_{(aq)} = HC^{13}N_{(g)} + C^{12}N^{-}_{(aq)}.$$

C^{13} has been used in a large number of researches in intermediary metabolism, particularly in multiple-labeling experiments. It provides an ideal stable partner for the long-lived radioisotope C^{14}. In general, its use as a tracer differs in no significant way from that of C^{14}. Synthetic procedures

[11] King, A. S., and Birge, R. T., *Nature* **124,** 127 (1929).
[12] Murphy, B. F., and Nier, A. O., *Phys. Rev.* **59,** 771 (1941).
[13] Taylor, T. I., and Bernstein, R. B., *J. Am. Chem. Soc.* **69,** 2076 (1947).
[14] Hutchinson, C. A., Stewart, D. W., and Urey, H. C., *J. Chem. Phys.* **8,** 532 (1940).

TABLE 27

SOME IONIC SPECIES DERIVED FROM CO_2

Mass Number	Ion
46	$(C^{12}O^{16}O^{18})^+$
45	$(C^{13}O^{16}O^{16})^+$ and $(C^{12}O^{16}O^{17})^+$
44	$(C^{12}O^{16}O^{16})^+$
30	$(C^{12}O^{18})^+$
29	$(C^{13}O^{16})^+$ and $(C^{12}O^{17})^+$
28	$(C^{12}O^{16})^+$
22	$(C^{12}O^{16}O^{16})^{++}$

for its incorporation into labeled molecules are the same as those used with C^{14} and so need not be discussed separately.

B. Assay

A mass spectrometer is needed. This isotope is usually assayed in the form of CO_2 because the gas is relatively easy to prepare pure by combustion techniques and it behaves well in the mass spectrometer. The following ions are all that need be considered (Table 27). Mass peaks at 45 and 44 are compared to establish the atom per cent C^{13}. The peak at 46 is 1/250 as high as that at 44 because O^{16}/O^{18} is about 500.

From Eq. 2 of Chapter IX, the atom per cent of C^{13} will be given by the expression $100R/(1 + R)$, where R is the ratio of the ion intensity at peak 45 to that at peak 44. The ratio observed must be corrected for the contribution of the ion $(C^{12}O^{16}O^{71})^+$, which makes the 46 peak too high by about 0.08% because the O^{17}/O^{16} ratio is about 1/2500.

Sometimes other molecular species are used for C^{13} assay because of special needs, e.g., if it is desired to ascertain whether labeling occurs in adjacent positions of a metabolite. One such research has been cited (see p. 158). This kind of research requires the use of C^{13} and provides an example of a rare instance in which C^{14} cannot be used interchangeably with C^{13}.

4. LONG-LIVED RADIOACTIVE CARBON, C^{14}

A. History

C^{14} is regarded generally as the most important single tracer isotope because it is the most versatile of all the isotopes of carbon, the element which is central to all biology. The story of its discovery illustrates well many of the general considerations about nuclear reactions and the production of radioactive isotopes presented in Chapter I and is included here to give the reader an added insight into the nature of research in this field.

Nowadays, when C^{14} is so readily available and so widely used, it may be startling to recall that in 1939 the existence of a useful radioactive carbon tracer was not considered likely. To understand this pessimism it is necessary to review some facts which had been established at that time. Kurie,[15] in 1934, showed that, when nitrogen was bombarded with fast neutrons, recoil particles which appeared to be protons were produced. He suggested the existence of the nuclear reaction

$$N^{14} + n \rightarrow C^{14} + H^{1} \tag{1}$$

Later, Bonner and Brubaker[16] noticed that when nitrogen was exposed to slow neutrons in a Wilson cloud chamber (a device for studying visually the tracks of recoil particles in nuclear reactions) there occasionally appeared short, heavy recoils which they took to be α particles. Burcham and Goldhaber,[17] however, showed that these particles were actually protons arising from reaction 1, a conclusion with which Bonner and Brubaker agreed.[18] It could be calculated from the energy of the proton recoils[18] that the unstable carbon, presumably C^{14}, formed in this reaction was about 0.17 Mev. unstable with respect to N^{14}.

Between 1935 and 1939, numerous efforts were made to detect a radioactivity which might be ascribed to C^{14} without success. By this time there was no question that such an isotope of carbon existed, and it seemed certain it was radioactive because of the estimated energy of reaction 1 and the additional fact that no pairs of neighboring stable isobars had been found to exist among the light elements. In other words, for a given mass there appeared to be only one stable nuclide among the elements in the first few rows of the periodic table. This meant that if the stable nuclide for mass 14 was nitrogen, as was evident from the existence of stable N^{14}, the nuclide of mass 14 corresponding to C^{14} would be expected to be radioactive. It would also be expected to decay by β^- emission to form N^{14}, as in the reaction.

$$C^{14} \rightarrow N^{14} + \beta^- \text{ (+ neutrino)} \tag{2}$$

The great difficulty arose in attempting to calculate the rate of this reaction. It was possible that the half-life for the hypothetical radioactivity to be expected with C^{14} might be either very short (of the order of fractions of a second) or very long (of the order of years). The latter alternative was considered highly unlikely on the basis of nuclear theory then prevalent. Study of a comparable pair of isobars, namely He^6 and Li^6, strengthened this conclusion. It was known that He^6 (analogous to C^{14}) had a very short

[15] Kurie, F. N. D., *Phys. Rev.* **45,** 904 (1934).
[16] Bonner, T. W., and Brubaker, W. M., *Phys. Rev.* **48,** 469 (1935).
[17] Burcham, W. E., and Goldhaber, M., *Proc. Cambridge Phil. Soc.* **32,** 632 (1936).
[18] Bonner, T. W., and Brubaker, W. M., *Phys. Rev.* **49,** 223 (1936).

half-life (<1 sec.) so that it was not conceivable that the half-life of C^{14} could be longer than a few days. The negative results of attempts to find radioactivity in carbon in the period 1935–1939 were thought to show conclusively that no radioactivity of half-life in the order of days existed. Therefore, it appeared that if C^{14} were radioactive its half-life would be too short for use as a tracer.

Equally pessimistic prospects appeared for other carbon isotopes. The possibilities were that: (1) isomers of stable C^{12} or C^{13} might exist which could be long-lived γ emitters; or (2) a heavier isotope of mass 15 might exist which would have better radioactive characteristics. The first possibility could be dismissed even in the primitive state of theory about nuclear isomers which existed in 1939. The existence of C^{15} as a long-lived isotope seemed no more probable than that of C^{14}. Arguments of a similar nature could be made against the existence of usable radioactive isotopes of nitrogen, oxygen, and hydrogen. In sum, it was felt that radioactive tracers might prove of limited usefulness in biology because of the lack of adequately long-lived radioactive isotopes for these important elements for which only stable tracers existed at the time.

Despite this bleak outlook it was desirable to make absolutely certain that no long-lived isotopes existed among these elements of major importance in biology, so that any possibility of extending the tracer technique to its ultimate limits in biology would not be overlooked.

The development of the internal "probe" target by Wilson and the writer in 1938[18a] made possible a new attack on the problem because it increased many hundredfold the intensity of bombardment possible with the apparatus then current. As these authors said, "Obviously, the method of internal targets should find its most important application in the preparation of radioactive isotopes as yet undiscovered."

Late in 1939, the writer began a systematic search for long-lived activities among the light elements. Carbon was chosen as the first element to investigate. All the reactions which could possibly lead to the production of an unstable carbon isotope were listed. These included reactions with protons of energy up to 8 Mev., deuterons of energy up to 16 Mev., α particles with energy up to 32 Mev., and neutrons both fast and slow. The reactions possible are shown in Table 28.

The reactions considered most promising were the deuteron reaction on C^{13} and the α-particle reactions on B^{11}. The α-particle reaction was first attempted by means of an external bell-jar chamber with B_2O_3 as target. Considerable knowledge had been accumulated about the recoil activities to be expected from the $B^{11}(\alpha,p)C^{14}$ reaction. It was known that recoil carbon emerged as $C^{11}O$ in almost 100% yield from B_2O_3 as a result

[18a] Wilson, R. R., and Kamen, M. D., *Phys. Rev.* **54**, 1031 (1938).

TABLE 28

POSSIBLE NUCLEAR REACTIONS LEADING TO PRODUCTION OF C^{14} OR OTHER POSSIBLE LONG-LIVED FORMS OF CARBON

Reaction	Target Material	Recoil Activity
$Be^9(\alpha,n)C^{12*}$	Beryllium metal	?
$B^{10}(\alpha,p)C^{13*}$	B_2O_3	C*O
$B^{11}(\alpha,p)C^{14}$	B_2O_3	C*O
$B^{11}(p,\gamma)C^{12*}$	B_2O_3	C*O
$B^{11}(d,n)C^{12*}$	B_2O_3	C*O
$C^{12}(d,p)C^{13*}$	Graphite	C*
$C^{13}(d,p)C^{14}$	Graphite	C*
$N^{14}(n,p)C^{14}$	Ammonium nitrate (aq.)	?
$N^{15}(n,p)C^{15}$	Ammonium nitrate (aq.)	?

of the (d,p) reaction on B^{10}, so that it was expected by analogy that in the $B^{11}(\alpha,p)C^{14}$ reaction any C^{14} formed could be found in the gas space surrounding the target after bombardment. The gas space from an intensive bombardment of B_2O_3 by 16-Mev. α particles was examined, but no long-lived activity was found ascribable to either C^{14} or to any new long-lived isomeric nuclei which might be formed by reactions such as $B^{11}(\alpha,p)C^{14}$, $B^{10}(\alpha,p)C^{13*}$, $B^{11}(\alpha,n)N^{14}$, and $B^{10}(\alpha,n)N^{13*}$.

Similar negative findings resulted when the gas-space products from deuteron bombardment at 18 Mev. and from α-particle bombardment at 32 Mev. were tested.[18b]

Finally, in experiments in collaboration with Dr. S. Ruben, bombardment of graphite by deuterons of 7 to 8 Mev. of energy on a probe target for a total of 4800 μa. hours produced a weak radioactivity which was shown to be isotopic with carbon.[19,20] Shortly thereafter, these bombardments were repeated using graphite enriched with C^{13}. Increased yields of radioactive carbon were noted, indicating that the new radioactivity was, in all likelihood, C^{14} produced by the (d,p) reaction on C^{13}. Calculations by the writer using the known cross section for this reaction and the yields of C^{14} from the C^{13}-enriched graphite indicated that the magnitude of the half-life was of the order of 10^3 years. The sign of the particles emitted during radioactive decay was ascertained to be negative, as had been expected.

The success of these experiments ended the search for other long-lived isotopes in elements like hydrogen and nitrogen. Almost simultaneously,

[18b] These experiments were performed by Dr. E. Segrè, personal communication 1940.

[19] Ruben, S., and Kamen, M. D., *Phys. Rev.* **57**, 549 (1940).

[20] Kamen, M. D., and Ruben, S., *Phys. Rev.* **58**, 194 (1940).

however, Alvarez and Cornog[21] both noted that a radioactive body isotopic with hydrogen arose from bombardment of water with deuterons and showed the new isotope to be H^3 (see p. 272 ff.). Thus, within a few months after the scientific world had somewhat ruefully concluded that development of tracer techniques would be seriously handicapped because useful radioactive tracers for carbon, hydrogen, oxygen, and nitrogen did not exist, C^{14} and H^3 were discovered and the situation greatly improved.

There remained, however, the problem of production. It was evident that to produce samples of C^{14} in quantities large enough to be of use as a tracer extremely long and expensive bombardments would be required. The basis for cheap production was found in an unexpected source—the $N^{14}(n,p)C^{14}$ reaction with slow neutrons. It had been thought that emission of protons would be favored over emission of γ rays in the competing reaction $N^{14}(n,\gamma)N^{15}$ only at high neutron energies. The cross section for fast neutron reactions was not high compared to a typical (d,p) reaction like $C^{13}(d,p)C^{14}$. Furthermore, not only would large amounts of nitrogenous material be required to capture the fast neutrons produced as secondary radiation during operation of the cyclotron, but it was unknown whether the recoiling carbon would emerge in a chemical form which could be extractable in the presence of very large amounts of target material. For these reasons, little attention was paid at first to the large bottle of ammonium nitrate solution which had been placed near the cyclotron as part of the general program for detecting C^{14}. However, when the radioactive vapor which was obtained from these solutions merely by aspirating CO_2-free air through the bottles and thence through a combustion tube to oxidize any radioactive oxides of carbon entrained in the air stream was examined, it was found,[19, 20] surprisingly, that samples of C^{14} with total activities many orders of magnitude greater than those found with the (d,p) reaction were obtained. It was also established that the $N^{14}(n,p)C^{14}$ reaction proceeded with high yield, even with slow neutrons. In fact, it could be estimated that its cross section was as high as many of the most productive (d,p) reactions. Further, the radioactive carbon could be recovered with great ease merely by aspiration of the ammonium nitrate solutions placed to capture the slow neutrons. This process was replaced after 1945 when the tremendous neutron fluxes of the nuclear reactor became available; C^{14} could be produced, with appropriate target materials like boron nitride, in quantities which ended further concern about shortages in its availability.

B. Preparation and Properties

As noted above, C^{14} is now produced almost entirely by means of the $N^{14}(n,p)C^{14}$ reaction in the uranium pile. The original target material was

[21] Alvarez, L., and Cornog, R., *Phys. Rev.* **56**, 613 (1939).

ammonium nitrate in solution, but this compound was rapidly discarded because of excessive decomposition induced by the great radiation flux in the pile. Procedures now are based on Be_3N_2 solid. Most of the recoil C^{14} can be recovered as methane and burned to CO_2. The isotope is supplied as $BaCO_3$ containing as much as 10 to 20% C^{14}.

It is of interest to calculate the specific activity of samples of carbon containing, for example, 10% C^{14}. The half-life is close to 5.0×10^3 yr. The number of active atoms required to give an activity of 1 mc. is $1.5 \times 10^{15} \times \tau_{1/2}$, if $\tau_{1/2}$ is expressed in years. One millicurie of C^{14} will require $1.5 \times 10^{15} \times 5.0 \times 10^3$, or 7.5×10^{18} atoms. One millimole of carbon, containing 10% C^{14}, weighs 12.2 mg. One milligram of such carbon contains $6.02 \times 10^{20}/12.2$, or 4.9×10^{19} atoms of which 10% are C^{14}. Hence, per milligram of carbon there are 4.9×10^{18} C^{14} atoms, or $4.9 \times 10^{18}/7.5 \times 10^{18} = 0.65$ mc. With G-M end-window counters as ordinarily used (geometry approximately 5%), such a sample could be diluted 10^6-fold and assayed with a precision of 5% in 1 hr. With instrumentation of higher efficiency (gas counting or scintillation detectors) this dilution factor could be increased up to twentyfold.

The reaction $N^{14}(n,p)C^{14}$ is of much theoretical interest because it is one of two known so far in which a proton is emitted in a slow neutron reaction. In general, as has been noted, emission of a heavy particle is favored over emission of a γ ray if the energy is available. This is true for all (n,p) reactions in which the product nucleus is a negative β-particle emitter with maximum β-ray energy less than the difference in mass between neutron and proton (see p. 31). The effect of the potential barrier, which opposes the escape of any positively charged heavy particle, renders proton emission unlikely unless the barrier is very low, as in nuclei of small atomic number. Thus, the only instances in which the conditions required for appreciable yields by an (n,p) process with slow neutrons are satisfied occur in the nitrogen and chlorine nuclei. It is of interest that these two reactions, $N^{14}(n,p)C^{14}$ and $Cl^{35}(n,p)S^{35}$, are of major importance in the formation of two of the most important tracer bodies, C^{14} and S^{35}.

When any nitrogenous material is irradiated in an environment containing oxygen atoms, it is found that the major fraction of the recoil C^{14} atoms appear as the simple oxides[22, 22a] of carbon, CO and CO_2. This is easily explicable if it is remembered that practically no combination likely to be formed will be more complicated than a free radical of the type C—H, C—OH, C═O, etc. All such combinations in a water environment

[22] See Yankwich, P. E., Rollefson, G. K., and Norris, T. H., *J. Chem. Phys.* **14**, 131 (1940), for an extended discussion of the recoil chemistry of C^{14} produced in the neutron irradiation of various nitrogenous materials.

[22a] See also Calvin, M., Heidelberger, C., Reid, J. C., Tolbert, B. M., and Yankwich, P. E., "Isotopic Carbon," pp. 5–7. Wiley, New York, 1949.

will tend to be hydrolyzed or oxidized to CO and CO_2. There is some possibility of hydration to formic acid also. On this basis, molecules such as CH_4 and CH_3OH would be much less frequently encountered as stable end products.

The half-life of C^{14} has been measured by a number of workers using absolute counting rates in conjunction with mass spectrometer determinations of C^{14} content in samples submitted for analysis. The grand average value appears to be 5568 yr.[23] The β-ray spectrum is simple, and γ-ray emission is not appreciable ($<5\%$ of the total radiation). The upper limit for the β-ray energy is given as 0.155 ± 0.0002 Mev.[24] This corresponds to a range of 28.0 ± 0.02 mg./cm.^{2}.

Natural carbon contains a very small but detectable quantity of C^{14} resulting from processes such as induced transmutation of atmospheric nitrogen by the cosmic ray neutron component. Thus methane from sewage (biological origin) shows a radioactivity[25] corresponding to 1×10^{-12} g./g. of carbon. This activity, although insignificant as a source of background has proved important in age determinations of biogeochemical material.[26]

C. Assay of C^{14}

The assay of C^{14} is totally unlike that of C^{11} because, although no decay corrections are needed, the softness of the C^{14} radiations makes corrections for absorption necessary. Material on the assay of C^{14} has already been presented to some extent in Chapter III, wherein data on the self-absorption of samples containing C^{14} have been shown (see Fig. 31).

C^{14} is well adapted to assay with electrometer ionization chambers of the conventional size because the β radiations emitted have an average range in air of approximately 4 to 6 cm. Consequently, these radiations can be assayed with high efficiency in the ionization volume of most ionization chambers. Similar remarks apply to electroscopes.[27]

The assay of C^{14} in solid samples can be carried out either internally or externally with the G-M tube. If a device such as the screen wall tube is

[23] See Hollander, J. M., Perlman, I., and Seaborg, G. T., *Revs. Mod. Phys.* **25,** 478 (1953). Later values appear to cluster around 5200 ± 200 yr. See Seliger, H. H., and Schwebel, A. *Nucleonics* **12,** No. 7, 54 (1954).

[24] Berggren, J. L., and Osborne, R. K., *Bull. Am. Phys. Soc.* **23,** 46 (1948); Levy, P. W., *Phys. Rev.* **72,** 248 (1947).

[25] Anderson, E. C., Libby, W. F., Weinhouse, S., Reid, A. F., Kirshenbaum, A. D., Grosse, A. V., *Phys. Rev.* **72,** 931 (1947).

[26] Libby, W. F., "Radiocarbon Dating". Univ. Chicago Press, Chicago, 1952.

[27] A modified Lauritsen electroscope of high sensitivity for assay of C^{14} has been described. See Henriques, F. C., Jr., and Margnetti, C., *Ind. Eng. Chem. Anal. Ed.* **18,** 417 (1946).

used,[28] the sample can be mounted on a cylinder, over the inside of which it is spread in a thin layer. In this fashion the maximal quantity of sample can be exposed to the sensitive volume of the tube counter with the minimum self-absorption. A number of devices for this purpose are commercially available. Most tracer samples can be assayed with sufficient precision and in comparatively short times by placement external to a bell-jar type of counter, despite the fact that the over-all sensitivity relative to internal counting is lower by a factor of 5 to 8.

It is a common procedure to mount C^{14} samples as solid preparations on a precision stage which can be brought into a fixed position close to the tube window. It is necessary also to know accurately the weight per unit area of the C^{14} sample assayed. In the preparation of "thin samples," i.e., samples of thickness less than the saturation value (see Chapter III, p. 89), the following procedure can be employed. The samples are mounted on thin aluminum or stainless-steel disks which are weighed accurately. The diameter of these disks is slightly less than that of the tube window. The carbon samples are prepared usually as $BaCO_3$, although they may also be assayed as Na_2CO_3 or as the metal salts of organic acids. The sample suspended as a thin paste in absolute alcohol is transferred to the weighed disk with an eye dropper and spread over the middle portion of the disk. The paste is smeared roughly homogeneously and dried in an oven at 100°C. The disk is then removed, the deposit cooled, moistened with absolute alcohol, and spread as homogeneously as possible with the ground end of a Pyrex stirring rod. The disk is placed in the oven and thoroughly dried. After being removed from the oven, it is allowed to equilibrate to room temperature and weighed. The procedure of drying and weighing is repeated until constant weight is obtained. The sample should be upended and tapped lightly to knock off any material which is held loosely, and which is in danger of falling off in subsequent operations, before endeavoring to attain constant weight. In general, the material is held quite firmly and can be removed only by direct rubbing.[29]

The area covered will, in general, be rather asymmetrical and must be estimated. There are a number of simple procedures possible. The sample disks can be ruled with standard coordinates and the area read off directly. A transparent disk with ruled coordinates can be placed directly above the disk and used in a similar fashion. The use of a planimeter has been found most convenient. In another procedure thin lens paper is used to absorb a small volume ($\sim$0.1–0.5 ml.) of the C^{14} solution. The area of the paper defines the area of the sample and the total weight is that of the sample

[28] Libby, W. F., and Lee, D. D., *Phys. Rev.* **55**, 245 (1939).

[29] See Chapter VI and VII in reference 22a for a complete discussion of sample preparation.

plus paper. There are many other procedures possible, of course. These mentioned have been tested in a variety of researches and found satisfactory.

Whenever sufficient material is available or the specific activity is high enough so that dilution with unlabeled material can be tolerated, it is best to resort to the counting of "thick" samples, i.e., samples thick enough (>25 mg./cm.^{2}), so that no appreciable increase in counting rate occurs with added sample thickness, the self-absorption correction is obviated, and uniformity in thickness is not critical (as discussed in Chapter III, p. 91). Numerous procedures are available for the preparation of thick samples [30-33] as $BaCO_3$, or for the strictly analogous situation of S^{35} assay with benzidine sulfate or barium sulfate.[34-36] As mentioned in Chapter V, these depend on centrifugation, evaporation or filtration of carbonate suspended in various media, usually 95% ethyl alcohol. The sample cup may be fashioned out of filter paper and the sample deposited by filtration through a Büchner-type filter; it may be placed in a pipe-and-screw-cap assembly and evaporated; or it may be fitted to the end of a pipe of such dimensions that the end of the pipe butts against the flange of the cup with a closure made by using rubber tubing after which the carbonate is deposited by centrifugation. Any of these types of procedure will prove satisfactory after a few preliminary practice preparations. Information on the use of materials other than carbonate is also available in the literature.[37]

In the use of carbonates as counting samples it should be remembered that, in the presence of moist CO_2-containing air labeled carbonate can be lost to the atmosphere through exchange reactions of the type

$$H_2C^{12}O_3 + BaC^{14}O_3 \rightarrow BaC^{12}O_3 + H_2C^{14}O_3$$

It has been shown that, after 65 hr. of exposure to moist CO_2 of 178 mg. of $BaCO_3$ containing 1800 ct./min., over 30% of activity is lost from the carbonate.[38] Elimination of moisture from the atmosphere obviates the exchange loss. Hence such samples should be stored in a desiccator after

[30] Hutchens, T. T., Claycomb, C. K., Cathey, W. J., and Van Bruggen, J. T., *Nucleonics* **7,** No. 3, 41 (1950).

[31] Armstrong, W. D., and Schubert J., *Anal. Chem.* **20,** 270 (1948).

[32] Dauben, W. G., Reid, J. C., and Yankwich, P. E., *Anal. Chem.* **19,** 828 (1947).

[33] Roberts, J. D., Bennett, W., Holroyd, E. W., and Fugitt, C. H., *Anal. Chem.* **20,** 904 (1948).

[34] Tarver, H., and Schmidt, C. L. A., *J. Biol. Chem.* **130,** 67 (1939).

[35] Henriques, F. C., Jr., Kistiakowsky, G. B., Margnetti, C., and Schneider, W. G., *Ind. Eng. Chem. Anal. Ed.* **18,** 349 (1946).

[36] Hendricks, R. H., Bryner, L. C., Thomas, M. D., and Ivie, J. O., *J. Phys. Chem.* **47,** 469 (1943).

[37] Hogness, J. R., Roth, L. J., Leifer, E., and Langham, W. H., *J. Am. Chem. Soc.* **70,** 3840 (1948).

[38] Armstrong, W. D., and Schubert, J., *Science* **106,** 403 (1947).

preparation. Some details on the evaporation loss from $BaCO_3$ samples during preparation in various atmospheres and by varying procedures have been reported.[39]

The problem of bringing the sample into a precisely fixed and uniform position with relation to the tube counter window can be solved in a variety of ways, depending on the manner of sample mounting. The simplest procedure is to construct a sturdy shelf arrangement on which the sample can be placed and slid into position under the window. Sometimes it is essential that the sample be as close to the window as possible without danger of contamination. The sample area must not exceed three-fourths of the window area and should be well centered to avoid errors arising from exposure of different portions of the sensitive volume with different samples. It should be emphasized that the G-M counter should be checked for variations in response on a day-to-day basis by means of a stable standard with a β-ray source such as C^{14}-labeled polystyrene, available from atomic energy establishments and commercial sources. The counting tube is calibrated with this sample in the standard position for every series of assays.

In Chapter V it was explained that assay of C^{14} in the gas phase can be used to advantage, especially when large amounts of material of low specific activity are encountered, or when very small quantities of labeled material must be handled. Various procedures for gas-phase assay of C^{14} are given in Table 29. Details can be found in references quoted or in the monograph by Glascock.[39a]

Counting of C^{14} in the liquid phase is most desirable in biological work. Unfortunately, this cannot be done with any of the instruments described above because they depend on detection of radiations which penetrate the sample material. The energy of the C^{14} radiation is so low that excessive self-absorption occurs in even the thinnest liquid samples which can be made suitable for use with ionization chambers or pulse counters.[47] The

[39] See pp. 122–126 in reference 22a.

[39a] Glascock, R. F., "Isotopic Gas Analysis for Biochemists." Academic Press, New York, 1954.

[40] Brownell, G. L., and Lockhart, H. S., *Nucleonics* **10,** 26 (1952).

[41] Bernstein, W., and Ballentine, R., *Rev. Sci. Instr.* **21,** 158 (1950).

[42] Van Slyke, D. D., Steele, R., and Plazin, J., *J. Biol. Chem.* **192,** 769 (1951).

[43] Bradley, J. E. S., Holloway, R. C., and MacFarlane, A. S., *Biochem. J.* **57,** 192 (1954).

[44] Barker, H. A., *Nature* **172,** 631 (1953).

[45] Crothorn, A. R., *Nature* **172,** 632 (1953).

[46] Brown, S. C., and Miller, W. W., *Rev. Sci. Instr.* **18,** 496 (1947).

[47] A procedure using formamide as a solvent for C^{14}-labeled compounds has been described by Schwebel, A., Isbell, H. S., and Karabinos, J. V., *Science* **113,** 465 (1951). This method is based on the principle of the "infinitely thick sample" and is claimed to be more rapid and less tedious than the plating methods based on the

TABLE 29

SOME METHODS FOR GAS-PHASE ASSAY OF C^{14}

Assay Instrument	Molecular Form	Sensitivity	Ref.
1. Ionization chamber and vibrating-reed electrometer	CO_2	Chamber holds 11 mM. of CO_2 and detects down to 5×10^{-5} μc.	40
2. Ionization chamber–Lauritsen electroscope	CO_2	Chamber holds 20 mM. of CO_2 and detects down to 5×10^{-5} μc.	27
3. Proportional counter, using counting mixture of methane and CO_2 at atmospheric pressure	CO_2	Sensitive volume, 100 ml.; maximum volume CO_2, 13 ml. (N.T.P.); counter background ~100 ct./min.; greater volumes possible	41, 42
4. Proportional counter, CO_2 used alone	CO_2	Similar to (3), maximum pressure of CO_2, 20 cm. Hg	43
5. Proportional counter with C^{14}-acetylene as counting gas	C_2H_2	Three-liter counters; efficiency ~75%; background ~30 ct./min., using anticoincidence setup	44, 45
6. G-M counter with CS_2-CO_2 counting mixture; Neher-Pickering quenching circuit	CO_2	CO_2 at 70 cm. Hg pressure; background ~50 ct./min.; similar to (3)	46

obvious instrument to exploit, then, is the scintillation detector which converts the soft C^{14} radiation in a volume of liquid to a more easily detectable radiation—visible light. The light so produced can be detected with the sensitive photomultiplier tube. It is probable that the scintillation method can be developed in the near future to provide a reasonably routine sensitive method for C^{14} assay in liquids. Some results in the literature[48] already indicate that, with suitable compounds and scintillation media, C^{14} can be counted in the scintillation detector with efficiencies between 60 and 70%.*

Radioautographic detection of C^{14} has also been reported.[48a]

use of $BaCO_3$. The sample geometry is given as about 1%. Another procedure with liquid samples in a stainless-steel cup covered with a thin-walled aluminum foil and placed inside a gas-flow counter has been described for S^{35}-labeled material [see Walser, M., Reid, A. F., and Seldin, D. W., *Arch. Biochem. and Biophys.* **45**, 91 (1953)]. This method could also be adapted to use with C^{14}-labeled compounds.

[48] Hayes, F. N., and Gould, R. G., *Science* **117**, 480 (1953).

[48a] Cobb, J., and Solomon, A. K., *Rev. Sci. Instr.* **19**, 441 (1948).

* See Appendix 5.

D. Synthesis of Organic Intermediates for Tracer Carbon Studies

The availability of the stable isotope C^{13} and the long-lived isotope C^{14} makes possible preparation of almost any carbon-labeled compound for tracer studies by means of a large number of well-known synthetic procedures. Modifications in the standard recipes may be occasioned by the necessity for conservation and recovery of labeled reagents and by the need for conducting reactions on a micro or semimicro scale.

In the last few years an extensive literature has accumulated bearing on useful syntheses for compounds in which tracer carbon is incorporated. Tables 30–32 are intended to provide a partial summary of this literature. The reactions noted can be extended *ad lib.* to synthesize a large variety of compounds in addition to those listed. Yields given are referred to labeled reagent.

E. Biosynthesis of Labeled Carbon Compounds

The advantages of biosynthetic methods lie in the great versatility of living organisms and in particular of microorganisms. The elaboration of simple labeled molecules like carbon dioxide and acetic acid into a great variety of labeled metabolites is possible. Furthermore, in many instances the use of living organisms represents the only efficient procedure available for making a given compound.

A number of problems arise when one attempts to design biosynthetic procedures for production of a given labeled compound. First, it is necessary to select an organism which can accumulate a practical quantity of the desired compound. Second, it is required that culture conditions be established for optimal yields. Third, and most important, one must devise procedures for isolation and purification of what may be small amounts of the compound in high yield. In most instances it is also necessary to determine the distribution of labeling isotope in the molecule. These considerations present formidable obstacles in the general development of biosynthetic procedures. Nevertheless many useful compounds have been prepared without difficulty by the use of organisms or enzymic extracts. A brief description of some of these syntheses is presented in this section.

1. Biosyntheses of Carbohydrates and of Organic Acids from CO_2. Hexoses labeled either uniformly or predominantly in certain positions can be made photosynthetically by higher plants from labeled CO_2. Starting with $C^{13}O_2$ containing 7.26 atom per cent excess C^{13}, starch containing 7.05 atom per cent excess C^{13} was obtained[49] from the mesophyll cells of bean leaves (*Phaseolus vulgaris*). The plants were depleted of starch by being kept in the dark for 90 hr. Then the leaves were severed from the plant at

[49] Livingston, L. G., and Medes, G., *J. Gen. Physiol.* **31,** 75 (1947).

the base of the petiole and immediately exposed under strong illumination for 48 hr. to the labeled CO_2. Most of the CO_2 was taken up, approximately 67% appearing in starch or polysaccharide and most of the remainder in soluble carbohydrate. Thus, practically all the CO_2 was converted with less than 3% isotopic dilution to polymerized D-glucose, from which isolation of pure D-glucose could easily be achieved by standard acid hydrolysis and crystallization procedures.

Detailed procedures have been described[50] for the preparation of labeled starch, glucose, fructose, and sucrose from Turkish tobacco leaves. After exposure to labeled CO_2, the leaves were extracted with 80% alcohol. The insoluble residue contained the starch which could be isolated, purified,

TABLE 30

SYNTHESES WITH LABELED CO_2 OR CARBONATE

A. One-carbon compounds

Compound	Synthesis Scheme	Typical Yield (%)	Ref.
Carbon monoxide	Reduction with Zn	~ 100	1
	CO_2-CO exchange over tungsten filament	Depends on equil. mixture	2
Potassium cyanide	Reduction with NaN_3	75	3
	Reduction at high temperature with Mg powder followed by treatment with NH_3 gas	72	4
	Reduction with Zn dust in stream of NH_3 gas at high temperature	90	5
Guanidine hydrochloride	Ammonation followed by reaction of intermediate BaNCN with excess NH_4NO_3	60	6, 7
Potassium formate	High-pressure reduction of $KHCO_3$ with H_2	98	8
Methyl alcohol	High-pressure reduction with H_2 (cat.); methyl formate prepared from HCO_2K may also be used[7]	85	9
	Reduction with $LiAlH_4$	80	10
Methane	Reduction with H_2 (cat.)	96	11a
Urea	Reduction with NH_3	80	11b
	Reduction of $BaCO_3$ with Ba metal at high temperature to carbide, via reaction with barium amide to form cyanimide, followed by acid hydrolysis	95	11c
Cyanimide	See urea synthesis		11c
	Also: Thermal reduction of labeled $BaCO_3$ at 850°C. with NH_3		7

[50] Putman, E. W., Hassid, W. Z., Krotkov, G., and Barker, H. A., *J. Biol. Chem.* **173,** 785 (1948).

TABLE 30 (CONTINUED)
*B. Carboxylic acids**

Compound	Group labeled	Synthesis Scheme	Typical Yield (%)	Ref.
Acetic acid	—COOH	Grignard reaction with CH_3MgI	50–90	12–16
Propionic acid	—COOH	Grignard reaction with C_2H_5MgI	50–60	12, 13
Butyric acid	—COOH	Grignard reaction with C_3H_7MgI	50–60	12, 13
Caproic acid	—COOH	Grignard reaction with $C_5H_{11}MgI$	50–60	13
Isovaleric acid	—COOH	Grignard reaction with *i*-C_4H_9MgBr	84	17
Octanoic acid	—COOH	Grignard reaction with $C_7H_{15}MgBr$	81	18
Undecanoic acid	—COOH	Grignard reaction with $C_{10}H_{21}MgBr$	85	19
Lauric acid	—COOH	Grignard reaction with $C_{11}H_{23}MgBr$	95	20
Palmitic acid	—COOH	Grignard reaction with $C_{15}H_{31}MgBr$	90	19
Benzoic acid	—COOH	Grignard reaction with C_6H_5MgBr	85	21a, 21b
p-Anisic acid	—COOH	Grignard reaction with $CH_3OC_6H_4$-MgBr	84	22
Phenylacetic acid	—COOH	Grignard reaction with $C_6H_5CH_2$-MgCl	88	23
1-Naphthoic acid	—COOH	Grignard reaction with $C_{10}H_7MgBr$	82	24
2-Naphthoic acid	—COOH	Grignard reaction with $C_{10}H_7MgBr$	73	25
Veratric acid	—COOH	Halogen-metal interconversion with C_4H_9Li and $(CH_3O)_2C_6H_3Br$	90	26
9-Fluorenecarboxylic acid	—COOH	Reaction with fluorene and $(C_6H_5)_3$-CNa	70	27
p-Toluylbutyric acid	—COOH	Grignard reaction with *p*-toluylpropionylmagnesium bromide	—	28
Oxalic acid	—COOH	Reduction with K	50	29
p-Aminobenzoic acid	—COOH	Halogen-metal interconversion with $LiC_6H_4NLi_2$	48	30
Nicotinic acid	—COOH	Halogen-metal interconversion with NC_6H_4Li	27	30

* The major compound of importance prepared from carbonate and falling outside the category of carboxyl-labeled substances is acetylene, C_2H_2, prepared either by fusion of barium carbonate with magnesium (references 3, 31) with yields varying from 60 to 75%, or by direct reduction of CO_2 with barium (reference 32) with yields of 90–98%. Other important compounds derived in 75% yield from $C^{14}O_2$ are benzene (see reference 33) and butadiene-2,3-C^{14} (see reference 34).

REFERENCES FOR TABLE 30

1. Weinhouse, S., *J. Am. Chem. Soc.*, **70,** 442 (1948).
2. Bernstein, R. B., and Taylor, T. I., *Science* **106,** 498 (1947).
3. Cramer, R. D., and Kistiakowsky, G. B., *J. Biol. Chem.* **137,** 549 (1941); Adamson, A. W., *J. Am. Chem. Soc.*, **69,** 2564 (1947).
4. Loftfield, R. B., *Nucleonics* **1,** No. 3, 54 (1947); Abrams, R., *J. Am. Chem. Soc.*, **71,** 3835 (1949).

5. Adamson, A. W., *J. Am. Chem. Soc.* **69,** 2564 (1947); McCarter, J. A., *J. Am. Chem Soc.* **73,** 483 (1951).
6. Marsh, N. H., Lane, L. C., and Salley, D. J., quoted in Calvin, M., Heidelberger, C., Reid, J. C., Tolbert, B. M., and Yankwich, P. E., "Isotopic Carbon," p. 158. Wiley, New York, 1949.
7. Zbarsky, S. H., and Fischer, I., *Can. J. Research* **27B,** 81 (1949).
8. Melville, D., Rachele, J. R., and Keller, E. B., *J. Biol. Chem.* **169,** 419 (1947).
9. Tolbert, B. M., *J. Am. Chem. Soc.*, **69,** 1529 (1947).
10. Nystrom, R. F., Yanko, W. H., and Brown, W. G., *J. Am. Chem. Soc.* **70,** 411 (1948).
11a. Beamer, W. H., *J. Am. Chem. Soc.* **70,** 3900 (1948).
11b. Bryan, C. E., U. S. Atomic Energy Comm., Isotopes Div., Circ. C-8 (1947).
11c. Murray, A., III, and Ronzio, A. R., *J. Am. Chem. Soc.* **71,** 2245 (1949).
12. Buchanan, J. M., Hastings, A. B., and Nesbett, F. B., *J. Biol. Chem.* **150,** 413 (1943).
13. Barker, H. A., Kamen, M. D., and Bornstein, B. T., *Proc. Natl. Acad. Sci. U. S.* **31,** 373 (1945).
14. Sakami, W., Evans, W. E., and Gurin, S., *J. Am. Chem. Soc.*, **69,** 1110 (1947).
15. Spector, L. B., U. S. Atomic Energy Comm., MDDC 532.
16. Lemmon, R. M., quoted, see p. 178 in reference 6.
17. Leslie, W. B., U. S. Atomic Energy Comm., MDDC 674.
18. Weinhouse, S., Medes, G., and Floyd, N. F., *J. Biol. Chem.* **155,** 143 (1944).
19. Dauben, W. G., *J. Am. Chem. Soc.* **70,** 1376 (1948).
20. Harwood, H. J., and Ralston, A. W., *J. Org. Chem.* **12,** 740 (1947).
21a. Dauben, W. G., Reid, J. C., and Yankwich, P. E., *Anal. Chem.* **19,** 828 (1947).
21b. Hussey, A. S., *J. Am. Chem. Soc.* **73,** 1364 (1951).
22. Reid, J. C., and Jones, H. B., *J. Biol. Chem.* **174,** 427 (1948).
23. Dauben, W. G., Reid, J. C., and Yankwich, P. E., see p. 180 in reference 6.
24. Dauben, W. G., *J. Org. Chem.* **13,** 313 (1948).
25. Heidelberger, C., Brewer, P., and Dauben, W. G., *J. Am. Chem. Soc.* **69,** 1389 (1947).
26. Reid, J. C., quoted on p. 183 in reference 6.
27. Collins, C. J., *J. Am. Chem. Soc.* **70,** 2418 (1948).
28. Collins, C. J., San Francisco Meeting, Div. Org. Chem. Abstracts, 1949, p. 54L.
29. Long, F. A., *J. Am. Chem. Soc.* **61,** 570 (1939).
30. Murray, A., III, Foreman, W. W., and Langham, W. H., *J. Am. Chem. Soc.* **70,** 1037 (1948).
31. Sakami, W., quoted, see p. 205 in reference 6.
32. Arrol, W. J., and Glascock, R., *Nature* **159,** 810 (1947).
33. Turner, H. S., and Warne, R. J., *J. Chem. Soc.* 789 (1953).
34. Mann, K. M., and Nystrom, R. F., *J. Am. Chem. Soc.* **73,** 5894 (1951).

and hydrolyzed to glucose. The soluble portion was worked over for glucose and fructose. The sucrose present was hydrolyzed to obtain optimal yields of glucose and fructose. Sucrose may best be made from the leaves of the plant *Canna indica*. Another plant which serves as a sucrose producer is the sugar beet.[50a]

[50a] See, also, Wick, A., Blackwell, M. E., and Hillyard, N., *Arch. Biochem. and Biophys.* **32,** 274 (1951), for a biosynthesis procedure with cantaloupe leaves.

TABLE 31

SYNTHESES WITH LABELED CARBON MONOXIDE* AND CYANIDE

Compound	Group Labeled	Synthesis Scheme	Typical Yield (%)	Ref.
Phosgene	—	Photochemical chlorination	90–100	1
Cyanogen chloride	—	Chlorination of AgCN	80	2
Formic acid	—	Hydrolysis of NaCN	50–60	3
Acetic acid	CH_3—	Reaction $Zn(CN)_2$ with phenol via hydroxybenzaldimine and *p*-cresol with oxidation to acetic acid	50	4
Malonic acid	Uniform	Hydrolysis of cyanoacetic acid	85	5
Succinic acid	—COOH	Reaction of NaCN with ethylene dichloride followed by acid hydrolysis	50	6
Lactic acid	—COOH	Reaction of HCN with acetaldehyde, followed by acid hydrolysis	95	7, 8
3-Hydroxypropionic acid	—COOH	Reaction of HCN with chloroethanol, followed by acid hydrolysis	30–40	9
Glycine	—COOH	Reaction of KCN with N-chloromethylphthalimide, followed by acid hydrolysis	81	8
		Strecker synthesis with KCN, formaldehyde, and NH_3	45	10
Alanine	—COOH	Strecker synthesis with KCN, acetaldehyde and NH_3	35	10
Ethylamine	—CH_2NH_2	Reaction of NaCN with CH_3I followed by reduction with H_2	70	11
Diazomethane		Hydrolysis of labeled nitrosomethyl urea from labeled cyanide via methylamine	30	12a, 12b

* Carbon monoxide is a more versatile reagent than indicated by the paucity of data available in labeling syntheses using it. The reader should consult the review article by Van Straten, S. F., and Nicholls, R. V. V., *Can. Chem. Processing* **31,** 947 (1947).

REFERENCES FOR TABLE 31

1. Huston, J. L., and Norris, T. H., *J. Am. Chem. Soc.* **70,** 1968 (1948).
2. Rice, C. N., and Yankwich, P. E., *in* Calvin, M., Heidelberger, C., Reid, J. C., Tolbert, B. M., and Yankwich, P. E., "Isotopic Carbon," p. 158. Wiley, New York, 1949.
3. Gurin, S., *in* "Symposium on Use of Isotopes in Biological Research," Univ. Chicago Press, 1947; see also p. 165 in reference 2.
4. Anker, H. S., *J. Biol. Chem.* **166,** 219 (1946).
5. Gal, E. M., and Shalgin, A. T., *J. Am. Chem. Soc.* **73,** 2398 (1951).
6. Allen, M. B., and Ruben, S., *J. Am. Chem. Soc.* **64,** 948 (1942).

7. Cramer, R. D., and Kistiakowsky, G. B., *J. Biol. Chem.* **137,** 549 (1941).
8. Sakami, W., Evans, W. E., and Gurin, S., *J. Am. Chem. Soc.* **69,** 1110 (1947).
9. Nahinsky, P., Rice, C. N., Ruben, S., and Kamen, M. D., *J. Am. Chem. Soc.* **64,** 2299 (1942).
10. Loftfield, R. B., *Nucleonics* **1,** No. 3, 54 (1947).
11. Kilmer, G. W., and du Vigneaud, V., *J. Biol. Chem.* **154,** 247 (1944).
12a. Hershberg, E. B., Schwenk, E., and Stahl, E., *Arch. Biochem.* **19,** 300 (1948).
12b. Heard, D. H., Jamieson, J. R., and Solomon, S., *J. Am. Chem. Soc.* **73,** 4985 (1951).

With labeled glucose and fructose available, a large number of labeled compounds can be synthesized. Thus, with the transglucosidase enzyme obtained from the microorganism *Pseudomonas saccharophila,*[51] it is possible to synthesize many disaccharides because this enzyme will couple glucose with almost any ketomonosaccharide as well as aldopentoses.[52] Successful syntheses of disaccharides consisting of glucose as one component with sugars such as ketoxylose and arabinose as the other have been reported. Thus, sucrose labeled either in the fructose or glucose moiety has been made by using the transglucosidase enzyme.[52a]

In *Neisseria meningitidis* there is found a maltose phosphorylase which catalyzes the reversible reaction

$$\text{Maltose} + \text{inorganic phosphate} \rightleftharpoons \beta\text{-D-Glucose-1-phosphate} + \text{D-glucose}$$

Using this enzyme it has been possible to prepare maltose with labeled carbon either in the reducing or in the nonreducing moiety.[52b]

Assimilation and degradation products made by living organisms from glucose, fructose, and sucrose can be synthesized efficiently once these labeled sugars are available. One may cite a few examples. The production of succinic acid by *Tetrahymena geleii* has been mentioned previously (see p. 294 *et seq.*). From labeled succinate, it is a simple matter to prepare in high yield labeled fumarate and malate, with the enzymes succinodehydrogenase and fumarase. The production of alcohol in 50% yield by yeast fermentation and of lactic acid in 70% yield by *Lactobacillus casei* is readily called to mind. Forty per cent recovery of isotopically labeled carbon in 2,3-butyleneglycol beginning with labeled glucose[53] has been achieved. It has been demonstrated[54] that good yields of fumaric acid can be ob-

[51] Doudoroff, M., Kaplan, N. O., and Hassid, W. Z., *J. Biol. Chem.* **148,** 67 (1943).

[52] Hassid, W. Z., Doudoroff, M., and Barker, H. A., *Arch. Biochem.* **14,** 29 (1947).

[52a] Wolochow, H., Putman, E. W., Doudoroff, M., Hassid, W. Z., and Barker, H. A., *J. Biol. Chem.* **180,** 1237 (1949).

[52b] Fitting, C., and Putman, E. W., *J. Biol. Chem.* **199,** 573 (1952).

[53] Adams, G. A., and Stanier, R. Y., *Can. J. Research* **23B,** 1 (1945).

[54] Foster, J. W., Carson, S. F., Ruben, S., and Kamen, M. D., *Proc. Natl. Acad. Sci. U.S.* **27,** 590 (1941).

TABLE 32
MISCELLANEOUS SYNTHESES

Compound	Carbon Labeled	Synthesis Scheme	Typical Yield (%)	Ref.
A. Alcohols, amides, halides, etc.				
Methanol	—	Methylation of potassium formate to methyl formate with methyl sulfate, followed by catalytic reduction with H_2	~ 50	1
Ethanol	1-C	Catalytic reduction with H_2 of carboxyl-labeled ethyl acetate	90–100	2
Ethanol	2-C	Catalytic reduction with H_2 of methyl-labeled ethyl acetate	90–100	2
Propanol	1-C	Reduction with H_2 of carboxyl-labeled ethyl propionate	95	2, 3
9-Fluorenylcarbinol	—CH_2OH	Reduction of methyl ester of carboxyl-labeled 9-fluorene carboxylic acid with $LiAlH_4$	80	4a
Glycerol	1,3-C	Condensation of nitromethane with C^{14}-paraformaldehyde, followed by reduction and diazotization	19	4b
Urea	—	Reaction of labeled phosgene with ammonia in toluene	90–100	5
Urethane	H_2N—C=O	Reaction of labeled urea nitrate with sodium nitrite and ethyl alcohol	40	6
Methyl iodide	—	Reaction of labeled methanol with hydriodic acid	85–90	1
Ethyl iodide	1-C	Reaction of 1-C-labeled ethanol with hydriodic acid	80–90	2
Ethyl iodide	2-C	Reaction of 2-C-labeled ethanol with hydriodic acid	80–90	2
Propyl iodide	1-C	Reaction of 1-C-labeled propanol with hydriodic acid	80–90	2
Propyl bromide	1-C	Reaction of 1-C-labeled propanol with phosphorus and bromine	76	3
Carbon tetrachloride	—	Photochlorination of labeled methane	92–94	7

TABLE 32 (CONTINUED)

Compound	Carbon Labeled	Synthesis Scheme	Typical Yield (%)	Ref.
		A. Alcohols, amides, halides, etc. (Continued)		
Ethylene dichloride	1,2-C	1-C-labeled ethyl amine (from reaction of labeled cyanide and methyl iodide followed by catalytic hydrogenation) converted to *quaternary* methyl ammonium base, pyrolyzed to 1,2-labeled ethylene and chlorinated	50	8
Acetamide	1-C	Carboxyl-labeled acetate converted to acetyl bromide followed by ammonation	50	9
Pyruvamide	1-C	Carboxyl-labeled acetyl bromide converted to acetyl cyanide with cuprous cyanide followed by acid hydrolysis	50	9
Nitromethane	—	Labeled methanol converted via methyl iodide by Victor Meyer reaction with silver nitrite; methyl nitrite (16% yield) formed as a side product	71	10a
Biotin	Ureido-C	C^{14}-phosgene reacted with diamino acid derived from biotin	~10	10b
		B. Aldehydes, ketones, and carbohydrates		
	Group Labeled			
Formaldehyde	—	(a) Catalytic oxidation of labeled methanol	45–55	11
		(b) Hydrolysis of chloro-C^{14}-methylacetate	60	12
Acetaldehyde	1,2-C	Acid hydrolysis of 1,2-labeled acetylene	75	13
p-Anisaldehyde	—C═O—H	Catalytic reduction of carboxyl-labeled anisoyl chloride	73	14
Veratraldehyde	—C═O—H	Catalytic reduction of carboxyl-labeled veratroyl chloride	73	15

TABLE 32 (CONTINUED)

Compound	Group Labeled	Synthesis Scheme	Typical Yield (%)	Ref.
B. Aldehydes, ketones, and carbohydrates (Continued)				
Acetone	—C=O—	Pyrolysis of carboxyl-labeled barium acetate	48	16, 17
Acetone	1,2,3-C	Pyrolysis of 1,2-labeled barium acetate	48	18a
1,3 Dihydroxyacetone	2-C^{14}	Nitromethane and labeled formaldehyde followed by Na reduction, bromination, and hydrolysis	22	18b
Diethyl ketone	—C=O—	Pyrolysis of carboxyl-labeled barium propionate	45	19, 20
Acetophenone	—C=O—	Reaction of acetic anhydride prepared from carboxyl-labeled sodium acetate with benzene in presence of aluminum chloride	90	19
		Reaction of diethyl sodium malonate with carboxyl-labeled benzyl chloride, followed by acid hydrolysis	54	20
		Reaction of carboxyl-labeled acetic acid with benzene in presence of aluminum chloride	79	21
Glucose	1-C	Labeled nitromethane via 1-nitro-1-deoxysorbitol (condensation with D-arabinose) to D-glucose	12–15 } 25%	10a*
Mannose	1-C	Same procedure as for 1-C-labeled glucose but via 1-nitro-1-deoxymannitol	5–8 } 25%	10a*
C. Various acids				
	Carbon Labeled			
Acetic	—CH_3	Reaction of CO_2 with methyl-labeled magnesium iodide, prepared with labeled methyl iodide	70–80	22
Acetic	—CH_3	C^{14}-methyl hydrogen sulfate reacted with cyanide to form nitrile which is then hydrolyzed	91	23a

* Both glucose and mannose are synthesized beginning with a single batch of nitromethane. The total yield of the nitro alcohols based on the nitromethane is 25%. Conversion of the nitro alcohols to the corresponding sugars is carried out by acid hydrolysis of the sodium salts of the nitro alcohols.

TABLE 32 (CONTINUED)

Compound	Carbon Labeled	Synthesis Scheme	Typical Yield (%)	Ref.
C. Various acids (Continued)				
Butyric	3-C	Reaction of 1-C-labeled ethyl iodide with diethyl sodium malonate, followed by hydrolysis	—	23b
Lactic	—COOH	Reaction of acetaldehyde with labeled cyanide	35–40	13
Lactic	2,3-C	Reaction of 1,2-labeled acetaldehyde (from labeled acetylene) with cyanide	40–50	13, 24
Pyruvic	—COOH	(a) Esterification of carboxyl-labeled lactate with *n*-butanol, followed by oxidation with acid permanganate to butyl pyruvate, followed by alkaline hydrolysis	60	24
	2,3-C	(b) As in (a) but with 2,3-labeled lactate	60	24
	2-C	(c) Carboxyl-labeled acetyl bromide reacted with cuprous cyanide to pyruvyl nitrile, followed by hydrolysis	45–55	24, 25a
	2-C	(d) 1-C^{14}-acetyl bromide reacted with cuprous cyanide to form nitrile, thence to pyruvoamide, and thence to pyruvate by hydrolysis	40	25b
Acetoacetic acid	1,3-C	(a) Carboxyl-labeled acetate condensed in presence of triphenylmethyl sodium via ethyl acetoacetate to sodium salt	25	24
	3-C	(b) Carboxyl-labeled methyl acetate reacted with magnesium and bromoethylacetate via ethylacetoacetate to sodium salt		20
	1-C	(c) As in (b) but with carboxyl-labeled bromoethylacetate from carboxyl-labeled ethyl acetate, treated with phosphorus and excess bromine	10	20

TABLE 32 (Continued)

Compound	Carbon Labeled	Synthesis Scheme	Typical Yield (%)	Ref.
C. Various acids (Continued)				
Mandelic acid	H–C–OH (carbon bearing H and OH)	(a) Carbonyl-labeled acetophenone oxidation with selenium dioxide via phenylglyoxal hydrate	75	21
		(b) Carboxyl-labeled benzoyl chloride reacted with diazomethane followed by bromination and hydrolysis	24	26
Glycine	—COOH	(a) Carboxyl-labeled acetate chlorination via chloroacetic and ammonation	70	27
	—H_2NCH_2	(b) As in (a) but with methyl-labeled acetate	70	27
	—COOH	(c) From carboxyl-labeled acetate to acetyl bromide via reaction with benzoyl bromide; then hydrolysis to bromoacetic acid and ammonation	55–60	28
Serine	—COOH	Carboxyl-labeled glycine via ethyl ester and benzoylation to ethyl hippurate, followed by condensation with ethyl formate, reduction with aluminum amalgam, and hydrolysis of N-benzoyl serine		29
Methionine	3,4-C	(a) 1,2-Labeled ethylene dichloride reacted with benzyl mercaptan and sodium, followed by reaction with ethyl sodiophthalimidomalonate, alkaline hydrolysis, and reduction with sodium	30	8
	—CH_3	(b) Methylation of 3-benzylhomocysteine with labeled methyl iodide in liquid ammonia	84	1

TABLE 32 (CONTINUED)

Compound	Carbon Labeled	Synthesis Scheme	Typical Yield (%)	Ref.
C. Various acids (Continued)				
Phenylalanine	1,2-C	1,2-Labeled glycine via hippurate and condensation with benzaldehyde to azolactone which is treated with phosphorus and hydriodic acid, yielding hydroiodide of phenylalanine; hydriodic acid removed with ammonia	50	29
3,4-Dihydroxyphenylalanine	—CH_2	Carbonyl-labeled vertraldehyde reacted with hydantoin, product reduced with sodium amalgam and hydrolyzed with barium hydroxide; demethylation with hydriodic acid	30	30
Tyrosine	—CH_2	Carbonyl-labeled *p*-anisaldehyde condensed with hydantoin, followed by reaction with phosphorus and hydriodic acid, and treatment with ammonia	35	14
Tryptophan	—CH_2	(a) Labeled formaldehyde reacted with dimethylindole, product condensed with acetaminomalonic ester in presence of dimethylsulfate followed by hydrolysis of ester with alkali, thermal decarboxylation, and acid hydrolysis	35	31
	—COOH	(b) 4-C-labeled hydantoin condensed with formaldehyde, reduced catalytically, and hydrolyzed	60	33
Lysine	2-C	Reductive acetylation of ethyl isonitrosocyanoacetate-2-C^{14}	70	32
Hydantoin	4-C	Carboxyl-labeled glycine, esterified, treated with potassium cyanate, followed by evaporation in presence of HCl	55	33

TABLE 32 (CONTINUED)

Compound	Carbon Labeled	Synthesis Scheme	Typical Yield (%)	Ref.
C. Various acids (Continued)				
Aminoadipic acid	6-C	Labeled cyanide converted to γ-chlorobutyronitrile and thence to aminoadipic acid via diethyl-(3-cyano-propylphthalimido) malonate		34
Alanine	2-C	From labeled cyanide, via chloroacetate, condensation with benzaldehyde, decarboxylation to cinnamic nitrile, hydrolysis to cinnamic acid, conversion via lithium cinnamate in presence of lithium methyl to benzylacetone, then via oxime to benzoylated amine, oxidation to benzoylalanine with resolution and hydrolysis	10–15	35
D. Various hydrocarbons, carcinogens, and steroids				
Toluene	1,3,5-C	From carbonyl-labeled pyruvate via condensation to tricarboxylic acid, decarboxylation first in acid and then with copper oxide and quinoline	45	36
Mesitylene	1,3,5-C	Carbonyl-labeled acetone condensed in sulfuric acid	10	16
1,2,5,6-Dibenzanthracene	9-C	Carboxyl-labeled 2-naphthoylchloride condensed with β-methylnaphthalene in presence of aluminum chloride, followed by Elbs pyrolysis	25	37
2-Methylcholanthrene	11-C	Carboxyl-labeled 1-naphthoylchloride condensed with diethyl cadmium compound (reaction product from reaction of 4-bromo-7-methylhydrindene, magnesium, and ethyl iodide); mixture treated with HCl followed by Elbs pyrolysis	38	38

TABLE 32 (CONTINUED)

Compound	Carbon Labeled	Synthesis Scheme	Typical Yield (%)	Ref.
D. Various hydrocarbons, carcinogens, and steroids (Continued)				
Phenanthrene	9-C	Wagner rearrangement of 10-C-labeled fluorenylcarbinol over phosphorus pentoxide		4
Cholestenone	3-C	Carboxyl-labeled phenyl acetate condensed with enol-lactone (from keto acid gotten by ozonization of cholestenone), followed by hydrolysis, decarboxylation, and cyclization in acid	51 (based on enol-lactone	39, 40
Testosterone	3-C	Same procedure as for cholestenone but beginning with ozonization of testosterone benzoate	48 (based on enol lactone)	40, 41
Dehydroisoandrosterone acetate	16-C	Introduction of labeled diazomethane in synthesis of Kuwada and Nakamura		42

REFERENCES FOR TABLE 32

1. Melville, D. B., Rachele, J. R., and Keller, E. B., *J. Biol. Chem.* **169,** 419 (1947).
2. Sakami, W., quoted in Calvin, M., Heidelberger, C., Reid, J. C., Tolbert, B. M., and Yankwich, P. E., "Isotopic Carbon," p. 202. Wiley, New York, 1949.
3. Fries, B. A., and Calvin, M., *J. Am. Chem. Soc.* **70,** 2235 (1948).
4a. Collins, C. J., *J. Am. Chem. Soc.* **70,** 2418 (1948).
4b. Schlenk, H., and De Haas, B. W., *J. Am. Chem. Soc.* **73,** 3921 (1951).
5. Yankwich, P. E., quoted in, see p. 158 in reference 2.
6. Skipper, H. E., Bryan, C. E., and Hutchinson, O. S., Atomic Energy Comm., Isotopes Div., Circular C-8, September, 1947.
7. Beamer, W. H., *J. Am. Chem. Soc.* **70,** 3900 (1948).
8. Kilmer, G. W., and du Vigneaud, V., *J. Biol. Chem.* **154,** 247 (1944).
9. Anker, H. S., *J. Biol. Chem.* **176,** 1333 (1948).
10a. Sowden, J. C., *Science* **109,** 229 (1949); also *J. Biol. Chem.* **180,** 55 (1949).
10b. Melville, D. B., Pierce, J. G., and Partridge, C. W. H., *J. Biol. Chem.* **180,** 299 (1949).
11. Tolbert, B. M., and Christenson, F., quoted in, see p. 166 in reference 2.
12. Jones, A. R., and Skraba, W. J., *Science* **110,** 332 (1949).
13. Cramer, R. D., and Kistiakowsky, G. B., *J. Biol. Chem.* **137,** 549 (1941).
14. Reid, J. C., and Jones, H. B., *J. Biol. Chem.* **174,** 427 (1948).
15. Reid, J. C., quoted in, see p. 199 in reference 2.

16. Grosse, A. V., and Weinhouse, S., *Science* **104,** 402 (1946).
17. Barker, H. A., and Kamen, M. D., *Proc. Natl. Acad. Sci. U. S.* **31,** 219 (1945).
18a. Wood, H. G., Werkman, C. H., Hemingway, A., Nier, A. O., and Stuckwisch, C. G., *J. Am. Chem. Soc.* **63,** 2140 (1941).
18b. Arnstein, H. R. V., and Bentley, R., *J. Chem. Soc.* 2385 (1951).
19. Shantz, E. M., and Rittenberg, D., *J. Am. Chem. Soc.* **68,** 2109 (1946).
20. Dauben, W. G., Reid, J. C., and Yankwich, P. E., quoted in, see p. 200 in reference 2.
21. Brown, W. G., and Neville, O. K., U. S. Atomic Energy Comm., MDDC 1168.
22. Tolbert, B. M., *J. Biol. Chem.* **173,** 205 (1948).
23a. Hess, D. N., *J. Am. Chem. Soc.* **73,** 4038 (1951).
23b. Sakami, W., quoted in, see p. 196 in reference 2.
24. Sakami, W., Evans, W. E., and Gurin, S., *J. Am. Chem. Soc.* **69,** 1110 (1947).
25a. Calvin, M., and Lemmon, R. M., *J. Am. Chem. Soc.*, **69,** 1232 (1947).
25b. Anker, H. S., *J. Biol. Chem.* **170,** 333 (1948).
26. Doering, W. v. E., Taylor, T. I., and Schoenewaldt, E. F., *J. Am. Chem. Soc.* **70,** 455 (1948).
27. Ostwald, R., *J. Biol. Chem.* **173,** 207 (1948).
28. Bloch, K., *J. Biol. Chem.* **179,** 1245 (1949).
29. Gurin, S., and Delluva, A. M., *J. Biol. Chem.* **170,** 545 (1947).
30. Reid, J. C., see p. 225 in reference 2.
31. Heidelberger, C., Gullberg, M. E., Morgan, A. F., and Lepkovsky, S., *J. Biol. Chem.* **175,** 471 (1948); **179,** 139 (1949).
32. Fields, M., Walz, D. E., and Rothchild, S., *J. Am. Chem. Soc.*, **73,** 1000 (1951).
33. Bond, H. W., *J. Biol. Chem.* **175,** 531 (1948).
34. Borsook, H., Deasy, C. L., Haagen-Smit, A. J., Keighley, G., and Lowry, P. H., *J. Biol. Chem.* **173,** 423 (1948); **176,** 1383 (1948).
35. Baddiley, J., Ehrensväard, G., and Nilsson, H., *J. Biol. Chem.* **178,** 399 (1949).
36. Hughes, D. M., and Reid, J. C., see p. 225 in reference 2.
37. Heidelberger, C., Brewer, P., and Dauben, W. G., *J. Am. Chem. Soc.* **69,** 1389 (1947).
38. Dauben, W. G., *J. Org. Chem.* **13,** 313 (1948).
39. Turner, R. B., *J. Am. Chem. Soc.* **69,** 726 (1947).
40. Turner, R. B., *J. Am. Chem. Soc.* **72,** 579 (1950).
41. Turner, R. B., *Science* **106,** 248 (1947).
42. Hershberg, E. B., Schwenk, E., and Stahl, E., *Arch. Biochem.* **19,** 300 (1948).

tained through the use of a special strain of the mold *Rhizopus nigricans*, and of citric acid by using another mold, *Aspergillus niger*.[55] Finally, there may be mentioned work[56] in which it has been shown that a strain of *Pseudomonas fluorescens* can produce α-ketoglutaric acid from glucose in such yields that one may predict 20% recovery of isotopic carbon in the acid after incubation with labeled glucose.

Most important, organic acids can be prepared from CO_2 with relative ease. Thus suspensions of *Escherichia coli* catalyze the reversible reaction

[55] Carson, S. F., Mosbach, E. H., and Phares, E. F., *J. Bacteriol.* **62,** 235 (1951).
[56] Lockwood, L. B., and Stodola, F. H., *J. Biol. Chem.* **164,** 81 (1946).

$H_2 + CO_2 \rightleftharpoons HCOOH$. At equilibrium with an atmosphere containing 95% H_2 and 5% CO_2, the concentration of formate is 0.003 M.[57] These organisms have been used to synthesize labeled formate from H_2 and labeled carbonate.[58]

Doubly labeled acetate can be prepared efficiently by exploiting *Clostridium aceticum* which catalyzes the reaction[59]

$$4H_2 + 2CO_2 \rightleftharpoons CH_3COOH + 2H_2O$$

Yields as high as 90% isotopically labeled acetate from labeled carbon dioxide are possible with practically no dilution. The formation of doubly labeled acetate during the fermentation of glucose in the presence of labeled carbonate by *Clostridium thermoaceticum*[60] has been mentioned previously (see p. 154ff.). This synthesis can be accomplished in 70 to 80% isotope yield with about twentyfold dilution. One may also recall the conversion of labeled CO_2 to doubly labeled acetate by *Butyribacterium rettgeri* during the fermentation of lactate.[61]

Preparation of dicarboxylic acids such as succinic, fumaric, malic, and the tricarboxylic acid, citric, from labeled CO_2, has been discussed above.

2. Biosyntheses of Higher Fatty Acids. There are available a number of organisms which carry on an efficient reductive condensation reaction between acetate on the one hand and a variety of fatty acids on the other to form higher homologs in the fatty acid series in good yields. Thus, beginning with uniformly labeled acetate, it is possible to derive uniformly labeled butyric and caproic acids during the fermentation of lactate by *Butyribacterium rettgeri*.[61] A better organism for this kind of synthesis is *Clostridium kluyveri*, which metabolizes ethanol and acetate, propionate or butyrate to *n*-butyrate, *n*-valerate, and *n*-caproate.[62] A wide diversity of labeled acids can be obtained with an appropriately labeled starting acid. Thus:

$$CH_3C^*OOH, \quad CH_3C^*H_2CH_2C^*OOH,$$
$$CH_3C^*H_2CH_2C^*H_2CH_2C^*OOH, \quad C^*H_3COOH,$$
$$C^*H_3CH_2C^*H_2COOH, \quad C^*H_3CH_2C^*H_2CH_2C^*H_2COOH,$$
$$C^*H_3C^*OOH, \quad C^*H_3C^*H_2C^*OOH, \quad C^*H_3C^*H_2C^*H_2C^*H_2C^*OOH,$$
$$CH_3CH_2CH_2C^*OOH, \quad CH_3CH_2CH_2C^*H_2CH_2COOH$$

[57] Woods, D. D., *Biochem. J.* **30**, 515 (1936).

[58] Utter, M. F., Lipmann, F., and Werkman, C. H., *J. Biol. Chem.* **158**, 521 (1945).

[59] Volcani, B. E., and Barker, H. A., personal communication; also Karlsson, J. L., Volcani, B. E., and Barker, H. A., *J. Bacteriol.* **56**, 781 (1948).

[60] Barker, H. A., and Kamen, M. D., *Proc. Natl. Acad. Sci. U.S.* **31**, 219 (1945).

[61] Barker, H. A., Kamen, M. D., and Haas, V., *Proc. Natl. Acad. Sci. U.S.* **31**, 355 (1945).

[62] Barker, H. A., Kamen, M. D., and Bornstein, B. T., *Proc. Natl. Acad. Sci. U.S.* **31**, 373 (1945).

Dilution of isotope need not be more than twofold, and with proper quantities of reagents yields of any particular acid up to 90% can be obtained. It should be noted that labeling is not entirely uniform. Thus, beginning with carboxyl-labeled acetate, one may find a labeled content slightly greater in the carboxyl of butyrate compared to the β carbon.

3. Biosynthesis of Amino Acids. Although a number of amino acids labeled in a variety of positions are available commercially, these preparations are quite expensive in many cases and of an inadequate specific activity for certain metabolic studies, particularly those undertaken with microorganisms. Uniformly labeled amino acids,[63] as well as amino acids labeled primarily in the carboxyl carbons,[64] can be made in good yield with relatively high specific activity, if microorganisms which derive all or a major part of their carbon from CO_2 are used. One specific biosynthetic procedure based on photoassimilation of CO_2 will be presented.

The green alga *Chlorella pyrenoidosa* grows well as a photoautotroph. The Carnegie group of investigators has obtained 1-mg. quantities of uniformly labeled amino acids with counting rates of 250 ct./sec./μg. after a 7-day growth in a labeled medium. Their procedure follows.[65]

The initial step was to prepare an inoculum of fresh, rapidly growing *Chlorella* adapted to a synthetic medium with CO_2 as the sole carbon source. A slant of *Chlorella pyrenoidosa*, A.T.C.C. No. 7516, was used to inoculate a medium composed of 2.5 g. of KNO_3, 5 g. of $MgSO_4 \cdot 7H_2O$, 2.5 g. of KH_2PO_4, 2.0 g. of NaCl, 5 mg. of $FeSO_4 \cdot H_2O$, and 1 l. of boiled tap water. After 2 weeks of growth with 5% CO_2 in air bubbled through the culture, and with good lighting, a culture density of 1.35 mg. dry weight per milliliter was obtained.

For growth with $C^{14}O_2$, a 250-ml. round flask sealed with a stopcock was prepared, containing 100 ml. of medium buffered at pH 5.5: 2.9 g. of KNO_3, 12 g. of KH_2PO_4, 1 g. of NaCl, 2 g. of Na_2SO_4, 10 mg. of $MgSO_4$, and 1 l. of boiled tap water. This buffer had sufficient capacity to keep the pH below 7 when the $C^{14}O_2$ was later added in a solution containing 3 mM. of NaOH. An inoculum of 5 mg. dry weight of *Chlorella* centrifuged from a sample of the stock culture and resuspended in the buffer was drawn into the partly evacuated flask, followed successively by 20 ml. of H_2O

[63] Frantz, I. D., Jr., Feigelman, H., Werner, A. S., and Smythe, M. P., *J. Biol. Chem.* **195,** 423 (1952), use the chemosynthetic sulfur oxidizer *T. oxidans*.

[64] *R. rubrum*, a photoheterotroph, grows well with CO_2 and ethanol. Amino acids produced will be labeled in carboxyl, or in other carbons, depending on whether CO_2 in the organic substrate is labeled. See Tarver, H., Tabachnick, M., Canellakis, E. S., Fraser, D., and Barker, H. A., *Arch. Biochem. and Biophys.* **41,** 1 (1952). See, also, Cutinelli, C., Ehrensvaard, G., Högstrom, G., Reis, L., Saluste, E., and Stjernholm, R., *Arkiv Kemi* **3,** 501 (1951).

[65] Roberts, R. B., Abelson, P. H., Cowie, D. B., Bolton, E. T., and Britten, R. J., *Carnegie Inst. Wash. Publ. No.* **607,** Chapter 4 (1951).

TABLE 33

TOTAL RADIOACTIVITY APPEARING IN THE FRACTIONS REMOVED DURING THE SEPARATION OF THE PROTEINS IN *Chlorella*[65]

Fraction	Counting Rate (ct./sec.)	Percentage of Original CO_2
	$\times 10^6$	
Original CO_2	11.9	100
Final suspension	11.9	100
Culture medium	1.93	16.2
Cold-TCA-soluble	1.15	9.7
Alcohol-soluble	1.86	15.8
Alcohol-ether-soluble	0.042	0.4
Hot-TCA-soluble	0.40	3.4
Acidified alcohol wash	0.11	0.9
Hydrolyzed protein	3.96	32.9
Total in fractions		79.3

containing 3 mM. of NaOH and 1.43 mM. of $C^{14}O_2$, and finally by 3 ml. of 1.0 *N* HCl to return the *p*H to 5.5.

The stopcock was carefully opened and the internal pressure was allowed to rise nearly to atmospheric pressure. The flask was then rotated at a few revolutions per minute about 6 inches from a 150-watt tungsten lamp with a $CuSO_4$ solution heat filter interposed. A fan was used to keep the temperature at about 25°C.

After a 7-day growth period the gas remaining in the flask was withdrawn and the residual CO_2 trapped in 0.2 *N* NaOH. Assay of the NaOH solution showed that only 0.1% of the original $C^{14}O_2$ remained free. As Table 33 shows, however, 16% of the original radioactivity remained in the supernatant fluid in nonvolatile form. This loss can be reduced by harvesting the cells before the $C^{14}O_2$ is completely utilized. Actively growing *Chlorella* release little organic material to the medium. In one case a mixed culture of yeast and *Chlorella* was used to recover the "waste" substances

The *Chlorella* were harvested by centrifugation and put through the following fractionation:

1. Ten to thirty milligrams wet weight of washed cells, usually as a pellet in a 50-ml. plastic centrifuge tube, was suspended in 4 ml. of 5% trichloroacetic acid (TCA) and transferred to small (8-ml.) glass test tubes. After 30 minutes at 5°C. the suspension was centrifuged and the supernatant fluid poured off. This was the *cold-TCA-soluble fraction.*

2. Any residual fluid was removed from the walls of the tube with a cotton swab, and the precipitate was suspended in 4 ml. of 75% ethanol. This suspension was kept for 30 minutes at 40° to 50°C. and then centrifuged. The supernatant fluid was poured off, giving the *alcohol-soluble fraction.*

3. The precipitate was suspended in 4 ml. of a solution containing 2 ml. of ether and 2 ml. of 75% ethanol. After 15 minutes at 40° to 50°C., the suspension was centrifuged. The supernatant fluid was the *alcohol-ether-soluble fraction* and could be combined with the alcohol-soluble fraction.

4. The precipitate was suspended in 4 ml. of 5% TCA and kept in a bath of boiling water for 30 minutes and centrifuged. The supernatant fluid was the *hot-TCA-soluble fraction.*

5. The remaining precipitate was washed free of residual TCA by suspending in acidified alcohol and centrifuging, then suspending in ether and centrifuging. The supernatant fluids from these two washes contained little radioactivity and were usually discarded. The precipitate was the principal *protein fraction.*

6. After a 1.0-ml. sample of the alcohol-soluble fraction was taken for radioactivity measurement, an equal volume (3 ml.) of ether was added to the remainder, then 3 ml. of water was added, causing a separation into two phases. The ether phase was taken off with a pipet. This extraction was repeated by adding 3 ml. more of ether, which was again removed after shaking and combined with the first ether fraction. This gave the *alcohol-soluble, ether-soluble fraction* and left the *alcohol-insoluble, ether-insoluble fraction.*

The original sample was chosen to give 10 to 30 mg. wet weight of cells. With samples smaller than 10 mg. the precipitates tended to break up and contaminate the supernatant fluids. In most cases samples larger than 30 mg. were not necessary and resulted in inefficient use of the radioactive material.

Table 33 shows the amounts of radioactivity appearing in the various fractions. All counting was done on 0.025-ml. samples taken with a Carlsberg constriction pipet. This type of pipet is sufficiently accurate and convenient for solutions of high radioactivity, since sampling losses are small.

It has been observed in every C^{14} *Chlorella* preparation, whether for production or for test, that a certain fraction of the radioactivity appearing in the final suspension is not accounted for by the total of the various fractions. It is suspected that this results from the destruction of carbohydrate materials present during hydrolysis of the "protein" fraction. Chromatography of the cold-TCA extract has shown that it contains principally glucose and fructose.

The alcohol extract contains all the strongly colored matter of the cell. It contains no appreciable quantity of protein (less than 1% of the total carbon). The amount of protein produced by *Chlorella* depends on the culture conditions and has varied in different runs. In one case the protein fraction contained 50% of the carbon of the cell.

As a test run, 70 mg. dry weight of *Chlorella* protein plus a small amount

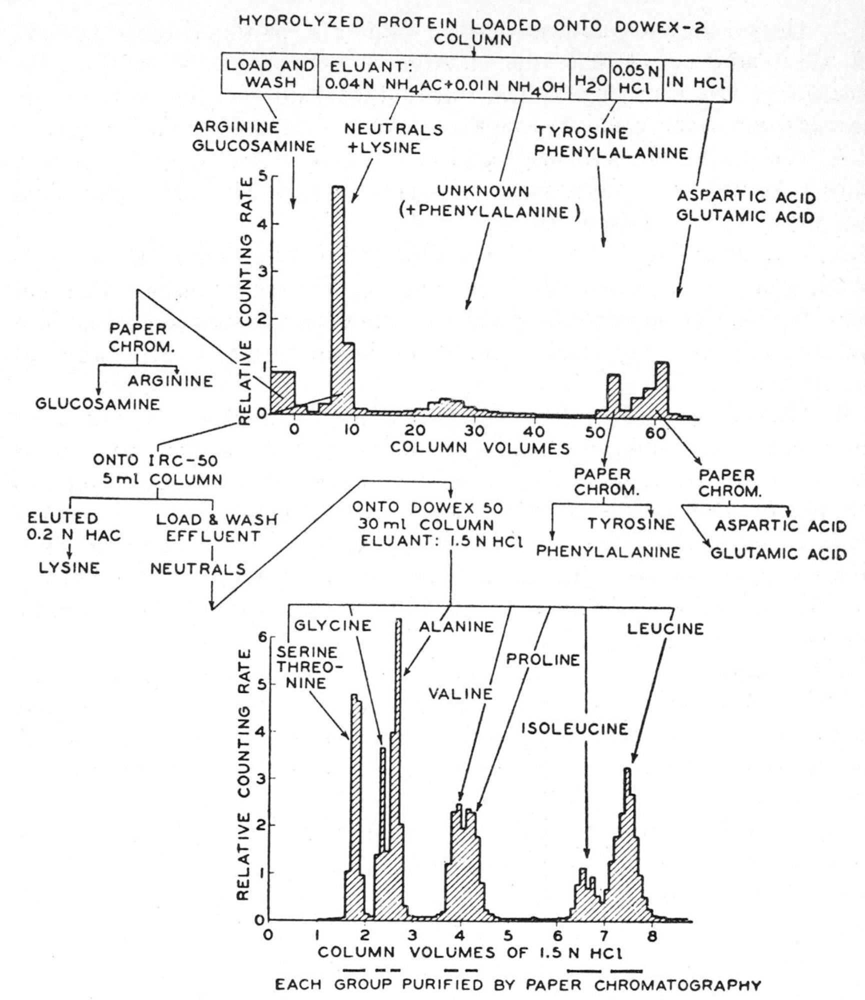

FIG. 67. Scheme for preparation of C^{14}-labeled amino acids.[65]

of radioactive protein was hydrolyzed and fractionated on ion-exchange columns (see Appendix 3).

Without serious loss of resolution in any of the separation steps, the amount of protein handled could be doubled by using the same volume of resin in the column steps. Thus, 50 mc. of C^{14} in the protein could be handled, with yields of about 3 mc. each of the more abundant amino acids.

The scheme for separation of the amino acids is shown in Fig. 67, along with histograms showing the degree of resolution obtained with the ion-exchange resin columns. The protein is hydrolyzed overnight in a sealed tube with 1 ml. of 6 *N* HCl at 100°C. After drying *in vacuo* to remove the

HCl, the hydrolyzate is taken up in 10 ml. of water and loaded on a 5-ml. Dowex 2 anion-exchange column prepared in the following way. Fine particles are removed by flotation, and the column is treated with 10 column volumes (c.v.) of 1 *N* NaOH. After washing with water to effluent neutrality, it is treated with 10 c.v. of a solution of 1 *N* sodium acetate containing 0.1 *N* NaOH, giving a column about half in the OH form and half in the acetate (Ac) form. The column is again washed with H_2O to effluent neutrality and finally loaded with hydrolyzate.

The effluent from the 10-ml. load solution and 10-ml. water wash contains 11% of the total radioactivity. This fraction is made up of arginine and hexosamine and a small percentage of unidentified materials. At this point elution is started with a solution 0.04 *N* in NH_4Ac and 0.01 *N* in NH_4OH. The first 10 to 20 c.v. of this eluant brings out the neutral group of amino acids: serine, glycine, threonine, alanine, proline, valine, leucine, isoleucine, and the basic lysine. These compounds account for 35% of the hydrolyzate. They are eluted in a band which varies in breadth and position from experiment to experiment. If the separation is not being followed on a fraction collector, it is well to collect 20 c.v. in this fraction. The next 20 c.v. brings out some phenylalanine and an unknown, ninhydrin-negative substance which is presumably a product of destruction of some of the amino acids during hydrolysis. At this point the column is washed with 2 c.v. of water to remove the NH_4Ac, and elution is begun with 0.05 *N* HCl. Ten column volumes of 0.05 *N* HCl removes phenylalanine and tyrosine. Elution with 1 *N* HCl then brings off aspartic acid and glutamic acid in 5 c.v. A clean-up with 6 *N* HCl brings off the remaining few per cent of the original radioactivity.

The particular advantages of this procedure for elution from Dowex 2 are: it avoids nonvolatile eluting agents; the small column requires only small volumes of solution; flow rates may be quite large without causing overlapping of groups; a fraction collector is not required, since the main groups come off with changes in eluant; and the aspartic-glutamic acid pair is almost free of other carbon compounds.

The solution containing the neutral group of amino acids is dried *in vacuo* to remove the NH_4Ac, brought up in 10 ml. of water, and loaded on a 5-ml. column of the carboxylic cation-exchange resin IRC-50. IRC-50 in the NH_4 form buffered to *p*H 4.7 holds up lysine and only traces of the other amino acids in this group. The lysine may be eluted with 0.2 *N* HAc and purified by paper chromatography. It is necessary to remove the lysine with this step, since it is so strongly bound to Dowex 50 that elution requires large volumes of strong acid eluant and is attended by appreciable losses.

Effluent from the IRC-50 load solution and a 10-ml. water wash is dried *in vacuo* and taken up in 3 ml. of 1.5 *N* HCl. This solution is loaded on a 30-ml. Dowex 50 column prepared in the following manner, essentially according to the prescription of Hirs *et al.*[65a]

Nearly colorless 200 to 400 mesh, 8% cross-linked Dowex 50[65b] is thoroughly washed with 2 *N* NaOH and heated to 70°C. in 2 *N* NaOH for several hours. It is then thoroughly washed with distilled water and treated with a large volume of 6 *N* HCl in a Büchner funnel. It is again thoroughly washed with distilled water and finally with 1.5 *N* HCl. After suspension in 1.5 *N* HCl, six 5-ml. portions of resin are poured into a 50-ml. burette of 1 $cm.^2$ cross-sectional area, and several column volumes of 1.5 *N* HCl are allowed to run through under gravity.

After the neutral group is loaded on in 3 ml. of 1.5 *N* HCl, the column is set up on an automatic fraction collector and 1.5 *N* HCl is allowed to run through at a rate of about 15 ml./hr. This flow rate is slow enough to achieve good resolution, and the complete separation can be performed conveniently overnight.

The neutral amino acids which are difficult to resolve completely by paper chromatography have the following R_F's in the phenol/water solvent: serine 0.32, glycine 0.40, threonine 0.51, and alanine 0.62. Fortunately, the first peak eluted from the Dowex 50 contains serine and threonine, and the second group glycine and alanine. As a result these pairs may be separated completely on a one-dimensional chromatogram, and the resolution of the column is more than adequate. The separation of leucine from isoleucine, however, depends entirely on the column.

No detailed study of the factors controlling the resolution of the Dowex 50 separation has been made. Experience indicates, however, that it is important to use a fine-grained resin (200 to 400 mesh) and low flow rates (½ c.v./hr. with a column of the dimensions given).[66]

4. Biosyntheses of Other Compounds. It is apparent that the possibilities of biosynthesis have only begun to be explored. There is little doubt that efficient syntheses from CO_2 of compounds such as chlorophyll, carotenoids, inositol, glycerol, and fats can be devised which far exceed in simplicity and yield comparable, if indeed possible, organic syntheses. There are rich possibilities in using various sugars and amino acids as starting materials in biosynthesis employing mutant strains which accumulate intermediary metabolites. Labeling of complex molecules such as many steroids, hormones, and pharmaceuticals may well require biosynthetic procedures for some time to come.

[65a] Hirs, C. H. W., S. Moore, and W. H. Stein, *J. Biol. Chem.* **195**, 669 (1952).
[65b] Lot L 2890-3 from the Dow Chemical Company, Midland, Michigan.
[66] Moore, S., and Stein, W. H., *J. Biol. Chem.* **192**, 663 (1951).

F. Degradation Methods

Procedures for establishing the placement of labeling isotope in organic molecules are of critical importance in most metabolic studies. A fairly large bibliography relating to methods for degradation of isotopic material has already accumulated. For the most part the substances involved have been carbohydrates and organic acids. In this section there will be given a short account of the procedures employed.

The carbohydrate most intensively studied has been glucose. Both chemical and biological methods have been employed. A chemical scheme based on oxidation with periodic acid will be described first.[67] In this procedure[68] compounds with adjacent hydroxyl groups or with an amino group adjacent to hydroxyl are cleaved as exemplified in the following equations:

$$RCHOHCHOHR' + HIO_4 \rightarrow RCHO + R'CHO + H_2O + HIO_3$$

$$RCHOHCHNH_2R' + HIO_4 \rightarrow RCHO + R'CHO + NH_3 + HIO_3$$

No such reaction occurs if the hydroxyl groups are not on adjacent carbon atoms. The reaction can also be extended to compounds with adjacent carbonyl groups, such as α-ketols, α-diketones, and α-ketonic aldehydes.

As applied to glucose, crystalline methyl-1-glucoside is prepared and treated with HIO_4. Carbon 3 appears as formic acid in the first oxidation. The remaining material is oxidized further. Carbon atoms 1, 2, 4, and 5 are recovered as formate, and carbon atom 6 appears as formaldehyde. The formaldehyde is removed continuously by steam distillation to minimize further oxidation to formate. The formaldehyde so isolated is oxidized in a separate step and recovered as CO_2. Another example of the use of HIO_4 is the degradation of adrenalin, the carbons of the $—CH_2NHCH_3$ group appearing as formaldehyde and formate.[69]

A biological procedure for degradation of glucose utilizing the lactic acid fermentation by *Lactobacillus casei* is available.[70, 71] In this procedure D-glucose is split to lactate. The lactate is then subjected to acid permanganate oxidation, the carboxyl groups, originally carbons 3 and 4 in the glucose appearing as CO_2. The other oxidation product is acetaldehyde in

[67] Wood, H. G., Lifson, N., and Lorber, V., *J. Biol. Chem.* **159,** 475 (1945).

[68] See review by Jackson, E. L., *in* "Organic Reactions" (R. Adams, ed.), Vol. 2, p. 341. Wiley, New York, 1944.

[69] Gurin, S., and Delluva, A. M., *J. Biol. Chem.* **170,** 545 (1947).

[70] Aronoff, S., Barker, H. A., and Calvin, M., *J. Biol. Chem.* **169,** 459 (1947).

[71] Shreeve, W. W., Feil, G. H., Lorber, V., and Wood, H. G., *J. Biol. Chem.* **177,** 679 (1949).

which the methyl group represents carbons 1 and 6 and the carbonyl group carbons 2 and 5. Oxidation of the acetaldehyde with sodium hypoiodite yields iodoform representing carbons 1 and 6 and formate containing carbons 2 and 5. Results obtained by this method have been checked by the chemical method based on HIO_4 oxidation. This procedure is a good example of the use of a specific biological degradation procedure. However, in this case as in others it is well to check the results by a purely chemical degradation scheme whenever possible.

Degradation of low-molecular-weight fatty acids has been accomplished invariably by pyrolysis or by wet combustion with permanganate, chromate, or hydrogen peroxide. Thus acetate,[72, 29] propionate,[73] and butyrate[74] have been converted to the barium salt and pyrolyzed to the corresponding ketone and carbonate. In this reaction, half of the original carboxyl appears as carbonate; the other half appears as the keto group in the ketone. Location of isotope in the ketone has been carried out, in the case of acetone formed from hydrolysis of barium acetate, by oxidation with hypoiodite to iodoform and acetate. Acid chromate oxidation of propionate results in acetic acid and CO_2, the latter originating from carboxyl of propionate, whereas alkaline permanganate oxidation of propionate, β-hydroxypropionate, or lactic acid results in a mixing of carbons with carboxyl contributing carbon to both reaction products,[73-76] carbonate and oxalate. Oxidation of butyric acid with peroxide[77] gives carbonate arising from carbon 1 and acetone from carbons 2, 3, and 4.

In addition to pyrolysis and oxidation procedures, there are available a number of decarboxylation methods. A general reaction involves the catalytic action of copper oxide, or copper chromate when boiling carboxylic acids in quinoline. Yields are variable[78] but may run as high as 95 %. Another reagent applicable in this connection is aniline citrate.[79] Alpha-amino acids are specifically decarboxylated by means of the well-known reagent triketohydrindene hydrate (ninhydrin).[80] An example of the use of

[72] Barker, H. A., and Kamen, M. D., *Proc. Natl. Acad. Sci. U.S.* **31,** 219 (1945).

[73] Wood, H. G., Werkman, C. H., Hemingway, A., Nier, A. O., and Stuckwisch, C. G., *J. Am. Chem. Soc.* **63,** 2140 (1941).

[74] Barker, H. A. Kamen, M. D., and Bornstein, B. T., *Proc. Natl. Acad. Sci. U.S.* **31,** 373 (1945).

[75] Nahinsky, P., and Ruben, S., *J. Am. Chem. Soc.* **63,** 2275 (1941).

[76] Nahinsky, P., Rice, C. N., Ruben, S., and Kamen, M. D., *J. Am. Chem. Soc.* **64,** 2299 (1942).

[77] Wood, H. G., Brown, R. W., Werkman, C. H., and Stuckwisch, C. G., *J. Am. Chem. Soc.* **66,** 1812 (1944).

[78] Dauben, W. G., Reid, J. C., Yankwich, P. E., and Calvin, M., *J. Am. Chem. Soc.* **68,** 2117 (1946).

[79] Edson, A. W., *Biochem. J.* **29,** 2082 (1935).

[80] Frantz, I. D., Loftfield, R. B., and Miller, W. W., *Science* **106,** 544 (1947).

hydrazoic acid (Schmidt reaction) for decarboxylation of benzoic and mandelic acids is available.[81] The use of bromine in decarboxylation of silver salts of fatty acids, i.e., acetate suspended in CCl_4, has been noted.[82]

The chemistry of benzimidazole derivatives of aliphatic acids can be applied to the elaboration of methods for characterization of isotopic fatty acids. In one such procedure,[82a] acetate is converted to 2-methylbenzimidazole by reaction with *o*-phenylenediamine. The imidazole derivative is degraded by heating with benzaldehyde, followed by treatment of the reaction product with permanganate. The original methyl group of acetate is recovered as CO_2, and the original carboxyl appears as benzimidazole.

Most of the methods described above, or variants thereof, are also applicable to hydroxy acids, polycarboxylic acids, keto acids, and esters. Thus lactic acid can be degraded by acid permanganate oxidation to acetaldehyde and CO_2, or by acid chromate oxidation to acetate and CO_2, the carboxyl carbon in either procedure giving rise to carbonate. Lactic acid may also be degraded by means of its benzimidazole derivative, as described for acetic acid.[82a] Acetoacetic acid when refluxed with mercuric sulfate in dilute acid solution is decomposed, giving carbonate from carboxyl carbon and acetone.[83] Pyruvic acid can be oxidized with ceric sulfate[84] to give carbonate from carboxyl carbon, or rather specifically by the use of the enzyme decarboxylase.[85] Alpha-ketoglutaric acid when oxidized with acid permanganate decarboxylates to carbonate and succinate, the carbonate being derived from the carboxyl group adjacent to the keto group.[86] A special procedure in the case of the ester of a keto acid is the thermal decomposition of ethyl pyruvate to carbon monoxide (derived from the carbethoxyl group) and ethylacetate.[87]

The important group of polycarboxylic acids involved in carbohydrate degradation, i.e., succinic, fumaric, malic, oxalacetic, and citric acids, has been worked over extensively by a variety of degradation procedures, many of which involve exploitation of specific enzyme systems. Thus, succinic dehydrogenase can be used to dehydrogenate succinate to fumarate, which can be converted in turn by fumarase to malic acid. Both fumaric and malic acids are easily oxidized by acid permanganate. Fuma-

[81] Doering, W. v. E., Taylor, T. I., and Schoenewaldt, E. F., *J. Am. Chem. Soc.* **70,** 455 (1948).

[82] Sprinson, D. B., *J. Biol. Chem.* **178,** 529 (1949).

[82a] Roseman, S., *J. Am. Chem. Soc.* **75,** 3854 (1953).

[83] Weinhouse, S., Medes, G., and Floyd, N. F., *J. Biol. Chem.* **155,** 143 (1944).

[84] Utter, M. F., Lipmann, F., and Werkman, C. H., *J. Biol. Chem.* **158,** 561 (1945).

[85] Cf. Carson, S. F., Ruben, S., Kamen, M. D., and Foster, J. W., *Proc. Natl. Acad. Sci. U.S.* **27,** 475 (1941).

[86] Evans, E. A., and Slotin, L., *J. Biol. Chem.* **141,** 439 (1941).

[87] Calvin, M., and Lemmon, R. M., *J. Am. Chem. Soc.* **69,** 1232 (1947).

rate forms carbonate and formate in a molar ratio 3:1, respectively, the formate deriving from one of the methylene carbons. Malic acid is converted to acetaldehyde and carbonate, the acetaldehyde representing the two central carbons of malate. By the use of malic dehydrogenase, malate can be converted to oxalacetate which in turn is easily decarboxylated, i.e., with aniline citrate, the carbonate deriving from the carboxyl group β to the keto group. All these methods are routine in biochemistry, and details can be found in most standard laboratory texts.

Chemical degradation of labeled succinic acid by means of the Curtius rearrangement has been reported.[29] In this reaction the diurethan is prepared via methylation with diazomethane, treatment with hydrazine, ethylation, and finally refluxing with hydrobromic acid which results in decarboxylation and formation of ethylene diamine. Pyrolysis of the barium salt *in vacuo* can be used to recover the carboxyl carbon as carbonate.[88]

Diazomethane is also a useful reagent in preparation of esters of fatty acids prior to stepwise degradation by the method of Barbier and Wieland.[89] In this procedure, which involves oxidative removal of one carbon at a time, beginning with carboxyl, the fatty acid is converted to the methyl ester using diazomethane. It is then caused to react with a Grignard reagent to form a substituted carbinol which is dehydrated to the corresponding unsaturated compounds. This can be oxidized with chromic oxide in glacial acetic acid to form the next lower fatty acid homolog and carbonate. The reaction sequence for caproic acid is:

$$\text{Caproate} \xrightarrow{CH_2N_2} \text{Methyl caproate} \xrightarrow{\text{2phenyl—Mg—Br}} \text{Amyl diphenyl carbinol} \xrightarrow{-H_2O} \text{1,1-Diphenyl hexene-1} \xrightarrow{CrO_3} \text{Valeric acid}$$

The carbonate formed derives from the original carboxyl. This reaction has been used to degrade labeled caproic acid.[90]

A good example of a degradative analysis for placement of label is afforded in researches on precursors in biosynthesis of uric acid.[91] After purification of uric acid isolated from pigeon excreta, alkaline MnO_2 is used to oxidize uric acid to CO_2, urea, and glyoxylic acid. In this degradation, CO_2 is derived from carbon 6, urea from carbons 2 and 8, and the aldehyde and carboxyl carbon of glyoxylic acid from carbons 4 and 5,

[88] Kushner, M., and Weinhouse, S., *J. Am. Chem. Soc.* **71**, 3358 (1949).

[89] Lane, J. F., and Wallis, E. S., *J. Am. Chem. Soc.* **63**, 1674 (1941).

[90] Stadtman, E. R., Stadtman, T. C., and Barker, H. A., *J. Biol. Chem.* **178**, 677 (1949).

[91] Sonne, J. C., Buchanan, J. M., and Delluva, A. M., *J. Biol. Chem.* **166**, 395 (1946); **173**, 69, 81 (1948).

respectively. The CO_2 is determined directly. The urea carbon is converted to carbonate by means of the enzyme urease; the glyoxylic acid, after isolation as semicarbazide, is oxidized by acid permanganate to carbonate (originally carbon 5) and formate (originally carbon 4). The formate is oxidized with mercuric oxide to carbonate. Another portion of the uric acid is split with potassium chlorate, forming urea (containing carbon 8) and alloxan. The alloxan is converted to alloxantin with H_2S which then can be oxidized to CO_2 (from carbons 4, 5, and 6) and urea (from carbon 2). In this way some cross checks on the isotopic assay of various positions may be obtained.

It is obvious that the variety of degradative procedures available is practically limitless and that these few remarks concerning application to labeled molecules constitute the sketchiest kind of presentation of possibilities. It can be expected that the literature relating to degradation methods will grow rapidly and to such an extent as to present a formidable problem to the reviewer in the very near future. The particular importance of tracer methods in this connection lies in the possibility of checking mechanisms postulated for degradation reactions. Thus, in the thermal decomposition of ethyl pyruvate it has been simple to show that the carbon monoxide formed originated from the carbethoxyl carbon atom rather than from the keto carbon.[87]

G. Radiation Hazards

The long half-life of C^{14} requires that it be handled as carefully as radium and other long-lived radioactive bodies. A general discussion of health physics as related to C^{14} is outside the scope of this book. It can be appreciated that tolerance activities depend on the form of carbon compound administered and its path of ingestion in so far as these govern the metabolic turnover and tissue concentrations of C^{14} which result. The isotope is most often supplied as $BaCO_3$, which is insoluble, and hence is a considerable hazard if it happens to be rubbed into skin or ingested by breathing into the lungs. Any such insoluble material should be handled with gloves in a place free of drafts and well enclosed.

Breathing of $C^{14}O_2$ does not appear to be a major hazard. A single inhalation of as much as 3 mc. would probably not be injurious if followed by immediate return to breathing of nonradioactive air.[92]

In tracer work, C^{14} is most often encountered as CO_2 or carbonate. Experiments have been made[93] on (1) the degree of exchange of inspired CO_2

[92] Brues, A. M., and Buchanan, D. L., *Cold Spring Harbor Symposia Quant. Biol.* **13,** 52 (1943).

[93] See Skipper, H. E., *Nucleonics* **10,** No. 2, 40 (1932), for a review of experimental data on hazards involved in the use of C^{14}.

with blood bicarbonate, (2) retention of bicarbonate carbon in blood and other tissues, (3) possible localization in sensitive portions of tissues, (4) stability of $BaCO_3$ particles retained in the lung after inhalation of $BaCO_3$ aerosols, and (5) possible carcinogenic effects arising from C^{14} radiation.

Adult mice were injected intraperitoneally with an amount of $NaHC^{14}O_3$ solution which corresponded to 50 mc. man-equivalents. From studies on retention of this material in mouse tissues it has been concluded that such a dose would probably be tolerable even though for a period of several months after injection certain bone-shaft cells might be receiving doses of 0.1 to 0.15 r./day which are upwards of three to four times the average tolerance dose recommended (0.05 r./day). High levels (50 mc. man-equivalents) as bicarbonate or 700 mc. man-equivalents as formate have shown no effects on the pattern of deaths from spontaneous leukemia.

Experiments with $BaCO_3$ aerosols indicate that C^{14} is eliminated quite rapidly and that solid $BaCO_3$ is less dangerous than might be expected. Loss of C^{14} is rapid because of exchange with lung CO_2 induced in the moist environment of the lung.

It seems safe to conclude that the use of C^{14} in tracer work is not attended by any appreciable hazards as long as the form in which it is encountered is primarily carbonate.[93a]

[93a] Ingestion of C^{14} as material used in synthesis of vital cell structures may be seriously hazardous. As an example, H. A. McQuade, M. Friedkin, and A. A. Atchison in *Nature* **175,** 1038 (1955) have noted significant increases in frequency of chromosome aberrations when onion root tips are treated with 2-C^{14}-thymine deoxyriboside, a precursor of deoxyribonucleic acid thymine.

CHAPTER XI

THE ISOTOPES OF OXYGEN, NITROGEN, PHOSPHORUS AND SULFUR

1. THE OXYGEN ISOTOPES

A. INTRODUCTION

The discovery of the rare isotopes O^{17} and O^{18} by Giauque and Johnson[1] in 1929 was crucial to the development of tracers because it led to a reinvestigation of the existence of the stable heavy-hydrogen isotope and accelerated the discovery of the rare stable isotopes of both carbon and nitrogen. The most recent measurements of the isotopic composition of oxygen[2] indicate the ratio O^{16}/O^{18} to be 489.2 ± 0.7 (corresponding to atom per cent abundance for O^{18} of 0.204) and the ratio O^{17}/O^{18} to be 2670 ±20. No radioactive isotope of oxygen is sufficiently long-lived to be useful in tracer work. One, O^{14}, is a positron emitter with $\tau_{1/2} = 76.5$ sec.; another, O^{15}, is a positron emitter with $\tau_{1/2} = 118$ sec.; the third, O^{19}, is a negatron emitter with $\tau_{1/2} = 29$ sec.[3] Only the isotope with mass 18 is useful as a tracer.[3a]

B. PREPARATION AND ASSAY

Fractional distillation[4] or electrolysis of water or heavy-water residues is the basis for production of O^{18}. The few isotope-exchange reactions which might be exploited have not come into general use. O^{18} is relatively expensive and difficult to obtain. The commercial sources vary from one country to another, but information can be obtained by writing the appropriate atomic energy establishment. In the United States, O^{18} water

[1] Giauque, W. F., and Johnson, H. L., *Nature* **123,** 318 (1929).

[2] Nier, A. O., *Phys. Rev.* **77,** 789 (1950).

[3] See Hollander, J. M., Perlman, I., and Seaborg, G. T., *Revs. Mod. Phys.* **25,** 469 (1953).

[3a] Some recent investigations of the chemical shift in the nuclear magnetic resonance of O^{17} indicate that it may be possible to use this isotope, despite its low abundance. O^{17} is the only stable oxygen isotope with a nuclear magnetic moment. Apparatus now appears to be available with sufficient sensitivity to exploit this property of O^{17}. See Weaver, H. E., Tolbert, B. M., and LaForce, R. C., *J. Chem. Phys.* **23,** 1956 (1955).

[4] Dostrovsky, I., Llewellyn, D. P., and Vroman, B. H., *J. Chem. Soc.* p. 3509 (1952).

with an O^{18} content of 10.3% and O^{17} content of 0.7% is available. These percentages represent an enrichment about fiftyfold over normal oxygen.

The assay of O^{18} can be based on density measurements of water, as discussed in Chapter III, but is best carried out with the mass spectrometer. The following ions are observed with molecular oxygen: $(O^{16}O^{18})^+$, $(O^{16}O^{17})^+$, and $(O^{16}O^{16})^+$, corresponding to mass peaks 34, 33, and 32, respectively. Mass peaks for the atomic ions at masses 18, 17, and 16 are not used because they contain contributions from ions produced from the water vapor which is likely to be present in the instrument. Water itself cannot be used because of the complicated pattern of ions it forms in the ion source and because of "memory" effects resulting from its tenacious persistence in the instrument from one run to the next.

Fortunately, it is possible to equilibrate water O^{18} with the oxygen of CO_2 by means of the exchange reaction[5]

$$H_2O^{18}{}_{(l)} + CO^{16}O^{18}{}_{(g)} \rightleftharpoons H_2O^{16}{}_{(l)} + CO^{16}O^{18}{}_{(g)} \quad (1)$$

This reaction proceeds rapidly enough at room temperature; at slightly acid pH it takes 4 hr. to reach equilibrium. By using a hot platinum wire the time required for equilibration can be shortened considerably.[6] Another method based on the use of sulfite to accelerate the exchange time has been described.[6a] Details can be found by consulting the literature.[6,7]

CO_2 is the most convenient molecular form to use because it can be easily separated from contaminating gases, particularly air, by condensation at low temperatures. It does not shorten filament life in the ion source of the mass spectrometer as does oxygen.

The ratio of peaks at mass 44, from the ion $(C^{12}O^{16}O^{16})^+$, to mass 46, from the ion $(C^{12}O^{16}O^{18})^+$ derived from CO_2 after equilibration, serves to determine the atom per cent excess of O^{18} in a sample of water prior to equilibration. The simplest procedure is to calibrate the mass spectrometer by determining ratios derived from water samples with known O^{18} content. The ratio, R, obtained from an unknown, is employed to calculate the atom per cent O^{18} by means of the equation $100R/(2 + R)$, (see Eq. *2*, Chapter IX). The atom per cent excess is obtained by subtracting the atom per cent O^{18} in ordinary water.

The atom fraction of O^{18} present in water before equilibration (N_0) can be calculated from measured or known values of atom fractions O^{18} in the water after equilibration (N_e) and in the CO_2 before and after equilibration (n_0 and n_e). If R_0 and R_e are the ratios of mass peak 44 to

[5] Cohn, M., and Urey, H. C., *J. Am. Chem. Soc.* **60,** 679 (1938).

[6] Dostrovsky, I., and Klein, F. S., *Anal. Chem.* **24,** 414 (1952).

[6a] Harrison, W. H., Boyer, P. D., and Falcone, A. B., *J. Biol. Chem.* **215,** 303 (1955).

[7] Cohn, M., *in* "Methods in Enzymology" (S. P. Colowick and N. O. Kaplan, eds.) Vol. IV. Academic Press, New York, in preparation.

TABLE 34

Methods for Assay of O^{18} in Various Compounds

Compound	Principle of method	Ref.
1. Inorganic phosphate	(a) Quantitative dehydration of KH_2PO_4 to metaphosphate and water; water equilibrated with CO_2 for assay	11
	(b) Reduction of $Ba_3(PO_4)_2$ with carbon to yield CO	12
2. Inorganic sulfate	Thermal reduction of $BaSO_4$ with carbon to CO_2 and assay of CO_2	13
3. Inorganic phosphate	Hydrolyze with alkaline phosphatase; bond ruptures between P and O, so that 3 of 4 oxygens come from organic phosphate and the 4th from solvent; assay resultant inorganic P as in (1).	14, 15
4. Organic compounds, general	(a) Pyrolysis over a cracking catalyst in H_2 to form water	16
	(b) Pyrolysis in N_2 over hot carbon to form CO which is converted to CO_2 with iodine pentoxide	17, 17a

46 before and after equilibration, then[6]

$$N_0 = \frac{1}{KR_e + 1} + \frac{b}{a}\left(\frac{1}{R_e + \frac{1}{2}} - \frac{1}{R_0 + \frac{1}{2}}\right)$$

where K is the equilibrium constant (2.088 at 25° C.[7]) for the exchange reaction (1), b is the number of millimoles of CO_2, and a is the number of millimoles of water taken for analysis.

There are small fluctuations in the normal abundance of O^{18} in water, depending on the source. Thus, sea water has a slightly higher O^{18} content than fresh water, owing to concentration during evaporation.[8] Water

[8] Gilfillan, E. S., Jr., *J. Am. Chem. Soc.* **56,** 406 (1934).

[9] Dole, M., *J. Chem. Phys.* **4,** 268 (1936).

[10] Dole, M., and Slobod, R. J., *J. Am. Chem. Soc.* **62,** 471 (1940).

[11] Cohn, M., *J. Biol. Chem.* **201,** 735 (1953).

[12] Cohn, M., and Drysdale, G. R., *J. Biol. Chem.* **216,** 831 (1955).

[13] Halperin, J., and Taube, H., *J. Am. Chem. Soc.* **74,** 375 (1952).

[14] Cohn, M., *J. Biol. Chem.* **180,** 771 (1949).

[15] Stern, S. S., and Koshland, D. E., Jr., *Arch. Biochem. and Biophys.* **39,** 229 (1952).

[16] Elving, P. J., and Ligett, W. B., *Chem. Revs.* **34,** 129 (1944).

[17] Doering, W. v. E., and Dorfman, F., *J. Am. Chem. Soc.* **75,** 5595 (1953).

[17a] Most recently, D. Rittenberg and L. Ponticorvo have described what appears to be a general method for conversion of oxygen in organic compounds based on heating *in vacuo* in the presence of mercuric chloride, cuprous chloride or chlorine, see *Intern. J. Appl. Radiation and Isotopes* **1,** No. 3, 208 (1956).

derived from atmospheric oxygen or from oxygen of carbonate rocks is more markedly higher in O^{18} content than fresh-water oxygen.[9, 10] These variations are of no practical significance in biological tracer work with O^{18} water, however.

In tracer research O^{18} may be used in a variety of compounds including inorganic phosphate and inorganic sulfate and in various organic compounds. The isotope must be in a form suitable for assay in the mass spectrometer with such compounds as the starting point. Details of procedures available are given by Cohn.[7] In Table 34 the various methods are summarized.

In addition to these procedures, a number of special methods have been developed using O^{18} for specific compounds, including decarboxylation of carboxylic acids,[6a, 18–23] isotopic equilibration of alcohol oxgen with CO_2 through mediation of water obtained by dehydration of the sample,[24] and assay of hydroxyl oxygen in glucose using the reaction with phenylenediamine to liberate the hydroxyl oxygen as water.[25] When dilution is not too great, methods based on the general ability of oxygen compounds like ketones, aldehydes, and carboxylic acids to exchange with water oxygen under suitable conditions can be employed.

C. The Use of O^{18} as a Tracer in Oxygen Metabolism

Oxygen is important biologically not only in its molecular form and as water but also in an endless variety of organic compounds. It can be expected that it will figure in almost as many tracer researches as do carbon and hydrogen. There is a major drawback: biochemical processes are all accomplished in a water environment. All atomic groupings of importance in biology (carboxyl, carbonyl, phosphate, sulfate, etc.) contain oxygen which can exchange with water oxygen. The use of tracer oxygen is predicated on the assumption that such exchange occurs at a low rate compared to the metabolic reaction being studied. Unfortunately, the literature is of little help in designing experiments with O^{18} in so far as predicting nonenzymic exchange rates for any given situation. Each experiment must include careful controls for actual determination of *in vitro* exchange and must not rely on assumptions based on findings in the literature.

Some notion of the inconsistencies in the published reports may be

[18] Bentley, R., *J. Am. Chem. Soc.* **71,** 2765 (1949).
[19] Bentley, R., and Rittenberg, D., *J. Am. Chem. Soc.* **76,** 4883 (1954).
[20] Doherty, D. G., and Vaslow, F., *J. Am. Chem. Soc.* **74,** 931 (1952).
[21] Hunsdiecker, H., and Hunsdiecker, C., *Ber.* **75,** 291 (1942).
[22] Sprinson, D. B., and Rittenberg, D., *Nature* **167,** 484 (1951).
[23] Dauben, W. G., Reid, J. C., Yankwich, P. E., and Calvin, M., *J. Am. Chem. Soc.* **68,** 2117 (1946).
[24] Anbar, M., Dostrovsky, I., Klein, F., and Samuel, D., *J. Chem. Soc.* 155 (1955).
[25] Koshland, D. E., Jr., and Stein, S. S., *J. Biol. Chem.* **208,** 139 (1954).

gained by comparing statements by different workers on exchange of oxygen in KH_2PO_4 with water oxygen under apparently identical conditions. Titani *et al.*[26] reported that complete exchange occurred in 100 hr. at 100°C., but Winter *et al.*[27] saw essentially no exchange under the same conditions in 41 hr. Similarly, Hall and Alexander[28] did not discern any effect of alkali in promoting exchange between chlorate ion and water, but Halperin and Taube,[13] in contrast, reported that there was complete exchange between chlorate ion and water in alkaline solution and no exchange in acid.

Despite these contradictions, there appear to be a number of general statements about exchange reactions in water which have validity.[29] Thus, carboxylic acids exchange both oxygen atoms, only the undissociated form of the acid being involved.[5] In contrast, no rapid hydroxyl oxygen exchange is found in alcohols,[5] with the exception of those containing strongly polar groups such as the tertiary alcohol trianisylmethanol. Aldehydes and ketones[5] show a tendency to exchange oxygen with water fairly rapidly. The exchange in the case of ketones does not appear to go by way of enolization.

From these statements it would be expected (and appears to be true) that only the carbonyl oxygen of sugars would be exchangeable[30] and, in proteins, only the oxygen of free carboxyl groups.[31] In the inorganic oxyanions, resistance to exchange seems to increase, the nearer the central element is to the upper right corner of the periodic system and the more oxygen atoms are coordinated.[28] Thus, iodate exchanges rapidly, but chlorate not at all under the same conditions (neutral or acid *p*H).[28] Sulfite exchanges rapidly, but sulfate does not.[13, 28]

Enzyme action frequently increases the *in vitro* rate of exchange in a given reaction. Thus, it has been noted[18] that the exchange of oxygen between the carboxyl of fatty acids and water at neutral *p*H is catalyzed by acetylcholinesterase. Similarly, chymotrypsin has been shown to catalyze exchange of oxygen between water and the carboxyl oxygen atoms of carbobenzyloxy-L-phenylalanine.[30] Catalysis of exchange has been observed between KH_2PO_4 and water oxygen in the presence of alkaline phosphatase.[15] These examples illustrate the basis for tracer research with O^{18} in showing clearly how exchange rates in the absence of enzymes or metabolic systems may be so slow that they cannot invalidate the use of O^{18} as a tracer for oxygen.

[26] Titani, T., Morita, N., and Goto, K., *Bull. Chem. Soc. Japan* **13,** 329 (1938).
[27] Winter, E. R. S., Carlton, M., and Briscoe, H. V. A., *J. Chem. Soc.* 131 (1940).
[28] Hall, N. F., and Alexander, O. R., *J. Am. Chem. Soc.* **62,** 3455 (1940).
[29] Reitz, O., *Z. Elektochem.* **45,** 101 (1939).
[30] Senkus, M., and Brown, W. G., *J. Org. Chem.* **2,** 569 (1938).
[31] Mears, W. H., *J. Chem. Phys.* **6,** 295 (1938).

The applications of O^{18} have been too many and varied to list here. Among some of the more interesting are researches on the effect of light in green plant respiration,[33] mechanism of the Hill reaction,[34] mechanism of phosphorylytic and phosphatatic cleavage,[25, 35] adenosine triphosphate formation in oxidative phosphorylation,[12] the source of the oxygen of respiratory CO_2,[36] and the mechanism of action of notatin[37] and uricase.[38] An interesting development is the demonstration that during oxidation reactions involving oxy-anions, like sulfite and chromate, there is actual transfer of atomic oxygen to the reducing agent.[13] One example of a similar mechanism in enzyme-catalyzed reactions is the case of *pyrocatechase*, the enzyme which splits the aromatic ring of catechol to *cis,cis*-muconic acid. When catechol is incubated with H_2O^{18} and unlabeled O_2 in the presence of the enzyme, no label is found in the product acid; however, when the oxygen is labeled and the water is not, the product acid is recovered with practically the same label content as that of the oxygen.[32]

In carrying out such researches it is often necessary to prepare O^{18}-labeled compounds. One of the most important of these is inorganic orthophosphate which is most easily prepared by heating a solution of KH_2PO_4 in water in a sealed tube at 120°C. for 8 days.[7] It will be noted that this procedure is based on the exchange of phosphate oxygen with water, a reaction about which there has been dispute.[26, 27] Organic phosphate compounds may be prepared by means of any reaction which incorporates inorganic phosphate without significant simultaneous exchange with water oxygen. Any of the intermediates in glycolysis can be prepared by enzymatic catalysis, as in the phosphorylase reaction or in the glyceraldehyde phosphate dehydrogenase reaction. Glucose-1-phosphate can be made enzymatically either by the use of muscle phosphorylase or by the exchange of the ester with inorganic phosphate in the presence of sucrose phosphorylase.[35] These are but a few examples of many which might be mentioned.

Probably the most widely used phosphate compound is adenosine triphosphate. Labeling with O^{18} in the terminal phosphate can be achieved by incubation of labeled inorganic phosphate and ATP with 3-phosphoglyceric acid, diphosphopyridine nucleotide, glyceraldehyde phosphate dehydrogenase, and 3-phosphoglycerate kinase.[11] Both of the two terminal groups can be labeled if adenylate kinase is also present in this mixture.[11]

[32] Hayaishi, O., Katagiri, M., and Rothberg, S., *J. Am. Chem. Soc.* **77,** 5450 (1955).

[33] Brown, A. H., *Am. J. Botany* **40,** 719 (1953).

[34] Brown, A. H., and Good, N., *Arch. Biochem. and Biophys.* **57,** 340 (1955).

[35] Cohn, M., *J. Biol. Chem.* **180,** 771 (1949).

[36] Lifson, N., Gordon, G. B., Visscher, M. B., and Nier, A. O., *J. Biol. Chem.* **180,** 803 (1949).

[37] Bentley, R., and Neuberger, A., *Biochem. J.* **45,** 584 (1949).

[8] Bentley, R., and Neuberger, A., *Biochem. J.* **52,** 694 (1952).

2. THE NITROGEN ISOTOPES

A. Introduction

Six isotopes of nitrogen are known, ranging in mass number from 12 to 17. N^{12} and N^{13} are positron emitters with half-lives of 0.0125 sec. and 9.93 min., respectively.[3] N^{16} and N^{17} are negatron emitters with half-lives of 7.35 and 4.14 sec., respectively.[3] N^{17} also emits neutrons.[39] The stable isotopes are N^{14} and N^{15}, the latter being present to the extent of 0.365 atom per cent.[2] The only radioactive isotope which could be or has been used[40, 41] in tracer research is N^{13}. It is made by a (*d,n*) reaction on carbon.[40] Most of the N^{13} remains trapped in the target and is recovered by combustion in a stream of oxygen. The gases obtained in this way include activity in the form of N_2 as well as oxides of nitrogen. The N_2 can be separated from the various radioactive oxides by condensation in liquid nitrogen. Procedures for quantitative recovery as NO_2 have been described.[41] Because of its short half-life and because it must be produced by a (*d,n*) reaction involving an installation like the cyclotron, N^{13} is too restricted in application to be considered of importance as a tracer for nitrogen.

The only practical tracer available is the rare stable isotope N^{15}, which was discovered by Naudé,[42] who observed isotope shifts in the band spectrum of NO corresponding to the existence of an isotope with mass 15.

B. Preparation and Assay of N^{15}

The commercial production of N^{15} has been based on the following exchange reaction between gaseous ammonia and solutions of ammonium salts, described by Thode and Urey.[43]

$$N^{15}H_{3\,(g)} + N^{14}H_4{}^+{}_{(aq)} \rightleftharpoons N^{14}H_{3\,(g)} + N^{15}H_4{}^+{}_{(aq)}$$

Later research has revealed another reaction which appears to be more efficient and which promises to make N^{15} cheaper to produce. This is the reaction between nitric oxide gas and nitric acid described by Spindel and Taylor.[44] The process involved is a good example of the use of exchange reactions in preparation of enriched stable isotopes and will be considered briefly.

The exchange column is shown diagrammatically in Fig. 68. Nitric acid entering the exchange vessel at the top percolates down over a packed

[39] Knable, K., Lawrence, E. O., Leeth, C. E., Moyer, B. J., and Thornton, R. L., *Phys. Rev.* **74,** 1217 (1948) (A).

[40] See, for instance, Ruben, S., Hassid, W. Z., and Kamen, M. D., *Science* **91,** 578 (1940).

[41] Ogg, R. A., Jr., *J. Chem. Phys.* **15,** 613 (1947).

[42] Naudé, S. M., *Phys. Rev.* **34,** 1498 (L) (1929).

[43] Thode, H. G., and Urey, H. C., *J. Chem. Phys.* **7,** 35 (1939).

[44] Spindel, W., and Taylor, T. I., *J. Chem. Phys.* **23,** 981 (1955).

column. It then drips into a refluxer in which it is converted to NO by reaction with SO_2. The SO_2 is oxidized to sulfuric acid in this reaction and accumulates in the refluxer. The NO gas rises up through the packed column, so that there is a flow of NO up and nitric acid down the column. Automatic regulation of the reaction between nitric acid and SO_2 is achieved by monitoring the position of the reaction zone with a photoelectric cell. The characteristic brown color of the NO_2 formed as a by-product of the reflux reaction is used to activate the photoelectric cell, which operates a relay to regulate the flow of SO_2. The regulation depends on the color developed in the reaction zone.

The nitric acid, enriched in N^{15}, which is produced is bled off at appropriate intervals. The by-product sulfuric acid can be saved and sold to recover some of the raw materials cost. Nitrous oxide, depleted in N^{15}, leaving the top of the column is oxidized to nitric acid in a waste refluxer and discarded. A number of these columns can be run in series to shorten the

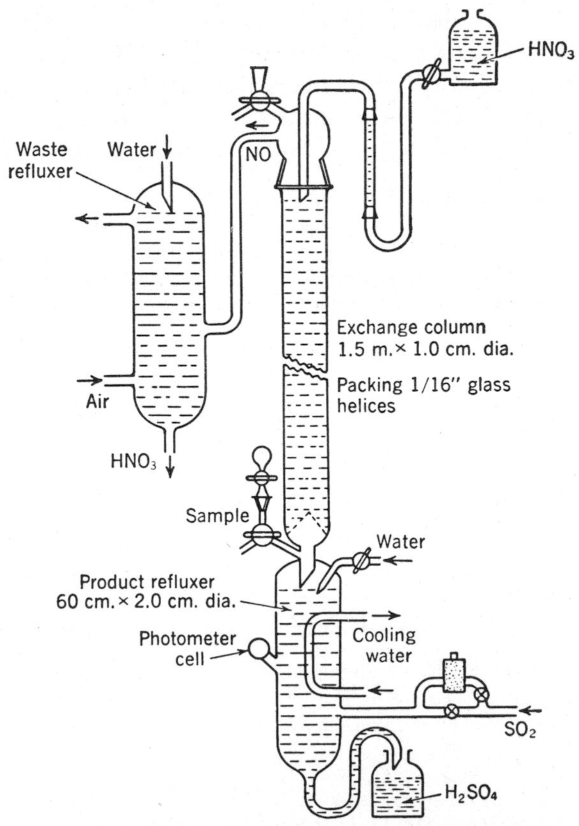

Fig. 68. Exchange system used for production of N^{15}. (After Spindel and Taylor.[44])

time required to reach equilibrium. With a column of the type shown in Fig. 68 the time needed is about 12 hr.

Assay of N^{15} requires a mass spectrometer. Nitrogen gas is the most satisfactory compound. Ion peaks at mass numbers 28, 29, and 30, corresponding to the ions $(N^{14}N^{14})^+$, $(N^{14}N^{15})^+$, and $(N^{15}N^{15})^+$, are used. The atom per cent of N^{15} can be determined from the ratio of peak 29 to 28, R, by the equation

$$\text{Atom per cent } N^{15} = 100R/(R + 2)$$

It is essential to ensure purity of the nitrogen; in particular, air must be excluded. Any leakage of air into the instrument will dilute the nitrogen and give a low reading. The presence of CO also must be avoided because it gives rise to the ions $(C^{12}O^{16})^+$ and $(C^{13}O^{16})^+$, which also have mass numbers 28 and 29. Hence, CO_2 must be excluded from the sample. The presence of CO_2 and air can be checked by examining peaks at 44 and 32, which arise from the ions $(C^{12}O^{16}O^{16+}$ and $(O^{16}O^{16})^+$. A peak at mass 40, due to argon, proves that air is present. It is best to calibrate the spectrometer for its response to air contamination by determining peak heights, using known amounts of air. The instrument can also be tested for proper functioning by measuring the intensity of peaks produced with highly enriched N^{15} at mass numbers 28, 29, and 30. If the intensity of the mass 28 peak is taken equal to unity, then the intensities of the peaks at mass 29 and 30 should be given by the expressions $[2P/(100 - P)]$ and $[P/(100 - P)]^2$ where P is the atom per cent N^{15}. The atom per cent in the sample for this test must be sufficiently high (25 to 75%) so that the peak at mass 30 can be measured readily.

In tracer researches, nitrogen is most often used in organic compounds like amino acids, purines, pyrimidines, quaternary bases, and porphyrins. For N^{15} assay the organic nitrogen must be converted to N_2. The process requires two steps. The first is the conversion of the organic nitrogen to ammonia. The second is the oxidation of ammonia to N_2. The first step can be accomplished by hydrolysis in some cases, but it is generally carried out by the Kjeldahl procedure. The ammonia solution produced by either treatment is allowed to react in the second step with alkaline hypobromite. The reaction follows:

$$2NH_3 + 3NaOBr \rightarrow N_2 + 3H_2O + 3NaBr$$

Details of these procedures are given in the literature.[45]

[45] See Rittenberg, D., *in* "Preparation and Measurement of Isotopic Tracers" (D. W. Wilson, A. O. Nier, and S. P. Reimann, eds.), pp. 31–39. J. W. Edwards, Ann Arbor, Michigan, 1946; also, Francis, G. E., Mulligan, W., and Wormall, A., "Isotopic Tracers," pp. 277–282. Athlone Press Univ. London, 1954.

TABLE 35

SOME LABELING PROCEDURES, STARTING WITH $N^{15}H_3$

Compound	Reaction	Ref.
Amino acids	(a) α-Keto acid reductively aminated with H_2 and $N^{15}H_3$ over palladium catalyst (*Knoop reaction*) (b) Phthalimide, made from $N^{15}H_3$ and phthallic acid (also available commercially), condensed with α-bromo acid, followed by hydrolysis to free amino acid (*Gabriel synthesis*)	47, 48
Creatine	Labeled cyanimide, from $N^{15}H_3$ and cyanogen bromide, condensed with sarcosine	49
Arginine	Labeled cyanimide converted to methyl isourea, which is condensed with ornithine (α-amino group protected by toluene sulfonyl radical) to give derivative of arginine which is freed by hydrolysis	50

C. THE USE OF N^{15} AS A TRACER FOR NITROGEN

A number of examples of the use of N^{15} have been described in previous chapters (see pp. 145–148).They indicate the importance of N^{15} as a tracer in biological researches. A discussion of many facets of research in protein metabolism illuminated by the use of N^{15} has been given by Sprinson.[46] In these researches, as in all others on nitrogen metabolism, it has usually been necessary to incorporate N^{15} into various organic compounds, beginning with $N^{15}H_3$. A large variety of N^{15}-labeled compounds are now available from commercial sources; a detailed discussion of synthetic procedures is, therefore, of only theoretical interest. A summary of some of the more useful reactions employed in labeling nitrogen compounds is given in Table 35. Many other procedures can be inferred from those presented in Tables 22 to 25 and 30 to 32 on the labeling of organic compounds with C^{14} and H^2. In addition, the biosynthetic procedures based on the use of microorganisms, such as those described in connection with C^{14}-labeling of amino acids (see p. 327ff.), can be used with no appreciable modifications for the preparation of N^{15}-labeled amino acids. It is necessary merely to use N^{15}-labeled ammonium salt in the medium employed to grow the organisms.

3. THE PHOSPHORUS ISOTOPES

A. PRODUCTION, PREPARATION, AND ASSAY

Phosphorus is a simple element with one stable isotope of mass number 31. Radioactive isotopes with mass numbers 29, 30, 32, 33, and 34 are

[46] Sprinson, D. B., *in* "Symposium on the Use of Isotopes in Biology and Medicine" (H. T. Clarke, ed.). pp. 182–209. Univ. Wisconsin Press, Madison, 1949.

known,[3] but only the isotopes with mass numbers 32 and 33 are sufficiently long-lived for use as tracers.

Both P^{32} and P^{33} are produced by the S(*n*,*p*)P reaction in the uranium pile. In one procedure,[51] the irradiated sulfur is fused and agitated with acetic acid. The resultant phosphoric acid is recovered from the melt by distillation. The residue is reworked for phosphate by leaching with dilute HCl. Purification is carried further by coprecipitation of phosphate with scavenger ferric hydroxide. The solids are dissolved in HCl, and the resulting solution is passed through a base-exchange resin column which removes cations. The effluent contains carrier-free phosphoric acid in dilute HCl.

P^{32} is a negative β emitter. Its upper energy limit is 1.701 Mev.,[3] and its average β energy is 0.70 Mev.[52] An aluminum foil about 0.5 mm. thick is required to cut the intensity of the radiations to half-value. No appreciable γ radiation is observed. The half-life is 14.30 days.[53] The assay of this isotope is uncomplicated by factors such as self-absorption, decay, and weak intensity and can be carried out with any of the instruments described in Chapter III. Sufficient sensitivity for practically all tracer researches is obtainable with a thin-window G-M counter or proportional flow counter. Self-absorption corrections do not become appreciable until thicknesses in excess of 50 mg./cm.^{2} are encountered. If assays are required on the same sample from day to day, corrections are best made with a standard P^{32} sample.

Samples of P^{32}, when first received from atomic-pile installations, need to be diluted before radioactivity assay can be attempted. These dilutions should not be made on so-called "carrier-free" phosphate samples using ordinary distilled water because appreciable amounts of P^{32} may be lost through adsorption on the walls of the container, precipitation as insoluble phosphate with traces of calcium or magnesium ions in the distilled water, and so on. Chemical-resistant glassware such as Pyrex or its equivalent should be used. A convenient diluting solution is 0.001 *M* H_3PO_4. Volatility of H_3PO_4 on subsequent evaporation under infrared lamps is not great enough to cause appreciable loss of P^{32}.

P^{33} is a negative β emitter with a half-life of 25 days.[54] The upper energy limit of its radiations is 0.27 Mev.[55] These radiations can be absorbed almost completely by 40 to 50 mg./cm.^{2} aluminum. No γ radiation in

[47] Schoenheimer, R., and Rittenberg, D., *J. Biol. Chem.* **127,** 285 (1939).
[48] Schoenheimer, R., and Ratner, S., *J. Biol. Chem.* **127,** 301 (1939).
[49] Bloch, K., Schoenheimer, R., and Rittenberg, D., *J. Biol. Chem.* **138,** 155 (1941).
[50] Bloch, K., and Schoenheimer, R., *J. Biol. Chem.* **138,** 167 (1941).
[51] Booth, A. H., quoted in *Chem. Eng. News* **27,** 1795 (1949).
[52] Caswell, R. S., *Phys. Rev.* **86,** 82 (1952).
[53] Cacciapuotti, B. N., *Nuovo cimento* **15,** 213 (1938).
[54] Jensen, E. N., and Nichols, R. T., *Phys. Rev.* **83,** 215 (1951).
[55] Sheline, R. K., Holtzman, R. B., and Fox, C. T., *Phys. Rev.* **83,** 919 (1951).

significant amounts can be detected. P^{33} accompanies P^{32} in pile-produced radioactive phosphorus. It is estimated that it accounts for about 1% of the initial radioactivity.[54] Because it has a longer half-life than P^{32}, the ratio of P^{33} to P^{32} increases with time, so that the apparent half-life of the radioactive phosphorus is greater than 14.3 days. If samples are measured with a thick enough foil to cut out the P^{33} radiations this uncertainty can be avoided. Decay corrections are best made when the standard is a sample prepared from the same material as that employed in the tracer research.

B. The Use of P^{32} as a Tracer

P^{32} was among the first of the radioactive isotopes to be discovered.[56] Because it was readily available and could be prepared and assayed with ease, it was used most extensively in early isotope work and even now is probably employed more frequently than any others. No attempt will be made to review the many applications to which it has been put. The reader can find a good summary in the monograph by Hevesy.[57]

The metabolic significance of phosphorus derives from its participation in esterification reactions (phosphorylation). These reactions may be used both for energy storage and for defining synthetic pathways. In general, it is necessary to have available a variety of P^{32}-labeled compounds for tracer studies on the mechanisms involved in phosphorus metabolism. With such labeled compounds, the interpretation of the results depends on the assumption that no exchange, other than that brought about metabolically, occurs between phosphate esters and inorganic phosphate. The validity of this assumption has been shown by a number of workers.[58,59] For instance, in one such research, Gourley[59] mixed P^{32}-labeled inorganic phosphate with such esters as glucose-1-phosphate-adenylic acid, 2-3,diphosphoglyceric acid, and adenosine triphosphate under conditions similar to those obtaining in blood plasma. No exchange was observed.

Useful procedures for synthesis of a variety of labeled phosphate compounds are summarized in Table 36.

Fractionation and isolation of phosphate esters and nucleosides are best accomplished by ion-exchange or chromatographic methods[61,70a,70b] or

[56] Amaldi, E., D'Agostino, O., Fermi, E., Pontecorvo, B., Rasetti, F., and Segrè, E., *Proc. Roy. Soc.* **149A,** 522 (1935).

[57] Hevesy, G., "Radioactive Indicators." Interscience, New York, 1948.

[58] Hevesy, G., and Aten, A. H. W., Jr., *Kgl. Danske Videnskab. Selskab. Biol. Medd.* **14,** 6 (1939).

[59] Gourley, D. R. H., *Nature* **169,** 192 (1952).

[60] Crane, R. K., and Lipmann, F., *J. Biol. Chem.* **201,** 235 (1953).

[61] Cohn, W. E., and Carter, C. E., *J. Am. Chem. Soc.* **72,** 4273 (1950).

[62] Kornberg, A., and Pricer, W. E., Jr., *J. Biol. Chem.* **191,** 535 (1951).

TABLE 36

INCORPORATION OF P^{32} IN VARIOUS PHOSPHORUS COMPOUNDS*

Compound	Procedure	Ref.
1. Adenosine triphosphate	(a) ARPPP*: phosphorylate ARPP using acetokinase and P^{32}-acetyl phosphate; purify by adsorption on Norite[60] or chromatography[61, 62]	63
	(b) ARPPP*: phosphorylate ARPP using P^{32}-phosphocreatine kinase	64
	(c) ARPP*P*: use either phosphorylation of AMP with rat liver mitochondria, or exchange with P*P* in aceto-coenzyme A-kinase system[65]	66
	(d) ARPP*P*: phosphorylate ARP as in 1(b)	64
	(e) ARPP*P: remove terminal P of ARPP*P* with hexokinase and excess glucose; phosphorylate product ARPP* with phosphopyruvate and corresponding kinase; purify on Norite[60]	66
	(f) ARPP*P: phosphorylate ARPP* with unlabeled phosphocreatine as in 1(b)	64
	(g) ARP*P*P*: phosphorylate ARP*P* as in 2(b)	64
2. Adenosine diphosphate	(a) ARP*P*: action of myosine on ARP*P*P*	64
	(b) ARPP*, or ARP*P: action of myosine on ARPP*P	64
3. Adenylic acid	ARP*: action of apyrase on ARP*P*P*	64
4. Acetyl phosphate	Reaction of labeled phosphate in pyridine with acetic anhydride; precipitation of acetyl phosphate at *p*H 7.5 using ethanol	66, 67
5. Phosphopyruvic acid	Phosphorylation of pyruvic acid in quinoline with P^{32}-labeled phosphorus oxychloride followed by alkaline hydrolysis and precipitation of barium salt in methanol	68
6. Inorganic pyrophosphate	Thermal dehydration of $Na_2HP^{32}O_4$	62
7. Inosine and uridine triphosphates	Transphosphorylation from ARPPP* to form URPPP* and IKPPP* using nucleoside diphosphokinase	69

* The following symbols are used: ARPPP for ATP, ARPP for ADP, ARP for AMP, and P for PO_4^{-3}. Asterisk (*) indicates labeled position.

[63] Rose, L. A., Greenberg-Manago, M., Korey, S. R., and Ochoa, S., *J. Biol. Chem.* **211,** 737 (1954).

[64] Rosenberg, H., *Australian J. Exptl. Biol. Med. Sci.* **33,** 17 (1955).

[65] Berg, P., *J. Am. Chem. Soc.* **77,** 3163 (1955).

[66] Kornberg, A., Kornberg, S., and Simms, E., *Biochim. et Biophys. Acta* **20,** 215 (1956).

[67] Avison, A. W. D., *J. Chem. Soc.* 732 (1955).

[68] Ohlmeyer, P., *J. Biol. Chem.* **190,** 21 (1951).

[69] Berg, P., and Joklik, W. K., *J. Biol. Chem.* **210,** 657 (1954).

[70a] Hummel, J. P., and Lindberg, O., *J. Biol. Chem.* **180,** 1 (1949).

[70b] Goodman, M., Benson, A. A., and Calvin, M., *J. Am. Chem. Soc.* **77,** 4257 (1955).

by procedures based on electrophoresis.[69] It is necessary to exercise care when determining specific activities of inorganic phosphate in the presence of organic phosphate. A procedure which appears to prevent the kind of contamination of inorganic phosphate that occurs with the usual precipitation method using magnesia mixture is based on formation of the phosphomolybdate complex and extraction of inorganic phosphate in this form with isobutanol.[71, 72] It is customary now to employ chromatographic procedures for isolation and purification of phosphate compounds.[61, 62, 69, 73]

C. Radiation Hazards

In estimating P^{32} dosages for human experimentation which will not exceed tolerance limits for long-range radiation damage, one may suppose only 10% of a given dose reaches bone where it is retained with an effective half-life of 13 days.[74] The other 90% is eliminated with an effective half-life of about 5 days. The bone dosage should not exceed a total of approximately 750 μc.[74] In a 70-kg. human being about 7 kg. is bone so that the specific dosage is about 100 μc./kg. of bone. If one adopts the notion that, instead of an average tolerance of 0.1 rep. per day, the total dosage should not exceed 1 rep., then the tolerance concentration is approximately 1.7 μc./kg. or a total of 12 μc. General dosage elsewhere in the body would be limited on a basis of 1 rep. total to about 4 μc./kg. Administered dosages totaling up to 5 mc. are considered permissible by most authorities.

4. THE SULFUR ISOTOPES

A. Preparation, Properties, and Assay

The normal isotopic composition of sulfur is S^{32} (95%), S^{33} (0.74%), S^{34} (4.2%), and S^{36} (0.016%). Isotopes available for radioactivity are, therefore, S^{31}, S^{35}, and S^{37}. Radioactive bodies assignable to all three of these isotopes have been found. S^{31} is a positron emitter with a half-life of 3.2 sec.[75] S^{35} is a negative β-ray emitter with a half-life of 87.1 days.[76-78]

[71] Ennor, A. H., and Stocken, L. A., *Nature* **168,** 199 (1951).

[72] Ennor, A. H., and Stocken, L. A., *Australian J. Exptl. Biol. Med. Sci.* **28,** 647 (1950).

[73] Davidson, J. N., and Smellie, R. M. S., *Biochem. J.* **62,** 594 (1952).

[74] Morgan, K. Z., *J. Phys. & Colloid Chem.* **51,** 1003 (1947).

[75] White, M. G., Creutz, E. C., Delsasso, L. A., and Wilson, R. R., *Phys. Rev.* **56,** 63 (1941).

[76] First observed by Andersen, E. B., *Z. physik. Chem.* **32B,** 237 (1936).

[77] Hendricks, R. H., Bryner, L. C., Thomas, M. V., and Ivie, J. O., *J. Phys. Chem.* **47,** 469 (1943).

[78] See also Kamen, M. D., *Phys. Rev.* **60,** 537 (1941), for a discussion of the history of isotopic assignment and production of radioactive sulfur.

Finally, S^{37} is a negative β-ray emitter, decaying with a half-life of 5.04 min.[79] As in C^{14}, no appreciable γ-ray emission is observed. It is apparent that S^{35} is the only radioactive isotope of sulfur suitable for biochemical investigations. The rare heavy isotopes S^{34} and S^{36} are also available as tracers, so that multiple labeling of sulfur is possible.

Neutron irradiation in the uranium pile, based on the reaction $Cl^{35}(n,p)S^{35}$, is the basis for production of S^{35}. Despite the rather high potential barrier for proton escape, the reaction proceeds with a good yield with slow neutrons.[78] As pointed out in the discussion of the $N^{14}(n,p)C^{14}$ reaction (pp. 31–305), the (n,p) reaction is exothermic when the negative β emitter formed has a maximum β-ray energy less than the mass difference of neutron and proton (0.8 Mev). The upper energy limit of the S^{35} β rays is only about 0.17 Mev. S^{35} can be obtained in a number of valence forms—sulfate, sulfur, and sulfide.

S^{35} exhibits radiation properties almost identical with those of C^{14}. The upper energy limit for the negative β particles emitted is reported as 0.1670 $\pm$0.0005 Mev.[80] The range in aluminum is given as 31.4 $\pm$0.5 mg./cm.^{2}.[81] The assay of S^{35} is accomplished by the same methods as those described for C^{14}. The samples are usually prepared as $BaSO_4$ or benzidine sulfate for counting.[82-84a] Benzidine hydrochloride is held to be a better precipitant for sulfate than barium chloride in that it gives more uniform precipitates. The preparation of benzidine sulfate follows standard procedures.[83-85a] The high specific activities in available S^{35} indicate no need for high-efficiency assay methods, such as gas counting.

B. Tracer Applications of S^{35}

Sulfur is usually present in one or more of several amino acids, such as cysteine and methionine, as well as in many other biologically important

[79] Zünti, W., and Bleuler, E., *Helv. Phys. Acta* **18,** 263 (1945).

[80] Price, H. C., Jr. Motz, J., and Langer, L. M., *Bull. Am. Phys. Soc.* **24,** 10 (1949).

[81] Solomon, A. K., Gould, R. G., and Anfinsen, C. G., *Phys. Rev.* **72,** 1097 (1947).

[82] See, however, Borsook, H., Buchman, E. R., Hatcher, J. B., Yost, D. M., and McMillan, E. M., *Proc. Natl. Acad. Sci. U.S.* **26,** 412 (1940), who use elementary sulfur which has the advantage of less absorption per unit weight but need be considered only when very weak samples are encountered. An arrangement using coincidence counting tubes is also mentioned.

[83] Henriques, F. C., Jr., Kistiakowsky, G. B., Margnetti, C., and Schneider, W. G., *Ind. Eng. Chem. Anal. Ed.* **18,** 349 (1946), have described an improved S^{35} assay, using a modified Lauritsen electroscope.

[84] Tarver, H., and Schmidt, C. L. A., *J. Biol. Chem.* **130,** 67 (1937).

[84a] Young, L., Edson, M., and McCarter, J. A., *Biochem. J.* **44,** 179 (1949); see also Schwebel, A., Isbell, H. S., and Karabinos, J. V., *Science* **113,** 456 (1951).

[85] Niederl, J. B., Baum, H., McCoy, J. S., Kuck, J. A., *Ind. Eng. Chem. Anal. Ed.* **12,** 428 (1940).

[85a] Tarver, H., *Advances in Biol. and Med. Phys.* **2,** 281 (1951).

compounds. The isotope can be used to advantage to follow the utilization and metabolism of these compounds in the animal body: one such example has been presented (p. 262). In particular, stably bound labeled sulfur may be introduced into protein molecules and used as a label for protein, thus supplying another auxiliary tracer for carbon compounds.

C. Distribution and Retention of Sulfur, with Remarks on Radiation Tolerance Dose

The distribution and retention of sulfur in the animal organism depends markedly on the form in which it is ingested. Thus the rat cannot use elementary sulfur in place of cystine or methionine for incorporation into tissue protein.[86] Ingestion as sulfate results in rapid excretion of most of the dose, a small fraction appearing in the animal, with the highest concentration in bone marrow. Sulfate also is not utilizable for amino acid synthesis. Ingestion as sulfide is rather inefficient, because most of the sulfur in this form is oxidized to sulfate in which form it undergoes the same fate as ingested sulfate.[87] Excretion by rats in urine and feces of S^{35}-labeled sodium sulfate is rapid and accounts for the major fraction of the excretion, approximately 95% of an intraperitoneal dose of 1 mg. being eliminated in 120 hr.[88]

It appears that the best way to incorporate sulfur into tissue, at least as protein, is introduction in the form of a thioamino acid, such as methionine or cysteine, because in this form it can participate in the metabolic pool reactions.[89] After feeding labeled methionine in small doses to bile fistula rats ($\sim$1 mg. per rat), it is found[86] that 56% of the methionine sulfur is incorporated into tissue protein. Of this tissue protein, 34% is in the carcass, 25% in the liver, 16% in the intestinal tract, 9% in the kidney, and smaller fractions in the genito-urinary tract, lungs, skin, hair, and bile. Similar distribution patterns are noted in the normal animal.[89-91] Selective localization of sulfur is not encountered, as with phosphorus when the latter is administered in inorganic form. To achieve such localization will probably require synthesis of a molecule, such as a sulfur-containing dye, which is known to deposit preferentially in some tissue.

Labeled sulfur injected as thiocyanate is converted to a small extent to sulfate.[92] Evidence that thiocyanate is actually metabolized as such has

[86] Tarver, H., and Schmidt, C. L. A., *J. Biol. Chem.* **146,** 69 (1942).
[87] Dziewiatkowski, D. D., *J. Biol. Chem.* **161,** 723 (1945).
[88] Dziewiatkowski, D. D., *J. Biol. Chem.* **178,** 197 (1949).
[89] Tarver, H., and Morse, L. M., *J. Biol. Chem.* **173,** 53 (1948).
[90] Maass, A. R., Larson, F. C., and Gordon, E. S., *J. Biol. Chem.* **177,** 209 (1949).
[91] Friedberg, F., Tarver, H., and Greenberg, D. M., *J. Biol. Chem.* **173,** 355 (1948).
[92] Wood, J. L., Williams, E. F., Jr., Kingsland, N., *J. Biol. Chem.* **170,** 251 (1947).

been presented in experiments[93] in which S^{35}-labeled thiocyanate is injected intraperitoneally into white rats and the resulting tissue distribution studied at 6- and 24-hr. intervals after injection. Some accumulation in thyroid water-soluble compounds is noted. The results are in agreement with the notion that thiocyanate may compete with iodine for the thyroid enzyme which mediates thyroxine formation.

Excretion data on sulfate indicate that no more than 5% of a 1-mg. dose given intraperitoneally must be considered as retained with a half-life given approximately by the natural decay rate (88 days). One may estimate a total-body tolerance dose for clinical experimentation as 100 μc. But if an amino acid such as methionine is used much higher incorporation occurs, and the tolerance dose may be only one-tenth as large, or less.[93a]

D. Synthesis of S^{35}-Labeled Compounds

Detailed descriptions of synthetic procedures for preparation of the naturally occurring sulfur amino acids as well as a variety of other compounds may be summarized briefly as follows.

1. Methionine and Homocysteine.[94] Beginning with labeled $BaSO_4$, labeled sulfide is obtained by fusion with iron filings. Labeled benzyl mercaptan is prepared in 75% yield by reaction of labeled sulfide with benzoyl chloride. The mercaptan is treated with γ-benzamido-γ-chlorobutyric acid ethyl ester to give S-benzylhomocysteine which may be converted to homocysteine by sodium reduction in liquid ammonia. Methylation with methyl iodide completes the synthesis. Over-all yield based on sulfur is about 20%.

Another procedure involves preparation of labeled benzyl mercaptan in approximately 88% yield using a Grignard reaction between benzylmagnesium chloride and elementary sulfur.[94, 95] Condensation of the mercaptan with excess ethylene dichloride results in formation of benzyl β-chloroethyl sulfide. Reaction of benzyl β-chloroethyl sulfide with ethyl sodiophthalimidomalonate[67, 96] in toluene results in a 70% yield of S-benzylhomocysteine. With carbon-labeled ethylene dichloride, one may prepare doubly labeled homocysteine or methionine.[97]

[93] Wood, J. L., and Williams, E. F., Jr., *J. Biol. Chem.* **177**, 59 (1949).

[93a] S^{35} with high specific activity has been shown to be a very efficient mutagen when incorporated into the protein of *Neurospora crassa*; see Hungate, F. P., and Mannell, T. J., *Genetics* **37**, 709 (1952).

[94] Seligman, A. M., Rutenberg, A. M., and Banks, H., *J. Clin. Invest.* **22**, 275 (1943).

[95] Kilmer, G. W., and du Vigneaud, V., *J. Biol. Chem.* **154**, 247 (1944).

[96] Snyder, H. R., and Chiddix, M. E., *J. Am. Chem. Soc.* **66**, 1000 (1944).

[97] Wood, J. L., and Gutmann, H. R., *J. Biol. Chem.* **179**, 535 (1949).

Benzylhomocysteine can also be prepared in 25% yield from benzyl mercaptan by reaction with 3,6-*bis*(β-chloroethyl)-2,5-diketopiperazine.

2. Cystine, Cysteine, Homocysteine, and Homocystine. Starting with S-benzylhomocysteine, homocysteine can be prepared by reduction with sodium in butyl alcohol.[94] Starting with $BaSO_4$ and proceeding to homocystine via S-benzylhomocysteine and homocysteine, the over-all yield is 24%.

To synthesize labeled cystine, labeled benzylthiomethyl chloride is first prepared from labeled benzyl mercaptan by reaction with polyoxymethylene in the presence of calcium chloride and dry hydrogen chloride. A malonic acid ester synthesis with sodiophthalimidomalonic ester follows, yielding S-benzylcysteine which can be converted to cystine by reduction with sodium in liquid ammonia to cysteine followed by oxidation with ferric chloride.[94, 98]

Another procedure employs condensation of labeled benzyl mercaptan with α-amino-β-chloropropionate or α-benzamido-β-chloropropionate[99] which results in 38 to 44% of benzylcysteine which after cleavage with sodium in liquid ammonia is oxidized in air to cystine in 77% yield.

Finally, syntheses of S^{35}-labeled cystine based on the conversion of serine to 2-phenyl-4-carboxymethyloxazoline have been reported. The hydrochloric acid salt of this compound is rearranged either to methyl α-benzamide-β-chloropropionate and the chlorine replaced directly by sulfhydryl, or it is rearranged as the thiobenzoic acid salt to give N,S-dibenzoylcysteine. The latter rearrangement gives 42% yields of optically active cystine based on sulfur.[100]

All labeled sulfur left in the various residues of such syntheses are recoverable by conversion to sulfate in the following manner.[84] The residues are fused with sodium peroxide, and the fusion mixture is neutralized to the methyl orange end point after solution in a large volume of water. The sulfate is then precipitated with a large excess of the Fiske benzidine reagent. The precipitate is filtered off, washed with acetone, and neutralized with 0.1 *N* alkali (phenol red end point). The benzidine is removed by successive evaporations and filtrations. Finally, the volume is brought to approximately 5 ml., and the last traces of benzidine removed by pouring this solution into 15 volumes of acetone. The sulfate is recovered from the acetone by allowing the solution to stand in the refrigerator until the salt is well crystallized. The sulfate is filtered off and dissolved in water.

3. Vesicants. Benzyl β-chloroethyl sulfide and *n*-butyl β-chloroethyl sulfide can be synthesized in good yields from the corresponding mercap-

[98] Wood, J. L., and du Vigneaud, V., *J. Biol. Chem.* **131,** 267 (1939).
[99] Melchior, J. B., and Tarver, H., *Arch. Biochem.* **12,** 301 (1947).
[100] Fry, E. M., *J. Org. Chem.* **15,** 433 (1950).

TABLE 37
IN VIVO UPTAKE OF S^{35}-LABELED AMINO ACIDS INTO PLASMA PROTEIN OF DOGS (AFTER SELIGMAN AND FINE[103])

Amino acid	Weight fed (mg.)	Total plasma removed (ml.)	Highest concen. of amino acid in protein fraction of plasma (mg./ml.)	Ingested Amino Acid: Incorporated in plasma protein (%)	Ingested Amino Acid: Excreted in urine as sulfate (%)	Diet supplements and remarks
L-Cystine	100	360	0.024	5.2	36†	Casein, tyrosine
L-Cystine	200	963	0.046	14.8	19†	Casein, tyrosine, tryptophan
DL-Homocystine	500	345	0.0032	0.15	66	Gelatin, betaine
DL-Homocystine	50*	310	0.0031	0.9	7	Casein hydrolyzate, infection
DL-Methionine	50	195	0.00034	0.06	18	Casein hydrolyzate, L-cystine
DL-Methionine	150*	66	0.0000	0.00	14	Gelatin, L-tyrosine

* Injected intravenously.
† Calculated on basis of DL-cystine.

tans, benzyl, and *n*-butyl mercaptan, by reaction with ethylene chlorhydrin in alkali as discussed previously (see p. 355).[101] The mode of action of these vesicants in insulin has been examined and the conclusion reached that a fraction of the vesicant becomes attached to a free amino group of the phenylalanine moieties in the insulin molecule.[102]

4. Biosyntheses. (*a*) *Protein.* Labeled plasma proteins can be prepared by *in vivo* synthesis from S-labeled amino acids. Thus, dogs made hypoproteinemic are fed labeled cystine, homocystine, or methionine with the result that up to 15% of administered sulfur can be recovered in the plasma protein.[103] A summary of typical experimental results is given in Table 37. It should be remarked (as discussed later in Section E, p. 359) that incorporation of amino acid, particularly cystine, may occur not by true peptide bond formation but by other mechanisms such as disulfide bond formation.

[101] Wood, J. L., Rachele, J. R., Stevens, C. M., Carpenter, F. H., and du Vigneaud, V., *J. Am. Chem. Soc.* **70,** 2547 (1948).
[102] Stevens, C. M., Wood, J. L., Rachele, J. R., and du Vigneaud, V., *J. Am. Chem. Soc.* **70,** 2554 (1948).
[103] Seligman, A. M., and Fine, J., *J. Clin. Invest.* **22,** 265 (1943).

Thus the high results with cystine may not be interpreted as owing to preferential high turnover of this amino acid.

Probably the most convenient procedures for securing S^{35}-labeled amino acids are based on the use of microorganisms. It has been shown[104, 105] that the yeast *Torula utilis* incorporates S^{35} efficiently into sulfur-containing amino acids when grown in media containing labeled sulfate. Suspensions of *E. coli* have been described incorporating as much as 23.4 mc. of organic sulfur in as little as 150 γ of packed cells.[106] The protein fraction of the cells is hydrolyzed and the various amino acids are separated and purified by paper-strip chromatography (see Appendix 3). Methionine and cystine can be labeled to the extent of 2×10^9 ct./sec./mg. sulfur, assayed with an end-window G-M counter.

(*b.*) *Antibiotics.* Considerable attention has been given to the use of labeled antibiotics in researches into biosynthesis and mechanism of action of these agents. It may be remarked in passing that S-labeled penicillin obtained by growing penicillium molds in nutrient media containing labeled sulfate is being used in investigations concerning the nature of the primary reaction between penicillin and the bacterial cell. As discussed previously (see p. 263ff.), preliminary results indicate that uptake is dependent on penicillin concentration in all organisms but that, at low concentrations, sensitive strains pick up more penicillin than resistant strains. Penicillin taken up is strongly and irreversibly attached and can be removed only by such drastic means as heating for 5 days at 60°C.[107-109]

E. Protein Turnover *in Vivo* and *in Vitro*

In early experiments with bile fistula dogs it was found that only a fraction of administered S^{35}-labeled methionine could be recovered in bile urine and feces.[84] This led to an examination of the general tissue protein. Tissues were removed from the animals as soon as possible after sacrifice, and proteins were separated by precipitation with 4% trichloroacetic acid. The total proteins were converted to sulfate by the method of Pirie.[110] It was found that in a fasting animal fed a very small amount of labeled methionine 56% of the labeled sulfur appeared in rat protein and 36% was oxidized to sulfate. A detailed analysis of the various organs showed the spe-

[104] Schlüssel, H., *Biochem. Z.* **321,** 421 (1951).

[105] Schlüssel, H., and Feinendegen, L., *Biochem. Z.* **321,** 533 (1951).

[106] Cowie, D. B., Bolton, E. T., and Sands, M. K., *Arch. Biochem. and Biophys.* **35,** 140 (1952).

[107] Rowley, D., Miller, J., Rowlands, S., and Lester-Smith, E., *Nature* **161,** 1009 (1948).

[108] Cooper, P. D., and Rowley, D., *Nature* **163,** 480 (1949).

[109] Maass, E. A., and Johnson, M. J., *J. Bacteriol.* **58,** 361 (1949).

[110] Pirie, N. W., *Biochem. J.* **26,** 2041 (1932).

cific activity in the total sulfur fraction from the different tissue proteins to be quite variable. The intestinal mucosa exhibited very high specific activities and the various muscle tissues very low specific activities. Liver protein, after treatment according to the method of Banga and Szent-Györgyi,[111] showed no variations in specific activity of the various fractions isolated. It was found that a rapid conversion of methionine sulfur to cystine sulfur occurred.

From these results it could be concluded that there was a rapid turnover of sulfur-containing proteins in many tissues which could be due either to exchange of sulfur-containing moieties in the administered methionine and cellular protein, or to cleavage and resynthesis of peptide bonds with consequent introduction of labeled sulfur. The latter alternative seemed much more likely in view of the researches of Schoenheimer and his colleagues,[112] who showed the same protein turnover, using amino acids labeled with deuterium and N^{15}. The animals accomplished this redistribution of sulfur under fasting conditions and after a single minimal dose of methionine. It was found that considerable methionine was introduced into tissue protein even after reserve protein was largely depleted by fasting. Thus, it was concluded, in accord with Schoenheimer's concept of the "metabolic pool," that cellular protein was labile. Furthermore, it was found that proteins of erythrocytes exchanged labeled sulfur very slowly in contrast with the rapid turnover in plasma proteins, a finding in harmony with those established by the N^{15} and H^2 work.

It is interesting to note that the use of the sulfur isotope as marker for amino acids led to substantially the same conclusions as were derived from researches with other labeling isotopes (H^2,C^{13},N^{15}).

The extension of labeling methods to study of the reversibility of biological degradation reactions *in vitro* has been discussed (Chapter VI, p. 167ff.). Among the many tracer experiments on protein "synthesis" or incorporation of labeled amino acid into protein in cell-free preparations one finds studies with S^{35}-labeled cystine and methionine which illustrate both advantages and limitations of the tracer method.[99, 113]

If a few micromoles of S^{35}-labeled cystine are incubated aerobically with rat liver slices or homogenates in Krebs saline solution for a few hours, then the protein fraction is found to contain considerable amounts of activity. However, boiled preparations also exhibit a high incorporation of amino acid sulfur. Reduction of the active protein with thioglycolic acid or cysteine effectively removes a large fraction of the activity, so that formation of disulfide bonds rather than peptide bonds appears to be in-

[111] Banga, I., and Szent-Györgyi, A., *Enzymologia* **9,** 111 (1940).

[112] Schoenheimer, R., "The Dynamic State of Body Constituents." Harvard U. P., Cambridge, Mass., 1946.

[113] Melchior, J. B., and Tarver, H., *Arch. Biochem.* **12,** 309 (1947).

volved. Thus labeled cystine cannot be used as a true test substance for occurrence of peptide bond synthesis.

When S^{35}-labeled methionine is incubated with liver slices a fraction of a per cent of the methionine present in the protein is replaced by labeled methionine. A large fraction (5 to 20%) of the methionine is converted to cystine, however, and a smaller fraction (1 to 9%) is oxidized to sulfate. This conversion to cystine results in a certain uptake of the spurious nature remarked on in the previous paragraph. Hence to establish methionine replacement, it is necessary to recover pure methionine from the protein and determine its labeled content. Results of these studies show that uptake of methionine with true peptide bond formation depends on the concentration of methionine over the range studied (4.2 to 42 μM./ml.). The percentage fixed decreases with increasing concentration, but the absolute amount increases with increasing concentration.

Chapter XII

VARIOUS RADIOACTIVE NUCLIDES OF IMPORTANCE IN BIOLOGY

1. INTRODUCTION

A general introduction to tracer methodology has been presented in the preceding chapters. In this concluding section, some useful data on isotopes of importance in biology will be outlined. These isotopes have been used in a great number of biological experiments and studies which in themselves comprise a large fraction of the literature on tracer applications in biology. A short bibliography of the more recent applications is included at the end of the chapter.

2. ALKALI METAL AND ALKALINE EARTH TRACERS

A. General Survey of Alkali Metal Tracers

Isotopes suitable for tracer studies are available for all the alkali metals with the exception of lithium. Sodium and potassium, however, are the two alkali metal elements of major biological interest. The function of these elements is concerned with mineral metabolism and, in particular, with water and electrolyte balance. Most tracer researches have dealt with the distribution and movement of these elements as ions in cellular fluids and tissues (see p. 253).

Medical studies with labeled sodium have been concerned mainly with determination of circulation time and intercellular space in a variety of heart and circulatory disturbances. A number of examples have already been discussed in Chapter VIII. From the standpoint of radiation dosage in tracer studies it is emphasized that the true biologic half-life of sodium is difficult to ascertain in abnormal subjects because it is not correlated in a simple way either with the time required to reduce plasma sodium to half value or with the time required to eliminate half of a given dose by way of the urine. In normal subjects the half value for plasma sodium is likely to be close to the true biologic half-life. Variations in the various half-lives determined either by plasma analysis or urinary excretion as a function of a number of heart and kidney disturbances have revealed a complex pattern from which ready generalizations are excluded.[1,2] In

[1] Threefoot, S. A., Burch, G., and Reaser, P., *J. Lab. Clin. Med.* **34,** 1 (1949).

[2] Burch, G. E., Threefoot, S. A., and Cronvich, J. A., *J. Lab. Clin. Med.* **34,** 14 (1949).

normal humans the half-life of sodium in the plasma is 14.6 mo.[3] A maximum estimate for Na^{24} dosage would appear to be 15 μc. body retention (70 kg.). Na^{22}, with its much longer half-life (2.6 yr. as compared to 15 hr. for Na^{24}), must be used with caution because of a small retention of sodium in bone.[4] Unfortunately, definitive data on this point do not appear to have been gathered. For K^{42}, maximal estimates of 20 μc. retained per 70-kg. body are given.[3]

B. Preparation, Properties, and Assay

1. Radioactive Sodium. Normal sodium consists of but one isotope, Na^{23}. Radioactive bodies with mass numbers 22 and 24 are available as tracers. The latter, which is relatively short-lived ($\tau_{1/2}$ = 15.06 hr.[5]) can be produced in very large quantities either by neutron irradiation of Na^{23} in the uranium pile, according to the reaction $Na^{23}(n,\gamma)Na^{24}$, or by deuteron bombardment in the cyclotron, as in $Na^{23}(d,p)Na^{24}$, or $Al^{27}(d,\alpha p)Na^{24}$. In addition, reactions such as $Mg^{24}(n,p)Na^{24}$ and $Al^{27}(n,\alpha)Na^{24}$ are available. Because of the short half-life, Na^{24} is procurable only by laboratories reasonably close to the production machines

The cross section for the $Na^{23}(d,p)Na^{24}$ is one of the largest known at moderate energies,[6] and yields reported range from 0.5 mc./μa.-hr. at 8 Mev. to values three to four times as great at 16 Mev. The element can be bombarded directly as metal in a bell-jar target or as a metaborate (45% sodium) on an internal target.

When the metaborate $NaBO_2$ is used,[7] approximately 100 mg. of the salt is fused on a knurled copper plate, by means of an oxygen torch. Typical dimensions of the target area are 1.5 x 1.2 x 0.1 inch. After bombardment, the target plate is digested (behind lead shielding) in 50 ml. of distilled water at about 70°C. for half an hour. The surface of the target is scraped with a rubber "policeman" or coarse toothbrush on a long handle to remove all target material. The solution is then filtered through a medium porosity sintered-glass funnel and the residue is discarded. Ten milliliters of 6 *N* HCl and an equal volume of methyl alcohol (about 60 ml.) are added to the solution. The solution is evaporated to dryness; the methyl borate is removed in the process. This procedure is repeated once. All HCl is removed by repeated evaporation with distilled water. The final NaCl residue is dissolved in a standard volume of distilled water, and the solution is filtered through a clean sintered-glass funnel. When metallic sodium is

[3] National Bureau of Standards (U.S.) Handbook No. **52** (1953).
[4] Berggren, H., *Acta Radiol.* **27,** 248 (1946).
[5] Sreb, J. H., *Phys. Rev.* **81,** 469 (1951).
[6] Clarke, E., and Irvine, J. W., Jr., *Phys. Rev.* **66,** 231 (1944).
[7] Reid, A. F., private communication, 1950.

bombarded, it may be removed from the target with water, neutralized with acid, and filtered free of all insoluble material. This treatment satisfactorily removes all contaminating radioactivity coming from target plate material.

If Na^{24} is prepared by the high-energy deuteron bombardment of aluminum, a suitable target is "aluminum alloy," with composition 95% aluminum, 2.25% magnesium, 0.25% chromium, and the rest iron and copper. The metal is dissolved in 6 N HCl; sodium (50 mg.) and copper (10 to 20 mg.) carrier are added. The aluminum is precipitated with a slight excess of ammonia. The precipitate is separated by centrifugation and decantation. It is worked over again by dissolving it in HCl and repeating the precipitation with ammonia. The second supernate is added to the first. The combined solutions are saturated with H_2S to precipitate copper and remaining aluminum. The filtrate is acidified with HCl and evaporated to dryness. The ammonium salts are destroyed by boiling with a 1:4 HCl-HNO_3 mixture. The residue contains the Na^{24} as a mixture of NaCl and Na_2SO_4.[7a]

Na^{24} decays almost entirely by emission of β^- particles, the maximum energy of which is given as 1.390 Mev.[8] Two γ rays are emitted in cascade in each disintegration. These rays have energies of 2.758 and 1.380 Mev., respectively.[8a] A very small fraction ($<0.01\%$) of the disintegrations proceed by emission of β^- rays with an upper energy limit of 4.17 Mev., followed by emission of the 1.380-Mev. γ ray.[8b] Assay of Na^{24} presents no problem because of the hardness of the radiations emitted.

The long-lived sodium isotope, Na^{22} ($\tau_{1/2} = 2.6$ yr.), can be produced only by cyclotron bombardment because no neutron reactions, excepting possibly $Mg^{24}(n,H^3)Na^{22}$, exist for its production by the uranium pile reactor. The best reaction is $Mg^{24}(d,\alpha)Na^{22}$. The target is either magnesium metal or magnesium oxide. After bombardment the radioactive sodium is easily extracted by successive leachings with hot water. Alternatively, the magnesia can be dissolved in a minimal quantity of dilute HCl. The magnesium is removed by precipitation as the hydroxide. The sodium is recovered in the filtrate. The best procedure for separation of high specific activity Na^{22} from magnesium appears to be based on the use of Dowex 50 ion-exchange resin.[9]

Yields are good, varying from 0.3 μc./μa.-hr. at 8 Mev. to 1 μc./μa.-hr. at 16 Mev. Activities as high as 3 to 5 mc. are available with recovery in

[7a] Stewart, D. B., U. S. Atomic Energy Comm. Declassified Doc. AECD-2738 (1949).

[8] Siegbahn, K., *Phys. Rev.* **70**, 127 (1946).

[8a] Bloom, S. D., *Phys. Rev.* **88**, 312 (1952).

[8b] Turner, J. F., and Cavanaugh, P. E., *Phil. Mag.* [7] **42**, 636 (1951).

[9] Linnenbom, U. J., *J. Chem. Phys.* **20**, 1657 (1952).

carrier-free amounts of sodium if the water-leaching method is used. Na^{22} is available from commercial sources.

Na^{22} emits positrons with an upper energy maximum of 0.575 Mev.[10] In addition, there is the accompanying annihilation γ radiation (0.5 Mev.), as well as a γ ray at 1.30 Mev.[11] Assay procedures are identical with those for C^{11}.

2. Radioactive Potassium. Normal potassium contains two stable isotopes, K^{39} and K^{41}, with relative abundances of 93.35% and 6.61%. In addition, there is a radioactive isotope, K^{40}, present to the small extent of 0.012% ($\tau_{1/2} = 4.5 \times 10^8$ yr.). This isotope is too long-lived to be useful as a tracer because, even with a reaction such as $Ca^{40}(n,p)K^{40}$ in the uranium pile reactor, only very low intensities are available. Furthermore, the maximum specific activity obtainable is limited by the long half-life (small disintegration constant) to **$\sim$20μc./g. K^{40}.**

The most available isotope for tracer purposes has been K^{42} ($\tau_{1/2} = 12.4$ hr.),[11a] which can be produced by any of the following reactions: (1) $K^{41}(d,p)K^{42}$, (2) $Sc^{45}(n,\alpha)K^{42}$, (3) $Ca^{42}(n,p)K^{42}$, and (4) $K^{41}(n,\gamma)K^{42}$.

Unit quantities up to 1 curie with specific activities greater than 200 mc./g. of element are produced by the uranium pile reactor, using capture reaction 4 on potassium. In theory, much lower total intensities and higher specific activities could be achieved from reactions 2 and 3. The cyclotron yields from reaction 1 are 350 to 500 μc./μa.-hr. at 12 Mev., rising by a factor of 2 at 16 to 20 Mev. Another reaction, $Ca^{44}(d,\alpha)K^{42}$, has a much lower yield (about 1 to 5% of the deuteron reaction on potassium) but can be used to achieve high specific activities. Laboratories possessing only weak neutron sources can produce tracer quantities of K^{42} by utilizing reactions 2 or 3.

K^{42} must be prepared free of the contamination with Na^{24} which arises from sodium impurities in the target potassium. It is difficult to be sure of how completely the two elements are separated because both K^{42} and Na^{24} have nearly identical half-lives, and the radiations they emit are not radically different.

Bombardment preparations are similar to those for sodium. The metal is removed after bombardment by solution in water. The solution is neutralized with a nominal quantity of HCl. Sodium is added as carrier, and several precipitations are carried out with the insoluble potassium cobaltinitrite. The precipitate is finally dissolved in weak acid, and cobalt is removed as the sulfide. The filtrate is freed of nitrite by boiling down with

[10] Laslett, L. J., *Phys. Rev.* **52,** 529 (1937); **76,** 858 (1949).

[11] Oppenheimer, F., and Tomlinson, E. P., *Phys. Rev.* **56,** 858 (1939); Good, W. M., Peaslee, D., and Deutsch, M., *ibid.* **69,** 313 (1946).

[11a] Hurst, D., and Walke, H., *Phys. Rev.* **51,** 1033 (1937).

HCl to dryness. Potassium may be rid of sodium also by precipitation with perchloric acid and ethyl alcohol. Chromatographic procedures should simplify the problem of preparation of sodium-free potassium.[11b]

K^{42} emits very high-energy negative β particles. Its spectrum reveals two energy components, one with an upper limit of 2.04 Mev. (25%) and the other 3.58 Mev. (75%).[12] The expected γ-ray component with energy of 1.51 (difference of two β energy upper limits) has also been reported.[12] It is assayed in the same manner as radioactive sodium.

It must be noted that natural potassium is itself radioactive because of the presence of K^{40}. In working with potassium, this natural contamination must be taken into account. The negative β radiation from K^{40} has an upper energy limit of 1.35 Mev. There is emission of 0.6 β particle per second from 1 cm.^{2} of normal potassium.[13] In addition to negative β emission, there is disintegration by K capture and an associated γ ray with $E = 1.5$ Mev. Most of the decay occurs by this process.[14] In all work with low-intensity radioactivity, reagents containing potassium should be avoided as much as possible because of its natural radioactivity.

It has been found that a potassium isotope, K^{43} ($\tau_{1/2} = 22.4$ hr.), can be produced by the cyclotron by means of the $A^{40}(\alpha,p)K^{43}$ reaction.[15] Two β components with upper energy limits of 0.25 Mev. and 0.8 Mev. have been reported. A γ component with energy of 0.5 Mev. has also been described. Exceptionally high yields of K^{42} can be achieved by means of 40-Mev. α-particle bombardment of argon. Yield figures quoted are 1 mc./μa.-hr. The use of argon as target material makes possible the production of K^{42} and K^{43} with high specific activity. The target chamber is filled with continuously circulated argon gas. Recoil potassium collects on the walls and in a glass-wool plug in the outlet tube from which it may be removed essentially carrier-free by treatment with hot distilled water.

C. The Alkaline Earth Tracers—Magnesium, Calcium, and Strontium

The alkaline earth metals magnesium and calcium exhibit a variety of functions. They activate a number of important enzyme systems in muscle metabolism. Magnesium is the essential mineral constituent of chlorophyll and other related pigments vital to plant function. Calcium is important in elaboration of skeletal structure as well as in regulation of protein reac-

[11b] See Frierson, W. J., and Jones, J. W., *Anal. Chem.* **22**, 1447 (1951). For a general text which gives details of chromatography of inorganic ions, see Smith, O. C., "Inorganic Chromatography." Van Nostrand, New York, 1953.

[12] Siegbahn, K., *Arkiv. Mat. Astron. Fysik* **34B**, No. 4 (1947).

[13] Dzelepow, B., Kopjowa, M., and Vorobjov, E., *Phys. Rev.* **69**, 538 (1946).

[14] Graf, T., *Phys. Rev.* **74**, 1199 (1948).

[15] Overstreet, R., Jacobson, L., and Stout, P. R., *Phys. Rev.* **75**, 231 (1949).

tivity. In mammalian tissue, both metals are distributed more or less parallel except that calcium is present in large amounts in bone. Tracers are available for all the alkaline earth metals.

Magnesium possesses three stable isotopes with mass numbers 24, 25, and 26. The lightest of these is the most abundant (78.6%). Two radioactive isotopes can be made, but only in a cyclotron or other high-energy, high-intensity particle accelerator. One of these isotopes, Mg^{27}, produced by the $Mg^{26}(d,p)Mg^{27}$ reaction, is too short lived ($\tau_{1/2} = 9.45$ min.)[16] for exploitation as a tracer, although some experiments with it on the role of chlorophyll in photosynthesis have been attempted.[17] The other, Mg^{28}, made in the cyclotron by the $Mg^{26}(\alpha,2p)Mg^{28}$ reaction at 39 Mev., has a half-life of 21.3 hr. It emits β^- rays with an upper limit of about 0.4 Mev.[18] It decays to Al^{28}, which in turn is radioactive, emitting β^- rays with an upper energy limit of 3 Mev. and γ rays with an energy of 1.8 Mev.

Natural calcium possesses stable isotopes with mass numbers 40, 42, 43, 44, 46, and 48. The light isotope (Ca^{40}) is by far the most abundant (96.96 atom per cent). Of the three radioactive isotopes available (mass numbers 41, 45, and 47), the one most suitable for tracer application is Ca^{45} ($\tau_{1/2} =$ 152 days).[19] It may be produced in the uranium pile reactor by the reactions (1) $Ca^{44}(n,\gamma)Ca^{45}$ or (2) $Sc^{45}(n,p)Ca^{45}$. The reaction on scandium is preferable because only Ca^{45} is produced, whereas in reaction 1 Ca^{41} ($\tau_{1/2} =$ 8.5 days) is formed simultaneously.

In the cyclotron the major reaction for production is $Ca^{44}(d,p)Ca^{45}$. Here again Ca^{41} is formed, also by $Ca^{40}(d,p)Ca^{41}$. Calcium targets must be allowed to age for some time before chemical separation. Conceivably the reaction $Ti^{47}(d,\alpha)Ca^{45}$ might be used to obtain only Ca^{45} in high specific activity. High-energy spallation of bismuth or copper can be used to produce radioactive calcium also.[20]

The target employed in the cyclotron is calcium metal or calcium oxide. In a helium atmosphere the metal is fused to the copper backing plate. Eutectic mixtures of calcium and magnesium seem to be more easily handled. Such a mixed target would, in fact, be advantageous, because Na^{22} would also be formed.

After bombardment of calcium, the target material is dissolved in dilute HCl. Carriers in 1-mg. amounts are added for contaminating activities (NaCl, $ScCl_3$, KCl). $Sc(OH)_3$ is precipitated by addition of a slight excess of carbonate-free ammonia. The filtrate is treated with scandium carrier in

[16] Ecklund, S., and Hole, N., *Arkiv Mat. Astron. Fysik* **29A,** No. 26 (1943).

[17] Ruben, S., Frenkel, A. W., and Kamen, M. D., *J. Phys. Chem.* **46,** 710 (1942).

[18] Sheline, R. K., and Johnson, N. R., *Phys. Rev.* **89,** 520 (1953).

[19] Matthews, D. W., and Pool, M. L., *Phys. Rev.* **72,** 163 (1947).

[20] See U. S. Atomic Energy Comm. Declassified Doc. AECD-2378, pp. 25, 26 (1949).

this fashion once again to remove the last traces of scandium activity. The filtrate then is heated on a water bath, and the calcium is precipitated as oxalate by adding excess ammonium oxalate. This precipitate is filtered, washed, and redissolved in dilute HCl. Carrier NaCl and KCl are added, and the oxalate precipitation is repeated. Yields are 0.1 to 0.5 μc./μa.-hr. at 12 to 14 Mev. With a bell-jar G-M tube with thin window, 1 μc. corresponds to approximately 1×10^5 ct./min.

Ca^{45} exhibits a simple negative β spectrum.[21] The upper energy limit is 0.260 Mev.[21] Its soft radiation requires an assay technique similar to that for C^{14}.

Natural strontium consists of four stable isotopes with mass numbers 84, 86, 87, and 88, the last named being the most abundant (82.56 atom per cent). An isotope well suited for tracer investigations is Sr^{89} ($\tau_{1/2}$ = 53 days).[22] Neutron reactions on strontium and yttrium are available.

Strontium metal is the target material usually employed. The metal is pressed or fused under helium into the copper cooling plate. It can be bombarded internally *in vacuo*, as well as in the bell-jar target in helium. The major contaminant produced in the reaction is the yttrium ($\tau_{1/2}$ = 87 days), which is a very intense γ-ray emitter. For this reason, separation must be carried out behind lead. The metal is dissolved in dilute HCl. Carrier yttrium is added and precipitated with excess ammonia. The filtrate, containing the strontium, also contains some copper. The strontium is freed from this contamination by precipitation with ammonium carbonate. The precipitate is dissolved in a slight excess of lactic acid. The isotope is usually administered in the form of lactate.

Yields vary from 8 to 11 μc./μa.-hr. at 12 to 14 Mev. A 2000-μa.-hr. bombardment will give a strontium sample containing approximately 2 g. of strontium with a Sr^{89} content of 20 mc. β equivalents. Only a slight amount of Sr^{85} ($\tau_{1/2}$ = 65 days) is formed, so it is not a complication. Its presence is detectable by its γ radiation (0.510 Mev.)[23] because Sr^{89} emits only negative β radiation. With Sr^{89} as tracer, only the β particles are to be taken as evidence for appearance of Sr^{89}. This can be ascertained by absorption measurements. Practically all the Sr^{89} radiation is absorbed by 700 mg./cm.^{2} aluminum.

The negative β spectrum of Sr^{89} is simple. Its upper energy limit is 1.463 Mev.[24] Assay procedures are similar to those for P^{32}. This isotope is

[21] Macklin, P. A., Feldman, L., Lidofsky, I., and Wu, C. S., *Phys. Rev.* **77,** 137 (1950).

[22] Glendenin, L. E., *in* "Radiochemical Studies: The Fission Products" (C. D. Coryell and N. Sugarman, eds.), Div. IV, Vol. 9, p. 10, National Nuclear Energy Series. McGraw-Hill, New York, 1946.

[23] Ter Pogossian, M., *Bull. Am. Phys. Soc.* **24,** 9 (1949).

[24] Langer, L. M., and Price, H. C., Jr., *Phys. Rev.* **51,** 713 (1949).

suitable for some investigations in calcium metabolism because it has been shown by Pecher[25] that the two elements are homologous to some extent. The fraction of tracer calcium retained in the tissues is higher for calcium than for strontium, but the distribution pattern is quite similar. Strontium is selectively absorbed by bone because there is a high calcium level in bone and the strontium is deposited by exchange with calcium. This observation might be exploited in selective radiation therapy of osteogenic tissue.[26]

Other strontium isotopes with mass numbers 90 and 91, produced mainly by fission processes in the uranium pile, are available in samples of very high specific activity. Sr^{90} ($\tau_{1/2}$ = 19.9 yr.)[27] is a β^- emitter which decays to radioactive Y^{90} ($\tau_{1/2}$ = 61 hr.),[28] also a β^- emitter. The upper energy limit for the Sr^{90} β^- rays is 0.61 Mev.[29] No γ radiation is detectable. The daughter Y^{90}, however, emits a hard β^- radiation, the upper energy limit of which is 2.18 Mev.[24] Sr^{91}, which is shorter lived ($\tau_{1/2}$ = 9.7 hr.),[30] emits a complex spectrum of β^- radiations and γ radiations. The β^--ray components have energies of 2.665, 2.03, 1.36, 1.09, and 0.62 Mev.[31] Corresponding γ radiations with energies of 0.551, 0.64, 0.66, 0.747, 1.025, and 1.413 Mev. are reported.[31]

Sr^{90} is a very dangerous isotope because it has such a long half-life and because it localizes in bone. Recommended tolerance dosages for body retention (70 kg.) are no more than 2 μc. for Sr^{89} and 1 μc. for Sr^{90} together with its daughter, Y^{90}. Ca^{45} can be retained to the extent of 65 μc. in the 70-kg. body.

3. TRACER ISOTOPES OF HALOGENS

A. Fluorine

Fluorine is characterized by its atomic number of 9. For this atomic number there is only one stable nucleus, that with mass number 19. Radioactive isotopes with mass numbers 17 and 20 have been noted but are too short-lived for use in biology. The isotope F^{18} is also short-lived[32] ($\tau_{1/2}$ = 112 min.) but can be applied in a limited way in biochemical investigations. The re-

[25] Pecher, C., *Proc. Soc. Exptl. Biol. Med.* **46,** 86 (1941).

[26] Treadwell, A. de G., Low-Beer, B. V. A., Friedell, H. L., and Lawrence, J. H., *Am. J. Med. Sci.* **204,** 521 (1942).

[27] Powers, R. I., and Voigt, A. F., *Phys. Rev.* **79,** 175 (1954).

[28] Bothe, W., *Z. Naturforsch.* **1A,** 179 (1946).

[29] Meyerhof, W. E., *Phys. Rev.* **74,** 621 (1948).

[30] Finkle, B., Katcoff, S., and Sugarman, N., *in* "Radiochemical Studies: The Fission Products" (C. D. Coryell and N. Sugarman, eds.), Div. IV, Vol. 9, p. 663, National Nuclear Energy Series, McGraw-Hill, New York, 1951.

[31] Ames, D. P., Bunker, M. E., Langer, L. M., and Sorenson, B. M., *Phys. Rev.* **91,** 68 (1953).

[32] Snell, A. H., *Phys. Rev.* **51,** 143 (1937).

actions for its production are $O^{17}(d,n)F^{18}$, $O^{18}(d,2n)F^{18}$, and $Ne^{20}(d,\alpha)F^{18}$. No suitable neutron reactions are available, so production must be carried out by a cyclotron or other high-energy charged-particle accelerator.

The deuteron-induced transmutation of oxygen is more convenient than that of neon because the latter can be bombarded in the gas phase only. Many oxygen-bearing materials can be used as target. Distilled water, lithium oxide, and boron oxide are the best compounds, from the standpoint of both oxygen content and ease of the chemical separation processes for recovery of radioactive fluorine.

Distilled water is bombarded in a bell-jar target. The water is contained in a small volume (2 to 5 ml.) behind a 1-mil aluminum window.[33] The only appreciable contaminant produced is recoil copper radioactivity if deuterons are used. This can be removed by addition of a little copper fluoride as carrier and precipitation of copper with H_2S. The filtrate is made up to the desired volume.

Lithium oxide targets contain most of the F^{18} as LiF, but a small quantity may also be found in the gas phase outside the target. This gaseous activity can be conducted through distilled water to which a small quantity of fluoride is added as carrier. The resultant solution is ready for use. The solid material is washed off the target plate with a small volume of water containing carrier fluoride (about 0.1 mg. in 2 to 3 ml.). The radioactive fluoride solution so obtained is shaken with a large volume of fluorine gas (50 to 100 cm.^{3}). Most of the F^{18} leaves the solution by equilibration with the gaseous fluorine through the electronic exchange reaction

$$F^{*-} + F_{(aq)} \rightleftharpoons F^{*}_{(aq)} + F^{-}$$

Removal of the F^{19} from the solution is virtually quantitative if the gas volume is much greater than the solution volume. The resultant gas can be recovered by solution in the appropriate volume of water.

When B_2O_3 is the target compound, the major fraction of the fluorine activity is found in the gas space. This is probably because recoil compounds formed are volatile borofluorides or oxyfluorides. When lithium oxide is used, a large probability exists for nonvolatile LiF formation whereas with boron oxide neither component reacts immediately to form nonvolatile compounds. Bombardment of boric oxide gives rise to C^{11}, as well as F^{18}, in the gas phase. The active gases, after addition of several cubic centimeters of carrier CO and HF, are led through warm acidified (HF) water, whereby the F^{18} is trapped and separated from the C^{11} present as $C^{11}O$. The small C^{11} contamination surviving this treatment decays

[33] Volker, J. E., Hodge, H. C., Wilson, H. J., and Van Voorhis, S. N., *J. Biol. Chem.* **134,** 543 (1940). These authors used the $O^{18}(p,n)F^{18}$ reaction which is formally equivalent to the $O^{18}(d,2n)F^{18}$ reaction.

sufficiently rapidly so that pure F^{18} decay is observed before one half-life has elapsed (2 hr.). Gas phase yields of 5 to 10 mc. of carrier-free ${}_9F^{18}$ are observed after a 25-μa.-hr. bombardment of B_2O_3 at 16 Mev.

F^{18} emits positrons with maximum energy of 0.635 ± 0.015 Mev.[34] No appreciable γ emission other than the usual annihilation radiation occurs.

Little application of F^{18} as a fluorine tracer has been made to biochemical or physiological investigations. The half-life of 2 hr. and the high yields available indicate the possibility of more extensive use. Among the fields of interest are insecticides, insect physiology, and studies of the mode of action of fluoride as an inhibitor of metal-activated enzyme systems such as phosphatases and carboxylases.

B. Chlorine

Natural chlorine has two isotopes with mass numbers 35 (75 atom per cent) and 37 (25 atom per cent). Three radioactive nuclei with mass numbers 34, 36, and 38 are available as tracers. The lightest and heaviest of these three isotopes, ${}_{17}Cl^{34}$ and ${}_{17}Cl^{35}$, have the short half-lives of 33 min.[35] and 38.5 min.,[36] respectively. Cl^{36} is extremely long lived ($\tau_{1/2} = 3.08 \times 10^5$ yr.).[37]

Cl^{34} is produced by an α-particle-induced transmutation of phosphorus, $P^{31}(\alpha,n)Cl^{34}$, as well as deuteron bombardment of sulfur, $S^{33}(d,n)Cl^{34}$. The radioactive isotope is recovered from target elementary phosphorus by dissolving the phosphorus in a few milliliters of warm 6 N HNO_3 after which the solution is shaken with chlorine gas as carrier, and the vapors are pumped into a vessel cooled with liquid air. A similar procedure may be used for extraction from sulfur. Cl^{34} is a positron emitter.

The positron radiation is complex. Its components are given at maximum energies of 4.5, 2.6, and 1.3 Mev.[38] Gamma radiations with energies of 3.22, 2.10, and 1.16 Mev. are listed,[39] as well as the annihilation radiation accompanying the positrons.

Cl^{36} is available in practical yields only from the uranium pile reactor. Its long half-life precludes production in the cyclotron, by reactions such as $Cl^{35}(d,p)Cl^{36}$, or $S^{36}(d,2n)Cl^{36}$. The uranium pile reactor process, based on the $Cl^{35}(n,\gamma)Cl^{36}$ reaction, can produce samples with specific Cl^{36} content greater than 0.1 mc./g. of chlorine.

${}_{17}Cl^{36}$ may decay either by positron emission or K capture to stable ${}_{16}S^{36}$ or by negatron emission to stable ${}_{18}A^{36}$. Work with highly active

[34] Blaser, J. P., Boehm, F., and Marmier, P., *Phys. Rev.* **75**, 1953 (1949).

[35] Sagane, R., *Phys. Rev.* **50**, 1141 (1936).

[36] Hole, N., and Siegbahn, K., *Arkiv Mat. Astron. Fysik* **33A**, No. 9 (1946).

[37] Bartholomew, R. M., *Can. J. Phys.* **33**, 43 (1955).

[38] Ruby, L., and Richardson, J. R., *Phys. Rev.* **83**, 698 (1951).

[39] Ticho, H. K., *Phys. Rev.* **84**, 847 (1951).

specimens indicates that little or no positron emission occurs.[39a] The only readily apparent radiations are negative β particles with an upper energy limit of 0.714 Mev.[40]

The assay of Cl^{36} depends mainly on the detection of these β particles, which are sufficiently hard for the use of the bell-jar type of thin-window G-M tube but not the thick-walled cylinder tube. The energy is low enough so that appreciable self-absorption occurs in sample materials with thickness greater than 8 to 10 mg./cm.^{2}. Because of the low yields, it is advantageous to develop the assay of the material in the gaseous state, using a high sensitivity electrometer ionization chamber or modified Lauritsen electroscope filled with a heavy gas such as argon.

Cl^{38} is available from neutron or deuteron activation of chlorine, as in $Cl^{37}(n,\gamma)Cl^{38}$ and $Cl^{37}(d,p)Cl^{38}$. To attain high specific activity from the neutron-induced reaction, a compound such as ethyl chloride must be used. The recoil Cl^{38} is separated as Cl^- by precipitation with silver ion. Another neutron reaction, $K^{41}(n,\alpha)Cl^{38}$, can be employed to produce samples with high specific activity. Potassium oxide is bombarded and Cl^{38} separated as AgCl after addition of carrier Cl^-.

Extremely good yields can be obtained in the cyclotron with the (d,p) reactions on chlorides. The best target is LiCl. After bombardment the LiCl (0.5 to 1.0 g.) is washed with a few milliliters of distilled water into a flask through which is aspirated 500 to 1000 cm.^{3} of chlorine gas. Equilibration of Cl^- and Cl_2 by electronic exchange reaction, i.e., $Cl^{*-} + Cl_2 \rightarrow Cl^- + Cl^*_2$, results in removal of Cl^{38} into the gas phase. The chlorine and solution vapors are condensed in a vessel cooled with liquid air. The amount of Cl^{38} recovered in the gaseous chlorine depends on the relative quantities of chlorine in the gas and solution phases. A 25-μa.-hr. bombardment at 16 Mev. yields some 500-mc. equivalents of Cl^{38}, recovered as chlorine gas by the above procedure.

Cl^{38} emits three distinct continuous spectra[41, 41a] of negative β particles, with maximum energies 1.16 (30.8%), 2.80 (15.8%), and 4.99 Mev. (53.4%). The 4.99-Mev. β rays are among the most energetic found in artificial radioactivity. Beta radiations with energies of 1.64 and 2.19 Mev. are also found. These are the energies expected, since they are differences of the β-spectra energies, i.e., $4.99 - 2.80 = 2.19$, and $2.80 - 1.16 = 1.64$.[41b] These radiations are readily detectable and assay procedures parallel those for C^{11}.

[39a] Johnston, F., and Willard, J. E., *Phys. Rev.* **75,** 528 (1949).

[40] Feldman, L., and Wu, C. S., *Phys. Rev.* **87,** 1091 (1952).

[41] Watase, Y., *Proc. Phys.-Math. Soc. Japan* **23,** 618 (1941).

[41a] Langer, L. M., *Bull. Am. Phys. Soc.* **24,** 8 (1949).

[41b] Similar data of more recent origin are given by Langer, L. M., *Phys. Rev.* **77,** 50 (1950).

TABLE 38
RADIOACTIVE ISOTOPES OF BROMINE

Mass Number	Half-life	Production Reactions	Radiations
78	6.4 min.[42]	$Se^{77}(d,n)Br^{78}$	Positrons, $E_{max} = 2.3$ Mev.; γ radiations: annihilation (0.5 Mev.), and components with $E = 0.46$ and 0.108 Mev.[43]
80	4.4 hr.[44]	$Br^{79}(n,\gamma)Br^{80}$ $Br^{79}(d,p)Br^{80}$ $Se^{80}(d,2n)Br^{80}$	Gamma-ray emitter, components with $E = 0.049$ and 0.037 Mev.; also[45] x-rays from K capture to Se^{80}.
	18 min.	Same as 4.4-hr. isomer	Negatrons, $E_{max} = 1.99$ and 1.1 Mev.[46]
82	35.87 hr.[47]	$Br^{81}(n,\gamma)Br^{82}$	Negatrons, $E_{max} = 0.465$ Mev.; each β disintegration associated with 3 γ rays in cascade, $E = 0.547$, 0.787, and 1.35 Mev.[48]
83	140 min.[49]	$Se^{82}(d,n)Br^{83}$ Fission product from uranium or Th+ neutrons	Negatrons, $E_{max} = 0.94$ Mev.,[50] no γ radiation.

Most researches with tracer chlorine have been carried out with Cl^{38}. These researches resemble those described for radioactive sodium, being concerned primarily with the distribution of chloride ion in intracellular and extracellular spaces of mammalian organisms.

C. BROMINE

The stable isotopes of bromine possess mass numbers of 79 and 81, the abundance of each being very close to 50 atom per cent. Some unstable isotopes, with their properties, are listed in Table 38. In addition to the isotopes shown, there are a number of bromine activities found in the fission products from uranium and thorium. These isotopes have mass numbers

[42] Snell, A. H., *Phys. Rev.* **52,** 1007 (1937).
[43] Valley, G. E., and McCreary, R. L., *Phys. Rev.* **56,** 863 (1939).
[44] Alichanian, A. I., Alichanov, A. I., and Dzelepow, B. S., *Phys. Z. Sowjetunion* **10,** 78 (1936).
[45] Siday, R. E., *Proc. Roy. Soc.* **178A,** 189 (1941).
[46] Laberrique-Frolow, L. J., *Compt. rend.* **234,** 2599 (1952).
[47] Cobble, J. W., and Atterbury, R. W., *Phys. Rev.* **80,** 917 (1950).
[48] Roberts, A., Downing, J. R., and Deutsch, M., *Phys. Rev.* **60,** 544 (1941).
[49] Langsdorf, A., and Segrè, E., *Phys. Rev.* **57,** 105 (1940).
[50] Duffield, R. B., and Langer, L. M., *Phys. Rev.* **81,** 203 (1951).

greater than 83 and are for the most part too short-lived for use as tracers.[51] A number of bromine isotopes with atomic numbers less than 78 appear to exist. These could be useful as tracers particularly because they can be produced in high specific activity from selenium or arsenic by α-particle bombardment.[52] Thus, Br^{77} ($\tau_{1/2} = 57$ hr.)[53] decays 95% by K capture and 5% by positron emission. The upper energy limit of the positrons is 0.336 Mev.[54] A complex γ spectrum is noted, in addition to the annihilation radiation, with energy components 0.813, 0.641, 0.520, 0.298, 0.284, 0.237, and 0.160 Mev.[54]

Of the isotopes listed, the most applicable is Br^{82}. Adequate yields of this isotope are obtainable with the cyclotron by means of a Szilard-Chalmers process based on neutron irradiation of organic bromides, such as tetrabromoethane, bromoform, or ethyl bromide. Excessive neutron intensities, as in the uranium pile reactor, require more stable materials such as metallic bromides.

The procedure for preparation and extraction of Br^{82} from an irradiated ethyl bromide target follows: 10 to 20 kg. of c.p. ethyl bromide is exposed to the maximal flux of neutrons from the $Be^9(d,n)B^{10}$ reaction in the cyclotron. A deuteron bombardment of beryllium for a few hundred microampere-hours suffices. The ethyl bromide is extracted three times in a large funnel with a small volume of water. No carrier bromide need be added because sufficient bromide is present from decomposition of the ethyl bromide. The volume of water should not exceed half that of the ethyl bromide because it is desirable to evaporate the water extract to a small volume. Five to ten millicuries of Br^{82} in as little as 100 mg. of bromide can be recovered in this fashion. A small quantity of sodium thiosulfate may be added to ensure conversion of all recoil Br^{82} to Br^-.

The labeled bromine obtained as bromide can be used directly for experiments on the distribution of bromide in intracellular and extracellular space, as in the experiments cited for sodium, potassium, and chloride. In other applications it is desirable to convert the Br^{82} to the form of elementary bromine. If very large samples are available in which dilution is not critical, this may be accomplished by shaking the slightly alkaline solution of bromide recovered in the Szilard-Chalmers process with liquid bromine. An exchange reaction between Br_2 and Br^- equilibrates the Br^{82} so that, on subsequent acidification, the bromine expelled will contain most of the Br^{82}. For the sample obtained as described above, a satisfactory

[51] Hollander, J. M., Perlman, I., and Seaborg, G. T., *Revs. Mod. Phys.* **25**, 469 (1953).

[52] Woodward, L. L., McCown, D. A., and Pool, M. L., *Phys. Rev.* **74**, 870 (1948).

[53] Hopkins, H. H., Jr., and Cunningham, B. B., quoted in reference 51.

[54] Canada, R., and Mitchell, A. C. G., *Phys. Rev.* **83**, 955 (1951).

procedure is to transfer the concentrated aqueous extract (up to 100 ml.) to a distilling flask. The bromide is oxidized to bromine by dropwise addition of MnO_2 suspended in H_2SO_4. The bromine is expelled with gentle heating and collected in a vessel cooled with dry ice-acetone mixture.[55]

D. Iodine

Iodine is a simple element with the stable isotope $_{53}I^{127}$. Some radioactive isotopes are shown in Table 39. The important isotopes are those with mass numbers 130 and 131 because they alone are available in sufficient yields for tracer purposes.

In the cyclotron, the iodine isotopes are produced by bombardment of tellurium with deuterons. I^{131} yields of 50 to 100 mc. can be obtained with bombardments of 5000 μa.-hr., at 16 Mev. Much higher yields of the shorter-lived I^{130} are obtained. Because the target is tellurium, these quantities can be obtained in carrier-free amounts.

Te^{131} is also produced and decays by an isomeric transition ($\tau_{1/2} = 30$ hr.) to its isomer, a negatron emitter ($\tau_{1/2} = 25$ min.) which decays to I^{131}. Hence tellurium targets continue to form I^{131} after bombardment. The maximum I^{131} activity is reached 3 to 4 days after the end of the bombardment.

Tellurium may be bombarded internally in the cyclotron as a component in a eutectic mixture of copper, cobalt, and tellurium, which assays 61% in tellurium.[56] More often it is bombarded as tellurium metal in helium in a bell-jar target. After bombardment, the tellurium metal powder is washed, with the aid of a stiff toothbrush, into an all-glass distilling apparatus. Carrier iodine (as iodide) is added, usually in quantities of 0.5 to 1.0 mg. From 20 to 25 ml. of dilute sulfuric acid (6 to 8 *N*) is added, and the suspension is brought slowly to the boiling point. Concentrated HNO_3 (25 to 40 ml.) is added slowly from a dropping funnel to oxidize the iodide to iodine. The vapors are caught in a receiving flask containing CCl_4. The CCl_4 solution of iodine, together with the condensed nitric acid phase, is transferred to a small separatory funnel, 25 ml. of distilled water is added, and the funnel is shaken vigorously. The CCl_4 is washed twice in this fashion. It is then shaken with about 25 ml. of water containing an excess of sodium bisulfite (approximately 2 equivalents per equivalent of iodine added as carrier). The iodine is reduced to iodide, which passes into the aqueous layer. The CCl_4 layer is washed twice with small portions of water. The washings are combined with the first water-sulfite extract and filtered to remove droplets of CCl_4. The procedure requires no more than 1 hr. after removal of tellurium from the target.

To recover the radioactive iodine carrier-free, a somewhat different

[55] Numerous syntheses of a variety of organic radiohalides can be found in the literature. See Bloch, H. S., and Ray, F. E., *J. Natl. Cancer Inst.* **7**, 61 (1946).

[56] Irvine, J. W., Jr., private communication, 1949.

TABLE 39

SOME RADIOACTIVE ISOTOPES OF IODINE

Mass Number	Half-life	Production Reactions	Radiations
124	4.5 days[57]	$Sb^{123}(d,n)I^{124}$ $Te^{124}(p,n)I^{124}$ [58]	Positrons, E_{max} = 2.20, 1.50, 0.67 Mev.; annihilation γ, also γ radiation, E = 0.603,[59] 0.73, 1.72, 1.95 Mev.
125	60.0 days[60]	$Te^{124}(d,n)I^{125}$	K capture (Te x-rays), E = 0.035 Mev.[60]
126	13.0 days[61]	$Te^{125}(d,n)I^{126}$ $Te^{126}(d,2n)I^{126}$	Negatrons, E_{max} = 1.268 and 0.085 Mev.; γ radiation, E = 0.395 Mev.[59]
128	24.99 min.[62]	$I^{127}(n,\gamma)I^{128}$ [63] $Te^{128}(d,2n)I^{128}$	Two negatrons, E_{max} = 1.59 and 2.02 Mev. (93%);[64] γ radiation, E = 0.4 Mev.
129	1.72×10^7 yr.[65]	Uranium fission	Negatrons, E_{max} = 0.12 Mev.;[66] γ radiation, E = 0.039 Mev.[66]
130	12.6 hr.[67]	$Te^{130}(d,2n)I^{130}$	Two negatrons, E_{max} = 0.61 and 1.03;[68] γ radiation, E = 0.417, 0.537, 0.667, 0.744 Mev.
131	8.075 days[69]	$Te^{130}(d,n)I^{131}$ Decay of Te^{131} Uranium fission	Four negatrons, E_{max} = 0.60 (87.2%), 0.335 (9.3%), 0.250 (2.6%), and 0.815 (0.7%) Mev.; γ radiations are listed[70] with energies of 0.720, 0.638, 0.364, 0.284, 0.160, and 0.080 Mev.; about 1% decay occurs to a 12-day isomeric state of Xe^{131}, which has a highly converted γ ray.[71]

[57] Marquez, L., and Perlman, I., *Phys. Rev.* **78,** 189 (1950).

[58] L. DuBridge, quoted by Seaborg, G. T., and Perlman, I., *Revs. Mod. Phys.* **20,** 617 (1948).

[59] Mitchell, A. C. G., Mei, J. Y., Maienschein, F. C., and Peacock, C. L., *Phys. Rev.* **76,** 1450 (1949).

[60] Friedlander, G., and Orr, W. C., *Phys. Rev.* **84,** 484 (1951).

[61] Livingood, J. J., and Seaborg, G. T., *Phys. Rev.* **54,** 775 (1938).

[62] Hull, D. E., and Seelig, H., *Phys. Rev.* **60,** 553 (1941).

[63] Amaldi, E., D'Agostino, O., Fermi, E., Pontecorvo, B., Rasetti, F., and Segrè, E., *Proc. Roy. Soc.* **149A,** 522 (1935).

[64] Siegbahn, K., and Hole, N., *Phys. Rev.* **76,** 133 (1946).

[65] Katcoff, S., *in* "Radiochemical Studies: The Fission Products" (C. D. Coryell and N. Sugarman, eds.), Div. IV, Vol. 9, p. 980, National Nuclear Energy Series. McGraw-Hill, New York, 1951.

[66] Butement, F. D. S., *Proc. Phys. Soc. (London)* **64A,** 395 (1951).

[67] Livingood, J. J., and Seaborg, G. T., *Phys. Rev.* **54,** 775 (1938).

[68] Roberts, A., Elliott, L. G., Downing, J. R., Peacock, W. C., and Deutsch, M., *Phys. Rev.* **64,** 268 (1943).

[69] Seliger, H. H., Cavallo, L., and Culpepper, S. V., *Phys. Rev.* **90,** 443 (1952).

[70] Bell, R. E., and Graham, R. L., *Phys. Rev.* **86,** 212 (1952).

[71] Graham, R. L., and Bell, R. E., *Phys. Rev.* **84,** 380 (1951).

procedure is recommended.[72] After transfer of the tellurium powder to the all-glass distilling apparatus, 15 ml. of 50% CrO_3 and 30 ml. of 18 N H_2SO_4 are added, and the mixture is refluxed for 0.5 hr.[73] The tellurium is brought into solution by this procedure, in which the iodine is oxidized to its higher valence states. The solution is cooled and excess oxalic acid (20 g.) added, along with 50 ml. of 18 N H_2SO_4. The solution is distilled into receivers containing 1 ml. of 1 N NaOH and 1 drop of 0.01 N sodium thiosulfate. Nearly ten fractions, each approximating 50 ml., are distilled before removal of iodine is complete. Water should be added to the boiling solution when the sulfuric acid becomes too concentrated. All these fractions are combined and concentrated to a small volume. The solution so obtained is transferred to a distilling flask, and 18 N H_2SO_4 is added until no further evolution of CO_2 can be detected. Finally, 25 ml. of 10 N H_2SO_4 is added, and six to eight fractions are distilled, each fraction being received in 1 ml. of 0.75 N NaOH containing 1 drop of 0.01 N sodium thiosulfate. In these operations, it is important that the sulfuric acid does not become so concentrated that it fumes and renders the distillate acid. The distillates are collected, concentrated, and made up to a suitable volume. The pH is adjusted with acid to the desired value.

I^{131} is a product of the fission of uranium. The uranium pile has become the major source of supply for this isotope. The 12.6-hr. iodine (I^{130}) must still be produced in the cyclotron, however.

When I^{131} samples are received in carrier-free form from uranium pile installations they require dilution for radioactive assay. It is necessary to keep assay solutions slightly alkaline to avoid loss of I^{131} by volatilization as I_2 formed by oxidation in air in heated neutral or acid solutions. A solution which is convenient for dilutions of carrier-free samples of high activity is made up with the following composition: NaOH, 0.8 g.; $NaHSO_3$, 0.44 g.; KI, 0.25 g.; water to make 1 liter. This solution contains a reducing agent in addition to alkali to ensure maintaining the I^{131} as iodide ion.

The 26-min. iodine (I^{128}) is obtained in good yields by irradiation of ethyl iodide and extraction of recoil iodide in a manner similar to that described for radiobromine. This isotope has been used in early investigations to show rapid selective uptake in rabbit thyroid.[74]

[72] Perlman, I., Morton, M. E., and Chaikoff, I. L., *Endocrinology* **30,** 487 (1942); see also Morton, M. E., Perlman, I., and Chaikoff, I. L., *J. Biol. Chem.* **140,** 603 (1941); Ballantine, D. S., and Cohn, W. E., U. S. Atomic Energy Comm. MDDC-1600 (1947).

[73] Levy, M., Keston, A. S., and Udenfriend, S., *J. Am. Chem. Soc.* **70,** 2289 (1948), use a procedure in which the tellurium metal is heated with dry alkali, followed by extraction and recovery of iodine.

[74] Hertz, S., Roberts, A., and Evans, R. D., *Proc. Soc. Exptl. Biol. Med.* **38,** 510 (1938).

By far the most extensive tracer investigations with the halogens have been carried out with the radioactive iodine isotopes I^{130} and I^{131}. A large bibliography exists on tracer iodine because of the importance of this element in thyroid function.

E. Radiation Hazards

It may be remarked that radiation hazards attendant on the clinical use of radiohalides can be estimated fairly well because of the large background of physiological research on these elements. Thus, fluoride, chloride, and bromide are largely extracellular and hence can be treated somewhat like sodium in so far as recommendations about tolerance dosage are concerned. Thus, F^{18} with its 2-hr. half-life and 0.7-Mev. positron has one-seventh the half-life of Na^{24} and half the upper energy of the Na^{24} β particle which allows an added increase by a factor of 10 to 14 in retained dosage. Thus, it would appear that, if total Na^{24} body retention allowed is 15 μc., then F^{18} can be retained in dosages of 150 μc. per 70 kg. Similarly, Cl^{36}, Cl^{38}, and Br^{82} would be allowed in the ranges 75, 150, and 15 μc. per 70 kg., respectively.

I^{131} is taken up with great efficiency by thyroid and requires different treatment from the health physics standpoint. If C is concentration of I^{131} in millicuries per gram thyroid, then the dosage in rep. is given by $158{,}000C$.[75] General experience indicates average retention of a given dose of I^{131} in thyroid of normal humans is 10 to 30%, with the average thyroid weight for a 70-kg. man being about 30 g.

Thyroid is quite radioresistant, and a total dose of 100 rep. is conservative for safety. According to the relation given above, it would take $100 \times 30/158{,}000 = 0.019$ mc. I^{131} retained to deliver 100 rep. to a 30-g. thyroid, assuming no excretion. Taking an average figure of 20% for retention of an administered dose, one arrives at a figure of 5×0.019 or 0.095 mc. I^{131} administered. Biological elimination through turnover adds a factor of safety of approximately 2, so that the total dosage figure is about 0.2 mc.[76]

4. TRACE ELEMENTS

A. Introduction

In this section are considered isotopic tracers of some elements present in living organisms in relatively low amounts, but essential to normal function. Among these trace elements—so called because they appear in such minute amounts—are manganese, iron, cobalt, copper, zinc, and molybdenum, in order of increasing atomic number. Other elements which are not dietary essentials but which are of interest because of their effects

[75] Marinelli, L. D., Quimby, E. H., and Hine, G. J., *Radiology* **51,** 564 (1948).

[76] More recent tolerance doses have been less conservative, with total safe retention in body being given as 0.3 μc.

in living tissue are also considered. These are arsenic, antimony, silver, gold, and mercury.

Data on metabolism and proved function of the trace elements are still meager. For the most part, tracer studies to date have dealt with simple distribution and retention of single or multiple doses of these elements at low concentrations. It has been possible to note where activity appears. Information on specific chemical compounds with which such activity is associated cannot be supplied until analytical methods appropriate for assay of both radioactive and carrier isotopes have been elaborated. In the discussion which follows, emphasis is placed on isotopes used so far; no attempt has been made to list all possible isotopes. A list can be found in the table in Appendix 4.

B. Manganese

Although little is known of its specific biological role, manganese has been found to serve as an activator of enzymes such as arginase, phosphoglucomutase, carboxylases, and peptidases. At the physiological level, studies on rats show that manganese aids lactation and prevents degeneration and atrophy of the testes. In the chick, it is effective in preventing the onset of the bone condition called perosis. The requirement for manganese in living organisms is very small; dosages must be administered with maximal specific activity. An example of research with labeled manganese has already been given (see p. 240). Information on the mode of retention and excretion of manganese administered in biological doses is available from a number of studies.

Natural manganese is a simple element, with one stable isotope of mass number 55. It has radioactive isotopes with mass numbers 51, 52, 54, and 56. The heaviest of these, Mn^{56}, has a half-life of only 2.6 hr. and is available only in laboratories close to a transmutation machine. Because Mn^{56} is the only manganese isotope available from slow neutron activation, the supply of manganese isotopes will continue to be furnished by the cyclotron or some other charged-particle accelerator. Of the isotopes available only two need be considered here, those with mass numbers 52 and 54.

Mn^{52} appears to exist in two isomeric states, with half-lives of 21.3 min. and 6.0 days.[77-79] Positrons are emitted by both isomers, the 21-min. positrons having an upper energy limit of 2.66 Mev.[80] A γ ray with energy of 1.46 Mev.[80] accompanies this positron as well as the usual annihilation radiation. The 6.0-day isomer[81] decays, by both positron emission and *K*

[77] Livingood, J. J., and Seaborg, G. T., *Phys. Rev.* **54,** 391 (1938).

[78] Hemmendinger, A., *Phys. Rev.* **58,** 929 (1940).

[79] Handley, T. H., quoted in reference 51.

[80] Osborne, R. K., and Deutsch, M., *Phys. Rev.* **71,** 467 (1947) (abstract).

[81] Good, W. M., Peaslee, D., and Deutsch, M., *Phys. Rev.* **69,** 680 (1946).

capture, to Cr^{52}. Sixty-five per cent of the disintegrations go by K capture. Positrons are emitted with a single upper energy of 0.582 Mev. Each positron is accompanied by a cascade of three γ rays with energies of 1.46, 0.73, and 0.94 Mev.[82]

Mn^{54}, which is long-lived ($\tau_{1/2}$ = 310 days),[77] decays by K capture to an excited state of Cr^{54}, which then drops to its ground state, emitting a γ ray with energy of 0.85 Mev.[83]

Both these manganese isotopes can be produced from iron by deuteron bombardment, according to the reactions $Fe^{54}(d,\alpha)Mn^{52}$ and $Fe^{56}(d,\alpha)Mn^{54}$. Because of the high abundance of Fe^{56} (91.6 atom per cent) compared to Fe^{54} (6.0 atom per cent), the production of Mn^{54} is favored. Mn^{54} can also be produced by the $Fe^{54}(n,p)Mn^{54}$ reaction, but the yield in this case is too low to warrant its use, except possibly in the uranium pile reactor. Typical yields for Mn^{52} are 50 to 100 μc. for 2000 μa.-hr. of deuteron bombardment of iron at 14 Mev. Longer bombardments can give Mn^{54} in yields up to 0.2 mc.

The iron target is dissolved in HCl, and the iron is removed by extraction with ether (see Section 4-C). The aqueous layers, containing cobalt, manganese, and phosphorus activities, are evaporated to dryness to remove ether and HCl. The residues are taken up in 0.5 M HNO_3, with the addition of some sodium phosphate as carrier. The phosphate is removed by two precipitations with bismuth in 0.5 M HNO_3. The filtrate is evaporated to dryness and dissolved in concentrated HNO_3. After addition of a minimal quantity of carrier manganese sulfate, the manganese is precipitated as MnO_2 by the addition of a few crystals of $KClO_3$. This precipitate is collected on a sintered-glass filter and redissolved with H_2O_2. Two more precipitations are required to free the manganese of traces of radioactive cobalt.

The final solution is made by dissolving the MnO_2 in dilute HCl, evaporating to dryness to get rid of excess acid, and then bringing the solution to the appropriate volume by addition of water to the residue. Greenberg and Campbell have given procedures for the preparation of manganese samples for assay under a variety of conditions.[84]

Mn^{52} free of Mn^{54} may easily be prepared in 100-mc. amounts by deuteron bombardment of chromium, according to the reaction $Cr^{52}(d,2n)Mn^{52}$. The copper target is plated with chromium for the bombardment. To extract the Mn^{52}, the chromium must be scraped off the copper and dissolved by fusion with sodium peroxide, followed by leaching with water. Solution with HCl is slow but is advantageous in minimizing copper contamination.

[82] Peacock, W. C., and Deutsch, M., *Phys. Rev.* **69**, 306 (1946).

[83] Deutsch, M., and Elliott, L. G., *Phys. Rev.* **65**, 211 (1944).

[84] Greenberg, D. M., and Campbell, W. W., *Proc. Natl. Acad. Sci. U. S.* **26**, 448 (1940).

The manganese can be recovered as the peroxide in a manner similar to the procedure described above. The use of Mn^{52}, free of Mn^{54}, is indicated in certain therapeutic applications in which the long half-life of the Mn^{54} is not advantageous. No estimates are available on tolerance dosage.

C. Iron

1. Tracer Researches. Iron is found in practically all tissues. Its most familiar function is as the central atom in ferriheme, which is the prosthetic group of the hemoglobin complex in mammalian erythrocytes and muscle tissue. In addition, ferriheme is present in all mammalian cells in minute amounts as the prosthetic group of catalase, peroxidases, and the cytochrome pigments. From these examples, iron is seen to occur mainly as an essential component of enzymes involved in oxidative processes.

The use of tracer iron to aid in unraveling the biochemical role of iron as well as the physiological aspects of iron absorption, mobilization, and excretion has been exploited extensively. The major finding has been that absorption of iron from the gut in dogs is regulated by the dog's iron reserves. If these are high, little or no iron is absorbed. This situation contrasts with that of other mineral constituents, such as phosphate, in which a continuous turnover between ingested and reservoir material is observed, with the over-all level being maintained by balance between absorption and excretion. The animal body is very efficient in utilizing its iron and excretion is maintained at a low level under normal conditions.

The nature of the reservoir iron in tissues has been established by the work of Granick and Michaelis.[85] Extending the observations of Laufberger,[86] these workers have identified the stored iron in mucosa and other tissues with a remarkable iron-containing protein, "ferritin," which has been found to contain as much as 23% by weight of iron. This protein has been isolated from bone marrow, spleen, and liver, as well as in gastrointestinal mucosa. Removal of iron can be accomplished without denaturing the protein moiety, called "apoferritin." With labeled iron, Hahn *et al.*[87] demonstrated the conversion of heme iron of the red blood cells into iron which could be isolated as a constituent of crystalline ferritin. On this basis, it has been possible to devise a scheme for the mechanism of iron absorption, transport, storage, and function. Briefly, the mechanism postulated is entry of iron in the ferric state into the gastrointestinal tract with the food. The iron is converted to the ferrous state by reducing agents in the food and in the tissues. Absorption of ferrous iron occurs mainly in the

[85] See Granick, S., *Chem. Revs.* **38,** 379 (1946), for a general discussion.

[86] Laufberger, V., *Bull. soc. chim. biol.* **19,** 1575 (1937).

[87] Hahn, P. F., Granick, S., Bale, W. F., and Michaelis, L., *J. Biol. Chem.* **150,** 407 (1943).

mucosal cells of the duodenum and jejunum. The presence of iron evokes an increase in concentration of apoferritin, which combines with the iron to form ferritin, the latter accumulating as reserve iron. Regulation of iron absorption is maintained by the equilibrium between ferritin, plasma iron in the blood serum and ferrous iron in the mucosa, the level of which is fixed by oxidation-reduction equilibria.

2. Tracer Isotopes. The isotopic composition of natural iron is Fe^{54} (6.0%), Fe^{56} (91.6%), Fe^{57} (2.1%), and Fe^{58} (0.28%). Radioactive isotopes are known with mass numbers 53, 55, and 59. Of these three, the two heavier isotopes, Fe^{55} and Fe^{59}, are available as tracers.

Most of the tracer research with iron has been accomplished with Fe^{59} ($\tau_{1/2}$ = 47 days).[88]

The β^- radiations consist of three groups with energy maxima given as 1.560 (0.3%), 0.462 (53.9%), and 0.271 (45.8%) Mev.[89]

Because of the large fraction of radiation which is of low energy, assay based on counting of β radiation involves preparation of thin samples to minimize self-absorption. Before the advent of the scintillation detector this type of procedure was most frequent. Detailed instructions for the preparation of Fe^{59} samples by ashing of biological material, solution, and electroplating were available.[90, 91] Now, however, it is no longer necessary to count the β radiations. Instead, the γ radiations can be used for assay by exploiting the highly efficient scintillation detectors. From the standpoint of the biologist, the direct counting of liquid samples with good efficiency makes the use of the scintillation detector the method of choice.

Fe^{59} is produced either by the deuteron reaction on the rare iron isotope Fe^{58} ($Fe^{58}(d,p)Fe^{59}$), or by neutron irradiation of cobalt according to the reaction $Co^{59}(n,p)Fe^{59}$. The neutron reaction has in its favor the 100% abundance of the cobalt isotope, but the cross section is low. Fe^{59} devoid of Fe^{55} can be produced only by transmutation of cobalt.

The metal can be bombarded internally or externally in the cyclotron. To obtain maximum specific activity, the amount of iron used is kept minimal. After bombardment, the surface layers of iron are dissolved in dilute HCl. Contaminating copper is removed by precipitation with H_2S. The filtrate is boiled to expel H_2S, and the iron is oxidized to the ferric state by careful addition of H_2O_2. The iron is precipitated as the hydroxide with ammonia, filtered, and redissolved in HCl to which carrier cobalt

[88] Livingood, J. J., and Seaborg, G. T., *Phys. Rev.* **54,** 51 (1938). Later values for the half-life of Fe^{59} are 47.1 and 45.1 days; see Tobailem, J., *Compt. rend.* **233,** 1360 (1951), and Schuman, R. P., and Camilli, A., *Phys. Rev.* **84,** 158 (1951).

[89] Metzger, F. R., *Phys. Rev.* **88,** 1360 (1952).

[90] Ross, J. F., and Chapin, M. A., *Rev. Sci. Instr.* **13,** 77 (1942).

[91] Dunn, R. W., *Nucleonics* **10,** No. 8, 40 (1952).

and manganese chlorides are added. The radioactive iron may be purified by extraction with diethyl ether in 6.4 *N* HCl[91] or by precipitation with pyridine.[92] Yields obtained vary between 0.03 and 0.05 μc./μa.-hr. at 12 to 16 Mev.

Production of Fe^{59} admixed with about four times as much of the long-lived Fe^{55} is based on neutron irradiation of iron in the uranium pile. Specific activities for Fe^{59} are quoted at greater than 1 mc./g. iron. Samples of Fe^{59} with less than 10% Fe^{55} contaminant can also be obtained by neutron activation of iron enriched in Fe^{58} isotope. These samples are quoted as having specific activities greater than 500 mc./g. iron and are obtainable in millicurie lots.

Procedures of the Szilard-Chalmers type for preparation of high-specific activity Fe^{59} from uranium pile irradiation of potassium ferrocyanide crystals or hydroferrocyanic acid[93] have been described.

In the case of the potassium ferrocyanide, 30 g. of crystals are bombarded, followed by solution of crystals in 250 ml. of cold distilled water and precipitation of 50 mg. of added aluminum nitrate carrier as hydroxide ($pH = 7.5$). After centrifugation, the active precipitate is washed twice with distilled water, redissolved in dilute HNO_3, and reprecipitated. The cycle is repeated three times.[94]

Hydroferrocyanic acid has a nominal advantage in that no K^{42} activity is formed. The procedure for extraction of the Fe^{59} involves merely extraction of the solid with dry, dilute ethereal HCl in a Soxhlet apparatus. Specific activity of samples obtained after a 1-month irradiation at relatively low neutron flux (2.5×10^{10}/cm.^{2}/sec.) is 1.5 mc./g.

Fe^{55} ($\tau_{1/2} = 2.94$ yr.)[95] disintegrates solely by *K* capture to Mn^{55}. Consequently the only radiations observed are the manganese x-rays and very soft conversion electrons.[96] The energy of the x-radiation is only 0.0065 Mev. This radiation is considerably softer than that of Fe^{59}, intensity being reduced to half value by approximately 10 mg. of ash.[97] Solid iron samples must be prepared for assay in the same manner[91] as for Fe^{59}, but the sample thicknesses must be kept to less than 1 to 2 mg./cm.^{2}.

Fe^{55} is produced in the cyclotron by deuteron transmutation of manganese, according to the reaction $Mn^{55}(d,2n)Fe^{55}$. The radioactive iron is obtained in higher specific activities than are available with Fe^{59} because

[92] Ray, T. W., *J. Lab. Clin. Med.* **25,** 745 (1940).

[93] Dewhurst, H., and Miller, N., Natl. Research Council (Canada), Atomic Energy Project Rept. CRC-351 (1947).

[94] Kenny, A. W., and Maton, W. R. E., *Nature* **162,** 567 (1948).

[95] Bouchez, R., deGroot, S. R., Nataf, B., and Tolhoek, H. A., *J. phys. radium* **11,** 105 (1950).

[96] Miller, F., Jr., *Phys. Rev.* **67,** 309, (1945).

[97] Peacock, W. C., Evans, R. D., Irvine, J. W., Jr., Good, W. M., Kip, A. F., Weiss, S., and Gibson, J. G., II, *J. Clin. Invest.* **25,** 605 (1946).

the target nucleus, manganese, is chemically separable. The target is usually manganese metal-bonded to a water-cooled copper surface. After irradiation, the manganese is scraped into a beaker, dissolved in 6 *N* HCl, and the resultant solution diluted to 1 *N* in HCl. Copper is removed by precipitation with H_2S. Carrier cobalt, zinc, and phosphate are added along with the desired quantity of iron. Iron is precipitated with cupferron in the cold,[98] leaving the contaminating radioactivities in the filtrate. The precipitate is ashed and redissolved in acid. The precipitation is repeated three or four times in the presence of carriers for manganese, cobalt, etc., until the sample obtained gives only the weak characteristic radiation of Fe^{55}. In an alternative procedure manganese is removed as the dioxide before separation of iron. Yields are in the range 0.015 to 0.020 μc./μa.-hr. at 12 to 16 Mev.

Fe^{55} with less than 10% Fe^{59} is available from uranium pile neutron activation of targets enriched in Fe^{54}. Millicurie lots are supplied with specific activities greater than 500 mc./g. iron.

The great difference in radiation characteristics of Fe^{55} and Fe^{59} provides a convenient basis for experiments involving double labeling of iron. Apparatus for distinguishing the two isotopes in a given sample has been described,[99] consisting of two types of G-M tube counters. One thin-window G-M tube counter filled to atmospheric pressure with helium is thirty times as sensitive to Fe^{59} as to Fe^{55}. This is owing, on the one hand, to the fact that helium is less sensitive to the soft x-radiation from Fe^{55} than to the more energetic β radiation of Fe^{59}, and, on the other hand, to preferential absorption of the Fe^{55} x-rays by the 2.7-mg./cm.^{2} mica window. Likewise, with a rather thick (0.76-mm.) beryllium window which transmits about 55% of the Fe^{55} x-rays (because of the low atomic number of beryllium) and only about 5% of the Fe^{59} radiation, and with a gas filling of high atomic number, i.e., argon at approximately atmospheric pressure, a G-M tube counter is produced which counts practically only the Fe^{55} radiation.

Total body retention for Fe^{59} is recommended not to exceed 11 μc.; for Fe^{55}, total retention should be no greater than 1 mc.[3]

D. Cobalt

Cobalt deficiencies are known to cause a variety of plant and animal diseases. The cobalt required to counteract disease manifestations is a trace amount. Little is known about the function of cobalt in metabolism. Numerous experiments indicate that the cobalt requirement is highest for ruminants. Even for ruminants this requirement is exceedingly small, being of the order of 0.1 part per million in the diet. In rabbits as little as

[98] Smith, G. F., *Cupferron and Neo-Cupferron*, Columbus, Ohio (1938).

[99] Peacock, W. C., and Good, W. M., *Rev. Sci. Instr.* **17**, 255 (1946).

0.0024 part per million fails to produce deficiency symptoms.[100] In ruminants the cobalt is thought to mirror a requirement for nutrition of microorganisms in the rumen.[101] Extensive studies on cobalt nutrition in sheep have been carried out in Australia, and the results appear to bear out this conclusion.[102, 103]

Cobalt appears to be uniquely associated[104] with Vitamin B_{12}, a fact probably of significance in relation to the well-known production of abnormalities in erythropoietic activity brought about by cobalt deficiency. That a polycythemia can be induced by low-cobalt diet is well authenticated.[105, 106] Cobalt activates the enzyme arginase.[107]

Large amounts of cobalt inhibit growth and respiration of microorganisms. Histidine and histidine-like compounds overcome this inhibition. It appears that this may be ascribed to the chelation of cobalt by histidine which results in a cobalt-histidine complex. This complex—in analogy to the iron porphyrin protein hemoglobin—shows the interesting property of combining reversibly with molecular oxygen.[108]

Application of tracer methods to cobalt metabolism has not been extensive, although adequate tracers have been available.

Considerable data are available on tissue distribution of cobalt administered to young calves orally or intravenously as cobaltous chloride.

Medical interest in cobalt is concerned mainly with its use as a radium substitute.[109, 110]

Natural cobalt has one stable isotope, Co^{59}. As shown in Appendix 4, a large number of radioactive isotopes are available for tracer studies. Good production of cobalt isotopes is obtained from deuteron bombardment

[100] Thompson, J. F., and Ellis, G. H., *J. Nutrition* **34,** 121 (1947). Miller, F., Jr., *Phys. Rev.* **67,** 309 (1945).

[101] McCance, R. A., and Widdowson, E. M., *Ann. Rev. Biochem.* **13,** 315 (1944).

[102] Marston, H. R., *Ann. Report, Div. of Biochem. and Gen. Nutrition* (Australia), 1947–1948; also *Ann. Rev. Biochem.* **8,** 557 (1939).

[103] Marston, H. R., Lee, H. C., and McDonald, I. W., *J. Agr. Sci.* (*Australia*) **38,** 216 (1949); **38,** 222 (1949).

[104] Rickes, E. L., Brink, N. G., Koniuszy, F. R., Wood, T. R., and Folkers, I. T., *Science* **108,** 134 (1948).

[105] Askew, H. O., and Watson, J., *New Zealand J. Sci. Technol.* **25A,** 81 (1943).

[106] Wintrobe, M. M., Grinstein, M., Dubash, J. J., Humphreys, S. R., Aschenbrucker, H., and Worth, W., *Blood* **2,** 323 (1947).

[107] Anderson, A. B., *Biochem. J.* **39,** 139 (1945).

[108] Burk, D., Hearon, J., Caroline, L., and Schade, A. L., *J. Biol. Chem.* **165,** 723 (1946).

[109] Myers, W. G., *Am. J. Roentgenol. Radium Therapy* **60,** 816 (1948).

[110] Other radium substitutes which are finding some use in therapy are Ir^{192}($\tau_{1/2}$ = 74.4 days) [see Freundlich, H. F., Haybittle, J. L., and Quick, R. S., *Acta Radiol.* **34,** 115 (1950)], and Cs^{137}($\tau_{1/2}$ = 37 yrs) [see Eastwood, W. S., *Nucleonics* **10,** No. 2 62 (1952)].

of iron, a mixture of isotopes being produced with half-lives ranging from 18 hr. to 270 days. The isotopes used are mainly Co^{56} and Co^{58}, as a mixture from bombarded iron targets.

The iron is separated from the target material by means of ether extractions in 6.4 *N* HCl. The residual aqueous layers containing carrier cobalt and manganese are treated with potassium nitrite to remove the cobalt as the insoluble potassium cobaltinitrite. This precipitate is dissolved in concentrated nitric acid, evaporated to dryness, redissolved in distilled water, and carrier iron and manganese are added. The cobalt is precipitated by the addition of 1 ml. of 1% silver nitrate in 6 *N* acetic acid and 2 ml. of 50% potassium nitrite.[92] The silver potassium cobaltinitrite is much more insoluble than the simpler potassium salt and so loss of cobalt is minimized. Four such precipitations suffice to effect complete purification of the cobalt. Very good yields are obtainable with the cyclotron bombardment of iron, Co^{56} and Co^{58} being observed to be formed in quantities of 4 μc./μa.-hr. The yield of Co^{57} is approximately 1 μc./μa.-hr.

The uranium pile reactor produces the 5.27-year Co^{60} with specific activities greater than 10 curies per gram of cobalt; Co^{58} ($\tau_{1/2}$ = 72 days) is also available from the $Ni^{58}(n,p)Co^{58}$ reaction. Availability of such samples should lead to widespread use of tracer techniques in the field of cobalt metabolism. Detailed descriptions of sample preparation and radioassay of cobalt isotopes can be found in the literature (see Bibliography).

E. Copper

Copper is an essential dietary constituent. Copper deficiencies in mammals result in inhibition of growth, erythropoiesis, and hemoglobin formation.[111] Copper requirements are very low, the average intake per day for the human adult being only 2 or 3 mg. Larger amounts are toxic. It is present in numerous protein and enzyme complexes, such as ascorbic acid oxidase and phenol oxidases. It appears to be involved in formation of catalase, peroxidases, cytochromes, and cytochrome oxidase.

In all experiments with copper, samples of high specific activity are required because of the low uptake and low copper requirements. Details of the preparation of copper samples for assay are given in the literature (see Bibliography).

Normal copper consists of two isotopes, Cu^{63} (68%) and Cu^{65} (32%). Radioactive isotopes suitable for biological investigations are those with mass numbers 61 and 64. These are made in high specific actvity by proton or deuteron bombardment of nickel according to the reactions

[111] Hart, E. B., Steenbock, H., Elvehjem, C. A., and Waddell, J., *J. Biol. Chem.* **65**, 67 (1925).

$Ni^{61,64}(p,n)Cu^{61,64}$, $Ni^{60}(d,n)Cu^{61}$, $Ni^{61}(d,2n)Cu^{61}$, and $Ni^{64}(d,2n)Cu^{64}$. Both isotopes are formed by nickel irradiation, so that the resultant decay curve is a composite of the two half-lives.

Cu^{61} ($\tau_{1/2}$ = 3.33 hr.)[112] decays both by positron emission (68%) and K capture (32%).[113] Two positron spectra are found, one with an upper energy limit of 1.205 Mev. (96%) and one with an upper energy limit of 0.55 Mev. (4%).[114] Gamma radiations, in addition to annihilation radiation, are detected with energies of 0.655, 0.284, and 0.076 Mev.[114] Cu^{64} ($\tau_{1/2}$ = 12.80 hr.)[115] has been discussed in Chapter I (see p. 21). It decays in three modes, i.e., by negatron emission (39%), by K capture (42%), and by positron emission (19%).[116] The β^- radiations have a maximum energy of 0.571 Mev., and the positrons have a maximum energy of 0.657 Mev.[117] A weak component (2.5%) with 1.35-Mev. energy is present in addition to the annihilation radiation.[118]

The nickel target, after irradiation, is dissolved in aqua regia, evaporated to dryness, taken up in water, and extracted with a CCl_4 solution of dithizone (diphenylthiocarbazone).[119] The reagent is present in a quantity of 1 mg./ml. of CCl_4. The CCl_4 extract is evaporated to dryness, ashed with 0.1 to 0.2 ml. of H_2SO_4 and a few drops of H_2O_2. The residue is taken up to a volume of 1 ml. in water. This solution, containing minimal quantities of copper, is neutralized with redistilled ammonia to the phenol red end point.

Cu^{61} is obtained in yields of 0.5 to 1.0 mc. in a few hundred microampere-hours of bombardment. The total amount of copper is governed by the purity of the nickel which usually assays 0.1% in copper. There will be obtained usually about 0.2 to 0.4 mg. of copper from a typical nickel target.

Cu^{64} can be made without Cu^{61} content by the $Cu^{63}(d,p)Cu^{64}$ or $Cu^{63}(n,\gamma)Cu^{64}$ reaction. Deuteron bombardment of copper results in very high yields, usually 3 mc./μa.-hr. Long-lived zinc, Zn^{65}, is also produced and may be separated from copper by precipitating the copper in strong acid with H_2S. Specific activities are low because of the high unlabeled copper content of the target material. The uranium pile reactor makes Cu^{64} by the (n,γ) reaction on Cu^{63} with specific activities greater than 3000

[112] Cook, C. S., and Langer, L. M., *Phys. Rev.* **74,** 227 (1948).

[113] Cook, C. S., *Phys. Rev.* **83,** 462 (1951).

[114] Owen, G. E., Cook, C. S., and Owen, P. H., *Phys. Rev.* **78,** 686 (1950).

[115] Rabinowicz, E., *Proc. Phys. Soc.* (*London*) **63A,** 1040 (1950).

[116] Average of various values reported calculated by Hollander, J. M., Perlman, I., and Seaborg, G. T., *Revs. Mod. Phys.* **25,** 469 (1953).

[117] Cook, C. S., and Langer, L. M., *Phys. Rev.* **73,** 601 (1948).

[118] Deutsch, M., *Phys. Rev.* **72,** 719 (1947).

[119] Schultze, M. O., and Simmons, S. J., *J. Biol. Chem.* **142,** 97 (1942).

mc./g. of copper. Main contaminants present are Zn^{65} and Ag^{110}, with small amounts of other activities.[120]

Production of Cu^{64} by fast neutron irradiation of $ZnCl_2$ and displacement of the active copper with zinc powder has been reported.[121] The best yield of Cu^{64} was achieved by electrolyzing the irradiated $ZnCl_2$ solution in 0.05 *N* H_2SO_4 (saturated with H_2) between polished platinum electrodes.

Total body retention is recommended not to exceed 150 μc.[3]

F. Zinc

Little is known about the physiological function of zinc. Deficiency of this mineral element retards growth in rats and mice. Various plant-deficiency diseases are related to zinc requirements. The major zinc-containing proteins are insulin, uricase, and carbonic anhydrase. Like the elements discussed in previous sections, zinc is needed in very small amounts for normal function.

Zinc is a complex of five stable isotopes, the two nuclei with largest abundance being Zn^{64} and Zn^{66}. The only radioactive isotope easily applicable to tracer research is Zn^{65} ($\tau_{1/2}$ = 250 days).[122] Another radioactive isotope, Zn^{63}, has the rather short half-life of 38 min.[123] The use of a third radioactive isotope, Zn^{69}, is complicated by occurrence of isomeric transitions.[124]

The long-lived zinc isotope is produced by neutron or deuteron activation of zinc and deuteron bombardment of copper according to the reactions $Zn^{64}(n,\gamma)Zn^{65}$, $Zn^{64}(d,p)Zn^{65}$, and $Cu^{65}(d,2n)Zn^{65}$. High specific activities are obtainable by using the deuteron reaction with copper. Old copper parts of the cyclotron, such as deflector plates, which have received intense bombardments for long periods, are good sources of radioactive zinc. The surfaces of these copper parts are rubbed with concentrated nitric acid with glass-fiber brushes. The resultant solution is freed of nitric acid by repeated evaporations with HCl. Copper may be removed electrolytically or by repeated precipitations with H_2S in 0.5 *N* HCl. The filtrate is freed of H_2S by boiling to dryness. The residue is taken up in a minimal quantity of 0.5 *N* HCl to which is added a little 20% citric acid.[125] The

[120] Frierson, W. J., Hood, S. L., Whitney, I. B., and Comar, C. L., *Arch. Biochem. and Biophys.* **38,** 397 (1952).

[121] Erbacher, O., Herr, W., and Egedi, U., *Z. anorg. Chem.* **256,** 41 (1948).

[122] Livingood, J. J., and Seaborg, G. T., *Phys. Rev.* **55,** 457 (1939).

[123] Delsasso, L. A., Ridenour, L., Sherr, R., and White, M. G., *Phys. Rev.* **55,** 113 (1939).

[124] Kennedy, J. W., Seaborg, G. T., and Segrè, E., *Phys. Rev.* **56,** 1995 (1939).

[125] Sheline, G. E., Chaikoff, I. L., Jones, H. B., and Montgomery, M. L., *J. Biol. Chem.* **147,** 409 (1943); **149,** 139 (1943).

solution is transferred to a separatory funnel and rendered slightly alkaline with ammonia. A fraction of a milligram of zinc chloride may be added as carrier if desired. Normally there is sufficient zinc present, originating as an impurity in the copper, to supply sufficient carrier. The alkaline aqueous layer is extracted with a chloroform solution of dithizone. Excess dithizone is removed from the chloroform layer by washing with dilute ammonia. The labeled zinc is extracted as chloride from the chloroform phase with 0.1 *N* HCl. Small amounts of contaminating radioactivities such as cadmium, cobalt, lead, or nickel may be present and should be tested for before the labeled zinc is used. This may be done by taking an aliquot of the solution, adding carriers for all the elements suspected, including zinc, and performing a qualitative gravimetric analysis for each element.

It is possible to recover as much as 10 mc. of Zn^{65} from discarded copper parts of the cyclotron. The total quantity of zinc in which this is contained need be no more than 1 mg. and can be considerably less. Yields of 0.5 to 0.7 μc./μa.-hr. are noted with 12- to 16-Mev. deuterons. The uranium pile reactor produces Zn^{65} by the $Zn^{64}(n,\gamma)Zn^{65}$ reaction in 15-mc. amounts. The specific activities quoted are greater than 75 mc./g. zinc.

Zn^{65} decays both by positron emission (2.5%) and *K* capture (97.5%).[126] The positrons have a maximum energy of 0.32 Mev.[127] In addition to the annihilation radiation, there is a γ ray reported with energy 1.120 Mev.[127]

Total body retention should not exceed 430 μc.[3]

G. Molybdenum, Vanadium, and Tungsten

The striking growth requirement of plants for exceedingly small quantities of molybdenum is well known but not understood. Application of a few grams of MoO_3 per acre to molybdenum-deficient soils in Australia has increased yield of legumes as much as four-fold.[128] Molybdenum nutrition in higher plants is closely allied to nitrogen assimilation. In rats it is established that molybdate in concentrations of 500 mg. per cent is toxic, and concentrations in the range 50 to 100 mg. per cent retard growth.[129]

High molybdenum content in soils causes a severe diarrhea in ruminants.[130] It is thought this may be due to the complexing action of molybdenum on catechol, thus reducing toxicity of catechol for bacteria in the rumen and allowing bacterial proliferation.[131]

[126] Yuasa, T., *Compt. rend.* **235,** 366 (1952).

[127] Mann, K. C., Rankin, D., and Daykin, F. M., *Phys. Rev.* **76,** 1719 (1952).

[128] Anderson, A. J., *J. Australian Inst. Agr. Sci.* **8,** 73 (1942).

[129] Neilands, J. B., Strong, F. M., and Elvehjem, C. A., *J. Biol. Chem.* **172,** 431 (1948).

[130] Ferguson, W. S., Lewis, A. H., and Watson, S. J., *Nature* **141,** 553 (1948).

[131] McGowan, J. C., Brian, P. W., and Blaschko, H., *Nature* **159,** 373 (1947); see, however, Comar, C. L., Singer, L., and Davis, G. K., *J. Biol. Chem.* **180,** 913 (1949).

Molybdenum has been shown to play a part in the reactions catalyzed by the enzyme hydrogenase.[132] It is suggested that the enzyme in the living cell is linked through a molybdenum bridge to flavin and cytochrome.[133] Molybdenum has also been found to be the activating metal for a number of flavoproteins, e.g., nitrate reductases linked with reduced diphosphopyridine or triphosphopyridine nucleotide,[134] xanthine oxidase, etc.[135]

Among the various radioactive isotopes (Appendix 4) only one, Mo^{99}, appears to be useful for tracer studies. Mo^{99} ($\tau_{1/2}$ = 67 hr.) decays by emission of negative β particles to unstable isomers of technetium (Tc^{99}).[136] About 10%[137] decays to a metastable state of Tc^{99} ($\tau_{1/2}$ = 6 hr.) and 90% to the ground state of Tc^{99} with a very long half-life ($\tau_{1/2} = 2.2 \times 10^5$ yr.)[138]. The negatron spectrum consists of two components, E_{max} = 0.24 and 1.03 Mev.[139]

A complex of γ radiations is associated with decay to Tc^{99}, which in turn has a complex spectrum of γ radiations preceding emission of a single β^- component. Disintegration schemes for the sequence Mo^{99}–Tc^{99}–Ru^{99} are given in the literature.[140] Mo^{99} must be used with the precaution that technetium activities in equilibrium with the parent Mo^{99} are separated or taken into account in the radioassay. The uranium pile reactor produces Mo^{99} with specific activity of greater than 10 mc./g. molybdenum.

Total body retention should not exceed 50 μc.[3]

Vanadium and tungsten appear to be involved as essential tracer elements in the metabolism of algae and plants. Certain ascidians, for instance, store large amounts of vanadium.[141] Recently it has been claimed that vanadium specifically increases quantum yields in photosynthesis by *Chlorella*.[142] It is possible that tracer researches with radioactive isotopes of these elements may be expanded in volume considerably in the next few years.

The isotope of vanadium most useful for tracer purposes is V^{48}. It is a

[132] Shug, A. L., Wilson, P. W., Green, D. E., and Mahler, H. R., *J. Am. Chem. Soc.* **76,** 3355 (1954).

[133] Mahler, H. R., and Green, D. E., *Science* **120,** 7 (1954).

[134] Nicholas, D. J. D., and Nason, A., *Arch. Biochem. and Biophys.* **51,** 310 (1954).

[135] Recently the role of molybdenum in the function of metalloflavoproteins has been reviewed exhaustively; see McElroy, W. D., and Glass, B. (eds.) "Inorganic Nitrogen Metabolism." Johns Hopkins Press, Baltimore, Maryland, 1955.

[136] Seaborg, G. T., and Segrè, E., *Phys. Rev.* **55,** 808 (1939).

[137] Mandeville, C. E., and Scherb, M. V., *Phys. Rev.* **73,** 848 (1948).

[138] Fried, S., Jaffey, A. H., Hall, N. F., and Glendennin, L. E., *Phys. Rev.* **87,** 252 (1952).

[139] Bunker, M. E., and Canada, R., *Phys. Rev.* **80,** 961 (1950).

[140] Goldhaber, M., and Hill, R. D., *Revs. Mod. Phys.* **24,** 179 (1952).

[141] Bertrand, D., *Bull. Am. Museum Nat. Hist.* **94,** 407 (1950).

[142] Warburg, O., Krippahl, G., and Buchholz, W., *Z. Naturforsch.* **10b,** 422 (1955).

positron emitter ($\tau_{1/2}$ = 16.0 days).[143] The upper limit for the positrons is 0.69 Mev.; there appears to be a weak positron component with an upper energy limit of 0.8 Mev.[144] About 58 % of the decay occurs by positron emission, and the rest by K capture.[145] In addition to the annihilation radiation, radiations with energies of 2.29, 1.32, and 0.99 Mev. are reported.[144] The isotope is made by deuteron bombardment of titanium metal, using the reactions $Ti^{47}(d,n)V^{48}$ and $Ti^{48}(d,2n)V^{48}$. Carrier-free radioactive vanadium is extracted[146] from the target material by fusion with a mixture of carbonate and sodium nitrate. The vanadium is leached from the melt with water. The alkaline solution is acidified with HCl, and the large excess of NaCl is removed by concentrating the solution to a small volume. The radioactive vanadium remains in the supernatant solution.

A tungsten isotope available for tracer researches is W^{185} ($\tau_{1/2}$ = 73.2 days).[147] It is a negatron emitter with an upper energy limit of 0.428 Mev.[148] and a weak γ emitter with an energy of 0.134 Mev.[149] W^{185}, made by the $W^{184}(n,\gamma)W^{185}$ reaction in the uranium pile, is available with specific activities greater than 100 mc./g. tungsten.

H. Arsenic

The metabolism of arsenicals is a problem well suited to the tracer approach. Little is known about the role of arsenic, which is always found in minute amounts in human tissues. Most of the pharmacological data available bear on the elimination of arsenites or arsenates administered in abnormal concentrations. The mechanism for adaptation to increasing amounts of arsenicals also requires elaboration. The effect of arsenate on the phosphorylative mechanisms in glycolysis is fairly well understood in terms of a simple competitive interaction between phosphate and arsenate for the triose substrate attacked by glyceraldehyde oxidase.

Arsenic is a simple element with one stable isotope, $_{33}As^{75}$. Numerous radioactive isotopes exist (see the Appendix 4) but those listed in Table 40 appear to be best suited for tracer investigations. Elaborating on data furnished in this table, it may be remarked that As^{76} is produced most usually by neutron irradiation of some organic arsenical which can be adapted to a Szilard-Chalmers process. The arsenic compound used in most

[143] Walke, H., *Phys. Rev.* **52,** 777 (1937).

[144] Roggenkamp, P. L., *Dissertation Abstr.* **13,** 571 (1953).

[145] Good, W. M., Peaslee, D., and Deutsch, M., *Phys. Rev.* **69,** 313 (1946).

[146] Haymond, H. R., Maxwell, R. D., Garrison, W. M., and Hamilton, J. G., *J. Chem. Phys.* **18,** 756 (1950).

[147] Saxon, D., *Phys. Rev.* **74,** 1264 (1948) (A).

[148] Shull, F. B., *Phys. Rev.* **74,** 917 (1948).

[149] Cork, J. M., Keller, H. B., and Stoddard, A. E., *Phys. Rev.* **76,** 575 (1948).

laboratories is cacodylic acid. The aqueous solution is irradiated with neutrons, after which a minimal quantity of $AsCl_3$ is added as carrier. The recoil As^{75} as As^{+++} is precipitated with H_2S. Specific activities of 1 to 2 mc./mg. arsenic are obtainable from a few hours of bombardment with neutrons from the $Be^9(d,n)B^{10}$ reaction in the cyclotron at 12 to 16 Mev. with deuteron beam currents of 100 μa. Other possible target materials are arsenites and arsenates. Directions for preparation of As^{76} by uranium pile irradiation of cacodylic acid may be found in the literature.[150] Yields up to 150 mc., with specific activities of 30 mc./mg., are reported with cacodylic acid.[150] Ordinary uranium pile irradiation of As_2O_3 as target results in production of units with specific activity greater than 4 curies per gram of arsenic.

In cyclotron bombardments producing As^{74} the production reaction in use for the most part is deuteron activation of germanium. The target may be a germanium metal eutectic with copper ($\sim$18% germanium), which can be bonded well to a cooling plate for internal bombardment,[151] or germanium oxide powder which must be bombarded as an external target. After irradiation, the target material is dissolved with aqua regia and a small quantity of arsenic chloride is added as carrier. The mixture is distilled several times to dryness with HCl, removing nitrate and most of the germanium. The arsenic which is in the pentavalent state is reduced to the volatile trivalent state by addition of HBr. More HCl is also added, and the mixture is distilled into water cooled in an ice bath, until bromine fumes appear. More HBr and HCl are added, and the distillation is repeated.

The radioactive arsenic is now present in the distillate. Irvine[151] recommends that it be precipitated as the metal by the addition of a few grams of solid ammonium hypophosphite and by warming the solution on a steam bath for a few minutes. The precipitated metal is filtered by means of a sintered-glass filter crucible, taking care that the precipitate is not allowed to be sucked dry because of danger of oxidation by aeration. The precipitate is washed with oxygen-free distilled water, and finally with acetone, to minimize oxidation.

Yields of 3 to 4 mc. are observed when GeO_2 is bombarded for 1000 μa.-hr. with deuterons at 12 to 16 Mev. Safe total-body retention of As^{76} is estimated to be no more than 10 μc.

[150] Straube, R. L., Neal, W. B., Jr., Kelly, T., and Ducoff, H. S., *Proc. Soc. Exptl. Biol. Med.* **69,** 270 (1948).

[151] See Irvine, J. W., Jr., *J. Phys. Chem.* **46,** 910 (1942), for a detailed description of the separation procedure for arsenic and germanium. Another procedure for separation of As^{76} in carrier-free form is given by Lanz, H. J., Wallace, P. C., and Hamilton, J. G., *Univ. Calif. (Berkeley) Publs. Pharmacol.* **2,** 263 (1950).

TABLE 40

RADIOACTIVE TRACER ISOTOPES OF ARSENIC

Mass Number	Half-life	Production Reactions	Radiations	Production Source and Yield
72	26 hr.[152]	$Ga^{69}(\alpha,n)As^{72}$ $Ge^{72}(d,2n)As^{72}$ $Se^{74}(d,\alpha)As^{72}$	Positrons, E_{max} = 3.34 (19%), 2.50 (62%), 1.84 (12%), 0.67 (5%), 0.29 (2%);[153] γ radiation, E = 0.835 Mev.;[153] and others up to 3.0 Mev.; annihilation radiation	Produced only by cyclotron
73	76 days[152]	$Ge^{72}(d,n)As^{73}$	K capture, e^-, E = 0.0135, 0.0539 Mev.[154]	Produced only by cylotron
74	17.5 days[155]	$Ge^{73}(d,n)As^{74}$	Negatrons (53%), E_{max} = 1.56 (51%), 0.69 (49%) Mev.; positrons (43%), E_{max} = 1.53 (11%), 0.92 (89%) Mev.;[156] γ radiation, E = 0.596, 0.635 Mev.;[156]	Best produced by cyclotron, 2 to 3 mc. in 0.5 to 2 mg. As[157]
76	26.8 hr.[158]	$As^{75}(n,\gamma)As^{76}$ $As^{75}(d,p)As^{76}$ $Ge^{76}(d,2n)As^{76}$ $Se^{76}(n,p)As^{76}$ $Se^{78}(d,\alpha)As^{76}$	Negatrons, E_{max} = 1.29 (15%), 2.49 (25%), 3.04 (60%) Mev.;[159] γ radiation, E = 0.557, 1.22, 1.78 (weak) Mev.[160]	See text
77	38 hr.[161]	Uranium fission	Negatrons, E_{max} = 0.700 Mev.[162]	Produced by pile, millicurie units, carrier free

[152] McCown, D. A., Woodward, L. L., and Pool, M. L., *Phys. Rev.* **74,** 1515 (1948).

[153] Mei, J. Y., Mitchell, A. C. G., and Huddleston, C. M., *Phys. Rev.* **79,** 19 (1950).

[154] Johansson, S., *in* Commemorative Volume M. Siegbahn, p. 183. Upsala (1951).

[155] McCown, D. A., Woodward, L. L., Pool, M. L., and Finston, H. L., *Phys. Rev.* **74,** 1248 (1948) (Abstracts).

[156] Johansson, S., Cauchois, Y., and Siegbahn, M., *Phys. Rev.* **82,** 275 (1951).

[157] Hunter, F. T., Kip, A. F., and Irvine, J. W., Jr., *J. Pharmacol. Exptl. Therap.* **76,** 207 (1942).

[158] Weil, G. L., *Phys. Rev.* **62,** 229 (1942).

[159] Siegbahn, K., *Arkiv Mat. Astron. Fysik* **34A,** No. 7 (1947).

[160] Wu, C. S., Havens, W. W., Jr., and Rainwater, L. J., *Phys. Rev.* **74,** 1248 (1948) (A).

[161] Heiman, W. J., and Voigt, A., Iowa State College Report ISC-89 (1951).

[162] Canada, R., and Mitchell, A. C. G., *Phys. Rev.* **81,** 485 (1951).

I. Selenium, Antimony, and Tellurium

The metabolism of elements of pharmacological importance has remained relatively unexplored owing mainly to lack of suitable microanalytical methods. Tracer methods supply a means of determining distribution and excretion patterns for single doses in microamounts.

The isotope of choice for selenium is Se^{75} ($\tau_{1/2} = 127$ days),[163] which decays by K capture with emission of complex x-ray and γ-ray spectra, including production of conversion electrons. Components with energies of 0.08, 0.1, 0.12, 0.14, 0.27, 0.28, and 0.40 Mev. have been reported.[164] Pile-produced Se^{75} is available in units with specific activity greater than 5000 mc./g. selenium.

Interest in antimony stems primarily from its use as tartar emetic (antimony potassium tartrate) in treatment of parasitic conditions such as those attending presence of schistosomes in humans. The isotopes of choice appear to be ${}_{51}Sb^{122}$ ($\tau_{1/2} = 2.8$ days), Sb^{124} ($\tau_{1/2} = 60$ days),[165] and Sb^{125} ($\tau_{1/2} = 2.7$ yr.).[166] They are available both from cyclotron bombardment of antimony with deuterons and from pile neutron irradiation of antimony.[165] The radiation characteristics of these and other antimony isotopes are given in Appendix 4. High-specific activity samples of all three antimony isotopes are obtainable from neutron activation of antimony in the uranium pile.

The radioactive isotopes of tellurium are many and for the most part are characterized by isomerism with attendant internal conversion. Thus limitations are placed on their use in metabolism studies. One isotope used in the work on distribution and excretion is ${}_{52}Te^{121}$, which exists in a number of metastable states. One isomer decays with a half-life of 143 days[167] and γ emission to another isomer with half-life of 17 days. The γ-ray energies associated with the 143-day isomer are reported as 0.082, 0.088, 0.159, and 0.213 Mev.[168] The 17-day isomer decays by K capture and emission of a 0.615-Mev. γ ray.[169] The 0.213-Mev. γ ray from the 143-day isomer is associated with a high percentage of conversion electrons, the detection of which is a problem closely analogous to detection of the S^{35} radiations.

[163] Cowart, W. S., Pool, M. L., McCown, D. A., and Woodward, L. L., *Phys. Rev.* **73,** 1454 (1948).

[164] Jensen, E. N., Laslett, L. J., and Pratt, W. W., *Phys. Rev.* **76,** 430 (1949).

[165] Livingood, J. J., and Seaborg, G. T., *Phys. Rev.* **55,** 414 (1939).

[166] Leader, D. R., and Sullivan, W. H, *in* "Radiochemical Studies: The Fission Products" (C. D. Coryell and N. Sugarman, eds.), Div. IV, Vol. 9, p. 934. National Nuclear Energy Series, McGraw-Hill, New York, 1952.

[167] Edwards, J. E., and Pool, M. L., *Phys. Rev.* **69,** 140 (1946).

[168] Hill, R. D., and Mihelich, J. W., quoted by Seaborg, G. T., and Perlman, I., *Revs. Mod. Phys.* **20,** 616 (1948).

[169] Kent, C. V., and Cork, J. M., *Phys. Rev.* **62,** 297 (1942).

A very short-lived isomer ($\tau_{1/2} = 3 \times 10^{-8}$ sec.) decaying by emission of a 0.23-Mev. γ ray to the 17-day isomer, has also been reported.[170]

The Te^{121} isomers are made by deuteron or α-particle bombardment of antimony in the cyclotron.

Pile-produced isotopes of tellurium are Te^{127}, Te^{129}, and Te^{131}. They are all negatron emitters with associated isomeric transition γ rays and conversion electrons (see Appendix 4).

J. Silver, Gold, and Mercury

Radioactive silver isotopes which may be useful for purposes of radiation therapy are available either from deuteron bombardment of palladium in the cyclotron or from uranium pile neutron irradiation of uranium or silver. Among these are:

1. Ag^{106} ($\tau_{1/2} = 8.2$ days),[171] which decays by K capture. Gamma radiations with energies of 1.55, 1.24, 1.04, 0.815, 0.717, 0.620, 0.511, 0.409, and 0.220 Mev. are found.[172]

2. Ag^{110} ($\tau_{1/2} = 270$ days),[173] which decays both by K capture and by negatron emission. Negative β-ray components are found with maximum energies of 0.087 (about 58%), 0.530 (about 35%), 2.12 (about 3%), and 2.86 (about 1%) Mev.[174] The γ-ray spectrum is exceedingly complex, with ten components ranging in energy from 0.115 to 1.516 Mev.[174]

3. Ag^{111} ($\tau_{1/2} = 7.6$ days),[175] which decays only by emission of negatrons of maximum energies of 1.04 (91%), 0.80 (1%), and 0.70 (8%) Mev.[176]

Other isotopes, mostly short-lived, are given in Appendix 4. Cyclotron production of Ag^{106} by deuteron bombardment of palladium yields 10 μc./μa.-hr. at 12 to 16 Mev. Carrier-free samples of Ag^{111} are available from the uranium pile reactor. Ag^{110} from the uranium pile reactor is quoted as available in 13-mc. units with specific activities greater than 100 mc./g. silver.

Of the various radioactive gold isotopes, those with mass numbers 195, 196, 198, and 199 are best suited for tracer work. Au^{195} decays by K capture with a half-life of 180 days.[177] Only γ radiations with energies of 0.0308,

[170] Bittencourt, P. T., and Goldhaber, M., *Phys. Rev.* **70**, 780 (1946).

[171] Pool, M. L., *Phys. Rev.* **52**, 239 (1938).

[172] Hayward, R. W., *Phys. Rev.* **85**, 760 (1952) (A).

[173] Gum, J. R., and Pool, M. L., *Phys. Rev.* **80**, 315 (1950).

[174] Siegbahn, K., *Phys. Rev.* **77**, 233 (1950).

[175] Steinberg, E. P., and Glendennin, L. E., *in* "Radiochemical Studies: The Fission Products" (C. D. Coryell and N. Sugarman, eds.), Div. IV, Vol. 9, p. 877 National Nuclear Energy Series. McGraw-Hill, New York, 1951.

[176] Johansson, S., *Phys. Rev.* **79**, 896 (1950).

[177] Steffen, R. M., Huber, O., and Humbel, F., *Helv. Phys. Acta* **22**, 167 (1949).

0.0990, and 0.130 Mev. are observed.[178] It can be made only by bombardment with high-energy particles by reactions such as $Ir^{193}(\alpha,2n)Au^{195}$ and $Pt^{194}(d,n)Au^{195}$. Au^{196} ($\tau_{1/2}$ = 5.55 days)[179] emits negatrons with maximum energy of 0.27 Mev.[180] It decays primarily by K capture (about 95%). Gamma radiations with energies of 0.352 and 0.354 Mev. are reported.[180] Au^{198} ($\tau_{1/2}$ = 2.698 days) emits a single β^- component with maximum energy of 0.97 Mev.[181]. A γ ray with 0.411-Mev. energy is reported. Au^{199} ($\tau_{1/2}$ = 3.15 days)[182] decays by negatron emission. Two β^- radiations are found with maximum energies of 0.460 and 0.250 Mev.[182] Gamma radiations with energies of 0.209, 0.159, and 0.050 Mev. are observed.[183] High specific activity samples of both Au^{198} and Au^{199} are available from neutron activation of gold in the uranium pile.

Mercury isotopes with mass numbers 197 and 203 are available. Hg^{197} exists in three isomeric states with half-lives of 7.0×10^{-9} sec.,[184] 23 hr., and 64 hr.,[185] decaying by K capture. Internal conversion electrons are present. The γ components for the 23-hr. isomer have energies of 0.164 and 0.133 Mev.; that for the 64-hr. form has energies of 0.077 and 0.191 Mev.[186] Hg^{203} ($\tau_{1/2}$ = 47.9 days)[187] is a negatron emitter with maximum energy of 0.208 Mev.[188] A γ ray of 0.279 Mev.[188] is also reported.

K. Concluding Remarks

The material of this chapter could be further elaborated by discussion of many other elements. One might, for instance, devote attention particularly to the heavy elements lead, bismuth, and polonium because of their historical importance in development of radiochemistry and tracer methodology. An arbitrary limitation of subject material is imperative in keeping with the stated objectives of this book, however.

[178] Douglas, D. G., Foster, J. S., and Thompson, A. L., quoted in reference 51.

[179] Wilkinson, G., *Phys. Rev.* **75**, 1019 (1949).

[180] Stahelin, P., *Phys. Rev.* **87**, 374 (1952).

[181] Seliger, H. H., and Schwebel, A., *Nucleonics* **12**, No. 7, 54 (1954). A very weak β^- component is claimed to exist (E_{max} = 1.37 Mev.), as well as a β^- component (about 1%) with E_{max} = 0.290 Mev. Some γ radiations in addition to the 0.411-Mev. component are also reported; see reference 51.

[182] Bell, R. E., Graham, R. L., and Yaffe, L., quoted in reference 51, p. 614. For information on the properties of Au^{199} in solution, see Schweitzer, G. K., and Bishop, W. N., *J. Am. Chem. Soc.* **75**, 6330 (1953).

[183] Sherk, P. M., and Hill, R. D., *Phys. Rev.* **83**, 1097 (1951).

[184] McGowan, F. M., *Phys. Rev.* **77**, 138 (1950).

[185] Friedlander, G., and Wu, C. S., *Phys. Rev.* **63**, 227 (1943).

[186] Huber, O., Humbel, F., Schneider, H., and deShalit, A., *Helv. Phys. Acta* **24**, 127, 629 (1951).

[187] Cork, J. M., Martin, D. W., LeBlanc, J. M., and Branyan, C. E., *Phys. Rev.* **85**, 386 (1952).

[188] Slätis, H., and Siegbahn, K., *Phys. Rev.* **75**, 319 (1949).

BIBLIOGRAPHY

A. General References

1. Maximum permissible amounts of radioisotopes in the human body and maximum permissible concentrations in air and water. Handbook 52. National Bureau of Standards, U. S. Dept. of Commerce, Government Printing Office, Washington, D. C. (1953).
2. Comar, C. L., Radioisotopes in nutritional trace element studies. *Nucleonics* **3,** No. 3, 32; No. 4, 30; No. 5, 34 (1948).
3. Garrison, W. M., and Hamilton, J. G., Production and isolation of carrier-free radioisotopes, *Chem. Revs.* **49,** 237 (1951).
4. McElroy, W. D., and Glass, B. (eds.), "Inorganic Nitrogen Metabolism." Johns Hopkins Press, Baltimore, Maryland, 1955.
5. McElroy, W. D., and Glass, B. (eds.), "Phosphorus Metabolism," Vols. 1 and 2. Johns Hopkins Press, Baltimore, Maryland, 1951–1952.
6. McElroy, W. D., and Glass, B. (eds.), "Copper Metabolism." Johns Hopkins Press, Baltimore, Maryland, 1950.

B. Antimony

1. Brady, F. J., Lawton, A. H., Cowie, D. B., Andrews, H. L., Ness, A. T., and Ogden, G. E., Localization of trivalent radioactive antimony following intravenous administration to dogs infected with dirofilaria immitis. *Am. J. Trop. Med.* **25** (2), 103 (1945).
2. Ness, A. T., Brady, F. J., Cowie, D. B., and Lawton, A. H., Anomalous distribution of antimony in white rats following the administration of tartar emetic. *J. Pharmacol. Exptl. Therap.* **90,** 174 (1947).
3. Bahner, C. T., Localization of antimony in blood. *Proc. Soc. Exptl. Biol. Med.* **86,** 371 (1954).

C. Arsenic

1. Norton, L. B., and Hansberry, R., Radioactive tracer methods for determination of disposition of arsenic in the silkworm. *J. Econ. Entomol.* **34,** 431 (1941).
2. DuPont, O., Ariel, I., and Warren, S. L., Distribution of radioactive arsenic in the normal and tumor-bearing (Brown-Pearce) rabbit. *Am. J. Syphilis, Gonorrhea, Venereal Diseases* **20,** 96 (1942).
3. Hunter, F. T., Kip, A. F., and Irvine, J. W., Jr., Radioactive tracer studies on arsenic injected as potassium arsenite: I. Excretion and localization in tissues. *J. Pharmacol. Exptl. Therap.* **76,** 207 (1942).
4. Lowry, O. H., Hunter, F. T., Kip, A. F., and Irvine, J. W., Jr., Radioactive tracer studies on arsenic injected as potassium arsenite: II. Chemical distribution in tissues. *J. Pharmacol. Exptl. Therap.* **76,** 221 (1942).
5. Lawton, A. H., Ness, A. T., Brady, F. J., and Cowie, D. B., Distribution of radioactive arsenic following intraperitoneal injection of sodium arsenite into cotton rats infected with *Litomosoides carinii*. *Science* **102,** 120 (1945).
6. Straube, R. L., Neal, W. B., Jr., Kelly, T., and Ducoff, H. S., Biological studies with As^{76}. I. Preparation of As^{76} by pile irradiation of cacodylic acid. *Proc. Soc. Exptl. Biol. Med.* **69,** 270 (1948).
7. Ducoff, H. J., Neal, W. B., Jr., Straube, R. L., Jacobson, L. V., and Bruess, A. V., Biological studies with As^{76}. II. Excretion and tissue localization. *Proc. Soc. Exptl. Biol. Med.* **69,** 548 (1948).

8. Morrison, F. O., and Oliver, W. F., Distribution of radioactive arsenic in the organs of poisoned insect larvae. *Can. J. Research* **D27,** 265 (1949).

D. Bromine

1. Moore, F. D., Tobin, L. H., and Aub, J. C., Distribution of radioactive dyes in tumor-bearing mice. *J. Clin. Invest.* **22,** 161 (1943).
2. Winteringham, F. P. W., Some chemical problems in the use, as a fumigant, of methyl bromide, labeled with Br^{82}. *J. Chem. Soc.* 416 (1949).
3. Broyer, T. C., Further observations on the absorption and translocation of inorganic solutes using radioactive isotopes with plants. *Plant Physiol.* **25,** 367 (1950).
4. Maurice, D. M., and Fridanza, A., Permeability of yolk of hen's egg to Br^{82}." *Nature* **170,** 546 (1952).
5. Forbes, G., Reid, A., Bondurant, J., and Ethridge, J., Estimation of total body chloride in young infants by radiobromide dilution. *Proc. Soc. Exptl. Biol. Med.* **83,** 871 (1953).

E. Calcium

1. Campbell, W. W., and Greenberg, D. M., Studies in calcium metabolism with the aid of its induced radioactive isotope. *Proc. Natl. Acad. Sci. U. S.* **26,** 176 (1940).
2. Pecher, C., Biological investigations with radioactive calcium and strontium. *Proc. Soc. Exptl. Biol. Med.* **46,** 86 (1941).
3. Harrison, H. E., and Harrison, H. C., Studies with radiocalcium: the intestinal absorption of calcium. *J. Biol. Chem.* **188,** 83 (1951).
4. Comar, C. L., Hansard, S. L., Hood, S. L., Plumlee, M. P., and Barrentine, B. F., Use of calcium-45 in biological studies. *Nucleonics* **8,** No. 3, 19 (1951).
5. Boyd, E. S., and Neuman, W. F., Surface chemistry of bone. V. The ion-binding properties of cartilage. *J. Biol. Chem.* **193,** 243 (1951).
6. Araki, M., Yonezawa, T., Chin, S., Kaza, M., Shimada, N., and Yoshioka, R., Investigation of calcium metabolism with the aid of radioactive Ca^{45}. *J. Kyoto Prefect. Coll. Med.* **49,** 294 (1951).
7. Hansard, S. L., Comar, C. L., and Plumlee, M. P., Absorption and tissue distribution of radiocalcium in cattle. *J. Animal Sci.* **11,** 524 (1952).
8. Ririe, D., and Toth, S. J., Plant studies with radioactive calcium. *Soil Sci.* **73,** 1 (1952).
9. Hevesy, G., Conservation of skeletal calcium atoms through life. *Kgl. Danske Videnskab. Selskab. Biol. Medd.* **66,** No. 9, 1 (1955).

F. Chlorine (see references T-2, T-3)

1. Burch, G. E., Threefoot, S. A., and Ray, C. T., Rates of turnover and biologic decay of chloride and chloride space in the dog determined with the long-lived isotope, Cl^{36}. *J. Lab. Clin. Med.* **35,** 331 (1950).
2. Ray, C. T., Burch, G. E., and Threefoot, S. A., Biologic decay rates of chloride in normal and diseased man, determined with long-lived radiochloride, Cl^{36}. *J. Lab. Clin. Med.* **39,** 673 (1952).
3. Threefoot, S. A., Burch, G. E., and Ray, C. T., Chloride "space" and total exchanging chloride in man measured with long-life radiochloride, Cl^{36}. *J. Lab. Clin. Med.* **42,** 16 (1953).
4. Toth, S. J., and Kretschmer, A. E., Plant studies with radioactive chlorine. *Soil Sci.* **77,** 293 (1954).

G. Cobalt (see reference N-2)

1. Copp, D. H., and Greenberg, D. M., Studies in mineral metabolism with the aid of artificial radioactive isotopes. VI. Cobalt. *Proc. Natl. Acad. Sci. U. S.* **27,** 153 (1941).
2. Comar, C. L., and Davis, G. K., Excretion and tissue distribution of radioactive cobalt administered to cows, rabbits, swine, and young calves. Papers III and IV in series "Cobalt Metabolism Studies." *J. Biol. Chem.* **170,** 379 (1947); *Arch. Biochem. and Biophys.* **12,** 257 (1947).
3. Braude, R., Distribution of radioactive cobalt in pigs. *Brit. J. Nutrition* **3,** 289 (1949).
4. Chaiet, L., Rosenblum, C., and Woodbury, D. T., Biosynthesis of radioactive vitamin B_{12} containing Co^{60}. *Science* **111,** 601 (1950).
5. Rosenfeld, I., and Tobias, C. A., Distribution of Co^{60}, Cu^{64}, and Zn^{65} in the cytoplasm and nuclei of tissues. *J. Biol. Chem.* **191,** 339 (1950).
6. Chow, B. F., Barrows, L., and Ling, C. T., The distribution of radioactivity in the organs of the fetus or of young rats born by mothers injected with vitamin B_{12} containing Co^{60}. *Arch. Biochem. and Biophys.* **34,** 151 (1951).
7. Callender, S. T., Turnbull, A., and Wakisaka, A., Estimation of intrinsic factor of Castle by use of radioactive vitamin B_{12}. *Brit. Med. J.* **I,** 10 (1954).
8. Ballentine, R., and Burford, D. D., Radiochemical assay of Co^{60}. *Anal. Chem.* **26,** 1031 (1954).
9. Johnson, R. R., Bentley, D. G., and Moxon, A. L., Synthesis *in vitro* and *in vivo* of Co^{60}-containing vitamin B_{12}-active substances by rumen microorganisms. *J. Biol. Chem.* **218,** 379 (1956).

H. Copper (see also references G-6, X-5)

1. Yoshikawa, N., Hahn, P. F., and Bale, W. F., Red cell and plasma radioactive copper in normal and anemic dogs. *J. Exptl. Med.* **75,** 489 (1952).
2. Schultze, M. O., and Simmons, S. J., Use of radioactive copper in studies on nutritional anemia of rats. *J. Biol. Chem.* **142,** 97 (1942).
3. Havinga, E., and Bykerk, R., Relation between the nature of the bond in copper salts used for intravenous injection and the rate of absorption of copper by liver. *Rec. trav. chim.* **66,** 184 (1947).
4. Schubert, G., Maurer, W., and Riezler, W., Orientation experiments with radioactive copper on animals with tuberculosis. *Klin. Wochschr.* **26,** 493 (1948); Orientation experiments with radioactive copper in rats with nutritional anemia. *ibid.* **26,** 555 (1948).
5. Comar, C. L., Davis, G. K., and Singer, L., The fate of radioactive copper administered to the bovine. *J. Biol. Chem.* **174,** 905 (1948).
6. Smith, E. E., and Gray, P., The distribution of Cu^{64} in early embryo chicks. *J. Exptl. Zool.* **107,** 183 (1948).
7. Joselow, M., and Dawson, C. L., The copper of ascorbic acid oxidase. Exchange studies with radioactive copper. *J. Biol. Chem.* **191,** 11 (1951).
8. Poulson, D. F., Bowen, V. T., Hilse, R. M., and Rubinson, A. C., The copper metabolism of *Drosophila*. *Proc. Natl. Acad. Sci. U. S.* **38,** 912 (1952).

I. Fluorine

1. Volker, J. E., Hodge, H. C., Wilson, H. J., and Van Voorhis, S. N., Adsorption of fluorides by enamel, dentine, bone, and hydroxy apatite. *J. Biol. Chem.* **134,** 543 (1940).

J. Gold

1. Sheppard, C. W., Wells, E. B., Hahn, P. F., and Goodell, J. P. B., Studies of the distribution of intravenously administered colloidal sols of manganese dioxide and gold in human beings and dogs using the radioactive isotopes. *J. Lab. Clin. Med.* **32,** 274 (1947).
2. Zilversmit, D. B., Boyd, G. A., and Brucer, M., The effect of particle size on blood clearance and tissue distribution of radioactive gold colloids. *J. Lab. Clin. Med.* **40,** 255 (1952).
3. Weiss, L. C., Steers, A. W., and Bollinger, H. M., The determination of radioactive gold in biological tissue. *Anal. Chem.* **26,** 586 (1954).
4. Ter-Pogossian, M., and Sherman, A. T., Radioactive gold for the intra-cavity treatment of carcinoma of the cervix. *Radiology* **65,** No. 5, 779 (1955).
5. Allen, W. M., Sherman, A. T., and Arneson, M. N., Further results obtained in the treatment of cancer of the cervix with radio gold. A progress report. *Am. J. Obstet. Gynecol.* **70,** 786 (1955).

K. Iodine

1. Chaikoff, I. L., and Taurog, A., Application of radioactive iodine to studies in iodine metabolism and thyroid function. *In* "Symposium on the Use of Isotopes in Biology and Medicine" (H. T. Clarke, ed.), p. 292. Univ. Wisconsin Press, Madison, 1949.
2. Hertz, S., Treatment of thyroid disease by means of radioactive iodine. *In* "Symposium on the Use of Isotopes in Biology and Medicine" (H. T. Clarke. ed.), p. 377. Univ. Wisconsin Press, Madison, 1949.
3. LeBlond, C. P., Iodine metabolism. *Advances in Biol. and Med. Phys.* **1,** 353 (1948).
4. Bruner, H. D., and Perkinson, J. O., Jr., A comparison of I^{131} counting methods. *Nucleonics* **10,** No. 10, 57 (1952); Preparation of tissues for I^{131} counting. *ibid.* **10,** No. 11, 66 (1952).
5. Bibliography of applications of labeled iodine. *Isotopics* **4** No. 1, (1954).
6. Weiner, S. C., ed. "The Thyroid," pp. 161–228. Holber-Harper Book Co., 1955.
7. Rall, J. E., The Role of Radioactive Iodine in Diagnosis of Thyroid Disease. *Am. J. Med.* **20,** 719–731 (1956).

L. Iron (see references N-2, X-5)

1. Austoni, M. E., and Greenberg, D. M., Iron metabolism-absorption, distribution and excretion of iron in the rat on normal and iron-deficient diets. *J. Biol. Chem.* **134,** 17 (1940).
2. Hahn, P. F., Bale, W. F., and Balfour, W. M., Radioactive iron used to study red blood cells over long periods. The constancy of the total blood volume in the dog. *Am. J. Physiol.* **135,** 600 (1942).
3. Wintrobe, M. M., Greenberg, G. R., Humphreys, S. R., Aschenbrucker, H., Worth, W., and Kramer, R., The anemia of infection. III. The uptake of radioactive iron in iron-deficient and in pyriodoxime-deficient pigs before and after acute inflammation. *J. Clin. Invest.* **26,** 103 (1947).
4. Gibson, J. G., II, Weiss, S., Evans, R. D., Peacock, W. C., Irvine, J. W., Jr., Good, W. M., and Kip, A. F., Measurement of the circulating red cell volume by means of two radioactive isotopes of iron. *J. Clin. Invest.* **25,** 616 (1946).
5. Dubach, R., Callender, S. T. E., and Moore, C. V., Studies in iron transportation and metabolism. VI. Absorption of radioactive iron in patients with fever and with anemias of varied etiology. *Blood* **3,** 526 (1948).

6. Beinert, H., and Maier-Leibnitz, H., Cytochrome C labeled with radioactive iron. *Science* **108,** 634 (1948).
7. Saylor, L., and Finch, C. A., Determination of iron absorption using two isotopes of iron. *Am. J. Physiol.* **172,** 372 (1953).

M. Magnesium

1. Ruben, S., Frenkel, A. W., and Kamen, M. D., Experiments on chlorophyll and photosynthesis using radioactive tracers. *J. Phys. Chem.* **46,** 710 (1942).
2. Sheline, R. K., and Johnson, N. R., New long-lived magnesium isotope. *Phys. Rev.* **89,** 520 (1953).
3. Becker, R. S., and Sheline, R. K., Preliminary experiments using Mg^{28} as a tracer with chlorophylls, plants, and their extracts. *J. Chem. Phys.* **21,** 946 (1953).

N. Manganese (see references J-1, X-5)

1. Greenberg, D. M., and Campbell, W. W., Studies in mineral metabolism with the aid of induced radioactive isotopes. IV. Manganese. *Proc. Natl. Acad. Sci. U. S.* **26,** 448 (1946).
2. Greenberg, D. M., Copp, D. H., and Cuthberton, E. M., Studies in mineral metabolism with the aid of artificial radioactive isotopes. VII. Distribution and excretion particularly by way of bile of iron, cobalt and manganese. *J. Biol. Chem.* **147,** 749 (1943).
3. Millikan, C. R., Radioautographs of manganese in plants. *Australian J. Sci. Research Ser.* **B4,** 28 (1951).
4. Burnett, W. T., Jr., Bigelow, R. R., Kimball, A. W., and Sheppard, C. W., Radiomanganese studies on the mouse, rat and pancreatic fistula dog. *Am. J. Physiol.* **168,** 620 (1952).
5. Romney, E. M., and Toth, S. J., Plant and soil studies with radioactive manganese. *Soil Sci.* **77,** 107 (1954).
6. Maynard, L. S., and Cotzias, G. C., The partition of manganese among organs and intracellular organelles of the rat. *J. Biol. Chem.* **214,** 487 (1955).

O. Mercury

1. Reaser, P. B., Burch, G. E., Threefoot, S. A., and Ray, C. T., Thermal separation of radiomercury from radiosodium. *Science* **109,** 198 (1949).
2. Lippman, R. W., Finkle, R. D., and Gillette, D., Effect of proteinuria on localization of radiomercury in rat kidney. *Proc. Soc. Exptl. Biol. Med.* **77,** 68 (1951).

P. Molybdenum

1. Neilands, J. B., Strong, F. M., and Elvehjem, C. A., Molybdenum in the nutrition of the rat. *J. Biol. Chem.* **172,** 431 (1948).
2. Stout, P. R., and Meagher, W. R., Studies of the molybdenum nutrition of plants with radioactive molybdenum. *Science* **108,** 471 (1948).
3. Comar, C. L., Singer, L., and Davis, G. K., Molybdenum metabolism and interrelationships with copper and phosphorus. *J. Biol. Chem.* **180,** 913 (1949).
4. Totter, J. R., Burnett, W. T., Jr., Monroe, R. A., Whitney, I. B., and Comar, C. L., Evidence that molybdenum is a non-dialyzable component of xanthine oxidase. *Science* **118,** 555 (1953).
5. Evans, H. J., Role of molybdenum in plant nutrition. *Soil Sci.* **81,** No. 3, 199 (1955).

Q. Potassium (see reference T-2)

1. Greenberg, D. M., Joseph, M., Cohn, W. E., and Tufts, E. V., Studies in the potassium metabolism of the animal body by means of its radioactive isotope. *Science* **87,** 438 (1938).
2. Hahn, L., Hevesy, G., and Rebbe, O., Permeability of corpuscles and muscle cells to potassium ions. *Biochem. J.* **33,** 1549 (1939).
3. Brooks, S. C., Penetration of radioactive KCl into living cells. *J. Cellular Comp. Physiol.* **11,** 247 (1938).
4. Greenberg, D. M., Aird, R. B., Boelter, M. D. D., Campbell, W. W., Cohn, W. E., and Murayama, M. M., A study with radioactive isotopes of the blood-cerebrospinal fluid barrier to ions. *Am. J. Physiol.* **140,** 47 (1943).
5. Jacobson, L., Overstreet, R., King, H. M., and Handley, R., A study of potassium absorption by barley roots. *Plant Physiol.* **25,** 639 (1950).
6. Krebs, H. A., Eggleston, L. V., and Terner, C., *In vitro* measurements of turnover rate of potassium in brain and retina. *Biochem. J.* **48,** 530 (1951).
7. Aikawa, J. K., Effect of starvation on exchangeable potassium and tissue K^{42} content in rabbits. *Proc. Soc. Exptl. Biol. Med.* **78,** 524 (1951).
8. Walker, W. G., and Wilde, W. S., Kinetics of radiopotassium in the circulation. *Am. J. Physiol.* **170,** 401 (1952).

R. Selenium

1. McConnell, K. P., Passage of selenium through the mammary glands of the white rat and the distribution of selenium in the milk proteins after subcutaneous injection of sodium selenate. *J. Biol. Chem.* **173,** 653 (1948).
2. McConnell, K. P., and Cooper, B. J., Distribution of selenium in serum proteins and red blood cells after subcutaneous injection of sodium selenate containing radioselenium. *J. Biol. Chem.* **183,** 459 (1950).
3. McConnell, F. P., Portmann, D. W., and Rigdon, R. H., Intravascular life span of the duck red blood cell as determined by radioactive selenium. *Proc. Soc. Exptl. Biol. Med.* **83,** 140 (1953).
4. Heinrich, M., Jr., and Kelsey, F. E., Selenium metabolism. Loss of selenium from mouse tissues on heating. *Federation Proc.* **13,** 364 (1954).

S. Silver (see also reference J-1)

1. Hahn, P. F., Goodell, J. P. B., Sheppard, C. W., Cannon, R. O., and Francis, H. C., Direct infiltration of radioactive isotopes as a means of delivering ionizing radiation to discrete tissues. *J. Lab. Clin. Med.* **32,** 1442 (1947).
2. West, H. D., Johnson, A. P., and Johnson, C. W., Concentration of Ag^{111} in spontaneous and experimentally induced abscesses. *J. Lab. Clin. Med.* **34,** 1376 (1949).
3. Gammill, J. C., Wheeler, B., Carothers, E. L., and Hahn, P. F., Distribution of radioactive silver colloids in tissues of rodents following injection by various routes. *Proc. Soc. Exptl. Biol. Med.* **74,** 691 (1950).
4. Scott, K. G., and Hamilton, J. G., Metabolism of silver in the rat with radiosilver used as indicator. *Univ. Calif. (Berkeley) Publs. Pharmacol.* **2,** 241 (1950).
5. West, H. D., Elliot, R. R., Johnson, A. P., and Johnson, C. W., "*In vivo* localization of radioactive silver at predetermined sites in tissues. *Am. J. Roentgenol. Radium Therapy* **64,** 831 (1950).

T. Sodium (see pp. 397, 400)

1. Griffiths, J. H. E., and Maegraith, B. G., Distribution of radioactive sodium after injection into the rabbit. *Nature* **143,** 159 (1939).
2. Finn, W. O., The distribution of radioactive potassium, sodium, and chloride in rats and rabbits. *J. Appl. Phys.* **12,** 316 (1941).
3. Wang, J. C., Penetration of radioactive sodium and chloride into cerebrospinal fluid and aqueous humor. *J. Gen. Physiol.* **31,** 259 (1948).
4. Threefoot, S. A., Burch, G. E., and Reaser, P., The biological decay periods of sodium in normal man, in patients with congestive heart failure and in patients with the nephrotic syndrome as determined by Na^{22} as the tracer. *J. Lab. Clin. Med.* **34,** 1 (1949).
5. Quimby, E. H., Radioactive isotopes in clinical diagnosis. *Advances in Biol. and Med. Phys.* **2,** 243 (1951).
6. Miller, H., and Wilson, G. M., The measurement of exchangeable sodium in man using the isotope Na^{24}. *Clin. Sci.* **12,** 97 (1953).

U. Strontium (see reference F-2)

1. Treadwell, A. de G., Low-Beer, B. V. A., Friedell, H. L., and Lawrence, J. H., Metabolic studies on neoplasm of bone with the aid of radioactive strontium. *Am. J. Med. Sci.* **204,** 521 (1942).
2. Bregher, J. C., and Taylor, M., Radiophosphorus and radiostrontium in mosquitos. Preliminary report. *Science* **110,** 146 (1949).
3. Schubert, J., and Wallace, H., Jr., The effect of zirconium and sodium citrate on the distribution and excretion of simultaneously injected thorium and radiostrontium. *J. Biol. Chem.* **183,** 157 (1950).
4. Jowsey, J., Owen, M., and Vaughan, J., Microradiographs and autoradiographs of cortical bone from monkeys injected with Sr^{90}. *Brit. J. Exptl. Pathol.* **34,** 661 (1953).
5. Turk, E., A modified radiochemical strontium procedure. U. S. Atomic Energy Comm. Argonne National Laboratory Rept. ANL-5184.

V. Tellurium

1. De Meio, B. H., and Henriques, F. C., Jr., Tellurium IV. Excretion and distribution in tissues studied with a radioactive isotope. *J. Biol. Chem.* **169,** 609 (1947).

W. Vanadium

1. Goldberg, E. D., McBlair, W., and Taylor, K. M., The uptake of vanadium by tunicates. *Biol. Bull.* **101,** 84 (1951).

X. Zinc (see reference G-6)

1. Cohn, E. J., Ferry, J. D., Livingood, J. J., and Blanchard, M. H., Physical chemistry of insulin. II. Crystallization of radioactive zinc insulin containing two or more zinc atoms. *J. Am. Chem. Soc.* **63,** 17 (1941).
2. Sheline, G. E., Chaikoff, I. L., Jones, H. B., and Laurence, M., Studies on the metabolism of zinc with the aid of its radioactive isotope. *J. Biol. Chem.* **147,** 409 (1943).
3. Bergh, H., Studies on the metabolism of zinc using radiozinc, Zn^{65}. *Kgl. Norske Videnskab. Selskab. Forh.* **23B** (5), 49 (1950).

4. Banks, T. E., Tupper, R. L. F., and Wormall, A., The fate of some intravenously injected zinc compounds. 1. The determination of Zn^{65} in tissues. 2. The fate of injected zinc carbonate and phosphate and a zinc dithizone complex. *Biochem. J.* **47,** 466 (1950).
5. Epstein, E., and Stout, P. R., The micronutrient cations iron, manganese, zinc and copper: their uptake by plants from the adsorbed state. *Soil Sci.* **72,** 47 (1951).
6. Shaw, E., Menzel, R. G., and Dean, L. A., Plant uptake of Zn^{65} from soils and fertilizers in the greenhouse. *Soil Sci.* **77,** 205 (1954).

APPENDIX 1

RADIOACTIVITY UNITS AND STANDARDS[1]

a. The *curie* is the amount of any radioactive isotope required to give 3.7×10^{10} disintegrations/sec. A *millicurie* (mc.) is 10^{-3} curie (3.7×10^{7} disintegrations/sec.). A *microcurie* (μc.) is 10^{-6} curie (3.7×10^{4} disintegrations/sec.). The *rutherford* (rd.)[2] is the amount of any radioactive isotope required to produce 10^{6} disintegrations/sec. The *millirutherford* (mrd.) and *microrutherford* (μrd.) are 10^{-3} and 10^{-6} rd., respectively.

Conversion Factors

To	From		
	Disintegrations/sec.	Rutherfords	Curies
Disintegrations/sec.	1	10^{6}	3.7×10^{10}
Rutherfords	10^{-6}	1	3.7×10^{4}
Curies	2.7×10^{-11}	2.7×10^{-5}	1

Thus, 1 rd. = 27 μc., or mc. = 37 rd.

In calculations involving these quantities it is convenient to have the following conversion factors for time units:

To	From		
	Hours	Days	Years
Seconds	3.60×10^{3}	8.64×10^{4}	3.16×10^{7}
Minutes	60.0	1.44×10^{3}	5.26×10^{5}

b. The *roentgen* (r.) is that quantity of photon radiation unspecified as to direction, intensity, or energy distribution which, when incident on a cubic centimeter of dry air at standard conditions, i.e., 0.001293 g. air, produces secondary corpuscular (electron) radiation which, if completely

[1] Condensed from values given by Evans, R. D., *Advances in Biol. and Med. Phys.* **1**, 176–192 (1940).

[2] Proposed by Condon, E. U., and Curtiss, L. F., *Science*, **103**, 712 (1946). This unit, as yet, has not been adopted widely.

absorbed in air, would give rise to a total charge of one statcoulomb (1 esu.). The charge on a single electron or positive ion is 4.80×10^{-10} esu. Hence 1 esu. corresponds to $1/4.80 \times 10^{-10} = 2.093 \times 10^{9}$ ion pairs. Thus, 1 r. produces $(2.083 \times 10^{9})/0.001293 = 1.61 \times 10^{12}$ ion pairs per gram air. Assuming the work required to produce an ion pair in air is on the average 32.5 ev., it follows that 1 r. corresponds to an energy absorbed in air of $32.5 \times 1.61 \times 10^{12} = 5.24 \times 10^{13}$ ev. Converting from electron volts to ergs, remembering that 1.60×10^{-12} erg = 1 ev., it is seen that 1 r. corresponds to an energy dissipation of 83.8 ergs per gram air. Summarizing:

$$\begin{aligned} 1 \text{ r.} &= 1 \text{ esu/cm.}^3 \text{ standard air} \\ &= 2.083 \times 10^{9} \text{ ion pairs/cm. standard air} \\ &= 1.61 \times 10^{12} \text{ ion pairs/g. air} \\ &= 5.24 \times 10^{13} \text{ ev./g. air} = 83.8 \text{ ergs/g. air} \end{aligned}$$

c. The *roentgen-equivalent-physical* (rep.) is that amount of radiation incident on tissue which results in the dissipation of 83.8 ergs/g. tissue. The energy dissipation associated with a roentgen of photons in air is not identical with the energy dissipation of a roentgen of photons in tissue; hence 1 r. photons = 1 rep. is not strictly true.

d. The *roentgen-equivalent-mammal* (rem.) is the energy dissipation in tissue biologically equivalent in man or mammals to 1 r. of γ or x-radiation. The rem. is related to the roentgen by the equation 1 rem. = 83.8/RBE ergs/g. tissue, where RBE denotes an empirical term, the "relative biological efficiency," which takes into account the differing effectiveness of different radiations in inducing a given amount of change or damage in tissue. A related unit is the "n" unit which attempts to provide a dosage measure for neutrons. One n of neutrons produces the same ionization, measured, say, in a Victoreen thimble chamber, as would 1 r. of γ rays. In tissue, however, 1 n unit appears to be 2 to 2.5 times as effective as 1 r. in producing ionization; i.e., 1 n = 2–2.5 rep. $\simeq$ 170–210 ergs/g. tissue.

e. The *roentgen per hour at* 1 *meter* (rhm.)[1] is that amount of unshielded photon radiation which produces 1 r./hr. of ionization in air at a distance of 1 meter from the source. It can be shown that 1 g. radium, unshielded, produces photon radiation equal to 0.969 rhm. Radium is ordinarily used inside platinum needles with 0.5-mm.-thick walls. Correcting for the absorption of the platinum it can be stated that 1 g. radium (usual conditions) = 0.84 ± 0.02 rhm. Other radiation equivalents as calculated by R. D. Evans are given in the table on the preceding page.

f. The *rad*. The International Commission on Radiological Protection

Milliroentgens per Hour at 1 Meter (Mrhm.) Produced by Radiation from Various Isotopes (est. Accuracy 1–3%)

Isotope	mrhm./100 rd.
Na^{22}	3.52
Na^{24}	5.20
Mn^{52}	5.23
Mn^{54}	1.31
Fe^{59}	1.76
Co^{58}	1.51
Co^{60}	3.52
Zn^{65}	0.80
Br^{82}	4.06
I^{128}	0.049
I^{130}	3.39
I^{131}	~0.640
RaB + C (0.5-mm. Pt filter)	2.64

met in Copenhagen in 1953 and decided that the absorbed dose of any ionizing radiation should be expressed in terms of a new unit, the *rad*. This unit of dosage is defined as that quantity of radiation which results in the dissipation of 100 ergs/g. tissue.

g. General recommendations for maximum permissible dose rates with either external or internal radiations, protection procedures and working conditions in radiological installations, and a complete summary of data on which such recommendations are based will be found in the authoritative report entitled "Recommendations of the International Commission on Radiological Protection" published as Supplement No. 6 of the *British Journal of Radiology*, 1955.

Appendix 2

SOME TYPICAL WORKING RULES FOR RADIOCHEMISTRY LABORATORIES

General Rules

These rules are taken from those in force at the Monsanto Radiochemical Laboratories, Washington University, St. Louis, Missouri. They are not to be construed as authoritative or definitive. It is not possible to present a final set of recommendations at this time. The working rules given in this Appendix are merely illustrative of those obtaining in one operating radiochemical laboratory with its own particular problems. For general recommendations for all types of installations using radiological equipment, see Section g, Appendix 1.

1. Before work is begun, and again at the conclusion of each experiment or preparative run, all desk and/or hood surfaces should be monitored and shown to be below the permissible level. During experiments requiring several days or involving special contamination hazard, spot monitoring should be done at more frequent intervals.

2. Samples that contain more than 1 rd. ($\sim$25 μc.) of activity associated with radiation capable of penetrating 40 mg./cm.^{2} should first be set in place and the working region checked in at least two pertinent places for radiation level, with an instrument having a wall thickness less than about 40 mg./cm.^{2}. Work should then be planned so that no part of the body, including hands, will receive more total dose than 100 mr./day or more instantaneous intensity than 100 mr./hr.

It should be remembered that ordinary chemical glassware will absorb most β particles. As a result, an open vessel such as a beaker may be safe when held in the hand, but the area above the open top may be dangerous.

Throughout the work a pocket ionization chamber should be worn, and the dosage read and recorded at the end of the day.

3. Any sample larger than 100 rd. ($\sim$2.5 mc.) should be handled over a stainless steel or other suitable safety tray in an operating hood.

Any sample larger than 1 rd. ($\sim$25 μc.) should be handled in a hood whenever there is danger of spray or the formation of volatile compounds.

Any long-lived isotope such as C^{14} in amounts larger than 1 rd. ($\sim$25 μc.) should be handled only over a safety tray in an operating hood.

Hoods containing active samples as defined by this rule should be kept operating illuminated, and if possible closed, even when work is suspended, for example, overnight.

4. Radiation levels and contaminated work benches and hoods, within the meaning of rules 2 and 3 above, must be marked by appropriate warning signs. Any other open radiation level greater than 10 mr./hr. should be marked.

Disposal and Storage

5. Active materials to be disposed of may be flushed down the sink if less than 1 rd. per flush and less than 100 rd. per week. Nonflushable material (filter paper, etc.) may be discarded in the building waste provided any one load contains less than 1 rd. and gives less than 1 mr./hr. immediately outside the container; for this procedure samples above about 1000 disintegrations/min. should be sealed, not just thrown in loose. More active wastes must be stored or have special handling.

6. When samples are stored, they should be shielded in such a way that the radiation level at any part of the outside of the shield is less than 5 mr./hr. The shield should be labeled as to the approximate number of rutherfords and type of material it contains and the name of the person to whom it belongs. If it is not possible to keep the radiation level outside the shield below 5 mr./hr., it should be put in the vault provided for strong samples. Any samples in the vault should be listed on the record attached to the vault door.

Long-lived samples of more than 1 rd. should be stored in double containers, labeled as above.

7. Transfer of samples to other buildings should be governed by these rules: No samples larger than 100 rd. ($\sim$2.5 mc.) should ever be transferred. Samples up to 100 rd. may be transferred only if their half-life is less than 10 days; 10 rd. if $\tau_{1/2} < 100$ days, 1 rd. if $\tau_{1/2} > 100$ days.

Personnel Protection

8. In all work with any samples larger than 1 rd. ($\sim$25 μc.) rubber surgeon's gloves, or thicker impervious gloves, should be worn. (The surgeon's gloves are thick enough, 40 mg./cm.^{2}, to stop C^{14} particles. Without gloves, a 1-rd. sample of C^{14} irradiates adjacent skin at the rate of $\sim$1 daily dose per second.) Before the worker leaves the laboratory room, rubber gloves must be carefully washed at the foot-operated sink and then removed, and the hands washed. The hands should be checked occasionally at the hand counter, at least before leaving for meals and at the end of the day.

9. During work with any samples larger than 1 rd. ($\sim$25 μc.), clean laboratory coats are to be worn. So long as there is no reason to believe that

these are contaminated, they may be kept in the lockers, folded inside out. Street coats, hats, etc., of laboratory workers covered by these rules should be left only in the lockers.

10. There shall be no smoking or eating in the laboratories devoted to radiochemistry.

11. When the meaning of a regulation is not clear, or a hazard not covered arises, a conservative course should be followed and more information obtained as soon as practicable.

The elements ordinary uranium and ordinary thorium are not to be included among the active samples in the above ten rules.

Counting Room Contamination

12. The counting room is to be used only for activity measurements on mounted samples and for closely associated functions, such as the recording of counting data. It is not a general workroom.

13. No radioactive substances except properly mounted samples and standards may be brought into the counting room.

14. Laboratory coats worn in the active laboratories are to be removed and left outside the counting room before entering.

Monitoring Instruments and Permissible Levels for Radiochemistry Laboratories

1. Radiation intensities ior penetrating radiation (range 40 mg./cm.^{2}) are best surveyed with a portable ionization chamber and d-c amplifier, calibrated directly in roentgen units, with an appropriate window thickness. Acceptable alternatives include: (1) a portable Geiger counter survey instrument with a moderately thin or thin-walled counter, calibrated for the particular type of radiation in roentgen units; (2) a Lauritsen electroscope with can modified to $\sim$40 mg./cm.^{2} and calibrated in roentgen units; (3) for γ radiation only a pocket-type ionization chamber supported in place and read after a suitable time. Tolerance daily dosage for humans is roughly 0.1 r. or 100 mr.

2. Laboratory benches, hoods, and other working surfaces may be monitored with a portable Geiger counter with a thin window (for C^{14} the window may not be more than a few milligrams per square centimeter). Activity levels below 50 ct./min. are permissible. Occasional activity levels up to 1000 ct./min. are permissible, provided the level cannot be reduced by washing and can be shown to be safe by the swipe monitoring method (see below).

3. In the swipe monitoring technique a piece of clean, lightly oiled filter paper approximately 1 inch in diameter is rubbed over the desk top (or other surface) for a path length of about 36 inches. It is then either counted

directly on the hand counter or fastened dirty side up on a standard cardboard mount and counted on one of the counters suitable for measuring the activity suspected. Counts below 50 per minute indicate permissible conditions.

4. In checking hands (or clothes, etc.) on the hand counter, a counting rate of greater than 50 per minute from any surface is not tolerable. The hands should be washed again, dried, and counted until the count is safe. Great care must always be used to avoid contaminating the counter itself or breaking the thin mica window; never touch the sensitive end of the counter tube with anything, and never touch any part of the entire instrument with any hand, glove, or other object that is contaminated.

Appendix 3

CHROMATOGRAPHY

The simplicity and wide applicability of chromatographic procedures warrants description of one set of procedures which has been used successfully in studies on biosynthesis in certain microorganisms, especially *Escherichia coli.**

1. PAPER CHROMATOGRAPHY

A. Technique

1. General Technique. A technique of paper chromatography well adapted to studies of biosynthesis uses Whatman No. 1 filter paper ("For Chromatography"). An organic liquid flowing by capillarity is allowed to ascend the sheet. The physical arrangement of the equipment used is shown in Fig. 69*A*.

For analytical purposes, two-dimensional chromatograms are usually made; for preparative purposes, both one- and two-dimensional chromatograms are used. For a two-dimensional chromatogram a chemical mixture is applied to an area about 0.5 inch in diameter, 1 inch from either edge at a corner of a sheet of paper cut to 9.5 by 10 inches. The sheet is then suspended, with the longer sides vertical, in a glass tank by means of stainless-steel clips affixed to a support rod. The lower edge of the sheet is allowed to dip about 0.5 inch into the first solvent. After the solvent has ascended to the top of the sheet, a process requiring from 6 to 24 hr., depending on the type of solvent employed, the chromatogram is dried in an air stream in a hood. The dirt-laden upper half-inch is then trimmed off to prevent interference with the migration of the second-dimension solvent. The trimmed edge is saved whenever it is necessary to reclaim compounds which migrate with the front. The chromatogram is turned through a right angle and run in a second solvent, after which it is thoroughly dried in a stream of air, with heating when necessary.

The chemicals which are separated may then be located on the paper by means of appropriate chemical tests, or by means of radioautography when members of a mixture contain radioactive tracer atoms.

* Reproduced with the permission of the Carnegie Institution of Washington from Chapter 3 of Publication No. 607, "Studies of Biosynthesis in *Escherichia coli*," R. B. Roberts, D. B. Cowie, P. H. Abelson, E. T. Bolton, and R. J. Britten (1955).

A list of the solvents most frequently used is given in Table 41. The system employed makes use of organic alcohols undersaturated with water, thus avoiding the necessity for a saturator device in the chromatographic tank. Such a procedure is desirable for the sake of simplicity. Because some latitude in solvent composition is tolerable, little variation in solute separation is noted from run to run.

The chief causes of anomalies in separations are: incomplete removal of HCl from protein hydrolyzates, the presence of trichloroacetic acid in nucleic acid hydrolyzates, the presence of high levels of salt in the materials to be separated, and overloading with compounds to be separated. The interfering acids may be removed by repeated vacuum distillation or by ether extraction, and salts may be removed by ion-exchange methods. Phosphate and sulfate salts cannot be tolerated above about 100 γ per chromatogram.

2. Radioautography. The routine preparation of radioautographs of radioactive chromatograms is an extremely useful method for locating radioactive substances. Location of radioactive amino acids is precise; amino acid spots separated by as little as a millimeter or two can be distinguished, and often the fibrous structure of the paper is recorded in the

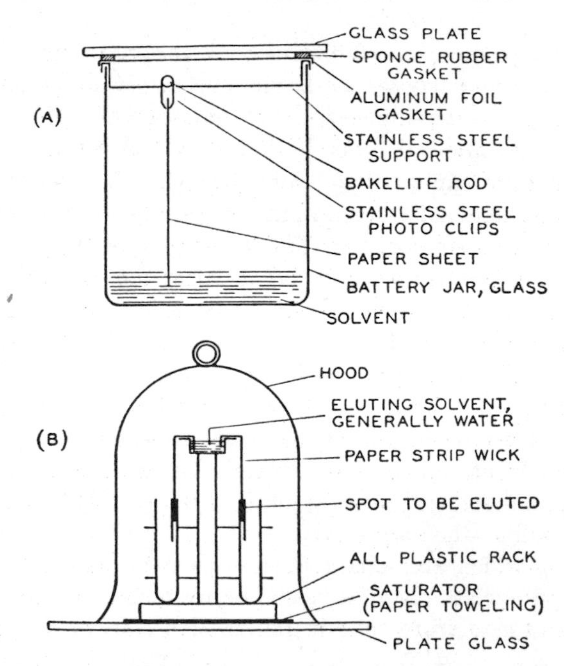

FIG. 69. Chromatographic apparatus. *A*, tank assembly. *B*, elution device. (After Roberts *et al.*, 1955.)

TABLE 41

COMPOSITION AND PRINCIPAL APPLICATIONS OF SOLVENTS FREQUENTLY USED IN PAPER CHROMATOGRAPHY

Solvent	Composition	Principal Applications	Comment
1. *sec*-Butyl alcohol Formic acid Water	70 ml. 10 ml. 20 ml.	Protein hydrolyzates; nucleic acid hydrolyzates; Krebs cycle components; preparation of labeled substrates	Runs 10 inches in 15 hr.; cleans paper efficiently; used in first dimension
2. Phenol (crystals) Concentrated NH_4OH Water	80 g. 0.3 ml. 20 ml.	Amino acids; peptides	Routinely used in second dimension for protein hydrolyzates
3. *tert*-Butyl alcohol 12 *N* HCl Water	70 ml. 6.7 ml. 23.3 ml.	Nucleic acid hydrolyzates	Used in first dimension; runs 10 inches in 24 hr.; cleans efficiently
4. Methanol Water saturated with $(NH_4)_2CO_3$	75 ml. 25 ml.	Krebs cycle components; culture fluids	Runs 10 inches in 6 hr.; used in first dimension
5. *sec*-Butyl alcohol Formic acid Water	85 ml. 5 ml. 10 ml.	Krebs cycle components; culture fluids	Used in second dimension; organic acids better resolved than in solvent 1
6. Isopropanol Formic acid Water	70 ml. 10 ml. 20 ml.	Peptides; oxidized sulfur compounds	Greater movement of strongly acidic materials than in solvent 1

darkened area of the film. The developed radioautograph provides a permanent record of the relative amounts of radioactivity distributed over the original chromatogram. A useful, although subjective, evaluation of the relative radioactivity may be made by simple visual inspection of the developed film. Often this evaluation is adequate to provide a reliable experimental result. When precise measurements are needed, direct counts of the chromatogram can be made, or elution and subsequent sampling into counting cups can be carried out.

The preparation of the radioautograph is simple and well adapted to the routine of analysis. Eastman "No-Screen" x-ray film 10 by 12 inches with emulsion on both sides is placed on top of the chromatogram and held securely by weights in complete darkness for an appropriate time. A legend written with radioactive ink on bond paper is taped to the chromatogram to serve as a positioning indicator as well as to identify the radioautograph. The exposure time varies with the amount of radioactivity and with the purpose of the radioautograph. For C^{14} or S^{35}, 50 to 60 ct./sec./cm.^{2} will give a weak but useful darkening of the film in 1 day.[1] Where weaker activities are to be detected, exposure times of weeks and even months are not impracticable. After exposure the film is developed in Kodak D-19 developer for 5 min. at 20°C. and treated in acid fixer (hypo) until clear. The developed film is then washed in running water and dried.

The amounts of radioactivity in the labeled compounds separated on the chromatograms may be determined directly by masking the chromatogram, leaving the radioactive area exposed (developed x-ray films containing holes make excellent masks for C^{14} or S^{35}). The G-M tube is placed directly over the hole, and a count is made. Since individual compounds (except for the minor constituents) occupy approximately the same area of paper, comparable measurements are made for each. It is necessary, however, to bear in mind that the appearance of the radioautograph may be deceptive in that on underexposed films the spots occupied by the amino acids appear smaller than is actually the case. Furthermore, the degree of blackening of the film is a true indicator of the amount of radioactivity on the chromatogram only over a limited range. Direct counting of the chromatogram may be of doubtful quantitative significance unlesss spot size, self-absorption, and counting geometry are evaluated, or kept controlled from area to area. In spite of these possible pitfalls, this very simple method consistently yields dependable results.

Determination of specific radioactivities by this simple method requires a minimal amount of manipulation, and the amino acids are still available for elution and sampling into counting cups in the usual way. The measurements made and the specific radioactivities calculated are valid as long as the relative proportions of the amino acids in the acid hydrolyzate remain fixed. In a large number of experiments where both the simple technique and a more elaborate method (the quantitative ninhydrin method of Moore and Stein[2] and determination of radioactivity in plastic cups under standard conditions) were used, the results from the two methods were in good agreement.

3. Elution. Removal of compounds from chromatograms is accomplished with the apparatus of Fig. 69*B*. Two-tenths milliliter of water or dilute acid will thoroughly wash 5 cm.^{2} of paper in 1 or 2 hr., carrying the compound

[1] That is, counts as measured by a thin (1.5 mg./cm.^{2}) mica end-window G-M counter tube in contact with the chromatogram. This quantity of radioactivity corresponds approximately to 0.05 μc./cm.^{2}.

[2] Moore, S., and Stein, W. H., *J. Biol. Chem.* **176**, 367 (1948).

to the bottom of the test-tube receiver. Recoveries are essentially quantitative. The eluted material may then be subjected to further analysis.

4. Chromatographic "Fingerprinting." The identification of an unknown substance produced by a living organism is often absolutely necessary for understanding how the organism carries out its life processes. When the substance can be labeled with isotopes, its positive identification, even though it is available only in trace quantities, is often possible by means of chromatographic fingerprinting.

In this method the unknown radioactive compound is mixed with a relatively large amount of a nonradioactive compound whose identity with the unknown substance is suspected. The mixture is then subjected to two-dimensional chromatography in a suitable pair of solvents. The location of the radioactive material is determined by means of a radioautograph. The location of the carrier substance is determined by a suitable chemical test and/or by visual or photographic inspection in visible or ultraviolet light. When enough carrier material is present, the spot it forms has a distinct character: it is circular or elliptical; it is compact or diffuse; or it is striated, forming fingers at its periphery. The spot formed on the radioautograph by the radioactive substance also has many distinguishing features. If the patterns formed by the chemical test on the paper and by the darkening of the film agree in every detail, then it may be concluded that the carrier and radioactive substances are identical. For the conclusion to be valid, it is required that the unknown substance make no significant contribution to the chemical test used to locate the carrier substance. The choice of the carrier substance is determined by the investigator's experience, and the ease with which an identification is made depends on the shrewdness of his choice.

B. Application

1. To Amino Acids. The amino acids of simple mixtures or of complex collections as found in protein hydrolyzates are conveniently separated by means of two-dimensional chromatography with *sec*-butyl alcohol/formic acid/water as the first solvent and phenol/ammonia/water as the second. Figure 70 shows a map of the location of amino acids run in this solvent pair. Phenylalanine and the leucines are not resolved. Peroxide treatment of the amino acid mixture yields methionine sulfoxide and sulfone, which are separated from valine. The position of diaminopimelic acid (DAP) is variable but runs alongside aspartic acid when no HCl is present on the starting spot. Cystine and homocystine and their oxidation products are crowded together in the region of the starting corner.

The sulfur-containing compounds can be resolved by the solvent pair isopropyl alcohol/water/formic acid and *tert*-butyl alcohol/HCl. Cysteine streaks annoyingly. Oxidation to the sulfonic acid and coupling with *n*-ethyl maleimide are useful steps to prevent streaking.

2. To Enzyme Reactions. Paper chromatography has also been used to analyze the products of the activities of enzymes. When enzyme reactions are carried out on a micro scale, it is often possible to separate the undegraded substrate from the reaction products and to estimate the extent of the reaction with little effort. Two examples are described below.

a. Decarboxylation of Aspartic Acid. *E. coli* grown in the presence of $C^{14}O_2$ contains labeled aspartic acid. Ninhydrin tests have demonstrated the C^{14} label to be in the carboxyl groups of the molecule. Because C^{14}

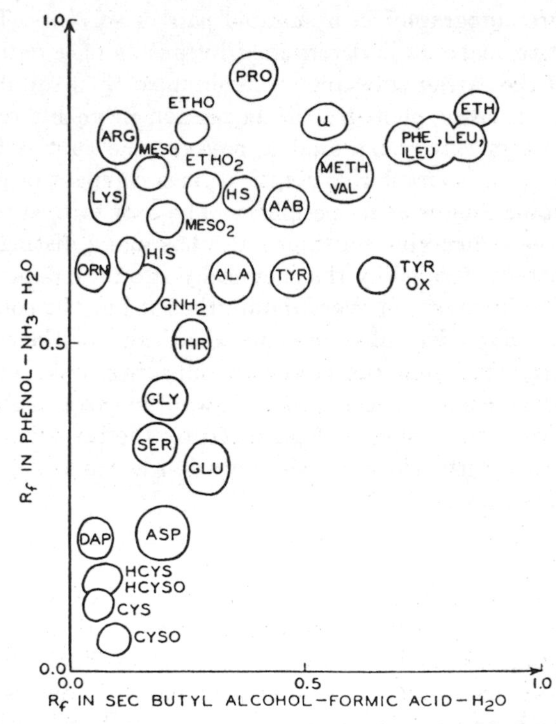

FIG. 70. Map of locations of amino compounds on chromatograms run in *sec*-butyl alcohol/formic acid/water; phenol/ammonia/water. AAB, α-aminobutyric acid; ALA, alanine; ARG, arginine; ASP, aspartic acid; CYS, cystine; CYSO, cysteic acid; DAP, diaminopimelic acid, ETH, ethionine; ETHO, ethionine sulfoxide; $ETHO_2$, ethionine sulfone; GLU, glutamic acid; GLY, glycine; GNH_2, glucosamine; HCYS, homocystine; HCYSO, homocysteic acid; HS, homoserine; HIS, histidine; ILEU, isoleucine; LEU, leucine, LYS, lysine; METH, methionine; MESO, methionine sulfoxide; $MESO_2$, methionine sulfone; ORN, ornithine; PHE, phenylalanine; PRO, proline; SER, serine; THR, threonine; TYR, tyrosine; TYROX, tyrosine (oxidized product); U, urea; VAL, valine. (After Roberts *et al.*, 1955.)

in both groups would indicate a symmetrical intermediate in the pathway for aspartic acid biosynthesis, it was necessary to determine the distribution of C^{14} between the carboxyl groups. For this purpose aspartic acid decarboxylase was allowed to act on the carboxyl-labeled aspartic acid. The reaction mixture was chromatographed in one dimension in phenol/water, with resulting separation of aspartic acid from the product alanine. Radioautographs were made. C^{14} determinations on the original aspartic acid, the undegraded residue, and the alanine product showed that about 30% of the original C^{14} was in C-1, i.e., in alanine, the remainder being in C-4 of aspartic acid. It was concluded, therefore, that a symmetrical intermediate is involved in aspartic acid synthesis. The chromatographic technique did not require the reaction to go to completion and avoided the necessity for analyzing the product $C^{14}O_2$ as the troublesome $BaC^{14}O_3$.

b. Arginase Action. Studies of arginine synthesis also required knowledge of the distribution of C^{14} within the molecule. Arginine is easily split into two fragments by means of arginase. The usual procedure to determine arginase activity is carried out in two steps: arginase digestion yielding ornithine and urea, and urease digestion yielding CO_2 and ammonia from urea. The product CO_2 is then analyzed for C^{14} content. Merely by carrying out the arginase digestion step, separating the components of the reaction mixture on a two-dimensional chromatogram (see Fig. 70), and directly counting the radioactivity in the residual arginine and the products ornithine and urea, an unequivocal determination of the C^{14} distribution can be made. It was shown that the amidine carbon of arginine is derived from the CO_2 supply of the culture medium, and the ornithine portion is derived from glutamic acid. The chromatographic procedure avoids gas-handling methods and gives an exact result without sacrificing the products in the course of measurement.

3. To Other Organic Acids. Paper chromatography of organic acids taking part in the operation of the Krebs cycle has been carried out with the solvent pair methanol/water/ammonium carbonate and *sec*-butyl alcohol/water/formic acid (see Table 41). The absolute R_F values of these compounds are somewhat variable; the pattern formed, however, is distinctive and reproducible. This system has been useful especially in the study of metabolic products released into culture fluids by growing bacteria.

4. To Purine and Pyrimidine Compounds. Compounds containing purine and pyrimidine rings are separated in the solvent pair *tert*-butyl alcohol/HCl and *sec*-butyl alcohol/water/formic acid. An acid hydrolyzate of bacterial nucleic acid contains only some of these compounds, and these are unambiguously separated in this solvent pair. The acid hydrolysis products are adenine, guanine, cytidylic acid, uridylic acid, deoxycytidylic acid, thymidylic acid, ribose, and orthophosphoric acid. Thymidylic acid is lost

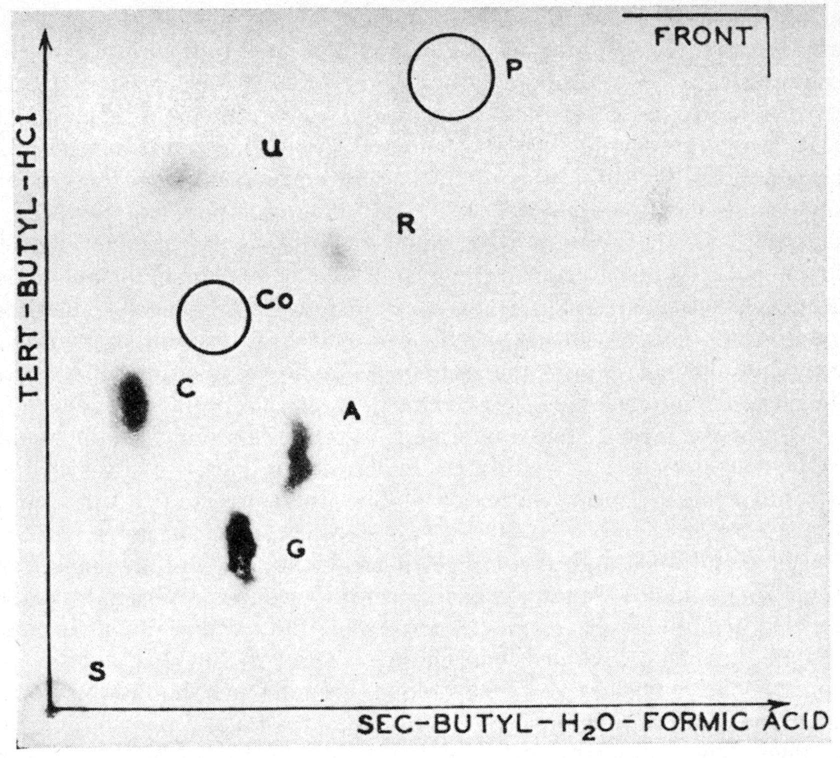

FIG. 71. Radioautograph of a two-dimensional chromatogram. Nucleic acid hydrolyzate from *E. coli* grown on C^{14} glucose. The locations of deoxycytidylic acid and orthophosphate are indicated by circles. A, adenine; C, cytidylic acid; CO, deoxycytidylic acid; G, guanine; P, orthophosphate; R, ribose; U, uridylic acid. (After Roberts *et al.*, 1955.)

in the front of the first-dimension solvent (*tert*-butyl alcohol/HCl). The regions occupied by the various ultraviolet-light-absorbing compounds are located with an ultraviolet lamp. Ribose is located by means of pentose tests or radioautographs when it is C^{14}-labeled. Phosphoric acid is located by means of phosphorus tests or radioautographs when it is P^{32}-labeled. The other materials are also located by means of radioautographs whenever they are labeled. In Fig. 71 is reproduced a radioautograph of an acid hydrolyzate of the nucleic acid of *E. coli* grown with C^{14} glucose as the sole carbon source. The location of phosphate is also shown.

2. ION-EXCHANGE CHROMATOGRAPHY

The studies of biosynthesis in *E. coli* have used principally three ion-exchange materials: Dowex 50, a sulfonic acid cation exchanger; Dowex 2, a polyamine anion exchanger; and IRC-50, a weak carboxylic acid cation exchanger. These materials have been applied mainly to the characterization of complex mixtures and to the purification of biologically synthesized chemicals.

The ion exchanger is poured as a slurry in water, acid, or alkali onto a pellet of glass wool in a buret in order to form a column. The exchanger is treated with a large excess of an acid, base, or salt to convert it to the form desired and to wash out impurities. It is then flushed thoroughly with water. The mixture of substances to be separated is poured into the buret and allowed to pass over the column. When the meniscus of the input solution reaches the top of the column bed, 1 or 2 column volumes of water are added to wash the column free of unadsorbed substances. The adsorbed materials are then eluted with an appropriate solution, usually a volatile acid. Ordinarily, small volumes (2 to 5 ml.) of resin and rapid flow rates (0.1 to 1 ml./min.) are employed. Such a procedure effects group separations depending on the ionic character of the solutes. Paper chromatography may be used to resolve the members of a group, provided their quantity is small enough to be accommodated. Alternatively, special columns of high resolving power and automatic fraction collecting may be employed.

GENERAL REFERENCES

1. Block, R. J., LeStrange, R., and Zweig, G., "Paper Chromatography: A Laboratory Manual." Academic Press, New York, 1952.
2. Brimely, R. C., and Barrett, F. C., "Practical Chromatography." Reinhold, New York, 1953.
3. Cramer, F., "Papier Chromatographie." Verlag Chemie, Weinheim, 1953.
4. Kunin, R., and Myers, R. J., "Ion Exchange Resins." Wiley, New York, 1950.
5. Nachod, F. C. (ed.), "Ion Exchange: Theory and Application." Academic Press, New York, 1949.
6. Pollard, F. H., and McOmie, J. F. W., "Chromatographic Methods of Inorganic Analysis." Academic Press, New York, 1953.
7. Samuelson, O., "Ion Exchangers in Analytical Chemistry." Wiley, New York, 1953.
8. Strain, H. H., "Chromatographic Adsorption Analysis." Interscience, New York, 1945.
9. Williams, T. I., "An Introduction to Chromatography." Chemical Publishing Co., New York, 1947.
10. Zechmeister, L., and Cholnoky, L., "Principles and Practice of Chromatography." Wiley, New York, 1944.

Appendix 4

RADIOACTIVE NUCLIDES OF INTEREST IN BIOLOGICAL TRACER RESEARCH

In the following table the nuclides are arranged according to atomic number (Z) and elementary symbol with mass number (A). Half-life values are in most cases averages of results reported by a number of investigators. Time units are abbreviated as follows: seconds (s), minutes (m), hours (h), days (d), and years (y).

The column headed "Radiation characteristics" summarizes modes of decay and energies of emitted radiation. The symbols used are:

α alpha particles
β^- negative beta particles
β^+ positive beta particles
γ gamma rays
K orbital electron capture (K, L . . . capture)
IT isomeric transition
e^- internal-conversion electrons
x x rays

The numbers immediately following the symbols indicate energies in million electron volts; these energies are maximum values for β^- and β^+ radiations. In the case of β^+ emission, annihilation radiation (γ, 0.5 Mev.) is always present. Isomers are indicated by underlining the mass number.

The reader will find a complete tabulation in Hollander, J. M., Perlman, I., and Seaborg, G. T., The table of isotopes, *Revs. Mod. Phys.* **25,** 469 (1953).

Z	Symbol and Mass Number (A)	Half-life	Radiation Characteristics
1	H^3	12.46 y	β^- 0.017–0.019
			No γ
4	Be^7	52.9 d	K, γ 0.478
6	C^{11}	20.35 m	β^+ 0.981
			No γ
	C^{14}	5700 y	No γ, β^- 0.155
9	F^{18}	112 m	β^+ 0.65
11	Na^{22}	2.6 y	β^+ 0.542
			γ 1.277
	Na^{24}	15.06 h	β^- 1.39
			γ 1.38, 2.76
12	Mg^{28}	21.2 h	β^- 0.3
			$\gamma < 0.1$
15	P^{32}	14.30 d	β^- 1.701
	P^{33}	25.4 d	β^- 0.27
			No γ
16	S^{35}	87.1 d	β^- 0.1670
			No γ
17	Cl^{36}	4.4×10^5 y	β^- 0.714, β^+
			K
	C^{38}	37.29 m	β^- 4.81 (53%), 2.77 (16%), 1.11 (31%)
			γ 2.15, 1.60
18	A^{37}	35.0 d	K
	A^{41}	109 m	β^- 1.245
			γ 1.37
19	K^{40}	4.5×10^8 y	β^- 1.3–1.9 (89%)
			K (11%)
			γ 1.46
	K^{42}	12.44 h	β^- 3.58 (75%), 2.04 (25%)
			γ 1.51
	K^{43}	22.4 h	β^- 0.25, 0.81
			γ 0.4
20	Ca^{45}	152 d	β^- 0.254
			No γ
23	V^{48}	16 d	β^+ 0.69 ($\sim$95%), 0.8 ($\sim$5%), K
			γ 1.32, 0.99
24	Cr^{51}	27.8 d	K, no β^+
			γ 0.32 (e^- 0.2)
25	Mn^{51}	44.3 m	β^+ 2.4
	Mn^{52}	6.0 d	K (65%), β^+ 0.58 (35%)
			γ 0.73, 0.94, 1.46
	$Mn^{\underline{52}}$	21.3 m	β^+ 2.66
			γ 1.46
			(IT, e^- 0.39)
	Mn^{54}	310 d	K, γ 0.84

Z	Symbol and Mass Number (A)	Half-life	Radiation Characteristics
25	Mn^{56}	2.576 h	β^- 2.81 (50%), 1.04 (30%), 0.65 (20%)
			γ 0.822, 1.77, 2.06
26	Fe^{52}	8.3 h	K (60%), β^+ 0.55 (40%)
	Fe^{55}	4 y	$K(\gamma\ e^-)$
	Fe^{59}	45.1 d	β^- 0.257, 0.460
			γ 1.097, 1.295
27	Co^{55}	18.2 h	β^+ 1.50 ($\sim$50%), 1.01 ($\sim$50%)
			γ 0.477, 0.535, 1.41
	Co^{56}	72 d	K, β^+ 1.50
			γ 0.845, 1.26, 1.74, 2.55, 3.25, 2.01
	Co^{57}	270 d	β^+ 0.26
			γ 0.117, 0.130
	Co^{58}	72 d	K, β^+ 0.470
			γ 0.805
	Co^{60}	5.27 y	β^- 0.306
			γ 1.1715, 1.3316
28	Ni^{57}	36 h	β^+ 0.835
			γ 1.38, 1.91
	Ni^{65}	2.564 h	β^- 2.10 (57%), 1.01 (14%), 0.60 (29%)
			γ 1.46, 1.12, 0.37
29	Cu^{60}	24.6 m	β^+ 1.8, 3.3
			γ 1.5
	Cu^{61}	3.33 h	K (32%), β^+ (68%), β^+ 1.205 (96%), 0.55 (4%)
			γ 0.076, 0.284, 0.655
	Cu^{64}	12.8 h	K (42%)
			β^- 0.571
			β^+ 0.657
			γ (1.34)
30	Zn^{62}	9.33 h	K ($\sim$90%), β^+ ($\sim$10%), 0.66
			γ 0.0418
	Zn^{63}	38.3 m	β^+ (93%) 2.36 (92%), 1.40 (7%), 0.5 ($\sim$1%), K (7%)
			γ 0.96, 1.89, 2.60
	Zn^{65}	250 d	K (97.5%), γ 1.120
			β^+ (2.5%) 0.325
	Zn^{69}	57 m	β^- 0.897
			No γ
	$Zn^{\underline{69}}$	13.8 h	IT 0.437
			γ 0.437, e^-
			Zn-x
	Zn^{72}	49 h	β^- 0.3 (95%), 1.6 (5%)
			γ
31	Ga^{66}	9.45 h	K (36%)
			β^+ (64%) 4.144 (87%), 1.4 (4%), 0.88 (7%), 0.40 (2%)
			γ 1.05–4.8
	Ga^{67}	77.9 h	K, γ 0.90–0.790
			e^-

Z	Symbol and Mass Number (A)	Half-life	Radiation Characteristics
31	Ga^{72}	14.3 h	β^- 3.15 (9%), 2.52 (8%), 1.5 (11%), 0.9 (32%), 0.6 (40%) γ 1.05–2.51
	Ga^{73}	5 h	β^- 1.4 No γ
33	As^{71}	60 h	K, β^+, γ 0.162
	As^{72}	26 h	K, β^+ 3.34 (19%), 2.50 (62%), 1.84 (12%), 0.67 (5%), 0.27 (2%) γ 0.835–3.0
	As^{73}	90 d	K, e^- γ 0.0529, 0.0135
	As^{74}	17.5 d	β^- (53%) 1.36 (51%), 0.69 (49%) β^+ (47%) 1.53 (11%), 0.92 (89%) γ 0.596, 0.635
	As^{76}	26.8 h	β^- 3.04 (60%), 2.49 (25%), 1.29 (15%) γ 0.557, 1.22
	As^{77}	38 h	β^- 0.760
	As^{78}	90 m	β^- 1.4 (~30%), 4.1 (~70%) γ 0.27
34	Se^{72}	9.7 d	K
	Se^{73}	7.1 h	β^+ 1.68 (1%), 1.318 (88%), 0.750 (10%), 0.25 (1%) K, x
	Se^{75}	127 d	K, γ 0.067–0.405 e^-
35	Br^{77}	2.4 d	K (95%) β^+ (5%) 0.336, γ 0.160–0.813
	Br^{80}	18 m	β^- (92%) 1.99 (85%), 1.1 (15%) K, β^+ (8%) 0.87 γ >0.6
	$Br^{\underline{80}}$	4.58 h	IT 0.049, 0.037 e^- γ 0.037, 0.049
	Br^{82}	35.87 h	β^- 0.465 γ 0.547–1.312
	Br^{83}	2.33 h	β^- 0.940 No γ
	Br^{84}	30 m	β^- 4.68 (40%), 3.56 (9%), 2.53 (16%), 1.72 (35%) γ 0.89, 1.89
36	Kr^{77}	65 m	K, β^+ 1.7 γ, x
	Kr^{79}	34.5 h	K (95%), β^+ (5%), 0.595 γ 0.263
	$Kr^{\underline{83}}$	113 m	IT, Kr-x 0.0322, 0.0093 e^-
	Kr^{85}	9.4 y	β^- 0.695 (99%), 0.15 (0.65%) γ 0.54

Z	Symbol and Mass Number (A)	Half-life	Radiation Characteristics
36	$K^{\underline{85}}$	4.36 h	K (23%), β^- (77%) 0.855 γ 0.150, 0.057
	Kr^{87}	75 m	β^- 3.63 (75%), 1.27 (25%) γ 0.41, 1.89, ~2.3
	Kr^{88}	2.77 h	β^- 2.8 (20%), 0.9 (12%), 0.52 (68%) γ 0.028
37	Rb^{81}	4.7 h	K (87%), β^+ (13%) 0.990 γ 0.95
	Rb^{82}	6.3 h	K (94%), β^+ (6%) 0.775 (76%), 0.175 (24%) γ 0.188–1.464
	Rb^{83}	83 d	K γ ~0.45, ~0.15
	Rb^{84}	34 d	K, β^+ 1.629 (39%), 0.822 (58%), others (?) γ 0.890
	Rb^{86}	19.5 d	β^- 1.82 (80%), 0.72 (20%) γ 1.075
	Rb^{87}	6.0×10^{10} y	β^- 0.275 No γ
38	$Sr^{\underline{85}}$	65 d	γ K, 0.513
	$Sr^{\underline{85}}$	70 m	IT (86%), K (14%) e^- γ 0.0095, 0.150, 0.233
	$Sr^{\underline{87}}$	2.8 h	IT, e^- γ 0.388
	Sr^{89}	53 d	β^- 1.461 No γ
	Sr^{90}	19.9 y	β^- 0.61 No γ
	Sr^{91}	9.7 h	β^- 2.665 (26%), 2.03 (4%), 1.36 (69%), 1.09 (33%) 0.62 (7%) γ 0.551–1.413
39	Y^{87}	80 h	K (99+%), β^+ (0.3%), 0.7 γ 0.685
	$Y^{\underline{87}}$	14 h	IT, e^- γ 0.381
	Y^{88}	104 d	K (99+%), β^+ (0.19%) 0.83 γ 0.91, 1.85, 2.76
	Y^{90}	61 h	β^- 2.18 No γ
	Y^{91}	61 d	β^- 1.537 γ 1.2, 0.2
	$Y^{\underline{91}}$	51 m	IT, e^- γ 0.551
	Y^{92}	3.6 h	β^- 3.60, 2.7, 1.3 γ 0.6
	Y^{93}	10 h	β^- 3.1 γ 0.7

Z	Symbol and Mass Number (A)	Half-life	Radiation Characteristics
42	Mo^{90}	5.7 h	K, β^+ γ ~0.1
	Mo^{93}	6.95 h	IT γ 0.262, 0.69, 1.51
	Mo^{93}	>2 y	K, Nb-$K\alpha$
	Mo^{99}	67 h	β^- 1.23 (80%), 0.45 (~20%) γ 0.040–0.780
47	Ag^{105}	40 d	K γ 0.0625–0.440
	Ag^{106}	8.2 d	K, γ 0.220–1.55 e^-
	$Ag^{\underline{106}}$	24.5 m	β^+ 1.95, 1.5 γ 0.5, >0.6
	$Ag^{\underline{110}}$	270 d	β^- 0.087 (~58%), 0.530 (~35%), 2.12 (~3%), 2.86 (~3%) γ 0.116–1.516
	Ag^{111}	7.6 d	β^- 1.04 (91%), 0.80 (1%), 0.70 (5%) γ 0.243
	Ag^{112}	3.2 h	β^- 4.2 γ 0.625, 1.40
48	Cd^{107}	6.7 h	K (99+%), β^+ (0.31%), 0.32 γ 0.846
	Cd^{109}	470 d	K, γ 0.0875
	$Cd^{\underline{111}}$	48.6 m	IT, e^- γ 0.150, 0.246
	Cd^{115}	53 h	β^- 1.11 (58%), 0.58 (42%) γ 0.335
	Cd^{115}	43 d	β^- 1.61 (98%), 0.7 (2%) γ 0.46–1.28
50	Sn^{113}	112 d	K, In-x e^- γ 0.393
	Sn^{117}	14 d	IT, e^- γ 0.159
	$Sn^{\underline{119}}$	~250 d	IT γ 0.0653, 0.0242
	$Sn^{\underline{121}}$	>400 d	β^- 0.42 No γ
	Sn^{123}	136 d	β^- 1.42 No γ
	Sn^{125}	9.4 d	β^- 2.37 (~95%), 0.40 (~5%) γ 1.90
51	Sb^{118}	5.1 h	K, x e^-, γ 1.5, 0.26
	Sb^{117}	2.8 h	K, x, e^- γ 0.156
	Sb^{119}	39 h	K, x

Z	Symbol and Mass Number (A)	Half-life	Radiation Characteristics
51	Sb^{120}	6.0 d	K, γ 1.1 Sn-x
	Sb^{122}	2.8 d	β^- 1.46, 1.94 γ 0.568, e^-
	Sb^{124}	60 d	β^- 2.291 (20%), 1.69 (7%), 0.95 (3%), 0.68 (26%), 0.50 (39%) γ 0.121–2.04
	Sb^{125}	2.7 y	β^- 0.616 (18%), 0.299 (49%), 0.128 (53%) γ 0.035–0.637
	Sb^{127}	93 h	β^- 1.2 γ 0.72
	Sb^{129}	4.2 h	β^-
52	Te^{118}	6.0 d	K
	Te^{119}	4.5 d	K, e^- 0.2, 0.5 γ 1.5
	Te^{121}	17 d	K, Sb-x γ 0.506, 0.573
	Te^{121}	154 d	IT, e^-, Te-x γ 0.082, 0.213
	$Te^{\underline{125}}$	58 d	IT 0.110, 0.0353
	Te^{127}	9.3 h	β^- 0.70 No γ
	$Te^{\underline{127}}$	115 d	IT 0.0885, Te-x
	Te^{129}	72 m	β^- 1.8 γ 0.3, 0.8 γ 0.030
	$Te^{\underline{129}}$	33.5 d	IT 0.1060 e^-
	Te^{131}	24.8 m	β^- 2.0 (~55%), 1.4 (~45%) γ 0.16, 0.7
	$Te^{\underline{131}}$	30 h	IT 0.177 e^-
	Te^{132}	77.7 h	β^- 0.22, e^-, x γ 0.231
53	I^{124}	4.5 d	β^+ 2.20 (51%), 1.50 (44%), 0.7 (5%) γ 0.603, 0.73, 1.72, 1.95
	I^{125}	60.0 d	K, x 0.035
	I^{126}	13.0 d	β^- 1.208 (27%), 0.85 (73%) γ 0.382
	I^{128}	24.99 m	β^- 202 γ 0.428
	I^{129}	1.72×10^7 y	β^- 0.12 γ 0.039
	I^{130}	12.6 h	β^- 1.03 (~60%), 0.61 (~40%) γ 0.744, 0.537, 0.417

Z	Symbol and Mass Number (A)	Half-life	Radiation Characteristics
53	I^{131}	8.141 d	β^- 0.815 (0.7%), 0.608 (87.2%), 0.335 (9.3%), 0.250 (2.8%) γ 0.080–0.722
	I^{132}	2.4 h	β^- 2.2, 0.9 γ 0.69, 1.41
	I^{133}	20.5 h	β^- 1.3 (~91%), 0.4 (~9%) γ 0.53, 0.85, 1.4
	I^{134}	52.5 m	β^- 1.6 (~70%), 2.8 (~30%) γ >2.2
	I^{135}	6.68 h	β^- 0.5 (35%), 1.0 (40%), 1.4 (25%) γ 1.8, 1.27
54	Xe^{127}	34 d	K, e^- γ 0.057–0.363
	$Xe^{\underline{129}}$	8.0 d	IT γ 0.196
	$Xe^{\underline{131}}$	12.0 d	IT γ 0.163
	$Xe^{\underline{133}}$	2.3 d	IT γ 0.233
	Xe^{133}	5.270 d	β^- 0.345 γ 0.081
	Xe^{135}	9.13 h	β^- 0.905 γ 0.25
55	Cs^{131}	9.6 d	K, Xe-x
	Cs^{132}	7.1 d	K γ 0.668
	Cs^{134}	2.3 y	β^- 0.648 (75%), 0.09 (25%) γ 0.561–1.164
	$Cs^{\underline{134}}$	3.1 h	IT 0.128, Cs-x
	Cs^{135}	3.0×10^6 y	β^- 0.21 No γ
	Cs^{136}	13.7 d	β^- 0.35 γ 0.9
	Cs^{137}	33 y	β^- 0.523 γ 0.6616
56	Ba^{131}	12.0 d	K, Cs-x γ 0.122–0.497
	Ba^{133}	~9.5 y	K, Cs-x γ 0.320, 0.085
	$Ba^{\underline{133}}$	38.8 h	IT e^- γ 0.276
	Ba^{135}	28.7 h	IT, Ba-x, e^- γ 0.269

Z	Symbol and Mass Number (A)	Half-life	Radiation Characteristics
56	Ba^{139}	85 m	β^- 2.27 γ 0.163, 0.20, 1.05
	Ba^{140}	12.8 d	β^- 1.022 (60%), 0.480 (40%) γ 0.0296–0.537
74	W^{181}	140 d	K γ 0.03–0.800
	W^{185}	73.2 d	β^- 0.428 γ 0.134
	W^{187}	24.1 h	β^- 0.63 (70%), 1.33 (30%) γ 0.0720–0.6189
76	Os^{185}	97 d	K, x γ 0.648, 0.878
	Os^{191}	16.0 d	β^- 0.143 γ 0.0417, 0.129
	Os^{193}	30.6 h	β^- 1.10 γ 0.066
77	Ir^{190}	12.6 d	K γ 0.2, 0.6
	Ir^{192}	74.37 d	β^- 0.66 γ 0.1364–0.6129
	Ir^{194}	19.0 h	β^- 2.18 γ 0.290, 0.326
78	Pt^{191}	3.00 d	K, x γ 0.083, 0.096, 0.173
	$Pt^{\underline{193}}$	4.33 d	K, x γ 0.135
	$Pt^{\underline{195}}$	3.5 d	IT γ 0.029–0.129
	Pt^{197}	18 h	β^- 0.670 γ 0.077, 0.191
	Pt^{199}	31 m	β^- 1.8
79	Au^{192}	5.0 h	β^+ $\sim$1.9 γ $\sim$2–3
	Au^{193}	15.8 h	K γ 0.051–0.235
	Au^{194}	39.5 h	K ($\sim$97%), β^+ ($\sim$3%) 1.8 γ 0.291–2.1
	Au^{195}	180 d	K, x γ 0.308, 0.099, 0.130
	Au^{196}	14 h	IT
	Au^{196}	5.5 d	K ($\sim$95%), β^- ($\sim$5%) γ 0.352, 0.332, 0.426
	Au^{198}	2.69 d	β^- 0.963, 0.290 γ 0.4118, 0.676
	Au^{199}	3.15 d	β^- 0.460 γ 0.050–0.62

Z	Symbol and Mass Number (A)	Half-life	Radiation Characteristics
80	Hg^{197}	65 h	K, Au-x
			γ 0.077
	$Hg^{\underline{197}}$	23 h	IT
			γ 0.133, 0.164
	Hg^{199}	44 m	IT, Hg-x
			e^-
			γ 0.155, 0.368
	Hg^{203}	47.9 d	β^- 0.208
			γ 0.279
81	Tl^{200}	27 h	K
			γ 0.365–1.210
	Tl^{202}	12.5 d	K
			γ 0.435
	Tl^{204}	3.5 y	β^- 0.765
			No γ
82	Pb^{203}	52 h	K
			γ 0.153, 0.269, 0.422
	Pb^{209}	3.22 h	β^- 0.635
			No γ
	$Pb^{210(RaD)}$	22 y	β^- 0.018
			γ 0.0465
	$Pb^{212(ThB)}$	10.6 h	β^- 0.355, 0.589
			γ 0.115–0.299
83	Bi^{204}	12 h	K, γ 0.217
			e^- 0.20, 0.75
	Bi^{206}	6.4 d	K, x
			γ 0.182–1.720
	Bi^{207}	~50 y	K, γ 0.064–2.49
	Bi^{208}	Long	K
	$Bi^{210(RaE)}$	5.02 d	β^- 99+%
			α (5 × 10^{-5}%) 4.87
	$Bi^{212(ThC)}$	60.5 m	β^- (66.2%) 2.256
			γ 0.15–2.20
			α 6.05, 6.08, 5.76, 5.60
			γ 0.04–0.472
	Bi^{213}	47 m	β^- (98%) 1.39 (68%), 0.959 (32%)
			α (2%) 5.86

Appendix 5

REMARKS ON THE USE OF LIQUID SCINTILLATION DETECTORS IN ASSAY OF BIOLOGICALLY IMPORTANT NUCLIDES

The most important tracer nuclides in biological research are C^{14} and H^3. Others which are critically essential, if perhaps of somewhat less importance are S^{35} and P^{32}. All of these, with the exception of P^{32}, require detection of low-energy beta radiations. Inasmuch as most tracer samples produced in the course of biochemical and physiological research are obtained in liquid form, radioactivity assay with apparatus based on the use of ionization chambers, or variants thereof, often necessitates processing of liquid samples so that the tracer material is obtained either in gaseous or solid form. This requirement is imposed by the self-absorption of the low-energy beta radiations characteristic of C^{14} and S^{35}, and especially H^3. The resultant inconvenience, loss of time, and increased hazard of sample loss hamper tracer research with low energy emitters.

In general, assay of liquid samples containing H^3, C^{14}, and S^{32} is feasible only by incorporation of the tracer materials as part of liquid scintillation media. The resultant increase in sensitivity, efficiency, and freedom from self-absorption effects, makes development of foolproof procedures based on liquid scintillation most desirable. While this has been possible in principle since 1950 (1, 2), it is only recently that commercial instrumentation which makes practicable the widespread routine use of liquid scintillation detectors has come on the market.

This development has been signalized by a growing literature which began to appear during the first printing of this edition. The occasion of a reprint of this edition affords an opportunity to include this brief summary of the present status of H^3 and C^{14} assay by liquid scintillation systems.

The major new development is that apparatus is now available which is sufficiently well designed and engineered to overcome a number of complications inherent in the use of phototube detectors. These difficulties are thermionic emission ("noise"), lack of reproducibility in performance of phototubes, and great sensitivity to normal radiation background. Thermionic emission, which is the result of spontaneous emission of electrons from the metal electrodes of the phototubes, can be effectively minimized by use of low temperatures as well as careful choice of tubes. In addition, the use of two phototubes in coincidence nearly eliminates whatever thermal noise is left after cooling, because it delivers to the amplifier system only those light pulses seen simultaneously by both phototubes. These are the true pulses resulting from the decay of the radioactive nuclide in the sample. Still further discrimination is obtained by using pulse-height selection which allows only pulses greater than a certain minimal height (voltage signal) to pass into the detector circuits. Because the pulses arising from thermionic emission are mostly very small by comparison with most of the pulses caused by the nuclide decay (even with as low-energy a beta emitter as H^3), it is possible in this way to select out for assay most of the nuclide radiation, while eliminating practically all the noise. On the other hand, by also using a discriminator circuit which cuts off all pulses *above* a certain height, it is possible to eliminate much of the cosmic ray background which consists mostly of relatively high-energy radiation.

The requirements in the circuitry become extreme only for H^3 assay. For C^{14} a single channel following the phototubes is sufficient and even refrigeration is not absolutely essential. With a high-energy beta emitter such as P^{32}, requirements are even less stringent. However, by designing an apparatus capable of handling H^3 routinely, the problem of assaying the other nuclides is solved simultaneously.

Such an instrument can also be adapted to simultaneous assay of pairs of nuclides. In addition, automatic sample changing can be incorporated. The result is an instrument sufficiently versatile to approximate closely the ideal of a universal assay system in one package.

In the arrangement most widely used at present,* there are two scaling units and two channels for passing different selected pulse heights. The circuitry involved is described in the literature (3). For the present it is sufficient to note one sequence of operations which is employed. The pulses from the phototubes are amplified to give a voltage drop which is fed into both of two parallel discriminator circuits. These measure the pulse height so obtained and decide whether the pulses on both sides exceed a set minimum. The coincidence circuits pass on those pulses which get through the two discriminators and which are in coincidence. The coincident pulses now go to another set of discriminators set to the proper values to fall inside a minimum and maximum voltage setting (channel width or "window") for one or the other of the two pulse height ranges appropriate for the two nuclide beta radiations to be assayed.

Each channel requires a rather careful but not overly taxing set of manipulations to achieve the proper maximal setting for the detection of

* This is known as the "Tricarb Scintillator" and is manufactured in the United States by the Packard Co., LaGrange, Illinois. Other models are now coming on the market.

the particular nuclide involved. The response of the phototube increases very rapidly with applied voltage (roughly as the eighth power). Thus the pulse heights also increase rapidly with the voltage across the phototube. Hence, there is a stringent requirement for very precise voltage regulation. The proper voltage is that which gives pulses of the correct size to be counted. This is determined by varying the voltage, holding the channel width constant, and observing the counting rate. At first, no counts are seen. Then, as the pulse heights begin to exceed the low value let through by the discriminator, the counting rate increases. Finally, the pulses begin to be too high to count, because they exceed the height set by the top setting of the channel. So a maximum in the counting rate is observed at any given value for the channel width. If the channel widths are varied by holding the low setting of the channel range constant and changing the high setting, a family of such counting curves is obtained. With any particular channel width, the proper voltage for the phototube operation is that which gives the maximum counting rate. It is important to arrange that the channel width can be varied easily. This is necessitated by the fact that changes in the scintillation recipe may require new determination of the characteristic counting curves.

It is obvious that the manipulations involved in operating a liquid scintillation system differ completely from those the reader may be accustomed to from his experience with Geiger-Muller counters, proportional counters, ionization chambers, and the like. However, the routine operation of liquid scintillators which are properly designed, as some of those now available commercially are, is well within the capabilities of most laboratory personnel.

What remains is the business of sample preparation for which a considerable exercise of judgment is still required. The reader is referred in the bibliography to two excellent articles (4, 5) dealing with the techniques involved in preparation of H^3 and C^{14} samples for assay in liquid scintillation media.

Briefly, the status of the art is somewhat as follows. Proper mixtures of solvent and phosphor cannot be prescribed in general for every compound to be tested. Each compound assayed must be investigated individually for its behavior in any given medium and the proper mixture determined by experiment. However, a great variety of compounds can be assayed using relatively few prescriptions.

In general the solvents most widely used are dry toluene and xylene. A large number of phosphors have been tested (6, 7). Among the phosphors, the most popular is 2,5-diphenyloxazole (PPO). Organic-soluble compounds, such as sterols, hydrocarbons, etc., are easily assayed using as solvent a 0.3 to 0.8% solution of PPO in toluene. Polar substances, such as carbonate, amino acids, and proteins, can be assayed by taking advantage of their tendency to form complexes or compounds with quaternary amines, particularly *p*-(diisobutylcresoxyethoxytheyl amine), or "Hyamine," as it is known commercially. Detailed procedures for the use of this material, as well as for a variety of other quaternary amines, are to be found in the articles cited below (5, 8).

The precautions involved in concocting scintillation mixtures are best understood in terms of what must be avoided. It is obvious that no material can be used which absorbs light in the wavelength region of optimum phototube response. Further, no substance which quenches fluorescence to any marked extent is utilizable. It is this latter limitation which renders prediction of performance for any given compound hazardous and which is responsible for the empirical nature of the procedures available. However, a large literature on this subject is available now. An introduction to this literature can be readily obtained by reference to the general reviews cited (4, 9).

It would not be surprising, in view of the present trend, to see the liquid scintillation detectors replace all others for much of the routine radioactivity assay in tracer research. There are no inherent disadvantages which militate appreciably against their use in the vast majority of assay situations and no single system is so versatile.

REFERENCES

1. Kallman, H., *Phys. Rev.* **78,** 62 (1950).
2. Reynolds, G. T., Harrison, F. B., and Salvini, G., *Phys. Rev.* **78,** 488 (1950).
3. Hiebert, R. D., and Watts, R. J., *Nucleonics*, **11,** (12), 38 (1953).
4. Davidson, J. D., and Reigelson, P., *Intern. J. Appl. Radiol. and Isotopes* **2,** 1 (1957).
5. Passman, J. M., Radin, N. S., and Cooper, J. A. D., *Anal. Chem.* **28,** 484 (1957).
6. Hayes, F. N., Ott, D. G., Kerr, V. N., and Rogers, B. S., *Nucleonics*, **13** (12), 38 (1955).
7. Hayes, F. N., Ott, D. G., and Kerr, V. N., *Nucleonics*, **14,** (1), 42 (1956).
8. Vaughan, M., Steinberg, D., and Logan, J., *Science*, **126,** 446 (1957).
9. Proc. Symposium on Tritium in Tracer Applications, sponsored by New England Nuclear Corp., Atomic Associates, Inc., and Packard Instrument Company, Inc., New York, November, 1957.

Author Index

The numbers in parentheses are footnote numbers and are inserted to enable the reader to locate a cross reference when the author's name does not appear at the point of reference in the text.

C

D

L

S

V

W

Y

Z

Subject Index

C

E

F

G

N

O

Y

Z